Selected Titles in This Series

Volume

6 **Kenneth R. Davidson**
C*-Algebras by example
1996

5 **A. Weiss**
Multiplicative Galois module structure
1996

4 **Gérard Besson, Joachim Lohkamp, Pierre Pansu, and Peter Peterson**
Miroslav Lovric, Manng Min-Oo, and McKenzie Y.-K. Wang, Editors
Riemannian geometry
1996

3 **Albrecht Böttcher, Aad Dijksma and Heinz Langer, Michael A. Dritschel and James Rovnyak, M. A. Kaashoek**
Peter Lancaster, Editor
Lectures on operator theory and its applications
1995

2 **Victor P. Snaith**
Galois module structure
1994

1 **Stephen Wiggins**
Global dynamics, phase space transport, orbits homoclinic to resonances, and applications
1993

C*-Algebras by Example

FIELDS INSTITUTE MONOGRAPHS

THE FIELDS INSTITUTE FOR RESEARCH IN MATHEMATICAL SCIENCES

C*-Algebras by Example

Kenneth R. Davidson

American Mathematical Society
Providence, Rhode Island

The Fields Institute
for Research in Mathematical Sciences

The Fields Institute is named in honour of the Canadian mathematician John Charles Fields (1863–1932). Fields was a remarkable man who received many honours for his scientific work, including election to the Royal Society of Canada in 1909 and to the Royal Society of London in 1913. Among other accomplishments in the service of the international mathematics community, Fields was responsible for establishing the world's most prestigious prize for mathematics research—the Fields Medal.

The Fields Institute for Research in Mathematical Sciences is supported by grants from the Ontario Ministry of Education and Training and the Natural Sciences and Engineering Research Council of Canada. The Institute is sponsored by McMaster University, the University of Toronto, and the University of Waterloo and has affiliated universities in Ontario and across Canada.

1991 *Mathematics Subject Classification.* Primary 46L05.

Library of Congress Cataloging-in-Publication Data
Davidson, Kenneth R.
   C*-algebras by example/Kenneth R. Davidson.
     p. cm. —(Fields Institute monographs, ISSN 1069-5273; 6)
   Includes bibliographical references (p. –) and index.
   ISBN 0-8218-0599-1 (alk. paper)
   1. C*-algebras. I. Title. II. Series.
QA326.D375 1996
512′.55–dc20
                                                    96-20184
                                                       CIP

Copying and reprinting. Individual readers of this publication, and nonprofit libraries acting for them, are permitted to make fair use of the material, such as to copy a chapter for use in teaching or research. Permission is granted to quote brief passages from this publication in reviews, provided the customary acknowledgment of the source is given.

Republication, systematic copying, or multiple reproduction of any material in this publication (including abstracts) is permitted only under license from the American Mathematical Society. Requests for such permission should be addressed to the Assistant to the Publisher, American Mathematical Society, P.O. Box 6248, Providence, Rhode Island 02940-6248. Requests can also be made by e-mail to reprint-permission@ams.org.

© Copyright 1996 by the American Mathematical Society. All rights reserved.
The American Mathematical Society retains all rights
except those granted to the United States Government.
Printed in the United States of America.
∞ The paper used in this book is acid-free and falls within the guidelines
established to ensure permanence and durability.
This publication was prepared by the Fields Institute.
10 9 8 7 6 5 4 3 2   01 00 99 98 97

To my sons, Colin and Stuart

The joy of doing any scientific research is in the quest for new ideas and deeper understanding. Finding something new *yourself* is exhilarating. Second to that, but still very important, is the pleasure of explaining it to someone else who cares.

In the pursuit of the unknown, you are never sure what you need to know. You can never know too much. Yet when approaching a new problem, you always know both too much (about the wrong things) and too little (about the question at hand). A big step towards the solution is figuring out exactly what it is that you really need to know.

Contents

Preface		xiii
I. The Basics of C*-algebras		**1**
I.1	Definitions	1
I.2	Banach Algebra Basics	3
I.3	Commutative C*-algebras	7
I.4	Positive Elements	9
I.5	Ideals, Quotients and Homomorphisms	12
I.6	Weak Topologies	15
I.7	The Density Theorems	19
I.8	Some Operator Theory	23
I.9	Representations of C*-algebras	26
I.10	C*-algebras of Compact Operators	36
	Exercises	40
II. Normal Operators and Abelian C*-algebras		**46**
II.1	Spectral Theory	46
II.2	The L^∞ Functional Calculus	48
II.3	Multiplicity Theory	53
II.4	The Weyl–von Neumann–Berg Theorem	57
II.5	Voiculescu's Theorem	64
	Exercises	72
III. AF C*-algebras		**74**
III.1	Finite Dimensional C*-algebras	74
III.2	AF Algebras	75
III.3	Perturbations	79
III.4	Ideals and Quotients	84
III.5	Examples	86
III.6	Extensions	91
	Exercises	95

IV. K-theory for AF C*-algebras — 97
- IV.1 Idempotents — 97
- IV.2 K_0 — 100
- IV.3 Dimension Groups — 102
- IV.4 Elliott's Theorem — 109
- IV.5 Applications — 112
- IV.6 Riesz groups — 118
- IV.7 The Effros–Handelman–Shen Theorem — 120
- IV.8 Blackadar's Simple Unital Projectionless C*-algebra — 124
- Exercises — 129

V. C*-Algebras of Isometries — 132
- V.1 Toeplitz Operators — 132
- V.2 Isometries — 136
- V.3 Bunce–Deddens Algebras — 137
- V.4 Cuntz Algebras — 144
- V.5 Simple Infinite C*-algebras — 147
- V.6 Classification of Cuntz Algebras — 150
- V.7 Real Rank Zero — 156
- Exercises — 162

VI. Irrational Rotation Algebras — 166
- VI.1 The algebras $\mathcal{A}_\theta$ — 166
- VI.2 Projections in $\mathcal{A}_\theta$ — 170
- VI.3 An AF algebra — 172
- VI.4 Berg's technique — 174
- VI.5 Imbedding $\mathcal{A}_\theta$ into $\mathfrak{A}_\theta$ — 177
- Exercises — 180

VII. Group C*-Algebras — 182
- VII.1 Group Representations — 182
- VII.2 Amenability — 185
- VII.3 Primitive Ideals — 190
- VII.4 A Crystallographic Group — 193
- VII.5 The Discrete Heisenberg Group — 200
- VII.6 The Free Group — 203
- VII.7 The Reduced C*-algebra of the Free Group — 206
- VII.8 $C_r^*(\mathbb{F}_2)$ is Projectionless — 210
- Exercises — 214

VIII. Discrete Crossed Products **216**
VIII.1	Crossed Products	216
VIII.2	Crossed Products by $\mathbb{Z}$	222
VIII.3	Minimal Dynamical Systems	223
VIII.4	Odometers	230
VIII.5	K-theory of Crossed Products	232
VIII.6	AF Subalgebras of Crossed Products	235
VIII.7	Crossed Product subalgebras of AF Algebras	238
VIII.8	Topological Stable Rank	244
VIII.9	An Order 2 Automorphism	247
	Exercises	250

IX. Brown–Douglas–Fillmore Theory **252**
IX.1	Extensions	252
IX.2	An Addition and Zero Element for $\mathrm{Ext}(X)$	254
IX.3	Some Special Cases	258
IX.4	Positive maps	259
IX.5	$\mathrm{Ext}(X)$ is a group	266
IX.6	First Topological Properties	268
IX.7	Ext for planar sets	273
IX.8	Quasidiagonality	281
IX.9	Homotopy Invariance	286
IX.10	The Mayer–Vietoris Sequence	289
IX.11	Examples	294
	Exercises	299

References 303

Index 307

Preface

These notes were developed in the fall of 1993 for a graduate course on C*-algebras. The subject of C*-algebras received a dramatic revitalization in the 1970s by the introduction of topological methods due the deep work of Brown, Douglas and Fillmore on extensions of C*-algebras and Elliott's use of K-theory to provide a useful classification of AF algebras and Kasparov's melding of the two into KK-theory. These results were the beginning of a marvelous new set of tools for analyzing concrete C*-algebras. The subject flourished by virtue of a rich and varied group of examples which were the perfect fodder for these new methods. Moreover, these examples served as a beacon for the kinds of general tools needed to study C*-algebras. Today, a student cannot get very far in the C*-algebra literature without being somewhat familiar with the lexicon of examples that now dot the landscape.

These notes are not intended as a systematic study of the general theory of C*-algebras, nor of K-theory. There are several excellent books on both of these aspects. Rather, I develop a modicum of the general theory for the sake of self-containment and then launch into a study of various important classes of examples.

A number of choices had to be made. I have developed the theory of AF algebras to quite an extent. However, in some of the other topics, such as irrational rotation algebras and Cuntz algebras, I have limited myself to the more modest goal of obtaining enough information to classify the algebras within their narrow context. Because this is not a course in K-theory, certain stronger results that require a more systematic study have been omitted. I have also given a detailed treatment of the Brown–Douglas–Fillmore theory. The discussion is limited to the commutative case as in their original work because of the technical simplifications. However the informed reader will notice that a number of the proofs have been influenced by the more general theory.

The hope is that these notes will open a student's eyes to some of the power, beauty and variety of these amazing algebras. This is just a glimpse into an exciting area of current research interest.

These notes were compiled during the author's participation in the special year on C*-algebras at the Fields Institute of Mathematics held in Waterloo, Canada during the 1994–95 academic year. The author wishes to thank the Fields Institute and the University of Waterloo for providing the release time that made this project possible. I also wish to thank everyone who provided feedback on the early drafts of these notes. I am especially indebted to Keith Taylor, who provided me with some wonderful material on the C*-algebras of crystallographic groups and Alan Paterson for advice on amenability. I am also grateful to Florin Boca, who provided me with detailed notes on showing that the irrational rotation algebras are limit circle algebras even though, in the end, I decided not to include them. I must also mention Ileana Ionascu and Raul Curto, who read the whole manuscript and made many detailed comments. In addition, I thank Ola Bratteli, Larry Brown, John Conway, Don Hadwin, Dick Kadison, Ian Putnam, Norberto Salinas, Jack Spielberg and Hong Sheng Yin who put me right at various points of the book. In spite of all their help, I know that I have been unable to get all the minor blemishes out. For this, I must bear the burden myself.

The text was prepared using $\mathcal{A}\mathcal{M}\mathcal{S}$-LaTeX, and typeset in 11pt Times roman. The commutative diagrams and figures were produced using the macro package Xy-pic version 3.2 written by Kris Rose and Ross Moore.

Kenneth R. Davidson
Waterloo, February, 1996

CHAPTER I

The Basics of C*-algebras

A **Banach algebra** is a complex normed algebra $\mathfrak{A}$ which is complete (as a topological space) and satisfies

$$\|AB\| \leq \|A\|\,\|B\| \quad \text{for all} \quad A, B \in \mathfrak{A}.$$

A **Banach ∗-algebra** is a complex Banach algebra $\mathfrak{A}$ with a conjugate linear involution ∗ (called the **adjoint**) which is an anti-isomorphism. That is, for all A, B in $\mathfrak{A}$ and λ in $\mathbb{C}$,

$$(A+B)^* = A^* + B^*$$
$$(\lambda A)^* = \overline{\lambda} A^*$$
$$A^{**} = A$$
$$(AB)^* = B^* A^*.$$

A **C*-algebra** is a Banach ∗-algebra with the additional norm condition

$$\|A^*A\| = \|A\|^2 \quad \text{for all} \quad A \in \mathfrak{A}.$$

Certain properties follow easily from these definitions. For example, the adjoint map is isometric. Indeed,

$$\|A\|^2 = \|A^*A\| \leq \|A^*\|\,\|A\|;$$

whence $\|A\| \leq \|A^*\| \leq \|A^{**}\| = \|A\|$. Other facts, such as some of the properties of positive elements, are surprisingly subtle.

Example I.1.1 The algebra of all bounded operators $\mathcal{B}(\mathcal{H})$ on a Hilbert space $\mathcal{H}$ is a C*-algebra with the usual adjoint operation. This follows from the well known identity

$$\|A^*A\| = \sup_{\|x\|=\|y\|=1} |(A^*Ax, y)| = \sup_{\|x\|=\|y\|=1} |(Ax, Ay)| = \|A\|^2.$$

It follows easily that a norm-closed subalgebra of a C*-algebra which is closed under adjoints is also a C*-algebra. A norm closed self-adjoint subalgebra of $\mathcal{B}(\mathcal{H})$ will be called a **concrete C*-algebra**.

The simplest example of a concrete C*-algebra is the algebra $\mathfrak{K}$ of all compact operators on a separable Hilbert space. Another important example is $C^*(T)$, the unital C*-subalgebra of $\mathcal{B}(\mathcal{H})$ generated by an operator T in $\mathcal{B}(\mathcal{H})$, which is the closure of all polynomials in T, T^* and I.

Example I.1.2 Let X be a locally compact Hausdorff space. Then $C_0(X)$, the space of all continuous functions on X vanishing at infinity, forms a C*-algebra with complex conjugation as the adjoint operation. Then

$$\|\bar{f}f\|_X = \sup_{x \in X} |\overline{f(x)}f(x)| = \sup_{x \in X} |f(x)|^2 = \|f\|_X^2.$$

Following the definitions for operators, we say that an element A of a C*-algebra $\mathfrak{A}$ is **self-adjoint** if $A^* = A$; N is **normal** if $N^*N = NN^*$; and U is **unitary** if $U^*U = I = UU^*$. We also define A to be **positive** if $A = A^*$ and the spectrum (see section I.2) $\sigma(A)$ is contained in the non-negative real line $[0, \infty)$.

It is often convenient to have a unit around, even when the algebra is not unital. So we show how to adjoin one while maintaining the C*-algebra structure. In fact, it is unique (see the exercises).

When a C*-algebra $\mathfrak{A}$ has an identity element I, compute

$$I^*A = (A^*I)^* = A^{**} = A \quad \text{for all} \quad A \in \mathfrak{A}.$$

So $I^* = I^*I = I$. Hence $\|I\|^2 = \|I^*I\| = \|I\|$. Since $I \neq 0$, this shows that $\|I\| = 1$.

Proposition I.1.3 *Every non-unital C*-algebra $\mathfrak{A}$ is contained in a unital C* algebra $\mathfrak{A}^\sim$ as a maximal ideal of codimension one.*

Proof. Form $\mathfrak{A}^\sim := \mathfrak{A} \oplus \mathbb{C}$ and define

$$(A, \lambda)(B, \mu) := (AB + \lambda B + \mu A, \lambda \mu)$$
$$(A, \lambda)^* := (A^*, \overline{\lambda})$$
$$\|(A, \lambda)\| := \sup_{\|B\| \leq 1} \|AB + \lambda B\|$$

This makes $\mathfrak{A}^\sim$ into a *-algebra. The norm is a Banach algebra norm because it is the norm induced from the space $\mathcal{B}(\mathfrak{A})$ of bounded operators on $\mathfrak{A}$ given by the *-algebra of operators $\{L_A + \lambda I : A \in \mathfrak{A}, \lambda \in \mathbb{C}\}$, where $L_A(B) = AB$ is a left multiplication operator. Thus this is a Banach *-algebra with unit $(0, 1)$. By design, $\mathfrak{A}$ is a maximal ideal of co-dimension one. The imbedding of $\mathfrak{A}$ into $\mathfrak{A}^\sim$ is isometric because

$$\|A\| = \|A(A^*/\|A\|)\| \leq \|(A, 0)\| = \sup_{\|B\| \leq 1} \|AB\| \leq \|A\|.$$

I.2. Banach Algebras Basics

It remains to verify the C*-algebra norm condition.

$$\begin{aligned}
\|(A,\lambda)\|^2 &= \sup_{\|B\|\leq 1} \|AB+\lambda B\|^2 \\
&= \sup_{\|B\|\leq 1} \|B^*A^*AB + \lambda B^*A^*B + \bar{\lambda}B^*AB + |\lambda|^2 B^*B\| \\
&\leq \sup_{\|B\|\leq 1} \|A^*AB + \lambda A^*B + \bar{\lambda}AB + |\lambda|^2 B\| \\
&= \|(A^*A + \lambda A^* + \bar{\lambda}A, |\lambda|^2)\| \\
&= \|(A,\lambda)^*(A,\lambda)\| \leq \|(A,\lambda)^*\|\,\|(A,\lambda)\|
\end{aligned}$$

Thus $\|(A,\lambda)\| \leq \|(A,\lambda)^*\|$. By symmetry, we have $\|(A,\lambda)\| = \|(A,\lambda)^*\|$. Hence the inequality above is an equality, and so

$$\|(A,\lambda)\|^2 = \|(A,\lambda)^*(A,\lambda)\|$$

as claimed. ∎

I.2 Banach Algebras Basics

For the convenience of the reader, we review the necessary background from Banach algebras that we need.

The **spectrum** of an element A of a unital Banach algebra $\mathfrak{A}$ is the set

$$\sigma(A) := \{\lambda \in \mathbb{C} : \lambda I - A \text{ is not invertible}\}.$$

The complement of the spectrum is called the **resolvent**, and $R_A(\lambda) = (\lambda I - A)^{-1}$ is the **resolvent function**.

Theorem I.2.1 *In any unital Banach algebra $\mathfrak{A}$, the spectrum of each A in $\mathfrak{A}$ is a non-empty compact set; and the resolvent function is analytic on $\mathbb{C} \setminus \sigma(A)$.*

Proof. If $|\lambda| > \|A\|$, then $\|\lambda^{-n}A^n\| \leq (|\lambda|^{-1}\|A\|)^n$ decreases geometrically fast; so the series

$$\sum_{n\geq 0} \lambda^{-n-1} A^n$$

is norm convergent. The limit is $(\lambda I - A)^{-1}$ since

$$(\lambda I - A)\sum_{n=0}^{k} \lambda^{-n-1} A^n = I - \lambda^{-k-2} A^{k+1}$$

which converges to I. Moreover, this shows that $R_A(\lambda)$ is analytic, and has a Laurent expansion about the point at infinity. Furthermore,

$$\lim_{|\lambda|\to\infty} \|R_A(\lambda)\| \leq \lim_{|\lambda|\to\infty} |\lambda|^{-1}(1 - |\lambda|^{-1}\|A\|)^{-1} = 0.$$

Similarly, if $\lambda_0 I - A$ is invertible and $|\lambda - \lambda_0| < \|(\lambda_0 I - A)^{-1}\|^{-1}$, then

$$(\lambda I - A)^{-1} = \sum_{n \geq 0} (\lambda - \lambda_0)^n (\lambda_0 I - A)^{-n-1}$$

is the Taylor expansion of R_A in a neighbourhood of λ_0. So the resolvent function is analytic on the complement of the spectrum, known as the resolvent of A. In particular, $f(R_A(\lambda))$ is an analytic scalar valued function for every continuous linear functional f on $\mathfrak{A}$. The resolvent is therefore an open set containing all λ in $\mathbb{C}$ with $|\lambda| > \|A\|$. So $\sigma(A)$ is a compact subset of $\{\lambda \in \mathbb{C} : |\lambda| \leq \|A\|\}$.

If the spectrum of A were empty, then $R_A(\lambda)$ would be a bounded entire function. By Liouville's Theorem, this leads to the absurd conclusion that R_A is the constant zero function. Indeed, for every functional f, $f(R_A(\lambda))$ is a scalar entire function vanishing at infinity; whence it is zero. So $R_A = 0$ by the Hahn–Banach Theorem. Hence the spectrum must be non-empty. ∎

Note that the power series technique of the proof provides the useful fact that if $\|A\| < 1$, then $I - A$ is invertible with inverse $\sum_{n \geq 0} A^n$. The following elementary functional property of the spectrum is known as the **spectral mapping property**.

Lemma I.2.2 *If p is a polynomial and $A \in \mathfrak{A}$ is an element of a unital Banach algebra, then*

$$\sigma(p(A)) = p(\sigma(A)).$$

Proof. For α in $\mathbb{C}$, factor $p - \alpha$ as

$$p(z) - \alpha = c \prod_{i=1}^{n} (z - \beta_i).$$

Then it follows that

$$p(A) - \alpha I = c \prod_{i=1}^{n} (A - \beta_i I).$$

Because all the terms commute, $p(A) - \alpha I$ is invertible if and only if $A - \beta_i I$ is invertible for every i. Thus α lies in $\sigma(p(A))$ if and only if there is a β_i in $\sigma(A)$ such that $p(\beta_i) = \alpha$. ∎

The **spectral radius** of A is defined to be

$$\mathrm{spr}(A) = \sup_{\lambda \in \sigma(A)} |\lambda|.$$

Proposition I.2.3 *For each A in a Banach algebra $\mathfrak{A}$, the spectral radius is determined by* $\mathrm{spr}(A) = \lim_{n \to \infty} \|A^n\|^{1/n}$.

Proof. Look more carefully at the Laurent series

$$(\lambda I - A)^{-1} = \sum_{n \geq 0} \lambda^{-n-1} A^n$$

I.2. Banach Algebras Basics

for R_A at infinity. Since R_A is analytic on $\{\lambda : |\lambda| > \mathrm{spr}(A)\}$, this series converges absolutely and uniformly for $|\lambda| \geq r > \mathrm{spr}(A)$. In particular, the Taylor series coefficients $r^{-n-1}\|A^n\|$ converge to 0 for $r > \mathrm{spr}(A)$ which implies that

$$\limsup_{n\to\infty} \|A^n\|^{1/n} \leq \mathrm{spr}(A).$$

On the other hand, there is a α in $\sigma(A)$ with $|\alpha| = \mathrm{spr}(A)$. By the spectral mapping property, α^n lies in the spectrum of A^n. Hence

$$\mathrm{spr}(A) = |\alpha^n|^{1/n} \leq \|A^n\|^{1/n} \quad \text{for all} \quad n \geq 1.$$

Thus

$$\limsup_{n\to\infty} \|A^n\|^{1/n} \leq \mathrm{spr}(A) \leq \inf_{n\geq 1} \|A^n\|^{1/n},$$

which shows that the limit exists. ∎

The rest of this section is devoted to the Gelfand theory of abelian Banach algebras.

Theorem I.2.4 *The only simple abelian unital Banach algebra is* $\mathbb{C}$.

Proof. Suppose that $\mathfrak{A}$ is a unital abelian Banach algebra containing an element A which is not scalar. Let α be an element of $\sigma(A)$, and consider the closed ideal $\mathfrak{J} = \overline{(A - \alpha I)\mathfrak{A}}$. No element of the form $(A - \alpha I)B$ is invertible, and thus

$$\|(A - \alpha I)B - I\| \geq 1.$$

Therefore I is not in the closure of such elements; whence $\mathfrak{J}$ is a proper ideal.

Thus when $\mathfrak{A}$ is simple, every element of $\mathfrak{A}$ must be scalar; and therefore $\mathfrak{A} = \mathbb{C}$. ∎

A **multiplicative linear functional** on a commutative Banach algebra $\mathfrak{A}$ is a non-zero homomorphism of $\mathfrak{A}$ into $\mathbb{C}$. The set $\mathcal{M}_\mathfrak{A}$ of all multiplicative linear functionals on $\mathfrak{A}$ is called the **maximal ideal space** of $\mathfrak{A}$. We shall show that multiplicative linear functionals have norm one. So $\mathcal{M}_\mathfrak{A}$ may be endowed with the topology induced by the weak-∗ topology on the Banach space dual of $\mathfrak{A}$.

Theorem I.2.5 *The multiplicative linear functionals on a unital abelian Banach algebra are continuous of norm 1. The map taking each multiplicative linear functional to its kernel is a bijection onto the set of maximal ideals of* $\mathfrak{A}$.

Proof. Suppose that φ is a multiplicative linear functional and $A \in \mathfrak{A}$ such that $\|A\| < 1 = \varphi(A)$. Let $B = \sum_{n\geq 1} A^n$. Then since $A + AB = B$,

$$\varphi(B) = \varphi(A) + \varphi(A)\varphi(B) = 1 + \varphi(B)$$

which is absurd. So $\|\varphi\| \leq 1$. Since $\varphi(I)$ must equal 1, this is an equality.

It follows that $M = \ker \varphi$ is a closed ideal of codimension 1 in $\mathfrak{A}$, and thus is maximal. Since φ is determined by M and the fact that $\varphi(I) = 1$, this correspondence is one to one. Conversely, if M is a maximal ideal of $\mathfrak{A}$, then $\mathrm{dist}(I, M) = 1$

because the unit ball about I consists of invertible elements. It follows that the closure of M still does not contain I. As this is a larger proper ideal, we deduce that M is closed. So the quotient algebra $\mathfrak{A}/M$ is a simple abelian unital Banach algebra. By Theorem I.2.4, this quotient is isomorphic to $\mathbb{C}$. So the quotient map φ is a continuous homomorphism of $\mathfrak{A}$ onto $\mathbb{C}$ with kernel M. ∎

Corollary I.2.6 *The maximal ideal space of a unital abelian Banach algebra is a compact Hausdorff space. If $\mathfrak{A}$ is abelian but not unital, then $\mathcal{M}_\mathfrak{A}$ is locally compact.*

Proof. It is clear that a weak-$*$ limit of multiplicative linear functionals is again multiplicative. So $\mathcal{M}_\mathfrak{A}$ is a weak-$*$ closed subset of the unit ball of the dual space of $\mathfrak{A}$. By the Banach–Alaoglu Theorem, it is weak-$*$ compact and Hausdorff.

In the non-unital case, form the algebra $\mathfrak{A}^\sim$ as in Proposition I.1.3 (which will be a Banach algebra but not a C*-algebra in this case). Then $\mathfrak{A}$ itself is a maximal ideal of $\mathfrak{A}^\sim$. This corresponds to the functional $\varphi_0(A + \lambda I) = \lambda$. Thus every other maximal ideal M of $\mathfrak{A}^\sim$ must have proper intersection with $\mathfrak{A}$; and therefore yields an ideal of codimension one in $\mathfrak{A}$. This then determines a non-zero homomorphism of $\mathfrak{A}$ into $\mathbb{C}$. Conversely, if φ is a non-zero multiplicative linear functional on $\mathfrak{A}$, it is clear that $\widetilde{\varphi}(A + \lambda I) = \varphi(A) + \lambda$ is the unique extension of φ to a multiplicative linear functional on $\mathfrak{A}^\sim$. (VERIFY!) So there is a bijective correspondence between $\mathcal{M}_\mathfrak{A}$ and $\mathcal{M}_{\mathfrak{A}^\sim} \setminus \{\varphi_0\}$. In particular, this is locally compact because $\mathcal{M}_{\mathfrak{A}^\sim}$ is the one-point compactification of $\mathcal{M}_\mathfrak{A}$ and it is Hausdorff. ∎

Now we define the **Gelfand transform** Γ of a commutative Banach algebra into $C_0(\mathcal{M}_\mathfrak{A})$ by $\Gamma(A) = \hat{A}$ where

$$\hat{A}(\varphi) = \varphi(A).$$

Theorem I.2.7 *The Gelfand transform is a contractive algebra homomorphism of an abelian Banach algebra $\mathfrak{A}$ into $C_0(\mathcal{M}_\mathfrak{A})$. The image algebra separates the points of $\mathcal{M}_\mathfrak{A}$.*

Proof. The functions $\hat{A}$ are continuous because of the definition of the weak-$*$ topology. In the non-unital case, $\mathcal{M}_\mathfrak{A} = \mathcal{M}_{\mathfrak{A}^\sim} \setminus \{\varphi_0\}$. Since $\hat{A}(\varphi_0) = \varphi_0(A) = 0$, it follows that $\hat{A}$ lies in $C_0(\mathcal{M}_\mathfrak{A})$. The map is contractive because each φ is contractive and the norm on $C_0(\mathcal{M}_\mathfrak{A})$ is the sup norm. Finally, $\Gamma(\mathfrak{A})$ separates points of $\mathcal{M}_\mathfrak{A}$ because the points correspond to distinct multiplicative linear functionals of $\mathfrak{A}$. ∎

Corollary I.2.8 *In a unital commutative Banach algebra $\mathfrak{A}$, A is invertible if and only if $\hat{A}$ is invertible, which is precisely when $\hat{A}$ does not vanish on $\mathcal{M}_\mathfrak{A}$. Thus*

$$\sigma(A) = \sigma(\hat{A}) = \{\varphi(A) : \varphi \in \mathcal{M}_\mathfrak{A}\}$$

and
$$\|\hat{A}\|_\infty = \mathrm{spr}(A).$$

I.3. Commutative C*-algebras

Proof. If A has an inverse, then since Γ is a homomorphism, $\Gamma(A^{-1}) = \Gamma(A)^{-1}$. Conversely, if A is not invertible, then as in the proof of Theorem I.2.4, the ideal $\mathfrak{J} = \overline{A\mathfrak{A}}$ is proper and thus is contained in a maximal ideal M. Let φ be the associated multiplicative linear functional. It follows that $\hat{A}(\varphi) = 0$ and thus $\hat{A}$ is not invertible in $C(\mathcal{M}_{\mathfrak{A}})$. It is now immediate that $\hat{A}$ has the same spectrum as A, and that this coincides with the range of $\hat{A}$. As the norm in $C(\mathcal{M}_{\mathfrak{A}})$ is the supremum norm and the range of $\hat{A}$ is $\sigma(A)$, we conclude that $\|\hat{A}\|_\infty = \mathrm{spr}(A)$. ∎

I.3 Commutative C*-algebras

We will show that every commutative C*-algebra has the form $C_0(X)$ as in Example I.1.2. This will be applied to the structure of normal operators.

Theorem I.3.1 *Let $\mathfrak{A}$ be an abelian C*-algebra. Then the Gelfand transform is an isometric $*$-isomorphism of $\mathfrak{A}$ onto $C_0(\mathcal{M}_{\mathfrak{A}})$.*

Proof. First assume that $\mathfrak{A}$ is unital. Let φ be a multiplicative linear functional on $\mathfrak{A}$. We will show that $\varphi(A^*) = \overline{\varphi(A)}$. Suppose first that $A = A^*$ is self-adjoint. Then form a family of unitary elements

$$U_t := e^{itA} = \sum_{n \geq 0} \frac{(itA)^n}{n!} \quad \text{for} \quad t \in \mathbb{R}.$$

Indeed,

$$U_t^* = \sum_{n \geq 0} \frac{(\overline{it}A)^n}{n!} = \sum_{n \geq 0} \frac{(-itA)^n}{n!} = e^{-itA} = U_t^{-1}.$$

So each U_t is unitary. Hence $\|U_t\|^2 = \|U_t^* U_t\| = \|I\| = 1$.
Therefore

$$1 \geq |\varphi(U_t)| = \left|\sum_{n \geq 0} \frac{(it\varphi(A))^n}{n!}\right| = |e^{it\varphi(A)}| = e^{-t\,\mathrm{Im}\,\varphi(A)}.$$

As this holds for all real t, we deduce that $\varphi(A)$ is real. Now if X is an arbitrary element of $\mathfrak{A}$, write it as $X = A + iB$ where

$$A := (X + X^*)/2 \quad \text{and} \quad B := (X - X^*)/2i$$

are the self-adjoint elements known as the **real** and **imaginary parts** of X. We know that $\varphi(A)$ and $\varphi(B)$ are real, and thus

$$\varphi(X^*) = \varphi(A - iB) = \varphi(A) - i\varphi(B) = \overline{\varphi(A) + i\varphi(B)} = \overline{\varphi(X)}.$$

So the Gelfand transform Γ satisfies $\widehat{A^*} = \hat{A}^*$. Therefore Γ is a $*$-homomorphism.
For $A = A^*$, we have $\|A\|^2 = \|A^*A\| = \|A^2\|$. Thus

$$\|\hat{A}\|_\infty = \mathrm{spr}(A) = \lim_{n \to \infty} \|A^{2^n}\|^{1/2^n} = \lim_{n \to \infty} (\|A\|^{2^n})^{1/2^n} = \|A\|.$$

So $\|\hat{A}\|_\infty = \|A\|$. For a general element T, we obtain
$$\|T\|^2 = \|T^*T\| = \|\widehat{T^*T}\|_\infty = \|\hat{T}^*\hat{T}\|_\infty = \|\hat{T}\|_\infty^2.$$
Consequently, the Gelfand transform is isometric.

Thus the image of $\mathfrak{A}$ under the Gelfand map is a unital norm closed self-adjoint subalgebra of $C(X)$ which separates points. By the Stone–Weierstrass Theorem, Γ is surjective and hence is a $*$-isomorphism.

Now consider the non-unital case. By Proposition I.1.3, $\mathfrak{A}$ is contained in a unital C*-algebra $\mathfrak{A}^\sim$ as a maximal ideal of co-dimension one. Clearly $\mathfrak{A}^\sim$ is also abelian. Moreover, from Corollary I.2.6, $\mathcal{M}_\mathfrak{A}$ is canonically homeomorphic to $\mathcal{M}_{\mathfrak{A}^\sim} \setminus \{\varphi_0\}$, where φ_0 is the multiplicative linear functional on $\mathfrak{A}^\sim$ with kernel $\mathfrak{A}$. Using the argument above, we see that the Gelfand transform is an isometric $*$-isomorphism of $\mathfrak{A}^\sim$ onto $C(\mathcal{M}_{\mathfrak{A}^\sim})$. This takes $\mathfrak{A}$ onto the ideal of functions vanishing at φ_0, which is naturally identified with $C_0(\mathcal{M}_\mathfrak{A})$. ∎

In particular, this applies to normal elements in C*-algebras.

Corollary I.3.2 *If N is a normal element of a unital C*-algebra $\mathfrak{A}$, then $C^*(N)$ is isometrically $*$-isomorphic to $C(\sigma(N))$, the continuous functions on the spectrum of N, via a map that takes N to the identity function $z(t) = t$. The (not necessarily unital) C*-algebra generated by N and N^* is carried onto $C_0(\sigma(N) \setminus \{0\})$.*

Proof. Since N is normal, $C^*(N)$ is abelian. It suffices to determine the maximal ideal space X. Note that a multiplicative linear functional φ in X is determined by $\varphi(N) = \lambda$ for then $\varphi(p(N, N^*)) = p(\lambda, \bar{\lambda})$ for every polynomial p. Thus the map from X into $\mathbb{C}$ taking φ to $\varphi(N)$ is a homeomorphism onto $\hat{N}(X)$. From the Gelfand theory, $\hat{N}(X) = \sigma(N)$. This map identifies $\hat{N}$ with the identity function z as desired. By Theorem I.3.1, this map is an isometric $*$-isomorphism.

When N is not invertible, the subalgebra generated by N and N^* (but not I) corresponds to the ideal of functions vanishing at 0. ∎

We obtain some useful consequences from this basic result. The first is an important fact known as the **continuous functional calculus** for a normal element. Many useful facts follow readily from this.

Corollary I.3.3 *If N is a normal element of a unital C*-algebra and f is a continuous function on $\sigma(N)$, the operator $f(N)$ is defined as the inverse of f under the Gelfand transform of $C^*(N)$. This map is an isometric $*$-isomorphism of $C(\sigma(N))$ onto $C^*(N)$. If $0 \in \sigma(N)$ and $f(0) = 0$, then $f(N)$ lies in the non-unital algebra generated by N and N^*.*

Moreover $\sigma(f(N)) = f(\sigma(N))$. Also if g is continuous on $f(\sigma(N))$, then
$$g(f(N)) = (g \circ f)(N).$$

Proof. The first part is immediate from Corollary I.3.2. We have
$$\sigma(f(N)) = \sigma(\widehat{f(N)}) = \sigma(f) = f(\sigma(N)).$$

I.4. Positive Elements

When p is a polynomial in z and $\bar{z}$, it is immediate from the fact that the functional calculus is a homomorphism that $p(f(N)) = (p \circ f)(N)$. The general case follows by approximating the continuous function g by polynomials. ∎

Corollary I.3.4 *Let $\mathfrak{A}$ be a C*-algebra.*
  (i) *If N in $\mathfrak{A}$ is normal, then $\|N\| = \operatorname{spr}(N)$.*
  (ii) *If $A = A^*$, then $\sigma(A)$ is real.*
  (iii) *If U in $\mathfrak{A}$ is unitary, then $\sigma(U)$ is contained in the unit circle.*

Proof. By Corollary I.3.2, we have
$$\|N\| = \|\hat{N}\|_{\sigma(N)} = \|z\|_{\sigma(N)} = \operatorname{spr}(N).$$
If A is self-adjoint, then $\overline{\hat{A}} = \hat{A}^* = \hat{A}$ is a real valued function. Since $\sigma(A)$ is the range of $\hat{A}$, part (ii) follows. Similarly, if U is unitary, then
$$|\hat{U}|^2 = \hat{U}^*\hat{U} = \hat{I} = 1. \qquad \blacksquare$$

I.4 Positive Elements

Positive elements play an important role in C*-algebras. They determine an **order** on the self-adjoint elements $\mathfrak{A}_{sa}$ of $\mathfrak{A}$ by setting $A \le B$ if $B - A$ is positive. Some of the properties of positive elements require non-trivial finesse. For convenience, we will work in unital algebras. This is no serious restriction because of Proposition I.1.3. We begin with another easy consequence of the continuous functional calculus.

Corollary I.4.1 *Each positive element A of a C*-algebra has a unique positive square root.*

Proof. The square root function $f(x) = \sqrt{x}$ is a continuous function on $[0, \|A\|]$ which contains $\sigma(A)$. Thus $B := f(A)$ is self-adjoint because f is real valued; and $\sigma(B) = f(\sigma(A))$ is contained in the positive real line. So B is positive. Moreover $B^2 = f^2(A) = z(A) = A$.

Suppose that C is another positive square root of A. Then by the functional calculus,
$$C = f(C^2) = f(A) = B. \qquad \blacksquare$$

Another useful application is the following analogue of the Hahn decomposition in measure theory.

Corollary I.4.2 *If A in $\mathfrak{A}$ is self-adjoint, then there are positive elements A_+ and A_- in $\mathfrak{A}$ such that $A = A_+ - A_-$ and $A_+ A_- = 0$.*

Proof. Let $f(x) = (x + |x|)/2$ and $g(x) = f(-x)$; so that f and g are positive, $fg = 0$ and $f(x) - g(x) = x$. Set $A_+ = f(A)$ and $A_- = A_+ - A = g(A)$. Then
$$A_+ A_- = f(A)g(A) = (fg)(A) = 0. \qquad \blacksquare$$

The following lemma contains some useful characterizations of positivity. However the main result about positivity is Theorem I.4.5 below.

Lemma I.4.3 *For $A = A^*$ in a C*-algebra $\mathfrak{A}$, the following are equivalent:*

(i) $A \geq 0$;
(ii) $A = B^2$ *for some* $B = B^*$;
(iii) $\|cI - A\| \leq c$ *for all* $c \geq \|A\|$;
(iv) $\|cI - A\| \leq c$ *for some* $c \geq \|A\|$.

Proof. (i) implies (ii) is Corollary I.4.1. Assuming (ii), there is a self-adjoint element B such that $A = f(B)$, where $f \in C(\sigma(B))$ is the function $f(x) = x^2$. Consequently $\|f\|_{\sigma(B)} = \|A\|$; and thus $0 \leq f \leq \|A\| \leq c$. Therefore $0 \leq c - f \leq c$. So

$$\|cI - A\| = \|(c - f)(B)\| = \|c - f\|_{\sigma(B)} \leq c.$$

This establishes (iii), which clearly implies (iv).

Assuming that (iv) holds for a particular value of c, we conclude that

$$c \geq \|cI - A\| = \|(c - z)(A)\| = \|c - z\|_{\sigma(A)}.$$

Thus the identity function z is non-negative on $\sigma(A)$. That is, $\sigma(A)$ is contained in $\mathbb{R}_+$; and thus A is positive. ∎

Corollary I.4.4 *If A and B are positive elements of $\mathfrak{A}$, then $A + B$ is also positive.*

Proof. Choose $R \geq \|A\|$ and $S \geq \|B\|$. Then $R + S \geq \|A + B\|$ and

$$\|(R + S)I - (A + B)\| \leq \|RI - A\| + \|SI - B\| \leq R + S.$$

So by part (iv) above, we see that $A + B$ is positive. ∎

We have accumulated enough tricks to tackle the main result. Notice that the proof is quite devious considering how straightforward this fact is for operators.

Theorem I.4.5 *If A belongs to a C*-algebra $\mathfrak{A}$, then A^*A is positive.*

Proof. Let $B = A^*A$. Since this is self-adjoint, Corollary I.4.2 allows us to write $B = B_+ - B_-$ where B_+ and B_- are positive and $B_+B_- = 0$. Let C be the positive square root of B_-, and set $T = AC$. Note that since C is a limit of polynomials in B_- with zero constant coefficient, we have $CB_+ = 0$. Then

$$-T^*T = -CA^*AC = -C(B_+ - B_-)C = CB_-C = B_-^2.$$

In particular, $-T^*T$ is positive, and thus has spectrum contained in $\mathbb{R}_+$.

Now write $T = X + iY$ where X and Y are the real and imaginary parts of T. Then

$$T^*T + TT^* = (X + iY)^*(X + iY) + (X + iY)(X + iY)^* = 2(X^2 + Y^2).$$

I.4. Positive Elements

This is a sum of positive elements, and hence is positive by the previous corollary. Consequently,

$$TT^* = (T^*T + TT^*) - T^*T = (T^*T + TT^*) + B_-^2$$

is a sum of positive elements, and hence is positive.

It is a well known ring theoretic fact (see Exercise I.4) that

$$\sigma(AB) \cup \{0\} = \sigma(BA) \cup \{0\}.$$

Since TT^* has non-negative spectrum, so does T^*T. Hence $\pm T^*T$ are both positive, and therefore $\sigma(T^*T) = \{0\}$. Since T^*T is self-adjoint, its norm equals its spectral radius; so $B_-^2 = -T^*T = 0$. As B_- is self-adjoint, we obtain $B_- = 0$; and so B is positive as required. ∎

We collect a couple of other results about the order structure that will be used frequently in the future.

Corollary I.4.6 *If $A \leq B$ in $\mathfrak{A}_{sa}$ and X belongs to $\mathfrak{A}$, then $X^*AX \leq X^*BX$.*

Proof. Let C be the positive square root of $B - A$. Then

$$X^*BX - X^*AX = X^*(B - A)X = (CX)^*(CX) \geq 0. \qquad \blacksquare$$

Lemma I.4.7 *If $0 \leq A \leq B$ are invertible in $\mathfrak{A}_{sa}$, then $B^{-1} \leq A^{-1}$.*

Proof. By the previous corollary,

$$I - B^{-1/2}AB^{-1/2} = B^{-1/2}(B - A)B^{-1/2} \geq 0.$$

Thus $(A^{1/2}B^{-1/2})^*(A^{1/2}B^{-1/2}) \leq I$; whence $\|A^{1/2}B^{-1/2}\| \leq 1$. The adjoint has the same norm, so

$$I \geq (A^{1/2}B^{-1/2})(A^{1/2}B^{-1/2})^* = A^{1/2}B^{-1}A^{1/2}.$$

Multiplying on both sides by $A^{-1/2}$ and applying the previous corollary yields

$$A^{-1} = A^{-1/2}IA^{-1/2} \geq A^{-1/2}(A^{1/2}B^{-1}A^{1/2})A^{-1/2} = B^{-1}. \qquad \blacksquare$$

An **approximate identity** for a Banach algebra $\mathfrak{A}$ is a net E_λ for $\lambda \in \Lambda$ which is bounded and satisfies

$$\lim_{\lambda \in \Lambda} E_\lambda A = \lim_{\lambda \in \Lambda} AE_\lambda = A \quad \text{for all} \quad A \in \mathfrak{A}.$$

In a C*-algebra, we further stipulate that $0 \leq E_\lambda$, $\|E_\lambda\| \leq 1$, and that $E_\lambda \leq E_\mu$ when $\lambda \leq \mu$. Since Λ is directed, for each λ and μ in Λ, there is an index $\nu \in \Lambda$ such that $E_\nu \geq E_\lambda$ and $E_\nu \geq E_\mu$.

Theorem I.4.8 *Every C*-algebra has an approximate identity.*

Proof. Let $\Lambda = \{A \in \mathfrak{A} : A \geq 0 \text{ and } \|A\| < 1\}$ ordered by the order on $\mathfrak{A}_{sa}$. First let us show that Λ is directed. Suppose that $A, B \in \Lambda$. Let $f(t) = t(1-t)^{-1}$ for $0 \leq t < 1$ and $g(t) = t(1+t)^{-1} = 1 - (1+t)^{-1}$ for $t \geq 0$. Note that $g(f(t)) = t$. Set $Y = f(A) + f(B)$ and $C = g(Y)$ (which lie in $\mathfrak{A}$ because $f(0) = g(0) = 0$). Since g is positive and $\|g\|_{\sigma(Y)} < 1$, we see that C belongs to Λ.

Since $X := f(A) \leq Y$, we have $I + X \leq I + Y$; and hence by Lemma I.4.7, $(I+X)^{-1} \geq (I+Y)^{-1}$. Thus
$$A = I - (I+X)^{-1} \leq I - (I+Y)^{-1} = C.$$
Similarly $B \leq C$. So Λ is directed.

Next we need to establish that approximation on a cofinal subset suffices. Care must be taken because $0 \leq A \leq B$ does not imply that $A^2 \leq B^2$ when A and B don't commute (see Exercise I.8). Now if $0 \leq A \leq B$ belong to Λ, and X belongs to $\mathfrak{A}$, compute
$$\|X - BX\|^2 = \|X^*(I-B)^2 X\| \leq \|X^*(I-B)X\| \leq \|X^*(I-A)X\|.$$
Similarly, $\|X - XB\|^2 \leq \|X(I-A)X^*\|$.

For any $X \geq 0$, let $A_n = g(nX)$ which lies in Λ by the argument in the first paragraph. Set $h(t) = t^2(1 - g(nt)) = t^2(1+nt)^{-1} \leq t/n$. Then
$$\|X(I-A_n)X\| = \|h(X)\| \leq \|h\|_{\sigma(X)} \leq \|X\|/n.$$
Hence
$$\lim_{B \in \Lambda} \|X - BX\|^2 \leq \lim_{n \to \infty} \sup_{\substack{B \in \Lambda \\ B \geq A_n}} \|X - BX\|^2$$
$$\leq \lim_{n \to \infty} \|X(I-A_n)X\| = 0.$$
Similarly, $\lim_{B \in \Lambda} \|X - XB\|^2 = 0$.

For arbitrary X,
$$\|X - XB\|^2 = \|(I-B)X^*X(I-B)\| \leq \|X^*X - X^*XB\|.$$
This tends to 0, as does $\|X - BX\|^2$. So Λ is an approximate identity. ∎

In the separable case, we can replace nets by sequences. The following corollary is left as an exercise.

Corollary I.4.9 *If $\mathfrak{A}$ is a separable C*-algebra, then there is an increasing sequence $0 \leq E_1 \leq E_2 \leq \ldots$ of positive norm-one elements which form an approximate identity for $\mathfrak{A}$.*

I.5 Ideals, Quotients and Homomorphisms

By an **ideal** in a C*-algebra, we will mean a norm-closed two-sided ideal. One-sided ideals will be specified as left or right ideals, and are also assumed to be closed unless otherwise specified.

Lemma I.5.1 *Every ideal of a C*-algebra is self-adjoint.*

I.5. Ideals, Quotients and Homomorphisms

Proof. Let $\mathfrak{J}$ be an ideal of $\mathfrak{A}$. Then $\mathfrak{B} := \mathfrak{J} \cap \mathfrak{J}^*$ is a C*-subalgebra of $\mathfrak{A}$. Notice that $\mathfrak{B}$ contains $\mathfrak{J}\mathfrak{J}^*$. By Theorem I.4.8, $\mathfrak{B}$ contains an approximate identity E_λ. Then for any J in $\mathfrak{J}$,
$$\lim_{\lambda \in \Lambda} \|J^* - J^* E_\lambda\|^2 = \lim_{\lambda \in \Lambda} \|(JJ^* - JJ^* E_\lambda) - E_\lambda(JJ^* - JJ^* E_\lambda)\| = 0.$$
As E_λ belongs to the ideal $\mathfrak{J}$, it follows that J^* also belongs to $\mathfrak{J}$. Therefore $\mathfrak{J} = \mathfrak{J}^*$ is self-adjoint. ∎

A subalgebra $\mathfrak{B}$ of a C*-algebra $\mathfrak{A}$ is called **hereditary** if $B \in \mathfrak{B}$ is positive, $A \in \mathfrak{A}$ and $0 \le A \le B$ implies that A belongs to $\mathfrak{B}$. It is an important fact that ideals are hereditary. The proof requires a factorization trick.

Lemma I.5.2 *If $X^*X \le A$ in a C*-algebra $\mathfrak{A}$, then there is an element B in $\mathfrak{A}$ with $\|B\| \le \|A\|^{1/4}$ such that $X = BA^{1/4}$.*

Proof. Define $B_n := X(A + \frac{1}{n}I)^{-1/2}A^{1/4}$. (This lies in $\mathfrak{A}$ even if it is not unital, but the computation is done in $\mathfrak{A}^\sim$ for convenience.) Set
$$D_{nm} := (A + \tfrac{1}{n}I)^{-1/2} - (A + \tfrac{1}{m}I)^{-1/2} \quad \text{and} \quad f_n(t) := t^{3/4}(t + \tfrac{1}{n})^{-1/2}.$$
Notice that f_n converges uniformly to $t^{1/4}$ on $K = [0, \|A\|]$. Then
$$\|B_n - B_m\|^2 = \|XD_{nm}A^{1/4}\|^2 = \|A^{1/4}D_{nm}X^*XD_{nm}A^{1/4}\|$$
$$\le \|A^{1/4}D_{nm}AD_{nm}A^{1/4}\|$$
$$= \|D_{nm}A^{3/4}\|^2 = \|f_n(A) - f_m(A)\|^2 \le \|f_n - f_m\|_K^2.$$
Since f_n is Cauchy, so is B_n.

Let $B = \lim_{n \to \infty} B_n$. Then
$$BA^{1/4} = \lim_{n \to \infty} B_n A^{1/4} = \lim_{n \to \infty} X(A + \tfrac{1}{n}I)^{-1/2}A^{1/2} = X. \qquad \blacksquare$$

Theorem I.5.3 *Suppose that $\mathfrak{J}$ is an ideal of a C*-algebra $\mathfrak{A}$. If J in $\mathfrak{J}$ is positive and $A^*A \le J$, then A belongs to $\mathfrak{J}$. In particular, ideals are hereditary.*

Proof. Factor $A = BJ^{1/4}$ using the previous lemma. Then $J^{1/4}$ belongs to the non-unital algebra generated by J and hence belongs to $\mathfrak{J}$. Thus so does A. ∎

Theorem I.5.4 *If $\mathfrak{J}$ is an ideal of a C*-algebra $\mathfrak{A}$, then the quotient algebra $\mathfrak{A}/\mathfrak{J}$ is a C*-algebra.*

Proof. Let us write $\dot{A}$ for the coset $A + \mathfrak{J}$. As usual, the adjoint is defined as $(A + \mathfrak{J})^* = A^* + \mathfrak{J}$; and the norm is defined as $\|\dot{A}\| := \inf_{J \in \mathfrak{J}} \|A - J\|$. Since $\mathfrak{J}$ is self-adjoint, $\|\dot{A}^*\| = \|\dot{A}\|$. The only thing that needs to be checked is the C*-norm condition.

Let E_λ be an approximate identity for $\mathfrak{J}$. We claim that
$$\|\dot{A}\| = \lim_{\lambda \in \Lambda} \|A - AE_\lambda\|.$$

Indeed, as AE_λ belongs to $\mathfrak{J}$, $\|A - AE_\lambda\| \geq \|\dot{A}\|$. On the other hand, for $\varepsilon > 0$ there is an element J in $\mathfrak{J}$ so that $\|A - J\| < \|\dot{A}\| + \varepsilon$. So

$$\lim_{\lambda \in \Lambda} \|A - AE_\lambda\| \leq \lim_{\lambda \in \Lambda} \|(A - J)(I - E_\lambda)\| + \|J - JE_\lambda\|$$
$$\leq \|A - J\| < \|\dot{A}\| + \varepsilon.$$

Letting ε decrease to 0 establishes the claim.

Now compute

$$\|\dot{A}^*\dot{A}\| = \lim_{\lambda \in \Lambda} \|A^*A(I - E_\lambda)\| \geq \lim_{\lambda \in \Lambda} \|(I - E_\lambda)A^*A(I - E_\lambda)\|$$
$$= \lim_{\lambda \in \Lambda} \|A(I - E_\lambda)\|^2 = \|\dot{A}\|^2 = \|\dot{A}^*\|\|\dot{A}\| \geq \|\dot{A}^*\dot{A}\|$$

So this is an equality. Hence the quotient norm is a C*-norm. ∎

Now we may deduce a fundamental fact about homomorphisms between C* algebras.

Theorem I.5.5 *Let π be a non-zero *-homomorphism of a C*-algebra $\mathfrak{A}$ into another C*-algebra $\mathfrak{B}$. Then $\|\pi\| = 1$ and $\pi(\mathfrak{A})$ is a C*-subalgebra of $\mathfrak{B}$. If π is injective, then it is isometric. So in general, π factors as $\dot{\pi}q$ where q is the quotient map of $\mathfrak{A}$ onto $\mathfrak{A}/\ker \pi$ and $\dot{\pi}$ is the induced isometric *-isomorphism of the quotient onto $\pi(\mathfrak{A})$.*

Proof. If $A = A^*$ in $\mathfrak{A}_{sa}$, then the spectrum $\sigma_\mathfrak{B}(\pi(A))$ is contained in $\sigma_\mathfrak{A}(A)$; and hence

$$\|\pi(A)\| = \mathrm{spr}(\pi(A)) \leq \mathrm{spr}(A) = \|A\|.$$

For general A in $\mathfrak{A}$,

$$\|\pi(A)\|^2 = \|\pi(A^*A)\| \leq \|A^*A\| = \|A\|^2.$$

Thus $\|\pi\| \leq 1$. In particular, π is continuous. So $\ker \pi$ is closed and π factors through the quotient $\mathfrak{A}/\ker \pi$. Write $\pi = \dot{\pi}q$ where q is the quotient map and $\dot{\pi}$ is the induced map on $\mathfrak{A}/\ker \pi$.

Now $\dot{\pi}$ is injective. If $\dot{\pi}$ were not isometric, there would be an element A such that $\|\dot{\pi}(A)\| < \|A\|$. Hence $r := \|\dot{\pi}(A^*A)\| < \|A^*A\| := s$. Let f in $C([0, s])$ be defined so that $f(t) = 0$ for $0 \leq t \leq r$ and $f(s) = 1$. Then by the continuous functional calculus Corollary I.3.3,

$$0 = f(\dot{\pi}(A^*A)) = \dot{\pi}(f(A^*A)).$$

But as f does not vanish on $\sigma(A^*A)$, Corollary I.3.2 shows that $f(A^*A) \neq 0$. This contradicts the injectivity of $\dot{\pi}$. Therefore $\dot{\pi}$ must be isometric. It follows that $\pi(\mathfrak{A})$ is closed, and hence is a C*-subalgebra of $\mathfrak{B}$. Finally, since $\|\dot{\pi}\| = 1$ and q is a quotient map, it follows that $\|\pi\| = 1$ as well. ∎

As a consequence, we recover a basic isomorphism theorem for rings in the C*-algebra context.

Corollary I.5.6 *Suppose that $\mathfrak{J}$ is an ideal of a C*-algebra $\mathfrak{A}$, and that $\mathfrak{B}$ is a C*-subalgebra of $\mathfrak{A}$. Then $\mathfrak{B} + \mathfrak{J}$ is a C*-algebra, and*

$$\mathfrak{B}/(\mathfrak{B} \cap \mathfrak{J}) \simeq (\mathfrak{B} + \mathfrak{J})/\mathfrak{J}$$

*is a *-isomorphism.*

Proof. Consider the map $\pi : \mathfrak{B} \to \mathfrak{A}/\mathfrak{J}$ given by $\pi(B) = B + \mathfrak{J}$. Clearly, $\ker \pi = \mathfrak{B} \cap \mathfrak{J}$. So we may factor $\pi = \dot{\pi} q$ through the quotient map $q : \mathfrak{B} \to \mathfrak{B}/(\mathfrak{B} \cap \mathfrak{J})$. However, we also have the factorization $\pi = Qj$ where j is the natural injection of $\mathfrak{B}$ into $\mathfrak{A}$ and Q is the quotient map of $\mathfrak{A}$ onto $\mathfrak{A}/\mathfrak{J}$.

Since $\pi(\mathfrak{B})$ is closed, so is $Q^{-1}(\pi(\mathfrak{B})) = \mathfrak{B} + \mathfrak{J}$. As $\mathfrak{B} + \mathfrak{J}$ is evidently a *-algebra, it is a C*-algebra. Applying Theorem I.5.5 to the restriction of Q to $\mathfrak{B} + \mathfrak{J}$ shows that there are isometric *-isomorphisms

$$(\mathfrak{B} + \mathfrak{J})/\mathfrak{J} \simeq \pi(\mathfrak{B}) \simeq \mathfrak{B}/(\mathfrak{B} \cap \mathfrak{J}). \qquad \blacksquare$$

Recall that an algebra $\mathfrak{A}$ is called **inverse closed** if, whenever $\mathfrak{A}$ is contained in a larger algebra $\mathfrak{B}$ (in the appropriate category of algebras), an element A in $\mathfrak{A}$ is invertible in $\mathfrak{B}$ only if it is already invertible in $\mathfrak{A}$.

Corollary I.5.7 *If $\mathfrak{A}$ is a unital C*-subalgebra of $\mathfrak{B}$ and $A \in \mathfrak{A}$, then*

$$\sigma_{\mathfrak{A}}(A) = \sigma_{\mathfrak{B}}(A).$$

So C-algebras are inverse closed.*

Proof. Since $\sigma_{\mathfrak{B}}(A) \subset \sigma_{\mathfrak{A}}(A)$, it suffices to show that if A in $\mathfrak{A}$ is invertible in $\mathfrak{B}$, then the inverse lies in $\mathfrak{A}$.

First suppose that $A = A^*$. Then $\mathfrak{C} := C^*(\{A, A^{-1}\})$ is a subalgebra of $\mathfrak{B}$ isomorphic to some $C(X)$ by Theorem I.3.1. Let $\hat{A}$ denote the image of A under this isomorphism. Then $\hat{A} \neq 0$ on X. So $0 \notin \sigma_{\mathfrak{C}}(A)$ which is a subset of the real line. Choose polynomials p_n such that $p_n(x)$ converges uniformly to x^{-1} on $\sigma_{\mathfrak{C}}(A)$. Then since $\widehat{A^{-1}} = \lim_{n\to\infty} p_n(\hat{A})$, we see that $A^{-1} = \lim_{n\to\infty} p_n(A)$ belongs to $C^*(A)$, which is contained in $\mathfrak{A}$.

For general A, if A^{-1} belongs to $\mathfrak{B}$, then $(A^*A)^{-1} = A^{-1}A^{-1*}$ lies in $\mathfrak{B}$ and therefore also in $\mathfrak{A}$. Thus $(A^*A)^{-1}A^* = A^{-1}$ belongs to $\mathfrak{A}$. $\blacksquare$

I.6 Weak Topologies

There are several important topologies on $\mathcal{B}(\mathcal{H})$ that are weaker than the norm topology. We will introduce two here, and develop some others in the exercises. These topologies will be applied to relate algebraic and analytic information about C*-algebras acting on Hilbert space.

The **weak operator topology** (WOT) on $\mathcal{B}(\mathcal{H})$ is defined as the weakest topology such that the sets
$$\mathcal{W}(T, x, y) := \{A \in \mathcal{B}(\mathcal{H}) : |((T - A)x, y)| < 1\}$$
are open. The sets
$$\mathcal{W}(T_i, x_i, y_i; 1 \leq i \leq n) := \bigcap_{i=1}^{n} \mathcal{W}(T_i, x_i, y_i)$$
form a base for the WOT topology. A net T_α converges WOT to an operator T (write $T_\alpha \xrightarrow{\text{WOT}} T$) if and only if
$$\lim_\alpha (T_\alpha x, y) = (Tx, y) \quad \text{for all} \quad x, y \in \mathcal{H}.$$

Analogously, the **strong operator topology** (SOT) is defined by the open sets
$$\mathcal{S}(T, x) := \{A \in \mathcal{B}(\mathcal{H}) : \|(T - A)x\| < 1\}.$$
A net T_α converges SOT to T (write $T_\alpha \xrightarrow{\text{SOT}} T$) if and only if
$$\lim_\alpha T_\alpha x = Tx \quad \text{for all} \quad x \in \mathcal{H}.$$

Clearly, the strong operator topology is stronger than the weak operator topology but weaker than the norm topology.

It is routine to verify that the adjoint is WOT-continuous. However, it is not SOT-continuous. Indeed, consider the unilateral shift S on ℓ^2 given by $Se_n = e_{n+1}$ on the orthonormal basis $\{e_n : n \geq 0\}$. The reader may readily verify that S^{*n} converges SOT (and thus also WOT) to 0. But S^n is isometric for all $n \geq 1$ and does not converge at all in the strong operator topology; while it converges WOT to 0.

Left and right multiplication by a fixed operator is continuous in both the WOT and SOT topologies. In other words, if $T_\alpha \xrightarrow{\text{SOT}} T$, then
$$AT_\alpha \xrightarrow{\text{SOT}} AT \quad \text{and} \quad T_\alpha A \xrightarrow{\text{SOT}} TA;$$
and if $T_\alpha \xrightarrow{\text{WOT}} T$, then
$$AT_\alpha \xrightarrow{\text{WOT}} AT \quad \text{and} \quad T_\alpha A \xrightarrow{\text{WOT}} TA.$$
However multiplication is not jointly continuous in either topology. It is not even WOT-continuous when restricted to the unit ball. For example, using the unilateral shift again, we have
$$\text{WOT--}\lim_{n \to \infty} S^{*n} = \text{WOT--}\lim_{n \to \infty} S^n = 0; \quad \text{but} \quad \text{WOT--}\lim_{n \to \infty} S^{*n} S^n = I.$$

An example showing that multiplication is not jointly SOT-continuous is more delicate, and will be left to an exercise. However, multiplication is SOT-continuous

I.6. Weak Topologies

on the unit ball. Indeed, suppose that
$$S_\alpha \xrightarrow{\text{SOT}} S \quad \text{and} \quad T_\alpha \xrightarrow{\text{SOT}} T,$$
and that $\|S_\alpha\| \leq 1$ for all α. Then
$$\|(ST - S_\alpha T_\alpha)x\| \leq \|(S - S_\alpha)Tx\| + \|S_\alpha\|\|(T - T_\alpha)x\|.$$
The right hand side tends to 0, and thus SOT–$\lim_\alpha S_\alpha T_\alpha = ST$.

It is an easy application of the uniform boundedness principle to show that a *sequence* which is converging WOT or SOT is necessarily bounded. The Banach–Alaoglu Theorem can be adapted to show that the unit ball of $\mathcal{B}(\mathcal{H})$ is compact in the weak operator topology. It is not compact in the strong operator topology, as a consideration of S^n again shows.

Proposition I.6.1 *The WOT-continuous linear functionals on $\mathcal{B}(\mathcal{H})$ and the SOT-continuous linear functionals coincide, and each functional has the form*
$$f(T) = \sum_{i=1}^n (Tx_i, y_i)$$
for a finite set of vectors $x_1, \ldots, x_n, y_1, \ldots, y_n$ in $\mathcal{H}$.

Proof. The functionals described are clearly WOT-continuous, and hence are also SOT-continuous. Conversely, suppose that f is a SOT-continuous linear functional. Then $f^{-1}(\mathbb{D})$ (where $\mathbb{D}$ is the unit disk) is SOT-open, and hence contains a basic neighbourhood of the origin
$$f^{-1}(\mathbb{D}) \supset \{T \in \mathcal{B}(\mathcal{H}) : \|Tx_i\| < 1, 1 \leq i \leq n\}$$
$$\supset \{T \in \mathcal{B}(\mathcal{H}) : \sum_{i=1}^n \|Tx_i\|^2 < 1\}$$
for certain vectors $x_1, \ldots, x_n$. Therefore
$$|f(T)| \leq \Big(\sum_{i=1}^n \|Tx_i\|^2\Big)^{1/2}.$$
Define a map Φ of $\mathcal{B}(\mathcal{H})$ into $\mathcal{H}^{(n)}$, the direct sum of n copies of $\mathcal{H}$, by the formula
$$\Phi(T) = (Tx_1, Tx_2, \ldots, Tx_n).$$
We may define a linear map F on the range of Φ by $F(\Phi(T)) = f(T)$. Since $\|F\| \leq 1$, we may extend it by the Hahn–Banach Theorem to a norm one linear functional $\widetilde{F}$ on $\mathcal{H}^{(n)}$. By the Riesz representation theorem for continuous linear functionals on Hilbert space, there are vectors $y_1, \ldots, y_n$ such that
$$\widetilde{F}((v_1, \ldots, v_n)) = \sum_{i=1}^n (v_i, y_i).$$
In particular,

$$f(T) = \widetilde{F}(\Phi(T)) = \sum_{i=1}^{n}(Tx_i, y_i).$$ ∎

Since closed convex sets are the intersection of closed half spaces (corresponding to continuous linear functionals), we obtain an immediate corollary.

Corollary I.6.2 $\mathcal{B}(\mathcal{H})$ *has the same closed convex sets in the weak operator and strong operator topologies.*

While $\mathcal{B}(\mathcal{H})$ is not metrizable in either of these two topologies, it is a useful fact that the unit ball is metrizable in both.

Proposition I.6.3 *If $\mathcal{H}$ is a separable Hilbert space, then the unit ball $\mathcal{B}(\mathcal{H})_1$ of $\mathcal{B}(\mathcal{H})$ is metrizable in both the weak and strong operator topologies.*

Proof. Choose a countable sequence $x_1, x_2, \ldots$ which is dense in the unit ball of $\mathcal{H}$. Define
$$d_W(S,T) := \sum_{i \geq 1, j \geq 1} 2^{-i-j} |((S-T)x_i, x_j)|$$
and
$$d_S(S,T) := \sum_{i \geq 1} 2^{-i} \|(S-T)x_i\|$$
It is left as an exercise to verify that these are indeed metrics equivalent to the weak operator and strong operator topologies respectively on the unit ball. ∎

The order structure is intimately related to the strong operator topology. Here is an easy lemma that will prove useful.

Lemma I.6.4 *Suppose that A_α, $\alpha \in \Lambda$, is an increasing net of self-adjoint operators which is bounded above by an operator M. Then $A = \text{SOT-}\lim_\alpha A_\alpha$ exists, and is the least upper bound for the net A_α. If A_α are all projections, then the limit is the projection onto the closure of the union of the ranges of A_α.*

Proof. For each x in $\mathcal{H}$, the net $(A_\alpha x, x)$ is an increasing sequence of real numbers bounded above by (Mx, x). Thus we may define a quadratic form
$$\Lambda(x) = \lim_\alpha (A_\alpha x, x).$$
Then a linear operator is defined by the *polarization identity*
$$(Ax, y) := \tfrac{1}{4}\left(\Lambda(x+y) - \Lambda(x-y) + i\Lambda(x+iy) - i\Lambda(x-iy)\right).$$
As the corresponding identity is valid for each A_α, it follows that
$$\lim_\alpha (A_\alpha x, y) = (Ax, y)$$
for every x, y in $\mathcal{H}$. So $A = \text{WOT-}\lim_\alpha A_\alpha$. Since
$$(Ax, x) = \lim_\alpha (A_\alpha x, x) = \sup_\alpha (A_\alpha x, x),$$

I.7. The Density Theorems

it is apparent that $A \geq A_\alpha$ for all α and that no smaller operator has this property. So $A = \sup_\alpha A_\alpha$.

To obtain strong convergence, notice that if $B \geq 0$, the Cauchy-Schwarz inequality for the form $\langle x, y \rangle := (Bx, y)$ yields

$$\|Bx\|^2 = \langle x, Bx \rangle \leq \langle x, x \rangle^{1/2} \langle Bx, Bx \rangle^{1/2} = (Bx, x)^{1/2}(B^3 x, x)^{1/2}.$$

Thus for x in $\mathcal{H}$, $\alpha_0 \in \Lambda$ and any $\alpha \geq \alpha_0$,

$$\|(A - A_\alpha)x\|^2 \leq ((A - A_\alpha)x, x)^{1/2}((A - A_\alpha)^3 x, x)^{1/2}$$
$$\leq \|A - A_{\alpha_0}\|^3 \|x\|^2 ((A - A_\alpha)x, x)^{1/2}$$

The right hand side converges to 0, and hence A_α converges to A strongly.

When A_α are all projections, the limit A is also a projection because the set of projections is SOT-closed. Indeed, P is a projection if and only if $2P - I$ is a self-adjoint isometry (a **symmetry**). Clearly both the set of self-adjoint operators and the set of isometries are SOT-closed. Thus the same is true for symmetries, and thus for projections. The limit projection A satisfies $Ax = x$ for all x in $\cup_\alpha \mathrm{Ran}(A_\alpha)$ and $Ax = 0$ for all x in $\cap_\alpha \ker(A_\alpha)$. This is evidently the projection onto $\overline{\cup_\alpha \mathrm{Ran}(A_\alpha)}$. ∎

I.7 The Density Theorems

A C*-subalgebra of $\mathcal{B}(\mathcal{H})$ which contains the identity operator and is closed in the weak operator topology is called a **von Neumann algebra**. If $\mathcal{S}$ is any subset of $\mathcal{B}(\mathcal{H})$, let the **commutant** of $\mathcal{S}$ be

$$\mathcal{S}' := \{T \in \mathcal{B}(\mathcal{H}) : ST = TS \text{ for all } S \in \mathcal{S}\}.$$

It is easy to verify that if $\mathcal{S}$ is self-adjoint, then $\mathcal{S}'$ is a self-adjoint unital algebra. Moreover, it is WOT-closed. Indeed, if $T_\alpha \in \mathcal{S}'$ and $T_\alpha \xrightarrow{\text{WOT}} T$, then for every S in $\mathcal{S}$

$$ST = \text{WOT--lim}_\alpha ST_\alpha = \text{WOT--lim}_\alpha T_\alpha S = TS.$$

In particular, the commutant of a C*-algebra is always a von Neumann algebra. The von Neumann algebra generated by an operator T is denoted by $W^*(T)$.

The following important result relating the **double commutant** $\mathfrak{A}'' := (\mathfrak{A}')'$ to the WOT-closure is called the **von Neumann Double Commutant Theorem**.

Theorem I.7.1 *Suppose that $\mathfrak{A}$ is a C*-subalgebra of $\mathcal{B}(\mathcal{H})$ with trivial null space. Then*

$$\mathfrak{A}'' = \overline{\mathfrak{A}}^{\text{WOT}} = \overline{\mathfrak{A}}^{\text{SOT}}.$$

Proof. Clearly $\overline{\mathfrak{A}}^{\text{SOT}} \subset \overline{\mathfrak{A}}^{\text{WOT}} \subset \mathfrak{A}''$ since the SOT is stronger than the WOT; and since $\mathfrak{A}''$ is WOT-closed and contains $\mathfrak{A}$. So fix an operator T in $\mathfrak{A}''$ and vectors $x_1, \ldots, x_n$. It suffices to find A in $\mathfrak{A}$ such that $\sum_{i=1}^n \|(T - A)x_i\|^2 < 1$ as this represents a basic SOT open neighbourhood of T.

First consider $n = 1$. Let P be the (orthogonal) projection onto the subspace $\overline{\mathfrak{A} x_1}$. Then P belongs to $\mathfrak{A}'$. Indeed, $\mathfrak{A} P \mathcal{H} \subset P \mathcal{H}$ and thus $PAP = AP$ for every A in $\mathfrak{A}$. Therefore
$$PA = (A^*P)^* = (PA^*P)^* = PAP = AP.$$
If $y = P^\perp x_1$, then $\mathfrak{A} y = \mathfrak{A} P^\perp x_1 = P^\perp \mathfrak{A} x_1 = 0$. As $\mathfrak{A}$ has trivial null space, $y = 0$; that is, x_1 lies in $\overline{\mathfrak{A} x_1}$. So $PT = TP$, and Tx_1 belongs to $\overline{\mathfrak{A} x_1}$. Therefore there is an operator A in $\mathfrak{A}$ such that $\|(T - A)x_1\| < 1$.

Now consider $n \geq 2$. Form the Hilbert space $\mathcal{H}^{(n)}$ which is a direct sum of n copies of $\mathcal{H}$. Let $A^{(n)}$ denote the operator on $\mathcal{H}^{(n)}$ given by
$$A^{(n)}(v_1, \ldots, v_n) = (Av_1, \ldots, Av_n).$$
This is the direct sum of n copies of A. Then let $\mathfrak{A}^{(n)} := \{A^{(n)} : A \in \mathfrak{A}\}$. An operator X in $\mathcal{B}(\mathcal{H}^{(n)})$ can be represented as an $n \times n$ matrix with coefficients X_{ij} in $\mathcal{B}(\mathcal{H})$.

We wish to compute $\mathfrak{A}^{(n)''}$. It is easy to see that an operator $X = [X_{ij}]$ lies in $\mathfrak{A}^{(n)'}$ if and only if each matrix entry X_{ij} belongs to $\mathfrak{A}'$. Hence an operator $T = [T_{ij}]$ in $\mathfrak{A}^{(n)''}$ must commute with each matrix unit E_{ij}, the operator with (i, j) coefficient equal to I and all other coefficients equal to 0. A routine calculation shows that this forces T to be diagonal with $T_{ii} = T_{11}$ for $1 \leq i \leq n$. That is, $T = T_{11}^{(n)}$. In addition, T commutes with $X^{(n)}$ for each X in $\mathfrak{A}'$, which means T_{11} belongs to $\mathfrak{A}''$. So $\mathfrak{A}^{(n)''} = (\mathfrak{A}'')^{(n)}$.

Now apply the $n = 1$ case to $T^{(n)}$ in $(\mathfrak{A}'')^{(n)}$ and $\mathbf{x} = (x_1, \ldots, x_n)$ to obtain an operator A in $\mathfrak{A}$ such that
$$1 > \|(T^{(n)} - A^{(n)})\mathbf{x}\|^2 = \sum_{i=1}^{n} \|(T - A)x_i\|^2.$$
Thus A lies in the given SOT neighbourhood of T. ∎

There is a minor deficiency in the Double Commutant Theorem because of the fact that convergent nets need not be bounded. In order to obtain a limit A, it may be necessary to use operators of arbitrarily large norm. This is remedied in **Kaplansky's Density Theorem**. First we need a technical result. Say that a function f on $\mathbb{R}$ is **strongly continuous** on $\mathcal{B}(\mathcal{H})_{sa}$ provided that whenever A_α is a net of self-adjoint operators with SOT–$\lim_\alpha A_\alpha = A$, then
$$\text{SOT–}\lim_\alpha f(A_\alpha) = f(A).$$

Lemma I.7.2 *Every continuous function f on $\mathbb{R}$ such that*
$$\limsup_{|t| \to \infty} \frac{|f(t)|}{|t|} < \infty$$
is strongly continuous on $\mathcal{B}(\mathcal{H})$.

I.7. The Density Theorems

Proof. Let $\mathcal{S}$ denote the family of strongly continuous functions on $\mathbb{R}$; and let $\mathcal{S}^b$ denote the set of bounded functions in $\mathcal{S}$. Since $\|f(A)\| \leq \|f\|_\infty$, it is easily verified that $\mathcal{S}$ and $\mathcal{S}^b$ are closed under uniform limits.

First we show that $\mathcal{S}^b \mathcal{S}$ is contained in $\mathcal{S}$. Indeed, suppose that f is in $\mathcal{S}^b$, g is in $\mathcal{S}$ and SOT–$\lim_\alpha A_\alpha = A$. Then for x in $\mathcal{H}$,

$$\|(fg(A_\alpha) - fg(A))x\| \leq \|f(A_\alpha)\| \|(g(A_\alpha) - g(A))x\|$$
$$+ \|(f(A_\alpha) - f(A))g(A)x\|.$$

Since $\|f(A_\alpha)\| \leq \|f\|_\infty$, both terms on the right tend to 0. Thus fg is strongly continuous. In particular, $\mathcal{S}^b$ is an algebra.

Next we show that $h(t) = (1+t^2)^{-1}$ belongs to $\mathcal{S}^b$. Indeed,

$$\|(h(A_\alpha) - h(A))x\| = \|((I+A_\alpha^2)^{-1}(A^2 - A_\alpha^2)(I+A^2)^{-1})x\|$$
$$= \|((I+A_\alpha^2)^{-1}(A_\alpha(A - A_\alpha) + (A - A_\alpha)A)(I+A^2)^{-1})x\|$$
$$\leq \|(I+A_\alpha^2)^{-1}A_\alpha\|\|(A - A_\alpha)(I+A^2)^{-1}x\|$$
$$+ \|(I+A_\alpha^2)^{-1}\|\|(A - A_\alpha)A(I+A^2)^{-1}x\|$$
$$\leq \|t(1+t^2)^{-1}\|_\infty \|(A - A_\alpha)y\| + \|(1+t^2)^{-1}\|_\infty \|(A - A_\alpha)z\|$$
$$= \tfrac{1}{2}\|(A - A_\alpha)y\| + \|(A - A_\alpha)z\|$$

where $y = (I+A^2)^{-1}x$ and $z = A(I+A^2)^{-1}x$. When SOT–$\lim_\alpha A_\alpha = A$, the right hand side converges to 0. Thus SOT–$\lim_\alpha h(A_\alpha) = h(A)$.

Clearly, $\mathcal{S}$ is invariant under translation and dilation by $\mathbb{R}$. Therefore the functions $h_s(t) = (1+s^2t^2)^{-1}$ belong to $\mathcal{S}^b$ for all real s. Since $z(t) = t$ belongs to $\mathcal{S}$, the multiplicative property shows that $k_s(t) = s t\, h_s(t)$ belongs to $\mathcal{S}^b$ for s in $\mathbb{R}$. In fact, these functions belong to $C_0(\mathbb{R})$. Moreover they are real valued and separate points in $\mathbb{R}$. Thus by the Stone–Weierstrass Theorem, they generate $C_0(\mathbb{R})$ as a uniformly closed algebra.

Now suppose that f is a continuous function on $\mathbb{R}$ such that $|f(t)| \leq C|t|$ for $|t| \geq 1$. Then $g(t) = (1+t^2)^{-1}f(t)$ belongs to $C_0(\mathbb{R})$ and hence lies in $\mathcal{S}^b$. Thus

$$zg(t) = t(1+t^2)^{-1}f(t)$$

belongs to $\mathcal{S}$ by the multiplicative property. However, this function is also bounded (by C); and so lies in $\mathcal{S}^b$. Thus multiplying by z again, we deduce that

$$z^2 g(t) = t^2(1+t^2)^{-1}f(t)$$

belongs to $\mathcal{S}$. Hence $g + z^2 g = f$ belongs to $\mathcal{S}$. In other words, f is strongly continuous. ∎

We are now ready to prove Kaplansky's density theorem.

Theorem I.7.3 *If $\mathfrak{A}$ is a C*-subalgebra of $\mathcal{B}(\mathcal{H})$ with trivial null space, then the unit ball of $\mathfrak{A}_{sa}$ is SOT-dense in the unit ball of $\mathfrak{A}''_{sa}$ and the unit ball of $\mathfrak{A}$ is SOT-dense in the unit ball of $\mathfrak{A}''$. Likewise, the positive operators in the unit ball of $\mathfrak{A}$*

are SOT-*dense in the positive contractions in* $\mathfrak{A}''$. *In addition, if* $\mathfrak{A}$ *is unital, the unitary group of* $\mathfrak{A}$ *is* SOT-*dense in the unitary group of* $\mathfrak{A}''$.

Proof. Suppose that T belongs to $\mathfrak{A}''_{sa}$ and $\|T\| \leq 1$. By the Double Commutant Theorem I.7.1, there is a net A_α in $\mathfrak{A}$ such that $T = \text{WOT-lim}_\alpha A_\alpha$. Since the adjoint is weakly continuous,

$$T = \text{WOT-lim}_\alpha (A_\alpha + A_\alpha^*)/2.$$

Since $\mathfrak{A}_{sa}$ is convex, Corollary I.6.2 shows that $\overline{\mathfrak{A}_{sa}}^{\text{SOT}} = \overline{\mathfrak{A}_{sa}}^{\text{WOT}}$. Hence there is a net B_β in $\mathfrak{A}_{sa}$ such that $T = \text{SOT-lim}_\beta B_\beta$.

Let

$$f(t) = (t \wedge 1) \vee (-1) = \begin{cases} 1 & \text{for} & t \geq 1 \\ t & \text{for} & -1 \leq t \leq 1 \\ -1 & \text{for} & t \leq -1 \end{cases}$$

By Lemma I.7.2, f is strongly continuous. Hence

$$T = f(T) = \text{SOT-lim}_\beta f(B_\beta).$$

Since $\|f(B_\beta)\| \leq \|f\|_\infty = 1$, this is the desired net in the unit ball of $\mathfrak{A}_{sa}$.

The same argument works for the positive part by using the function

$$g(t) = (t \wedge 1) \vee 0.$$

Similarly, when $\mathfrak{A}$ is unital, suppose that U is a unitary operator in $\mathfrak{A}''$. By the spectral theorem (see the next section), we can write $U = e^{iT}$ for some T in $\mathfrak{A}''_{sa}$. Let $T = \text{SOT-lim}_\beta B_\beta$ be a limit of self-adjoint elements. Since $h(t) = e^{it}$ is strongly continuous by Lemma I.7.2, we obtain

$$U = h(T) = \text{SOT-lim}_\beta h(B_\beta).$$

Since the $h(B_\beta)$ are unitary in $\mathfrak{A}$, the result follows.

Finally consider the case of the whole unit ball. Let X be an element of $\mathfrak{A}''$ with $\|X\| \leq 1$. Form the C*-algebra $\mathcal{M}_2(\mathfrak{A})$ of 2×2 matrices with coefficients in $\mathfrak{A}$ acting on $\mathcal{H} \oplus \mathcal{H}$. It is readily apparent that

$$\mathcal{M}_2(\mathfrak{A})'' = \overline{\mathcal{M}_2(\mathfrak{A})}^{\text{WOT}} = \mathcal{M}_2(\overline{\mathfrak{A}}^{\text{WOT}}) = \mathcal{M}_2(\mathfrak{A}'').$$

In particular, $T = \begin{bmatrix} 0 & X \\ X^* & 0 \end{bmatrix}$ belongs to the unit ball of $\mathcal{M}_2(\mathfrak{A}'')_{sa}$. Hence it is the SOT-limit of a net in the unit ball of $\mathcal{M}_2(\mathfrak{A})_{sa}$. For such a net to converge, each matrix coefficient must converge SOT. Thus the $(1,2)$ entries form a net in the unit ball of $\mathfrak{A}$ converging SOT to X. ∎

Remark I.7.4 When $\mathcal{H}$ is separable, the unit ball of $\mathcal{B}(\mathcal{H})$ in the strong operator topology is metrizable by Proposition I.6.3. Thus the nets in the Kaplansky Density Theorem may be replaced by sequences.

I.8 Some Operator Theory

Recall that a **partial isometry** is an operator U such that $U = UU^*U$. Associated to a partial isometry U is its **initial projection** $P = U^*U$ and it **range projection** $Q = UU^*$. Then U maps $P\mathcal{H}$ isometrically onto $Q\mathcal{H}$, and vanishes on $P^\perp \mathcal{H}$. The **polar decomposition** of an operator T in $\mathcal{B}(\mathcal{H})$ is a factorization $T = UA$ where A is positive and U is a partial isometry with initial space $\overline{\text{Ran}(A)}$ and range space $\overline{\text{Ran}(T)}$.

Theorem I.8.1 *Every operator T on a Hilbert space $\mathcal{H}$ has a unique polar decomposition $T = UA$. The positive operator $A = |T| := (T^*T)^{1/2}$ lies in $C^*(T)$; and the partial isometry U belongs to $W^*(T)$. If T is invertible, then U is a unitary element of $C^*(T)$.*

Proof. If $T = UA$, then $T^*T = A^*U^*UA = A^*A = A^2$. By Corollary I.4.1, A is the unique positive square root of T^*T, known as $|T|$. If x is in $\mathcal{H}$,

$$\|Ax\|^2 = (Ax, Ax) = (A^*Ax, x) = (T^*Tx, x) = (Tx, Tx) = \|Tx\|^2.$$

Thus we may define an isometric operator U on $\text{Ran}(A)$ by $U(Ax) = Tx$. Clearly, the range is precisely $\text{Ran}(T)$. Extend U by continuity to the closure $\overline{\text{Ran}(A)}$. Then define U to be 0 on $\text{Ran}(A)^\perp = \ker(A) = \ker(T)$ and extend by linearity to all of $\mathcal{H}$. By construction, U is a partial isometry with initial space equal to $\overline{\text{Ran}(A)}$ and range $\overline{\text{Ran}(T)}$. Moreover, the choice of U is uniquely determined since we require that $U(Ax) = Tx$ and that the range of U^*U equals $\overline{\text{Ran}(A)}$.

To verify that U belongs to $W^*(T)$, the double commutant theorem states that it suffices to show that U belongs to $C^*(T)''$. For X in $C^*(T)'$ and x in $\ker(T)$, it follows that $TXx = XTx = 0$; so Xx belongs to $\ker(T)$ as well. Therefore $UXx = 0 = XUx$ for every x in $\ker(T)$. Also if $x = Ay$ lies in $\text{Ran}(A)$,

$$UXx = UXAy = (UA)Xy = TXy = XTy = XUAy = XUx.$$

By continuity and linearity, it follows that X and U commute.

When T is invertible, we have the identity $U = T(T^*T)^{-1/2}$ which lies in $C^*(T)$. Also

$$U^*U = (T^*T)^{-1/2}T^*T(T^*T)^{-1/2} = I.$$

Since U is invertible, this shows that U is unitary. ∎

Corollary I.8.2 *If T is an operator on a Hilbert space $\mathcal{H}$, then the projection $[\text{Ran }T]$ onto the closure of the range of T and the projection $[\ker(T)]$ onto the kernel of T belong to $W^*(T)$.*

Proof. If $T = U|T|$ is the polar decomposition, then

$$[\text{Ran }T] = UU^* \quad \text{and} \quad [\ker(T)] = I - U^*U.$$

∎

In fact, von Neumann algebras contain a plentiful collection of projections. Recall that the spectral theorem for self-adjoint operators expresses an operator $A = A^*$ as an integral $A = \int_{\sigma(A)} \lambda \, E_A(d\lambda)$ where E_A is the spectral measure for A. We will prove the spectral theorem in Chapter II. For the moment, we just recover a few of its components that we need at this stage.

Let $\mathcal{O}$ be an open subset of $\mathbb{R}$. We will show that the spectral projection $E_A(\mathcal{O})$ belongs to $W^*(A)$. Indeed, consider the collection $\mathcal{F}_\mathcal{O}$ of all non-negative continuous functions f of compact support in $\mathbb{R}$ such that $f \leq \chi_\mathcal{O}$. This set is upwards directed in the usual order. Let $P = \text{SOT-lim}_{f \in \mathcal{F}_\mathcal{O}} f(A)$, which exists by Lemma I.6.4. We will show that $P = E_A(\mathcal{O})$.

First note that P is a projection. Since $f(A) \leq I$ for all f in $\mathcal{F}_\mathcal{O}$, we have $P \leq I$. Since P lies in $W^*(A)$ by construction, it commutes with each $f(A)$. For each f in $\mathcal{F}_\mathcal{O}$, one also has $f^{1/2}$ in $\mathcal{F}_\mathcal{O}$. Hence $P^2 \geq f^{1/2}(A)^2 = f(A)$. (Careful, this uses commutativity!) As this is true for all $f \in \mathcal{F}_\mathcal{O}$, we have $I \geq P^2 \geq P$. Thus $\sigma(P)$ is contained in $\{0, 1\}$; and so P is a projection.

If g in $C(\mathbb{R})$ has compact support in $\mathcal{O}$, then there is a function f in $\mathcal{F}_\mathcal{O}$ such that $g = fg$. Hence (again using commutativity)

$$g(A) \geq Pg(A) \geq f(A)g(A) = g(A).$$

Taking strong limits shows that $Pg(A) = g(A)$ whenever $g = g\chi_\mathcal{O}$. On the other hand, if $g|_\mathcal{O} = 0$, then $f(A)g(A) = 0$ for every f in $\mathcal{F}_\mathcal{O}$. Thus $Pg(A) = 0$. Consequently, P is the projection onto the closed union of the ranges of all $g(A)$ with support contained in $\mathcal{O}$. This is the spectral projection $E_A(\mathcal{O})$.

Now if $\mathcal{X}$ is a G_δ subset of $\mathbb{R}$, then it is the intersection of all open sets containing it. Thus we may similarly show that $E_A(\mathcal{X}) = \inf_{\mathcal{O} \supset \mathcal{X}} E_A(\mathcal{O})$. Again this spectral projection lies in $W^*(A)$. Since every Borel measurable set agrees with a G_δ set up to a null set, this shows that every spectral projection of A belongs to $W^*(A)$.

It is not difficult to recover the integral formula for A. For convenience assume that $\|A\| \leq 1$. It is an exercise to show that the Riemann sums satisfy

$$\sum_{k=-n}^{n} \tfrac{k}{n} E_A([\tfrac{k}{n}, \tfrac{k+1}{n})) \leq A \leq \sum_{k=-n}^{n} \tfrac{k+1}{n} E_A((\tfrac{k}{n}, \tfrac{k+1}{n}])$$

and that both sides converge to A.

We will use these spectral projections to show that there are abundant projections in every von Neumann algebra.

Theorem I.8.3 *If $\mathfrak{A}$ is a von Neumann algebra, then the closed convex hull of the set of projections in $\mathfrak{A}$ equals the set $\mathfrak{A}_{+1}$ of positive contractions in $\mathfrak{A}$; and the closed convex hull of the set of symmetries equals the set $\mathfrak{A}_{sa1}$ of self-adjoint contractions in $\mathfrak{A}$.*

I.8. Some Operator Theory

Proof. Let $0 \leq A \leq I$ be a positive contraction in $\mathfrak{A}$. Then

$$A = \lim_{n\to\infty} \frac{1}{n} \sum_{i=1}^{n} E_A((\tfrac{k}{n}, 1]).$$

Analogously, if $-I \leq B \leq I$, then

$$B = \lim_{n\to\infty} \frac{1}{n} \sum_{i=1}^{n} 2 E_B((\tfrac{2k}{n} - 1, 1]) - I. \qquad \blacksquare$$

Since a C*-algebra may not have any projections at all, this result does not generalize to arbitrary C*-algebras. Surprisingly, we are able to show that the unit ball of a unital C*-algebra is always the closed convex hull of the unitary elements. This is true even though they are not the full set of extreme points, nor is there necessarily any compactness of the unit ball in some weak topology. This result is known as the **Russo–Dye Theorem**; however, the proof given here is a stronger variant of the original.

Theorem I.8.4 *In any unital C*-algebra $\mathfrak{A}$, the closed convex hull of the set of unitary elements is the whole unit ball.*

Proof. It suffices to show that if $\|A\| < 1$, then A is in the convex hull of a set of unitary operators. When A is invertible, this is easy. For in this case, the polar decomposition $A = U|A|$ lies in $\mathfrak{A}$ by Theorem I.8.1. It is a simple computation to verify that $V_\pm := |A| \pm i(I - A^*A)^{1/2}$ are both unitary. Then

$$A = (UV_+ + UV_-)/2.$$

Next we show that if $\|A\| < 1$ and U is unitary, then $U + A$ is the sum of two unitaries. Indeed, $(U + A)/2 = U(I + U^*A)/2$ is an invertible contraction since $\|U^*A\| < 1$. So this is the average of two unitaries U_1 and V_1 by the previous paragraph. That is

$$U + A = U_1 + V_1.$$

By induction, we will show that there are unitaries such that

$$U + nA = U_1 + \cdots + U_n + V_n.$$

Indeed, assuming this holds, find unitaries U_{n+1} and V_{n+1} such that

$$V_n + A = U_{n+1} + V_{n+1}.$$

Then

$$U + (n+1)A = U_1 + \cdots + U_n + (V_n + A)$$
$$= U_1 + \cdots + U_n + U_{n+1} + V_{n+1}.$$

Now suppose that $\|T\| < 1 - \frac{2}{n}$. Then let $A = \frac{n}{n-1}T - \frac{1}{n-1}I$ and $U = I$. Since

$$\|A\| < \tfrac{n}{n-1} \tfrac{n-2}{n} + \tfrac{1}{n-1} = 1,$$

we obtain unitaries $U_1, \ldots, U_{n-1}$ and V_{n-1} such that

$$nT = I + (n-1)A = U_1 + \cdots + U_{n-1} + V_{n-1}.$$

Hence T is the average of n unitaries.

It follows that the convex hull of the unitary group includes the whole open ball. So the norm closure is equal to the closed unit ball. ∎

I.9 Representations of C*-algebras

In this section, we will develop the basic properties of representations of C*-algebras. This will culminate in a proof of the fundamental theorem of Gelfand and Naimark that every C*-algebra is isomorphic to a concrete C*-algebra of operators.

A **representation** π of a C*-algebra $\mathfrak{A}$ on a Hilbert space $\mathcal{H}$ is a $*$-homomorphism of $\mathfrak{A}$ into $\mathcal{B}(\mathcal{H})$. We say that π is **topologically irreducible** if $\pi(\mathfrak{A})$ has no proper closed invariant subspaces. It is called **algebraically irreducible** if it has no proper invariant manifolds (subspaces that are not necessarily closed). Our first theorem shows that these two notions coincide for C*-algebras. So we will call π **irreducible** when this condition holds.

Lemma I.9.1 *Let π be a representation of a C*-algebra $\mathfrak{A}$ on a Hilbert space $\mathcal{H}$. Then $\pi(\mathfrak{A})$ is topologically irreducible if and only if $\pi(\mathfrak{A})' = \mathbb{C}I$.*

Proof. If $\pi(\mathfrak{A})'$ is larger than the scalars, then it contains a non-scalar positive operator; and thus by Theorem I.8.3, it contains a proper projection P. Thus $P\mathcal{H}$ is an invariant subspace for $\pi(\mathfrak{A})$.

Conversely, suppose that $\mathcal{M}$ is a proper invariant subspace for $\pi(\mathfrak{A})$; and let P be the orthogonal projection onto $\mathcal{M}$. Invariance is expressed algebraically as $\pi(A)P = P\pi(A)P$ for every A in $\mathfrak{A}$. However, it then follows that

$$P\pi(A) = (\pi(A^*)P)^* = (P\pi(A^*)P)^* = P\pi(A)P = \pi(A)P$$

for every A in $\mathfrak{A}$. Thus P is a non-scalar operator in $\pi(\mathfrak{A})'$. ∎

Lemma I.9.2 *Let π be a topologically irreducible representation of a C*-algebra $\mathfrak{A}$ on a Hilbert space $\mathcal{H}$. Suppose that T in $\mathcal{B}(\mathcal{H})$, a finite dimensional subspace $\mathcal{K} \subset \mathcal{H}$ and $\varepsilon > 0$ are given. Then there is an element A in $\mathfrak{A}$ such that*

$$\|A\| \leq \|T|_\mathcal{K}\| \quad \text{and} \quad \|(\pi(A) - T)|_\mathcal{K}\| < \varepsilon.$$

Proof. Since $\pi(\mathfrak{A})$ is irreducible, Lemma I.9.1 shows that the commutant of $\pi(\mathfrak{A})$ is just the scalars. Hence $\pi(\mathfrak{A})'' = \mathcal{B}(\mathcal{H})$. By the Double Commutant Theorem I.7.1, $\overline{\pi(\mathfrak{A})}^{\text{SOT}} = \mathcal{B}(\mathcal{H})$.

Assume that $\|T|_\mathcal{K}\| = 1$ and let $S = TP_\mathcal{K}$. By Kaplansky's Density Theorem I.7.3, there is an element A_1 in $\mathfrak{A}$ with $\|\pi(A_1)\| \leq 1$ such that

$$\|(\pi(A_1) - S)P_\mathcal{K}\| < \varepsilon/2.$$

I.9. Representations of C*-algebras

By Theorem I.5.5, there is an element A_2 in $\mathfrak{A}$ such that $\|A_2\| < (1-\varepsilon/2)^{-1}$ such that $\pi(A_2) = \pi(A_1)$. Let $A = (1-\varepsilon/2)A_2$. Then $\|A\| \leq 1$ and

$$\|(\pi(A) - T)|_\mathcal{K}\| \leq \|(\pi(A_1) - S)P_\mathcal{K}\| + \varepsilon/2\|\pi(A_1)\| < \varepsilon. \qquad \blacksquare$$

Theorem I.9.3 *Every topologically irreducible representation π of a C*-algebra $\mathfrak{A}$ is algebraically irreducible.*

To prove this theorem, it suffices to show that if x and y are unit vectors in $\mathcal{H}$, then there is an element A in $\mathfrak{A}$ such that $\pi(A)x = y$. So this result clearly follows from the following stronger variant known as **Kadison's Transitivity Theorem**, which is the C*-algebra version of Jacobson's density theorem for rings.

Theorem I.9.4 *Let π be a topologically irreducible representation of a C*-algebra $\mathfrak{A}$ on a Hilbert space $\mathcal{H}$. Suppose that T in $\mathcal{B}(\mathcal{H})$, a finite dimensional subspace $\mathcal{K} \subset \mathcal{H}$ and $\varepsilon > 0$ are given. Then there is an element A in $\mathfrak{A}$ such that*

$$\pi(A)|_\mathcal{K} = T|_\mathcal{K} \quad \text{and} \quad \|A\| \leq \|T\| + \varepsilon.$$

Proof. Use Lemma I.9.2 to find A_0 in $\mathfrak{A}$ such that

$$\|A_0\| \leq \|T\| \quad \text{and} \quad \|(\pi(A_0) - T)P_\mathcal{K}\| < \varepsilon/2.$$

Recursively find A_n in $\mathfrak{A}$ such that $\|A_n\| < 2^{-n}\varepsilon$ such that

$$\left\|\left(\sum_{k=0}^n \pi(A_k) - T\right)P_\mathcal{K}\right\| < 2^{-n-1}\varepsilon.$$

Indeed, suppose that this holds for n. Apply Lemma I.9.2 to the operator

$$S = \sum_{k=0}^n \pi(A_k) - T,$$

the subspace $\mathcal{K}$ and $2^{-n-2}\varepsilon$ to obtain an approximant A_{n+1} in $\mathfrak{A}$ with

$$\|A_{n+1}\| \leq 2^{-n-1}\varepsilon \quad \text{and} \quad \left\|\left(\sum_{k=0}^{n+1} \pi(A_k) - T\right)P_\mathcal{K}\right\| < 2^{-n-2}\varepsilon.$$

Let $A = \sum_{k=0}^\infty A_k$. It is clear that $\pi(A)|_\mathcal{K} = T|_\mathcal{K}$ and

$$\|A\| \leq \|T\| + \sum_{n=1}^\infty 2^{-n}\varepsilon = \|T\| + \varepsilon. \qquad \blacksquare$$

A **positive linear functional** on a C*-algebra is a linear functional such that $f(A) \geq 0$ whenever $A \geq 0$. A **state** is a positive linear functional of norm 1.

For example, if π is a representation of $\mathfrak{A}$ on a Hilbert space $\mathcal{H}$ and x is a vector in $\mathcal{H}$, then

$$f(A) := (\pi(A)x, x)$$

is positive. Indeed, if $A \geq 0$, then
$$f(A) = (\pi(A^{1/2})^2 x, x) = \|\pi(A^{1/2})x\|^2 \geq 0.$$
In the unital case, this is a state when $\|x\| = 1$.

Associated to any positive linear functional is a positive semidefinite sesquilinear form on $\mathfrak{A}$ given by
$$[A, B] := f(B^*A).$$
That is, $[\cdot, \cdot]$ is linear in the first variable, conjugate linear in the second, and $[A, A] \geq 0$ for all A in $\mathfrak{A}$. Hence it satisfies the **Cauchy-Schwarz inequality**
$$|[A, B]|^2 \leq [A, A][B, B]$$
or equivalently
$$|f(B^*A)|^2 \leq f(A^*A)f(B^*B).$$

Lemma I.9.5 *Positive linear functionals are continuous. If E_λ is an approximate unit for $\mathfrak{A}$, then $\|f\| = \lim_\lambda f(E_\lambda)$. In particular, when $\mathfrak{A}$ is unital, $\|f\| = f(I)$.*

Proof. As the unital case is more straightforward, we prove it first. Positivity implies that if $A \leq B$, then $f(A) \leq f(B)$. So for $0 \leq A \leq I$, we have
$$0 \leq f(A) \leq f(I).$$
Now if X is in $\mathfrak{A}$ with $\|X\| \leq 1$, then $0 \leq X^*X \leq I$ and so
$$|f(X)|^2 = |f(I^*X)|^2 \leq f(X^*X)f(I) \leq f(I)^2.$$

In general, suppose that f is not bounded on $\mathfrak{A}_{+1}$. Then there are A_k in $\mathfrak{A}_{+1}$ such that $f(A_k) > 2^k$ for $k \geq 1$. Let $A = \sum_{k \geq 1} 2^{-k} A_k$. Then
$$\infty > f(A) \geq f\left(\sum_{k=1}^n 2^{-k} A_k\right) = \sum_{k=1}^n 2^{-k} f(A_k) > n.$$
This holds for all n which is absurd; thus f is bounded on $\mathfrak{A}_{+1}$. However, an arbitrary element X in $\mathfrak{A}$ may be written as
$$X = (X + X^*)/2 + i(X - X^*)/2i = A_1 - A_2 + iA_3 - iA_4$$
where $(X + X^*)/2 = A_1 - A_2$ and $(X - X^*)/2i = A_3 - A_4$ are the Hahn decompositions of Corollary I.4.2. Thus $\|A_i\| \leq \|X\|$ for $1 \leq i \leq 4$. Consequently, we deduce that f is bounded on the whole unit ball of $\mathfrak{A}$.

Let $M = \lim_\lambda f(E_\lambda)$, which exists because this is an increasing net bounded above by $\|f\|$. Then for any X in $\mathfrak{A}$ with $\|X\| \leq 1$,
$$|f(X)|^2 = \lim_\lambda |f(E_\lambda X)|^2 \leq \lim_\lambda f(E_\lambda^2) f(X^*X) \leq M\|f\|.$$
Choosing X so that $|f(X)|$ approximates the norm of f yields $\|f\|^2 \leq M\|f\|$, or equivalently, $\|f\| \leq M$. ∎

I.9. Representations of C*-algebras

The key to representing a C*-algebra on a Hilbert space is to build representations from states. This important procedure is called the **GNS construction** named after **Gelfand, Naimark** and **Segal**.

Theorem I.9.6 *Let f be a positive linear functional on $\mathfrak{A}$. Then there is a representation π_f of $\mathfrak{A}$ on a Hilbert space $\mathcal{H}$ and a vector x_f in $\mathcal{H}$ which is a cyclic vector for $\pi(\mathfrak{A})$ such that $\|x_f\|^2 = \|f\|$ and*

$$f(A) = (\pi_f(A)x_f, x_f) \quad \text{for all} \quad A \in \mathfrak{A}.$$

Proof. Let $\mathcal{N} = \{A \in \mathfrak{A} : f(A^*A) = 0\}$. Then

$$\mathcal{N} = \{A \in \mathfrak{A} : f(B^*A) = 0 \text{ for all } B \in \mathfrak{A}\}$$

because

$$|f(B^*A)|^2 \leq f(B^*B)f(A^*A) = 0 \quad \text{for all} \quad A \in \mathcal{N},\ B \in \mathfrak{A}.$$

Hence $\mathcal{N}$ is a closed subspace. Moreover, $\mathcal{N}$ is a left ideal since if N lies in $\mathcal{N}$ and A, B belong to $\mathfrak{A}$, then

$$f(B^*(AN)) = f((A^*B)^*N) = 0 \quad \text{for all} \quad A, B \in \mathfrak{A}.$$

Whence, AN belongs to $\mathcal{N}$.

Define a positive definite inner product on $\mathfrak{A}/\mathcal{N}$ by

$$(\dot{X}, \dot{Y}) := f(Y^*X)$$

where $\dot{X}$ denotes $X + \mathcal{N}$. This is well defined because if N_1, N_2 are in $\mathcal{N}$, then

$$f((Y + N_2)^*(X + N_1)) = f(Y^*X) + f((Y + N_2)^*N_1) + \overline{f(X^*N_2)}$$
$$= f(Y^*X).$$

Let $\mathcal{H}$ denote the Hilbert space obtained by completing $\mathfrak{A}/\mathcal{N}$ in the inner product norm.

Let π_0 denote the left regular representation of $\mathfrak{A}$ on $\mathfrak{A}/\mathcal{N}$:

$$\pi_0(A)\dot{X} := \dot{AX}$$

which is well defined because $\mathcal{N}$ is a left ideal. This is a $*$-representation because

$$(\pi_0(A)\dot{X}, \dot{Y}) = f(Y^*AX) = f((A^*Y)^*X)$$
$$= (\dot{X}, \pi_0(A^*)\dot{Y}) = (\pi_0(A^*)^*\dot{X}, \dot{Y}).$$

Hence $\pi_0(A^*) = \pi_0(A)^*$. Moreover, $\|\pi_0\| \leq 1$ because

$$\|\pi_0(A)\|^2 = \sup_{\|\dot{X}\| \leq 1} \|\pi_0(A)\dot{X}\|^2 = \sup_{\|\dot{X}\| \leq 1} f(X^*A^*AX)$$
$$\leq \sup_{\|\dot{X}\| \leq 1} \|A^*A\| f(X^*X) = \|A\|^2.$$

Thus π_0 extends by continuity to a $*$-representation π_f of $\mathfrak{A}$ on $\mathcal{H}$.

In the unital case, let $x_f = \dot{I}$. Then
$$(\pi_f(A)x_f, x_f) = f(I^*A) = f(A).$$
The vector x_f is cyclic because $\pi_f(\mathfrak{A})x_f = \mathfrak{A}/\mathcal{N}$ which is dense in $\mathcal{H}$. And
$$\|f\| = f(I) = \|x_f\|^2.$$

Otherwise, let E_λ be an approximate identity. The net $\dot{E}_\lambda$ is Cauchy because $\lim_\lambda f(E_\lambda) = \|f\|$ and thus there are indices $\alpha \leq \beta$ such that
$$f(E_\alpha) > \|f\| - \varepsilon \quad \text{and} \quad \|E_\lambda E_\alpha - E_\alpha\| < \varepsilon \quad \text{for all} \quad \lambda \geq \beta.$$
Hence if $\lambda \geq \beta$,
$$\operatorname{Re} f(E_\lambda E_\alpha) = f(E_\alpha) + \operatorname{Re} f(E_\lambda E_\alpha - E_\alpha) \geq \|f\| - 2\varepsilon.$$
Therefore
$$\|\dot{E}_\lambda - \dot{E}_\alpha\|^2 = f((E_\lambda - E_\alpha)^2) = f(E_\lambda^2) + f(E_\alpha^2) - 2\operatorname{Re} f(E_\lambda E_\alpha)$$
$$\leq f(E_\lambda) + f(E_\alpha) - 2(\|f\| - 2\varepsilon) \leq 4\varepsilon.$$
Thus for $\lambda, \mu \geq \beta$,
$$\|\dot{E}_\lambda - \dot{E}_\mu\| \leq \|\dot{E}_\lambda - \dot{E}_\alpha\| + \|\dot{E}_\alpha - \dot{E}_\mu\| \leq 4\varepsilon^{1/2}.$$
Let $x_f = \lim_\lambda \dot{E}_\lambda$. Then
$$(\pi_f(A)x_f, x_f) = \lim_\lambda (\pi_f(A)\dot{E}_\lambda, \dot{E}_\lambda) = \lim_\lambda f(E_\lambda A E_\lambda) = f(A).$$
This vector is cyclic because the vectors $\pi_f(A)x_f = \dot{A}$ form a dense subset of $\mathcal{H}$. And since
$$\lim_\lambda \pi_f(E_\lambda)\dot{A} = \lim_\lambda \pi_f(E_\lambda)\pi_f(A)x_f = \pi_f(A)x_f = \dot{A}$$
for this dense set of vectors, it follows that $\text{SOT-}\lim_\lambda \pi_f(E_\lambda) = I$. Hence
$$\|f\| = \lim_\lambda |f(E_\lambda)| = \lim_\lambda (\pi_f(E_\lambda)x_f, x_f) = \|x_f\|^2. \qquad \blacksquare$$

Corollary I.9.7 *Every state on $\mathfrak{A}$ has a unique extension to a state on $\mathfrak{A}^\sim$.*

Proof. Existence follows since the representation π_f constructed above extends to $\mathfrak{A}^\sim$ by setting $\pi_f(I) = I$. Thus we define $\tilde{f}(A) = (\pi_f(A)x_f, x_f)$ for all A in $\mathfrak{A}^\sim$. Clearly this is a positive linear functional of norm $\|x_f\|^2 = 1$. Uniqueness follows because $\tilde{f}(I) = \|\tilde{f}\| = 1$ by Lemma I.9.5. $\blacksquare$

Let $\mathcal{S}(\mathfrak{A})$ denote the set of all states on $\mathfrak{A}$, known as the **state space** of $\mathfrak{A}$. A state is called **pure** if it is an extreme point of $\mathcal{S}(\mathfrak{A})$.

Theorem I.9.8 *Let π be a representation of a C*-algebra $\mathfrak{A}$ with a cyclic unit vector x. Then the state $f(A) := (\pi(A)x, x)$ is pure if and only if π is irreducible.*

I.9. Representations of C*-algebras

Proof. Assume that $\pi(\mathfrak{A})$ is not irreducible. Then by Lemma I.9.1, there is a proper projection P in $\pi(\mathfrak{A})'$. Let $t = \|Px\|^2$. If $t = 1$, this means $Px = x$ and thus $P\pi(A)x = \pi(A)Px = \pi(A)x$ for all A in $\mathfrak{A}$. As x is cyclic, these vectors are dense in $\mathcal{H}$ and hence $P = I$, contrary to fact. Likewise if $t = 0$, we conclude that $P^\perp = I$. So $0 < t < 1$.

Let $g_1(A) = t^{-1}(\pi(A)Px, Px)$ and $g_2(A) = (1-t)^{-1}(\pi(A)P^\perp x, P^\perp x)$. These are states because
$$g_1(I) = t^{-1}\|Px\|^2 = 1 = (1-t)^{-1}\|P^\perp x\|^2 = g_2(I).$$
And since $P^\perp \pi(A) P = 0$ for all A in $\mathfrak{A}$, we obtain
$$(tg_1 + (1-t)g_2)(A) = (\pi(A)Px, Px) + (\pi(A)P^\perp x, P^\perp x)$$
$$= (\pi(A)x, x) = f(A).$$
So f is a convex combination of g_1 and g_2. If f were pure, then $g_1 = g_2$. But this leads to an absurdity. Indeed, suppose that
$$0 = g_1(A) - g_2(A) = t^{-1}(\pi(A)Px, Px) - (1-t)^{-1}(\pi(A)P^\perp x, P^\perp x)$$
$$= (\pi(A)x, t^{-1}Px - (1-t)^{-1}P^\perp x).$$
Then $y = t^{-1}Px - (1-t)^{-1}P^\perp x$ is orthogonal to $\pi(\mathfrak{A})x$ which is dense in $\mathcal{H}$; whence $y = 0$. This is false, and thus f is not pure.

Conversely, suppose that $f = tg_1 + (1-t)g_2$ for $g_1 \neq g_2$ and some $0 < t < 1$. Define a sesquilinear form on $\pi(\mathfrak{A})x$ by
$$[\pi(A)x, \pi(B)x] := tg_1(B^*A).$$
Then
$$0 \leq [\pi(A)x, \pi(A)x] = tg_1(A^*A) \leq f(A^*A) = \|\pi(A)x\|^2.$$
In particular, this form is well defined because $\pi(A)x = 0$ implies that
$$0 = g_1(A^*A) = f(A^*A).$$
So it is a positive semi-definite sesquilinear form of norm at most 1. Thus there is an operator H with $0 \leq H \leq I$ such that $[u, v] = (u, Hv)$. Now calculate for A, X, Y in $\mathfrak{A}$,
$$((\pi(A)H - H\pi(A))\pi(X)x, \pi(Y)x)$$
$$= (\pi(X)x, H\pi(A^*Y)x) - (\pi(AX)x, H\pi(Y)x)$$
$$= tg_1(Y^*AX) - tg_1(Y^*AX) = 0.$$
Hence H commutes with $\pi(\mathfrak{A})$.

If π were irreducible, then H would be scalar, say $H = cI$. Then
$$tg_1(B^*A) = c(\pi(A)x, \pi(B)x) = cf(B^*A).$$
Both f and g_1 are states; thus $c = t$ and $g_1 = f$, contrary to fact. Therefore the representation π is reducible. ■

Now we return to the proof of the **Gelfand–Naimark Theorem** that states that every C*-algebra can be isometrically represented as a concrete C*-algebra of operators. In view of the GNS construction, it is sufficient to show that there are enough states to determine the norm. We need some more results about states.

Lemma I.9.9 *Let f be a continuous linear functional on $\mathfrak{A}$ such that for some approximate identity E_λ,*
$$\|f\| = 1 = \lim_\lambda f(E_\lambda).$$
Then f is a state.

Proof. First let us reduce to the unital case. Let $\tilde{f}$ be any Hahn–Banach extension of f to $\mathfrak{A}^\sim$. Let $\tilde{f}(I) = a$. Since $\|\tilde{f}\| = \|f\| = 1$, we have $|a| \leq 1$. Also $\|2E_\lambda - I\| \leq 1$ and hence
$$1 \geq \lim_\lambda |\tilde{f}(2E_\lambda - I)| = |2 - a|.$$

Together, these inequalities imply that $\tilde{f}(I) = 1$. So we may assume that $\mathfrak{A}$ is unital and $f(I) = 1$.

Next we show that f is self-adjoint meaning that if $A = A^*$, then $f(A)$ is real. Let A be a self-adjoint element of norm one. Notice that $\|A \pm inI\|^2 = n^2 + 1$. Hence
$$|f(A) \pm ni| \leq \sqrt{n^2 + 1} \quad \text{for all} \quad n \in \mathbb{N}.$$
That is, $f(A)$ lies in the intersection of all disks centred at $\pm ni$ of radius $\sqrt{n^2 + 1}$. This intersection is the interval $[-1, 1]$.

Now if $0 \leq A \leq I$, then $\|2A - I\| \leq 1$. Hence $-1 \leq 2f(A) - 1 \leq 1$, and thus $0 \leq f(A) \leq 1$. So f is positive. ∎

Now the key step in constructing sufficiently many states uses the Hahn–Banach Theorem again.

Lemma I.9.10 *Let A be a self-adjoint element of $\mathfrak{A}$. Then there exists a pure state f on $\mathfrak{A}$ such that $|f(A)| = \|A\|$.*

Proof. We will work in $\mathfrak{A}^\sim$ if $\mathfrak{A}$ is not unital. Then $C^*(A)$ is a subalgebra of $\mathfrak{A}^\sim$. Since $\|A\| = \mathrm{spr}(A)$, there is a multiplicative linear functional φ in the maximal ideal space of $C^*(A)$ such that $|\varphi(A)| = \|A\|$. It is always the case that $\|\varphi\| = 1 = \varphi(I)$. Let Φ be any Hahn–Banach extension of φ to a functional on $\mathfrak{A}^\sim$. Then since $\|\Phi\| = 1 = \Phi(I)$, the previous lemma shows that Φ is a state on $\mathfrak{A}^\sim$ and thus restricts to a state on $\mathfrak{A}$.

Let $\mathcal{F}$ be the set consisting of all states f on $\mathfrak{A}$ such that $f(A) = \Phi(A)$. This is a weak-∗ closed bounded convex subset of the dual space of $\mathfrak{A}$. By the Banach–Alaoglu Theorem, $\mathcal{F}$ is weak-∗ compact. So by the Krein-Milman Theorem, it has an extreme point f_0. To see that f_0 is an extreme point of $\mathcal{S}(\mathfrak{A})$ and hence is pure,

I.9. Representations of C*-algebras

it suffices to show that $\mathcal{F}$ is a face of $\mathcal{S}(\mathfrak{A})$; for this implies that extreme points of $\mathcal{F}$ are extreme points of $\mathcal{S}(\mathfrak{A})$.

But this follows from the fact that $f(A) = \Phi(A) = \pm\|A\|$ for f in $\mathcal{F}$. So if $f = (g_1 + g_2)/2$ for f in $\mathcal{F}$ and g_i in $\mathcal{S}(\mathfrak{A})$, then

$$\|A\| = |f(A)| \leq (|g_1(A)| + |g_2(A)|)/2 \leq \|A\|.$$

This can only occur if $g_1(A) = g_2(A) = f(A) = \Phi(A)$. Hence g_i belongs to $\mathcal{F}$; which shows that $\mathcal{F}$ is a face. ∎

Corollary I.9.11 *If A is an element of $\mathfrak{A}$, then there exists an irreducible representation π of $\mathfrak{A}$ and a unit vector x such that $\|\pi(A)x\| = \|A\|$.*

Proof. Apply Lemma I.9.10 to A^*A to obtain a pure state f with $f(A^*A) = \|A\|^2$. Then let π_f and x_f be obtained from the GNS construction applied to f. By Theorem I.9.8, π_f is irreducible. Finally,

$$\|\pi_f(A)x_f\|^2 = (\pi(A^*A)x_f, x_f) = f(A^*A) = \|A\|^2. \qquad \blacksquare$$

Finally we have all the pieces to complete the proof of the Gelfand–Naimark Theorem.

Theorem I.9.12 *Every abstract C*-algebra $\mathfrak{A}$ is isometrically *-isomorphic to a concrete C*-algebra of operators. If $\mathfrak{A}$ is separable, then one may take the Hilbert space to be separable.*

Proof. Take $\pi := \sum \oplus_{f \in \mathcal{S}(\mathfrak{A})} \pi_f$. Then $\|\pi(A)\| = \|A\|$ for every A in $\mathfrak{A}$ by the previous corollary. When $\mathfrak{A}$ is separable, it suffices to choose one representation π_n for each element A_n of a countable dense subset of $\mathfrak{A}$ such that $\|\pi_n(A_n)\| = \|A_n\|$. Then $\sum \oplus_n \pi_n$ works and is separably acting. ∎

The representation used in this proof is called the **universal representation** of $\mathfrak{A}$. Note that every state on $\pi(\mathfrak{A})$ is a **vector state** in this representation; meaning that for each state f, there is a unit vector x such that $f(\pi(A)) = (\pi(A)x, x)$. As every linear functional on $\mathfrak{A}$ is a linear combination of states (see Exercise I.32), it follows that every functional f on $\mathfrak{A}$ may be represented by $f(A) = (\pi(A)x, y)$ for certain vectors x and y. Thus they extend to WOT-continuous linear functionals on $\pi(\mathfrak{A})''$.

The **Jacobson radical** rad($\mathcal{A}$) of a Banach algebra $\mathcal{A}$ is the intersection of the kernels of all algebraically irreducible representations. An algebra is called **semi-simple** if the Jacobson radical is $\{0\}$.

Corollary I.9.13 *C*-algebras are semi-simple.*

Proof. By Theorem I.9.3, topologically irreducible representations of C*-algebras are algebraically irreducible. So this applies to π_f when f is a pure state. Since the pure states separate points in $\mathfrak{A}$ by Corollary I.9.11, the representation

$$\sum_{\substack{f \in \mathcal{S}(\mathfrak{A}) \\ f \text{ pure}}} \oplus \, \pi_f$$

is isometric. Hence

$$\operatorname{rad}(\mathfrak{A}) = \bigcap_{\substack{f \in \mathcal{S}(\mathfrak{A}) \\ f \text{ pure}}} \ker \pi_f = \{0\}.$$

Therefore $\mathfrak{A}$ is semi-simple. ∎

We conclude this section with a couple of results about the relationship between representations of a C*-algebra and its ideals. A representation π is **non-degenerate** if $\pi(\mathfrak{A})\mathcal{H}$ is dense in $\mathcal{H}$.

Lemma I.9.14 *Suppose that $\mathfrak{I}$ is an ideal of a C*-algebra $\mathfrak{A}$, and that π is a non-degenerate representation of $\mathfrak{I}$ on a Hilbert space $\mathcal{H}$. Then there is a unique representation $\tilde{\pi}$ of $\mathfrak{A}$ on $\mathcal{H}$ extending π. Moreover, $\tilde{\pi}$ is irreducible if and only if π is irreducible.*

Proof. For A in $\mathfrak{A}$, J in $\mathfrak{I}$ and x in $\mathcal{H}$, define

$$\tilde{\pi}(A)(\pi(J)x) := \pi(AJ)x. \tag{1}$$

To verify that this is well defined, suppose that $\pi(J_1)x_1 = \pi(J_2)x_2$. Let E_λ be an approximate unit for $\mathfrak{I}$. Then

$$\pi(AJ_1)x_1 = \lim_\lambda \pi(AE_\lambda J_1)x_1 = \lim_\lambda \pi(AE_\lambda)\pi(J_1)x_1$$
$$= \lim_\lambda \pi(AE_\lambda)\pi(J_2)x_2 = \lim_\lambda \pi(AE_\lambda J_2)x_2 = \pi(AJ_2)x_2.$$

Hence $\tilde{\pi}(A)$ is well defined on a dense subset of $\mathcal{H}$. Moreover

$$\|\tilde{\pi}(A)(\pi(J)x)\| \le \lim_\lambda \|\pi(AE_\lambda)\| \, \|\pi(J)x\| \le \|A\| \, \|\pi(J)x\|,$$

and thus $\tilde{\pi}(A)$ extends uniquely to a bounded operator in $\mathcal{B}(\mathcal{H})$.

To see that this is a *-representation, one should verify that $\tilde{\pi}$ is multiplicative and self-adjoint. We do the latter only. For A in $\mathfrak{A}$ and J_1, J_2 in $\mathfrak{I}$,

$$(\tilde{\pi}(A^*)\pi(J_1)x_1, \pi(J_2)x_2) = (\pi(A^*J_1)x_1, \pi(J_2)x_2) = (\pi(J_2^*A^*J_1)x_1, x_2)$$
$$= (\pi(J_1)x_1, \pi(AJ_2)x_2) = (\pi(J_1)x_1, \tilde{\pi}(A)\pi(J_2)x_2)$$
$$= (\tilde{\pi}(A)^*\pi(J_1)x_1, \pi(J_2)x_2).$$

Hence $\tilde{\pi}(A^*) = \tilde{\pi}(A)^*$. The uniqueness follows from the fact that any extension of π must satisfy (1) which determines π on a dense set.

Suppose that π is not irreducible, and thus there is a proper invariant subspace $\mathcal{M}$ for $\pi(\mathfrak{I})$. Then since π is non-degenerate

$$\mathcal{H} = \overline{\pi(\mathfrak{I})(\mathcal{M} + \mathcal{M}^\perp)} \subset \overline{\pi(\mathfrak{I})\mathcal{M}} + \overline{\pi(\mathfrak{I})\mathcal{M}^\perp}.$$

I.9. Representations of C*-algebras

Since $\pi(\mathfrak{J})\mathcal{M}^\perp \subset \mathcal{M}^\perp$, we must have $\mathcal{M} = \overline{\pi(\mathfrak{J})\mathcal{M}}$. Therefore
$$\tilde{\pi}(\mathfrak{A})\mathcal{M} = \overline{\tilde{\pi}(\mathfrak{A})\pi(\mathfrak{J})\mathcal{M}} = \overline{\pi(\mathfrak{J})\mathcal{M}} = \mathcal{M}.$$
So $\tilde{\pi}$ is not irreducible either. The other direction is trivial. ∎

Our other result in this direction turns the tables and considers the restriction of a representations of $\mathfrak{A}$ to an ideal.

Lemma I.9.15 *Suppose that π is a representation of $\mathfrak{A}$ on a Hilbert space $\mathcal{H}$. Let $\mathfrak{J}$ be an ideal of $\mathfrak{A}$. Then the projection P onto $\overline{\pi(\mathfrak{J})\mathcal{H}}$ lies in the centre of $\pi(\mathfrak{A})''$. If π is irreducible and $\pi(\mathfrak{J}) \neq 0$, then the restriction $\pi|\mathfrak{J}$ is also irreducible.*

Proof. Since $\pi(\mathfrak{A})\pi(\mathfrak{J})\mathcal{H} = \pi(\mathfrak{J})\mathcal{H}$, it follows that $\overline{\pi(\mathfrak{J})\mathcal{H}}$ is invariant for $\pi(\mathfrak{A})$ and thus P lies in $\pi(\mathfrak{A})'$. If X lies in $\pi(\mathfrak{J})'$, then $X\pi(J)x = \pi(J)(Xx)$ belongs to $\pi(\mathfrak{J})\mathcal{H}$ for every J in $\mathfrak{J}$ and x in $\mathcal{H}$. Thus $P\mathcal{H}$ is invariant for $\pi(\mathfrak{J})'$; whence
$$P \in \pi(\mathfrak{J})'' \cap \pi(\mathfrak{A})' \subset \pi(\mathfrak{A})'' \cap \pi(\mathfrak{A})'$$
which is the centre of $\pi(\mathfrak{A})''$.

If π is irreducible, then P is scalar. As $\pi(\mathfrak{J}) \neq 0$, we have $P = I$. Thus $\pi|\mathfrak{J}$ is non-degenerate. So it is irreducible by the previous lemma. ∎

Using these results, we will derive a useful extension of Theorem I.4.8. A **quasicentral approximate unit** for an ideal $\mathfrak{J}$ of a C*-algebra $\mathfrak{A}$ is an approximate unit E_λ such that
$$\lim_\lambda \|E_\lambda A - AE_\lambda\| = 0 \text{ for all } A \in \mathfrak{A}.$$

Theorem I.9.16 *Every ideal $\mathfrak{J}$ of a C*-algebra $\mathfrak{A}$ has a quasicentral approximate unit.*

Proof. Let $E_\lambda, \lambda \in \Lambda$, be an approximate unit for $\mathfrak{J}$ given by Theorem I.4.8. Recall from the proof of that theorem that if $E_\lambda \leq E \leq I$, then
$$\|X - XE\| \leq \|X^*(I - E_\lambda)X\|^{1/2} \quad \text{for all} \quad X \in \mathfrak{J}.$$
Thus the convex hull $\mathcal{E}$ of $\{E_\lambda : \lambda \in \Lambda\}$ is directed and forms an approximate unit for $\mathfrak{J}$, again as in the proof of Theorem I.4.8. Therefore, it suffices to show that for $\lambda \in \Lambda$ and $A_1, \ldots, A_n$ in $\mathfrak{A}$, then there is an element F in $\mathcal{E}$ such that $F \geq E_\lambda$ and $\|A_iF - FA_i\| < 1/n$ for $1 \leq i \leq n$.

To construct F, form the C*-algebra $\mathcal{M}_n(\mathfrak{A})$. Notice that $E_\lambda^{(n)}$, the direct sum of n copies of E_λ, forms an approximate unit for the ideal $\mathcal{M}_n(\mathfrak{J})$. Let
$$A = A_1 \oplus \cdots \oplus A_n$$
be the diagonal operator with diagonal entries A_i. First we show that 0 is in the closure of $\mathcal{S} := \{AF^{(n)} - F^{(n)}A : F \in \mathcal{F}\}$ where $\mathcal{F} = \text{conv}\{E_\mu : \mu \geq \lambda\}$. Indeed, $\mathcal{S}$ is a convex set. So if 0 is not in $\overline{\mathcal{S}}$, the separation theorem version of the Hahn–Banach Theorem shows that there is a linear functional f so that
$$\text{Re } f(AE_\mu^{(n)} - E_\mu^{(n)}A) \geq 1 \quad \text{for all} \quad \mu \geq \lambda.$$

Let π be the universal representation of $\mathfrak{A}$, and let vectors x and y be chosen so that $f(A) = (\pi(A)x, y)$. Let P be the projection onto $\overline{\pi(\mathfrak{J})\mathcal{H}}$. Then it is easy to verify that

$$\text{SOT--}\lim_{\lambda \in \Lambda} \pi(E_\lambda) = P.$$

But P belongs to $\pi(\mathfrak{A})'$ by Lemma I.9.15. Hence

$$\lim_{\lambda \in \Lambda} f(AE_\lambda - E_\lambda A) = \lim_{\lambda \in \Lambda} (\pi(AE_\lambda - E_\lambda A)x, y)$$
$$= \lim_{\lambda \in \Lambda} (\pi(A)P - P\pi(A)x, y) = 0.$$

This contradicts the assumption that 0 is not in the closure of $\mathcal{S}$. Thus there is an element F in $\mathcal{F}$ so that

$$\|AF^{(n)} - F^{(n)}A\| = \max_{1 \le i \le n} \|A_i F - F A_i\| < 1/n.$$

Now define $F = F_{\mathcal{A},\lambda}$ indexed by finite subsets $\mathcal{A}$ of $\mathfrak{A}$ and $\lambda \in \Lambda$. To see that it is upward directed, suppose that $F_{\mathcal{A},\lambda}$ and $F_{\mathcal{B},\mu}$ are two such elements. Since they lie in $\mathcal{E}$, we have

$$F_{\mathcal{A},\lambda} = \sum_{i=1}^{k} s_i E_{\lambda_i} \quad \text{and} \quad F_{\mathcal{B},\mu} = \sum_{i=j}^{\ell} t_j E_{\mu_j}.$$

Let $\mathcal{C} = \mathcal{A} \cup \mathcal{B}$ and choose ν in Λ such that $\nu \ge \lambda_i$ for $1 \le i \le k$ and $\nu \ge \mu_j$ for $1 \le j \le \ell$. Then it is easily seen that $F_{\mathcal{C},\nu}$ dominates both $F_{\mathcal{A},\lambda}$ and $F_{\mathcal{B},\mu}$. It is routine to verify that it is a quasicentral approximate unit. ∎

I.10 C*-algebras of Compact Operators

In this section, we develop the complete structure theory for C*-subalgebras of the compact operators. This applies, in particular, to finite dimensional C*-algebras which are the basic building blocks used in Chapter III. The C*-algebra algebra $\mathfrak{K}$ of compact operators is one of the simplest and most basic C*-algebras. It will play a central role in the theory. As for matrix algebras, the existence of minimal projections in $\mathfrak{K}$ is crucial to the analysis.

Lemma I.10.1 *If $\mathfrak{A}$ is a non-zero C*-algebra of compact operators, then it contains a minimal projection E. Moreover, $E\mathfrak{A}E = \mathbb{C}E$.*

Furthermore, if π is a non-zero representation of $\mathfrak{A}$, then there is a minimal projection E of $\mathfrak{A}$ such that $\pi(E) \ne 0$. When π is irreducible, $\pi(E)$ has rank one.

Proof. As $\mathfrak{A}$ is non-zero, it contains a non-zero positive operator A. Because A is compact, its spectrum is a countable set with 0 as the only possible cluster point. From the continuous functional calculus Theorem I.3.3, the spectral projection P corresponding to a non-zero eigenvalue of A lies in $\mathfrak{A}$. Since P is compact, it has finite rank. Clearly, P dominates a projection E in $\mathfrak{A}$ with minimal positive rank. This projection is therefore minimal in $\mathfrak{A}$. If $E\mathfrak{A}E$ contained a self-adjoint

I.10. C*-algebras of Compact Operators

element T which is not a scalar multiple of E, then the same argument produces a spectral projection of T which is dominated by E, and so has smaller positive rank, contrary to fact. Hence $E\mathfrak{A}E = \mathbb{C}E$.

Suppose that π is a non-zero representation. Then there is a positive element A such that $\pi(A) \neq 0$. From the functional calculus, we deduce that there is a finite rank spectral projection P of A such that $\pi(P) \neq 0$. Now P is the sum of minimal projections of $\mathfrak{A}$ because it is finite rank. So there is a minimal projection E such that $\pi(E) \neq 0$.

If $\pi(E)$ has rank greater than 1, choose orthogonal unit vectors x and y in the range of $\pi(E)$. For any A in $\mathfrak{A}$, there is a scalar λ so that $EAE = \lambda E$. Hence

$$(\pi(A)x, y) = (\pi(EAE)x, y) = (\lambda x, y) = 0.$$

Thus $\overline{\pi(\mathfrak{A})x}$ is a proper invariant subspace, and so π is not irreducible. ∎

We write $\mathfrak{K}(\mathcal{H})$ to denote the C*-algebra of compact operators on $\mathcal{H}$, which is isomorphic to $\mathcal{M}_n$ if $\mathcal{H}$ is n-dimensional, and to $\mathfrak{K}$ when $\mathcal{H}$ is separable and infinite dimensional.

Lemma I.10.2 *Let $\mathcal{H}$ be a Hilbert space. Then the only irreducible C*-subalgebra of $\mathfrak{K}(\mathcal{H})$ is itself.*

Proof. Suppose that $\mathfrak{A}$ is a non-zero irreducible subalgebra of $\mathfrak{K}(\mathcal{H})$ acting on a Hilbert space $\mathcal{H}$ of the appropriate dimension. Let E be a minimal projection in $\mathfrak{A}$ provided by Lemma I.10.1 such that E is rank one. Then there is a unit vector e so that $E = ee^*$. Since $\mathfrak{A}$ is irreducible, it is algebraically irreducible by Theorem I.9.3. So if x and y are vectors in $\mathcal{H}$, one may choose elements A and B in $\mathfrak{A}$ so that $Ae = x$ and $Be = y$. Then $\mathfrak{A}$ contains

$$AEB^* = Aee^*B^* = (Ae)(Be)^* = xy^*.$$

Thus $\mathfrak{A}$ contains every rank one operator. As these operators span $\mathfrak{K}$, we obtain $\mathfrak{A} = \mathfrak{K}$. ∎

Before going on, we collect two useful corollaries.

Corollary I.10.3 *The C*-algebra of compact operators is simple.*

Proof. As $\mathfrak{K}$ is irreducible, any non-zero ideal of $\mathfrak{K}$ is irreducible by Lemma I.9.15. Hence the only non-zero closed ideal is all of $\mathfrak{K}$ by Lemma I.10.2. ∎

Corollary I.10.4 *If $\mathfrak{B}$ is an irreducible C*-subalgebra of $\mathcal{B}(\mathcal{H})$ which contains a non-zero compact operator, then $\mathfrak{B}$ contains $\mathfrak{K}$.*

Proof. Since $\mathfrak{B} \cap \mathfrak{K}$ is a non-zero ideal of $\mathfrak{B}$ by hypothesis, it is irreducible by Lemma I.9.15. Hence it is all of $\mathfrak{K}$ by Lemma I.10.2. ∎

This next lemma contains the key idea. A minimal idempotent which is not in the kernel of a representation σ generates a cyclic irreducible representation both for $\mathfrak{A}$ and for $\sigma(\mathfrak{A})$; and they are unitarily equivalent.

Lemma I.10.5 *Every non-degenerate representation of a C*-algebra $\mathfrak{A}$ of compact operators has an irreducible subrepresentation which is unitarily equivalent to the restriction of $\mathfrak{A}$ to a (minimal) reducing subspace.*

Proof. Let σ be a non-degenerate representation of $\mathfrak{A}$. By Lemma I.10.1, there is a minimal projection E such that $P = \sigma(E)$ is non-zero. Let f be a unit vector in the range of P; and let $\mathcal{H}_f^\sigma = \overline{\sigma(\mathfrak{A})f}$. Also let e be a unit vector in the range of E; and let $\mathcal{H}_e = \overline{\mathfrak{A}e}$. The subspaces $\mathcal{H}_e$ and $\mathcal{H}_f^\sigma$ are invariant for $\mathfrak{A}$ and $\sigma(\mathfrak{A})$ respectively. Since $E\mathfrak{A}E = \mathbb{C}E$, the state on $\mathfrak{A}$ given by $\varphi(A) = (Ae, e)$ satisfies

$$EAE = \varphi(A)E \quad \text{for all} \quad A \in \mathfrak{A}.$$

Define a linear map U from $\mathcal{H}_e$ to $\mathcal{H}_f^\sigma$ by the formula

$$U(Ae) := \sigma(A)f = \sigma(AE)f.$$

A simple computation shows that for A, B in $\mathfrak{A}$

$$(UAe, UBe) = (\sigma(AE)f, \sigma(BE)f) = (\sigma(EB^*AE)f, f)$$
$$= \varphi(B^*A)(Pf, f) = (B^*Ae, e) = (Ae, Be).$$

Consequently, U is an isometry from $\mathcal{H}_e$ onto $\mathcal{H}_f^\sigma$. In particular, U is well defined.

For A, B in $\mathfrak{A}$, we have

$$\sigma(A)(UBe) = \sigma(A)\sigma(B)f = \sigma(AB)f = U(ABe) = UAU^*(UBe).$$

Thus $\sigma(A)|_{\mathcal{H}_f^\sigma} = UA|_{\mathcal{H}_e}U^*$ for every A in $\mathfrak{A}$. So $\sigma|_{\mathcal{H}_f^\sigma}$ is unitarily equivalent to the restriction of $\mathfrak{A}$ to $\mathcal{H}_e$.

Let π_e denote the restriction map of $\mathfrak{A}$ to $\mathcal{H}_e$. Notice that $\pi_e(E) = ee^*$ is rank one. Indeed, $EAe = EAEe$ belongs to $\mathbb{C}e$ for every A in $\mathfrak{A}$, and thus $\mathbb{C}e$ is the range of $\pi_e(E)$. Suppose that P is a projection in $\mathcal{B}(\mathcal{H}_e)$ which commutes with $\pi_e(\mathfrak{A})$. Then

$$Pe = P\pi_e(E)e = \pi_e(E)Pe$$

is a multiple of e. As $Pe = P^2e$, this multiple is 0 or 1. By considering $P^\perp$ if necessary, we may suppose that $Pe = 0$. But then

$$PAe = P\pi_e(A)e = \pi_e(A)Pe = 0 \quad \text{for all} \quad A \in \mathfrak{A};$$

whence $P = 0$. Therefore $\pi_e(\mathfrak{A})' = \mathbb{C}I$; and so π_e is irreducible. Hence the subspace $\mathcal{H}_e$ is minimal. ∎

Corollary I.10.6 *Every irreducible representation of $\mathcal{M}_n$ or $\mathfrak{K}$ is unitarily equivalent to the identity representation.*

Proof. The only invariant subspace for $\mathcal{M}_n$ or $\mathfrak{K}$ is the whole space. ∎

I.10. C*-algebras of Compact Operators

It remains to show that representations of these algebras are direct sums of irreducible ones, not something more complicated. As we shall see in the next chapter, the situation is not as simple even for most abelian C*-algebras.

Theorem I.10.7 *Let $\mathfrak{A}$ be a non-degenerate C*-subalgebra of the compact operators. Then every representation σ of $\mathfrak{A}$ is the direct sum of irreducible representations which are unitarily equivalent to subrepresentations of the identity representation.*

Proof. By Lemma I.10.5, $\mathcal{H}_\sigma$ contains a subspace $\mathcal{H}_f^\sigma$ such that the restriction of $\pi(\mathfrak{A})$ to $\mathcal{H}_f^\sigma$ is an irreducible representation which is unitarily equivalent to a subrepresentation π_e of the identity representation. Choose a maximal family $\{\mathcal{H}_n^\sigma\}$ of pairwise orthogonal reducing subspaces with this property. Then $\mathcal{H}_\sigma$ is spanned by the H_n^σ's. For otherwise, the complement $(\sum_n \mathcal{H}_n^\sigma)^\perp$ would dominate another such subspace $\mathcal{H}_f^\sigma$ by Lemma I.10.5 again. This would contradict the maximality of the original family. Let σ_n be the restriction of σ to $\mathcal{H}_n^\sigma$. Then we obtain the decomposition $\sigma = \sum_n \oplus \sigma_n$, and each σ_n is equivalent to an irreducible subrepresentation of the identity. ∎

Finally, we apply this to the identity representation itself. Recall that $\mathcal{H}^{(k)}$ denotes the Hilbert space direct sum of k copies of $\mathcal{H}$. And for each A in $\mathcal{B}(\mathcal{H})$, $A^{(k)}$ denotes the operator on $\mathcal{H}^{(n)}$ given by $A^{(k)}(x_1, \ldots, x_k) = (Ax_1, \ldots, Ax_n)$.

Theorem I.10.8 *Let $\mathfrak{A}$ be a C*-subalgebra of the compact operators $\mathfrak{K}$. Then there are Hilbert spaces $\mathcal{H}_i$ of dimension n_i for $i \geq 0$ and non-negative integers k_i so that*

$$\mathcal{H} \simeq \mathcal{H}_0 \oplus \sum_{i \geq 1} \oplus \mathcal{H}_i^{(k_i)} \quad \text{and} \quad \mathfrak{A} \simeq 0 \oplus \sum_{i \geq 1} \oplus \mathfrak{K}(\mathcal{H}_i)^{(k_i)}.$$

Proof. Let $\mathcal{H}_0 = \ker \mathfrak{A}$. By Theorem I.10.7, the Hilbert space $\mathcal{H}_0^\perp$ may be decomposed as a direct sum of reducing subspaces $\mathcal{K}_j$ such that the restriction σ_j of $\mathfrak{A}$ to $\mathcal{K}_j$ is irreducible. By Lemma I.10.2, $\sigma_j(\mathfrak{A}) = \mathfrak{K}(\mathcal{K}_j)$. Collect together all equivalent representations into classes $\{\sigma_j : j \in \mathcal{S}_i\}$. Let $\mathcal{H}_i$ be a Hilbert space of dimension $n_i = \dim \mathcal{K}_j$ for j in $\mathcal{S}_i$; let k_i be the cardinality of $\mathcal{S}_i$; and let π_i denote the identity representation of $\mathfrak{K}(\mathcal{H}_i)$. Then it is evident that

$$\mathcal{H} \simeq \mathcal{H}_0 \oplus \sum_{i \geq 1} \oplus \mathcal{H}_i^{(k_i)} \quad \text{and} \quad \text{id} \simeq 0 \oplus \sum_i \oplus \pi_i^{(k_i)}.$$

Finally, notice that a rank one projection E_i in $\mathfrak{K}(\mathcal{H}_i)$ is mapped onto a projection of rank k_i. As this projection is compact, it follows that k_i is finite for all i. ∎

Exercises

I.1 Show that if S and T are two normal operators, then there is a $*$-isomorphism π from $C^*(S)$ onto $C^*(T)$ such that $\pi(S) = T$ if and only if S and T have the same spectrum.

I.2 Show that $C_0(X, \mathfrak{A})$, the space of continuous functions vanishing at infinity on a compact Hausdorff space X with values in a C*-algebra $\mathfrak{A}$, is a C*-algebra with the operations $f^*(x) = f(x)^*$ and $\|f\| = \sup_{x \in X} \|f(x)\|$.

I.3 If $\mathfrak{A}_n$ is a sequence of C*-algebras for $n \geq 1$, form the infinite product $\mathfrak{M} = \prod_{n \geq 1} \mathfrak{A}_n$ of all bounded sequences (A_n) with A_n in $\mathfrak{A}_n$ and the infinite sum $\mathfrak{J} = \sum \oplus_{n \geq 1} \mathfrak{A}_n$ consisting of such sequences which converge to 0. With the supremum norm and point-wise multiplication and conjugation, these become Banach $*$-algebras. Show that they are C*-algebras, and that $\mathfrak{J}$ is an ideal of $\mathfrak{M}$.

I.4 Show that in any algebra,
$$\left(A(BA - \lambda I)^{-1}B - I\right)(AB - \lambda I) = \lambda I.$$
Hence show that $\sigma(AB) \cup \{0\} = \sigma(BA) \cup \{0\}$.

I.5 Prove Corollary I.4.9.
HINT: List a dense subset A_n of $\mathfrak{A}$. Then choose an increasing sequence of elements E_n in Λ (from Theorem I.4.8) such that $\|A_k E_n - A_k\| < 1/n$ and $\|E_n A_k - A_k\| < 1/n$ for all $1 \leq k \leq n$.

I.6 Use the identity $\|A\|^2 = \|A^*A\| = \mathrm{spr}(A^*A)$ to prove that there is a unique C*-algebra norm on a given $*$-algebra. In particular, show that the unitization $\widetilde{\mathfrak{A}}$ of a non-unital C*-algebra $\mathfrak{A}$ is unique.

I.7 Show that there is a C*-algebra norm on $\mathcal{M}_n(\mathfrak{A})$.
HINT: Prove it first for concrete C*-algebras.

I.8 Show that there are positive 2×2 matrices A and B such that $A \leq B$ but $A^2 \not\leq B^2$. Prove that the inequality is valid in every C*-algebra when A and B commute.

I.9 Show that if $0 \leq A \leq B$, then $0 \leq A^{1/2} \leq B^{1/2}$.
HINT: Use Lemma I.4.7 to show that $\|A^{1/2}B^{-1/2}\| \leq 1$. Then use Exercise I.4 to show that $\mathrm{spr}(B^{-1/4}A^{1/2}B^{-1/4}) \leq 1$.

I.10 A function on $\mathbb{R}_+$ is called **operator monotone** if $0 \leq A \leq B$ implies that $f(A) \leq f(B)$.
(a) Use Lemma I.4.7 to show that $f_s(t) = st(1 + st)^{-1}$ is operator monotone for $s > 0$.
(b) Show that an integral of f_s by a positive measure is also operator mono-

Exercises

tone. Apply this to $\int_0^\infty f_s(t)s^{-\beta}\,ds$ to show that $f(t) = t^\beta$ is operator monotone for $0 < \beta < 1$.

(c) Show that $f(t) = \min\{t,1\}$ is not operator monotone on $\mathcal{M}_2$.

I.11 If A is a positive element in $\mathfrak{A}$, show that $\overline{A\mathfrak{A}A}$ contains A. Hence this is the hereditary C*-subalgebra of $\mathfrak{A}$ generated by A. Show that every separable hereditary subalgebra has this form.

I.12 For a separable Hilbert space $\mathcal{H}$, show that the ideal $\mathfrak{K}$ of compact operators is the unique proper ideal of $\mathcal{B}(\mathcal{H})$.

I.13 Let $\omega = \{n^{-1} : n \geq 1\} \cup \{0\}$. Find all ideals of the C*-algebra $\mathrm{C}(\omega, \mathcal{M}_2)$.

I.14 Show that every closed ideal of $\mathrm{C}_0(X)$ has the form
$$\mathfrak{J}_E = \{f \in \mathrm{C}_0(X) : f|_E = 0\}$$
for a closed subset E of X.

HINT: Use the Stone–Weierstrass theorem.

I.15 Verify the integral formula of the spectral theorem following the outline in section I.8.

I.16 For a von Neumann algebra $\mathfrak{A}$, show that the extreme points of $\mathfrak{A}_{sa1}$ are precisely the set of symmetries; and the extreme points of $\mathfrak{A}_{+1}$ are the projections. Show that for $\mathcal{B}(\mathcal{H})$, the set of extreme points of the unit ball is strictly larger than the set of unitaries.

I.17 Use the polar decomposition of a compact operator K to show that it may be written as $K = \sum_{n \geq 1} s_n e_n f_n^*$ where s_n is a sequence of real numbers decreasing to 0, and $\{e_n\}$ and $\{f_n\}$ are orthonormal sequences, where ef^* denotes the rank one operator $ef^*(x) = (x, f)e$. The sequence $s_n(K)$ are called the **singular values** of K.

I.18 A compact operator K in $\mathfrak{K}$ is **trace class** if
$$\|K\|_1 := \sum_{n \geq 1} s_n(K) < \infty.$$

The collection of all trace class operators is denoted by $\mathcal{C}_1$.

(a) Show that if x_n and y_n are orthonormal sequences, then
$$\sum_{n \geq 1} |(Kx_n, y_n)| \leq \|K\|_1.$$

(b) Show that $\mathcal{C}_1$ is complete in the trace norm, and that the ideal of finite rank operators is dense in $\mathcal{C}_1$.

(c) Show that $\|AKB\|_1 \leq \|A\|\,\|K\|_1\,\|B\|$. Hence deduce that $\mathcal{C}_1$ is an ideal of $\mathcal{B}(\mathcal{H})$.

(d) Fix an orthonormal basis $\{e_n\}$ and define the **trace** by
$$\mathrm{Tr}(K) = \sum_{n \geq 1}(Ke_n, e_n).$$
Show that $\mathrm{Tr}(KT) = \mathrm{Tr}(TK)$ for all K in $\mathcal{C}_1$ and T in $\mathcal{B}(\mathcal{H})$. Hence deduce that Tr is independent of the choice of basis.

(e) Each T in $\mathcal{B}(\mathcal{H})$ defines a linear functional φ_T on $\mathcal{C}_1$ by $\varphi_T(K) = \mathrm{Tr}(TK)$. Show that $\|\varphi_T\| = \|T\|$.

(f) Show that if φ is a linear functional on $\mathcal{C}_1$, then the sesquilinear form $\langle x, y \rangle := \varphi(xy^*)$ determines a bounded linear operator T such that $\varphi = \varphi_T$.

(g) Deduce that $\mathcal{B}(\mathcal{H})$ is the dual space of $\mathcal{C}_1$. This defines the **weak-∗ topology** on $\mathcal{B}(\mathcal{H})$ given by the functionals φ_K on $\mathcal{B}(\mathcal{H})$ for K in $\mathcal{C}_1$.

I.19 (a) Show that the weak operator topology corresponds to the functionals φ_F for finite rank F. Hence deduce that the weak-∗ topology is stronger than the WOT.

(b) Show that every von Neumann algebra is a dual space.

I.20 Show that the dual of the compact operators is $\mathcal{C}_1$.

I.21 The strong-∗ topology on $\mathcal{B}(\mathcal{H})$ is the weakest topology such that $T \to Tx$ for x in $\mathcal{H}$ and $T \to T^*$ are continuous maps. Show that the functionals which are strong-∗ continuous on $\mathcal{B}(\mathcal{H})$ coincide with the WOT-continuous functionals.

I.22 Show that multiplication is not jointly continuous in the strong operator topology.
HINT: Let Λ consist of ordered pairs (M, x) where M is a finite dimensional subspace and x is a unit vector orthogonal to M ordered by the relation $(M, x) < (N, y)$ if M and x are both contained in N. Let e be a fixed unit vector. Set $A_{(M,x)} = (\dim M)ex^*$ and $B_{(M,x)} = (\dim M)^{-1}xe^*$.

I.23 Complete the proof of Proposition I.6.3.

I.24 Prove the converse of Lemma I.7.2.
HINT: Let Λ be a net consisting of ordered pairs (M, x), where M is a finite dimensional subspace and x is a unit vector which is *not* orthogonal to M, and ordered by the relation $(M, x) < (N, y)$ if M and x are both contained in N. Set $A_{(M,x)} = (\dim M \|P_M x\|)^{-1}xx^*$. Show that this net converges strongly to 0, but that $\limsup_{(M,x)} \|f(A_{(M,x)}y\| = \infty$ for every $y \neq 0$ and f in $C(\mathbb{R})$ such that $\limsup_{|t| \to \infty} |f(t)/t| = \infty$.

I.25 Suppose that A belongs to a concrete C*-algebra $\mathfrak{A} \subset \mathcal{B}(\mathcal{H})$, and has a polar decomposition $A = U|A|$. Show that $Uf(|A|)$ belongs to $\mathfrak{A}$ provided that $f(0) = 0$.

Exercises

I.26 Show that if $\mathfrak{J}$ is an ideal of a C*-algebra $\mathfrak{A}$ and A lies in $\mathfrak{A}$, then there is an element J in $\mathfrak{J}$ such that $\|A - J\| = \mathrm{dist}(A, \mathfrak{J})$.
HINT: Use Corollary I.4.2 on $|A| - \|A + \mathfrak{J}\|I$.

I.27 If a C*-algebra $\mathfrak{A}$ contains a non-unitary isometry S, show that
$$\|S - A\| \geq \tfrac{1}{2n}$$
for every $A = \sum_{i=1}^{n} \lambda_i U_i$ which is the convex combination of n unitaries.
HINT: You may suppose that $\lambda_1 \geq 1/n$. Estimate $\|U_1^* S - \lambda_1 I\|$ and compare it with the fact that $\sigma(U_1^* S) = \overline{\mathbb{D}}$.

I.28 Show that in $C(\overline{\mathbb{D}})$, the function $f(z) = (1 - \tfrac{1}{p})z$ is not a convex combination of fewer than p unitaries.
HINT: If $f = \sum_{i=1}^{n} \lambda_i u_i$ is a convex combination of unitaries, show that $f - \lambda_i u_i$ maps $\overline{\mathbb{D}}$ into itself, and thus has a fixed point. Hence deduce that $\lambda_i \leq 1/p$.

I.29 Show that states on $C_0(X)$ correspond to regular Borel probability measures on X. Which are the pure states?

I.30 Show that irreducible representations of abelian C*-algebras are one dimensional.
HINT: Use the spectral theorem.

I.31 Show that every linear functional on a C*-algebra decomposes as a sum of a self-adjoint functional and a skew-adjoint functional.

I.32 Show that every self-adjoint functional f decomposes as $f = g_1 - g_2$ where g_i are positive linear functionals such that
$$\|g_1\| + \|g_2\| = \|f\|.$$
HINT: Mimic the proof that a real measure decomposes as the difference of two positive measures supported on disjoint sets. This is known as the **Jordan decomposition**.

I.33 Show that every irreducible representation comes from a (pure) state.
HINT: Use a state of the form $f(A) = (\pi(A)x, x)$.

I.34 Show that if $\mathfrak{J}$ is an ideal of a separable C*-algebra $\mathfrak{A}$, then there is a sequential quasi-central approximate unit for $\mathfrak{J}$ relative to $\mathfrak{A}$. Then show that this is still true even if $\mathfrak{J}$ is not closed (but is still self-adjoint).

I.35 Suppose that ρ is a representation of a C*-algebra $\mathfrak{A}$ on a Hilbert space $\mathcal{H}$ with a unit cyclic vector x. Define the state $f(A) = (\rho(A)x, x)$. Show that ρ is unitarily equivalent to the GNS construct π_f via a unitary operator U such that $Ux_f = x$. This shows that the GNS construction is unique.
HINT: Set $U\dot{X} = \rho(X)x$.

I.36 Show that every finite dimensional C*-algebra may be faithfully represented on a finite dimensional Hilbert space.
HINT: Show that finitely many states suffice for the Gelfand–Naimark Theorem in this context.

I.37 Suppose that $\mathfrak{A}$ is a C*-algebra containing the compact operators. Show that every faithful irreducible representation is unitarily equivalent to the identity representation.

I.38 A representation σ is a **subrepresentation** of a representation ρ if there is a central projection P in $\rho(\mathfrak{A})'$ such that $\sigma = P\rho|P\mathcal{H}$. Hence ρ splits as a direct sum $\rho \simeq \sigma \oplus \rho'$. Show that the *Schroeder–Bernstein* Theorem is valid for representations. That is, if σ and ρ are each unitarily equivalent to subrepresentations of the other, then they are unitarily equivalent.
HINT: (a) Show that $\rho \simeq \rho \oplus (\rho' \oplus \sigma')$.
(b) By iterating (a), show that there are countably many pairwise orthogonal projections P_n in $\rho(\mathfrak{A})'$ such that $P_n\rho|P_n\mathcal{H}$ are each unitarily equivalent to $\rho' \oplus \sigma'$.
(c) Hence deduce that
$$\rho \simeq \rho'' \oplus (\rho' \oplus \sigma')^{(\infty)} \simeq \rho \oplus (\rho' \oplus \sigma')^{(\infty)}.$$
(d) Hence show that
$$\rho \simeq (\rho \oplus \sigma') \oplus (\rho' \oplus \sigma')^{(\infty)} \simeq \sigma \oplus (\rho' \oplus \sigma')^{(\infty)}.$$

I.39 (a) Suppose that ρ and σ are representations of a C*-algebra $\mathfrak{A}$ on Hilbert spaces $\mathcal{H}_\rho$ and $\mathcal{H}_\sigma$. Suppose that T in $\mathcal{B}(\mathcal{H}_\sigma, \mathcal{H}_\rho)$ is a non-zero operator such that
$$\rho(A)T = T\sigma(A) \quad \text{for all} \quad A \in \mathfrak{A}.$$
Show that the partial isometry U in the polar decomposition of T also intertwines ρ and σ. Moreover, the range and domain projections are reducing subspaces for $\sigma(\mathfrak{A})$ and $\rho(\mathfrak{A})$ respectively. The restrictions to these subspaces are unitarily equivalent.
(b) If σ is irreducible in the above situation, then σ is unitarily equivalent to a subrepresentation of ρ. If both ρ and σ are irreducible, then they are unitarily equivalent.

Notes and Remarks.

Commutative Banach algebras and the Gelfand transform were introduced by Gelfand [1941]. The notion of an abstract C*-algebra is due to Gelfand and Naimark [1943], who proved Theorem I.9.12 that C*-algebras can be isometrically imbedded into $\mathcal{B}(\mathcal{H})$. Their definition of a C*-algebra was somewhat stronger than the present-day definition. This gap (Theorem I.4.5) was filled by Fukamiya [1952] and Kaplansky (unpublished). Segal [1947] proved the existence of an approximate identity, and further refined the GNS construction proving Theorem I.9.8. Theorem I.5.4 on ideals is also due to Segal [1949] and to Kaplansky [1949]. Weak topologies and the double commutant Theorem I.7.1 are due to von Neumann [1929]. Kaplansky [1951] extended this to his density Theorem I.7.3. Kadison [1955] proved the transitivity Theorem I.9.4. Theorem I.8.4 is due to Russo and Dye [1966], but the proof presented here is from Kadison and Pedersen [1985]. Quasicentral approximate units were introduced by Arveson [1977] and Ackemann and Pedersen [1977].

CHAPTER II

Normal Operators and Abelian C*-algebras

II.1 Spectral Theory

Early in Chapter I, we learned that if N is a normal operator with spectrum X, then the continuous functional calculus Corollary I.3.3 provides a $*$-isomorphism of $C(X)$ onto $C^*(N)$. So to classify normal operators, we need to classify the $*$-representations of $C(X)$ for a compact metric spaces X.

Two normal operators M and N are **unitarily equivalent** if there is a unitary operator U such that $N = UMU^*$. This is the most rigid notion of equivalence, as a unitary operator preserves the complete spatial structure of a Hilbert space. Likewise, two representations ρ and σ of $C(X)$ are unitarily equivalent if there is a unitary operator U such that $\sigma = \operatorname{Ad} U \rho$, where $\operatorname{Ad} U(X) = UXU^*$.

Recall that a representation ρ of a C*-algebra $\mathfrak{A}$ on $\mathcal{H}$ is **cyclic** if there is a **cyclic vector** x such that $\rho(\mathfrak{A})x$ is dense in $\mathcal{H}$. Associated to a unit cyclic vector is the state

$$\varphi(A) := (\rho(A)x, x).$$

By Exercise I.35, ρ is unitarily equivalent to π_φ, the representation obtained from φ by the GNS construction. In the commutative case, this is made even more explicit.

Theorem II.1.1 *Let X be a compact metric space. Every cyclic representation ρ of $C(X)$ is unitarily equivalent to a representation given by $\sigma_\mu(f) = M_f^\mu$, the operator of multiplication by f on $L^2(X, \mu)$, where μ is a regular Borel probability measure on X.*

Proof. Let x be a unit cyclic vector for ρ, and consider the state on $C(X)$ given by $\varphi(g) := (\rho(g)x, x)$. By the Riesz Representation Theorem for positive linear functionals on $C(X)$, there is a regular Borel probability measure μ on X such that

$$\varphi(g) = \int_X g \, d\mu.$$

Define U from $C(X)$ into $\mathcal{H}$ by $Ug = \rho(g)x$. Then calculate

$$\|Ug\|^2 = (\rho(g)x, \rho(g)x) = (\rho(g)^*\rho(g)x, x)$$
$$= \varphi(|g|^2) = \int_X |g|^2 \, d\mu = \|g\|^2_{L^2(\mu)}$$

II.1. Spectral Theory

As $C(X)$ is dense in $L^2(\mu)$ and $\rho(C(X))x$ is dense in $\mathcal{H}$, this operator extends by continuity to a unitary operator of $L^2(\mu)$ onto $\mathcal{H}$.

For f, g in $C(X)$, compute
$$\rho(f)Ug = \rho(f)\rho(g)x = \rho(fg)x = U(fg) = UM_f^\mu g.$$
Again by the density of $C(X)$ in $L^2(\mu)$, we deduce that $\rho(f) = UM_f^\mu U^*$, and hence that $\rho = \operatorname{Ad} U \, \sigma_\mu$. ∎

Of course, not every representation is cyclic. However, we have the next best thing.

Proposition II.1.2 *Every non-degenerate representation σ of a C*-algebra $\mathfrak{A}$ is the direct sum of cyclic representations.*

Proof. If x in $\mathcal{H}$ is a unit vector, let $\mathcal{H}_x = \overline{\sigma(\mathfrak{A})x}$. Since $\sigma(\mathfrak{A})$ has non-trivial null space, it follows that x belongs to $\mathcal{H}_x$. (See the proof of Theorem I.7.1.) It is evident that this subspace is invariant for $\sigma(\mathfrak{A})$ and the restriction σ_x of σ to $\mathcal{H}_x$ is cyclic with cyclic vector x. Then we apply Zorn's Lemma to obtain a maximal family $\{x_\alpha : \alpha \in \mathcal{A}\}$ of unit vectors with the property that $\mathcal{H}_{x_\alpha}$ are pairwise orthogonal. It suffices to show that $\mathcal{H} = \sum \oplus_{\alpha \in \mathcal{A}} \mathcal{H}_{x_\alpha}$. Indeed, the right hand side is a direct sum which forms a subspace of $\mathcal{H}$. If it is proper, choose a unit vector y orthogonal to it. For any α, and any A, B in $\mathfrak{A}$,
$$(\sigma(A)y, \sigma(B)x_\alpha) = (y, \sigma(A^*B)x_\alpha) = 0.$$
Thus $\mathcal{H}_y$ is orthogonal to each $\mathcal{H}_{x_\alpha}$. This contradicts the maximality of the family constructed above. Therefore $\mathcal{H}$ must be the direct sum of a family of cyclic subspaces. ∎

Applying this to the abelian case yields one version of the Spectral Theorem.

Theorem II.1.3 *Every normal operator on a separable Hilbert space is unitarily equivalent to a multiplication operator M_f on some $L^2(\mu)$ space.*

Proof. Let N be a normal operator with spectrum X; and let $\rho(f) = f(N)$ be the associated representation of $C(X)$. By the preceding proposition, we may decompose ρ as a direct sum of cyclic representations. As $\mathcal{H}$ is separable, this is a countable direct sum. So we may write $\rho = \sum \oplus_{n \geq 1} \rho_n$. Then by Theorem II.1.1, there are regular Borel probability measures μ_n such that $\rho_n \simeq \sigma_{\mu_n}$ and
$$N|_{\mathcal{H}_n} = \rho_n(z) \simeq M_z^{\mu_n}.$$
Form the space $X_\infty = X \times \mathbb{N}$ of countably many disjoint copies of X, and define a measure μ on X_∞ by putting the measure μ_n on $X \times \{n\}$. Then it is evident that $L^2(\mu) \simeq \sum \oplus_{n \geq 1} L^2(\mu_n)$. If Z denotes the coordinate function $Z(x, n) = x$, then
$$M_Z \simeq \sum_{n \geq 1} \oplus M_z^{\mu_n} \simeq N.$$
∎

II.2 The L^∞ Functional Calculus

Most formulations of the spectral theorem include more information, specifically about a more powerful measurable functional calculus. Part of this can be deduced immediately, but first we make some additional preparations.

A vector x is called **separating** for a C*-algebra $\mathfrak{A}$ of operators if A in $\mathfrak{A}$ and $Ax = 0$ implies that $A = 0$. It is an elementary fact that *if x is a cyclic vector for a C*-subalgebra $\mathfrak{A}$ of $\mathcal{B}(\mathcal{H})$, then x is a separating vector for $\mathfrak{A}'$.* Indeed, suppose that T is in $\mathfrak{A}'$ and $Tx = 0$. Then for every A in $\mathfrak{A}$, one has $TAx = ATx = 0$. But $\mathfrak{A}x$ is dense in $\mathcal{H}$, whence $T = 0$.

Proposition II.2.1 *Let $\mathfrak{A}$ be an abelian subalgebra of $\mathcal{B}(\mathcal{H})$ for a separable Hilbert space $\mathcal{H}$. Then $\mathfrak{A}'$ has a cyclic vector; whence $\mathfrak{A}''$ has a separating vector.*

Proof. By Proposition II.1.2, $\mathcal{H} = \sum \oplus_{n \geq 1} \mathcal{H}_{x_n}$ is the sum of cyclic subspaces for $\mathfrak{A}'$. The subspace $\mathcal{H}_{x_n}$ is invariant for $\mathfrak{A}'$. Thus the projection P_n onto $\mathcal{H}_{x_n}$ commutes with $\mathfrak{A}'$, and so lies in $\mathfrak{A}''$. By the Double Commutant Theorem I.7.1, $\mathfrak{A}''$ is the weak operator topology closure of $\mathfrak{A}$; and thus is abelian. Hence $\mathfrak{A}' = \mathfrak{A}'''$ contains $\mathfrak{A}''$. Let $x = \sum_{n \geq 1} 2^{-n} x_n$. Then $\mathfrak{A}'x$ contains $\overline{\mathfrak{A}' 2^n P_n x} = \overline{\mathfrak{A}' x_n} = \mathcal{H}_{x_n}$ for all $n \geq 1$. So x is cyclic for $\mathfrak{A}'$. ∎

Say that a representation ρ of $C(X)$ is **multiplicity free** if $\rho(C(X))'$ is abelian. An abelian subalgebra of $\mathcal{B}(\mathcal{H})$ is called **maximal abelian** if it is not contained in any larger abelian subalgebra. A maximal abelian von Neumann algebra is often abbreviated as a **masa**, which stands for maximal abelian self-adjoint algebra.

Theorem II.2.2 *Let ρ be a representation of $C(X)$ on a separable Hilbert space $\mathcal{H}$. Then the following are equivalent.*

(i) *$\rho(C(X))$ has a cyclic vector.*
(ii) *ρ is multiplicity free.*
(iii) *$\rho(C(X))''$ is maximal abelian.*
(iv) *$\rho(C(X))''$ is unitarily equivalent to the algebra $L^\infty(\mu)$ acting by multiplication on $L^2(\mu)$ for some regular Borel probability measure μ on X.*

Proof. Assume that (i) holds, and apply Theorem II.1.1 to show that $\rho = \operatorname{Ad} U \, \sigma_\mu$ for a unitary operator U and a Borel probability measure μ on X. So for convenience, we may suppose that $\rho = \sigma_\mu$. Suppose that T commutes with $\rho(C(X))$. Let $f = T1$ where 1 represents the constant function 1 in $L^2(\mu)$. Then for g in $C(X)$,

$$Tg = TM_g^\mu 1 = M_g^\mu T1 = M_g^\mu f = fg.$$

Moreover, since $\|fg\|_2 = \|Tg\|_2 \leq \|T\| \|g\|_2$ for all $g \in C(X)$ which is dense in $L^2(\mu)$, it follows that $\|f\|_\infty \leq \|T\|$. Consequently, $T = M_f^\mu$ is a multiplication operator by a bounded measurable function f in $\mathcal{L}^\infty(\mu)$. Conversely, every such multiplication operator commutes with $\rho(C(X))$. Thus $\rho(C(X))' = L^\infty(\mu)$ is abelian; whence (ii) ρ is multiplicity free and (iv) holds.

II.2. The L^∞ Functional Calculus

When $\rho(\mathrm{C}(X))$ is abelian, we have
$$\rho(\mathrm{C}(X))'' = \overline{\rho(\mathrm{C}(X))}^{\mathrm{WOT}} \subset \rho(\mathrm{C}(X))'.$$
But (ii) implies that $\rho(\mathrm{C}(X))' \subset \rho(\mathrm{C}(X))''$. Thus $\rho(\mathrm{C}(X))''$ is equal to its own commutant, so it is maximal abelian. Hence (ii) implies (iii).

Now (iii) together with Proposition II.2.1 shows that $\rho(\mathrm{C}(X))''$ has a cyclic vector x. But
$$\overline{\rho(\mathrm{C}(X))x} = \overline{\rho(\mathrm{C}(X))''x} = \mathcal{H},$$
so ρ is cyclic. Hence (iii) implies (i).

Finally if (iv) holds, then by Theorem II.1.1, ρ is cyclic; and therefore (iv) implies (i). ∎

Of course, most representations of $\mathrm{C}(X)$ are not cyclic, but the general structure can be obtained by applying Proposition II.1.2. In order to get the full power of the result, we need a couple of preparatory lemmas. Recall that $L^1(\mu)^* = L^\infty(\mu)$, and that this pairing determines the weak-$*$ topology on $L^\infty(\mu)$.

Lemma II.2.3 *Let μ be a Borel probability measure on a compact metric space X. Suppose that f_α is a net of functions in $L^\infty(\mu)$ converging in the weak-$*$ topology to f. Then $M^\mu_{f_\alpha}$ converges in the weak operator topology to M^μ_f.*

Proof. Fix vectors x, y in $L^2(\mu)$. Then $h = x\overline{y}$ belongs to $L^1(\mu)$. Hence
$$\lim_\alpha (M^\mu_{f_\alpha} x, y) = \lim_\alpha \int f_\alpha h\, d\mu = \lim_\alpha \int f h\, d\mu = \lim_\alpha (M^\mu_f x, y). \quad \blacksquare$$

Lemma II.2.4 *Let μ and ν be regular Borel probability measures on a compact metric space X. Suppose that σ is a $*$-isomorphism of $L^\infty(\mu)$ onto $L^\infty(\nu)$ which is the identity on $\mathrm{C}(X)$. Then μ and ν are equivalent measures (so that $L^\infty(\mu) = L^\infty(\nu)$), and σ is the identity map.*

Proof. Since σ is a $*$-isomorphism, it preserves order; and therefore it preserves suprema. Let $\mathcal{O}$ be an open subset of X. Then $\chi_\mathcal{O}$ is the supremum of an increasing sequence of continuous functions. For example, let
$$f_k(\xi) = \begin{cases} 1 & \text{if } \operatorname{dist}(\xi, \mathcal{O}^c) \geq 2^{-k} \\ 2^k \operatorname{dist}(\xi, \mathcal{O}^c) & \text{if } \operatorname{dist}(\xi, \mathcal{O}^c) < 2^{-k} \end{cases}.$$
Hence
$$\sigma(\chi_\mathcal{O}) = \sup_{k \geq 1} \sigma(f_k) = \sup_{k \geq 1} f_k = \chi_\mathcal{O}.$$
If $\sigma(\chi_E) = \chi_E$ and $\sigma(\chi_F) = \chi_F$, then
$$\sigma(\chi_{E^c}) = \sigma(1 - \chi_E) = 1 - \sigma(\chi_E) = 1 - \chi_E = \chi_{E^c},$$

and

$$\sigma(\chi_{E\cap F}) = \sigma(\chi_E \chi_F) = \sigma(\chi_E)\sigma(\chi_F) = \chi_E \chi_F = \chi_{E\cap F}.$$

The result for unions follows from these two relations. Finally, if $\sigma(\chi_{E_k}) = \chi_{E_k}$ for $k \geq 1$, then

$$\sigma(\chi_{\bigcup_{k\geq 1} E_k}) = \sigma(\sup_{n\geq 1} \chi_{\bigcup_{k=1}^n E_k}) = \sup_{n\geq 1} \sigma(\chi_{\bigcup_{k=1}^n E_k})$$
$$= \sup_{n\geq 1} \chi_{\bigcup_{k=1}^n E_k} = \chi_{\bigcup_{k\geq 1} E_k}.$$

This shows that the set $\{E : \sigma(\chi_E) = \chi_E\}$ is a sigma algebra containing the open sets. So it contains all Borel sets, which yields all measurable sets up to subsets which are measure zero relative to $\mu + \nu$. In particular, $\mu(E) = 0$ if and only if $\chi_E = 0$ in $L^\infty(\mu)$; which is therefore if and only if $\chi_E = \sigma(\chi_E) = 0$ in $L^\infty(\nu)$; which is if and only if $\nu(E) = 0$. So μ and ν are equivalent measures. Finally, σ is the identity on the span of the characteristic functions. As this space is norm-dense in $L^\infty(\mu)$, it follows that σ is the identity map. ∎

With these two measure theoretic lemmas in hand, we can prove the main result of this section.

Theorem II.2.5 *Let ρ be a representation of $C(X)$ on a separable Hilbert space $\mathcal{H}$; and let $\mathfrak{M} = \rho(C(X))''$. Then there is a regular Borel measure μ on X so that $\mathfrak{M}$ is $*$-isomorphic to $L^\infty(\mu)$. Moreover, this map is a homeomorphism from the weak operator topology on $\mathfrak{M}$ to the weak-$*$ topology on $L^\infty(\mu)$.*

Proof. By Proposition II.2.1, $\mathfrak{M}$ has a separating vector x. Let $\mathcal{K}$ be the cyclic subspace generated by x. Consider the restriction map $\pi(A) = A|_{\mathcal{K}}$ for A in $\mathfrak{M}$. Clearly, this is a $*$-homomorphism. Since x is a separating vector, π is an isomorphism. Moreover, this map is obviously continuous in the weak operator topology.

Now $\pi(\mathfrak{M})$ has a cyclic vector x, whence by Theorem II.2.2, there is a regular Borel probability measure ν on X so that $\mathfrak{M}$ is unitarily equivalent to $L^\infty(\nu)$ acting on $L^2(\nu)$. Note that the weak operator topology on $L^\infty(\nu)$ coincides with the weak-$*$ topology it inherits as the dual of $L^1(\nu)$. Indeed, if h_1 and h_2 are functions in $L^2(\nu)$, then the WOT-continuous linear functional

$$\varphi(f) = (M_f^\nu h_1, h_2) = \int_X f(h_1 \overline{h_2})\, d\nu$$

is integration against the $L^1(\nu)$ function $h = h_1 \overline{h_2}$, and therefore is weak-$*$ continuous. Conversely, if h is in $L^1(\nu)$, then we set $h_2 = |h|^{1/2}$ and $h_1 = h/h_2$. Clearly $\|h_i\|_2^2 = \|h\|_1$; so they lie in $L^2(\nu)$. Thus the weak-$*$ continuous functional

$$\psi(f) = \int_X fh\, d\nu = \int_X f(h_1 \overline{h_2})\, d\nu = (M_f^\nu h_1, h_2)$$

is WOT-continuous.

II.2. The L^∞ Functional Calculus

Now turn the tables and work from the other direction. Apply Proposition II.1.2 to ρ to decompose it as a direct sum $\rho = \sum \oplus \rho_n$ of at most countably many cyclic representations ρ_n. By Theorem II.1.1, there are Borel probability measures μ_n on X so that ρ_n is unitarily equivalent to σ_{μ_n}. Let $\mu = \sum_n 2^{-n} \mu_n$, which is a measure equivalent to the family $\{\mu_n\}$. For each bounded Borel function f on X, choose a bounded sequence f_k in $C(X)$ converging to f a.e. $d\mu$ (which is possible by Egoroff's Theorem). Then f_k converges to f in the weak-$*$ topology on $L^\infty(\mu)$ by the Lebesgue Dominated Convergence Theorem; and thus also converges weak-$*$ to f in each $L^\infty(\mu_n)$. So by Lemma II.2.3,

$$\sum_n \oplus M_f^{\mu_n} = \text{WOT-}\lim_{k\to\infty} \sum_n \oplus M_{f_k}^{\mu_n}$$

belongs to $\mathfrak{M}$. Moreover, the same lemma now shows that the $*$-homomorphism $\widetilde{\rho}$ from $L^\infty(\mu)$ to $\mathfrak{M}$ given by $\widetilde{\rho}(f) = \sum_n \oplus M_f^{\mu_n}$ is continuous from the weak-$*$ topology on $L^\infty(\mu)$ to the weak operator topology on $\mathfrak{M}$.

The map $\widetilde{\rho}$ is injective. Indeed, if $\widetilde{\rho}(f) = 0$ for a non-zero function f in $L^\infty(\mu)$, then there is a function g in $L^\infty(\mu)$ such that $fg = \chi_E$ for a set E of positive measure. But then

$$\sum_n \oplus M_{\chi_E}^{\mu_n} = \widetilde{\rho}(\chi_E) = \widetilde{\rho}(f)\widetilde{\rho}(g) = 0.$$

This implies that $\mu_n(E) = 0$ for all n; whence $\mu(E) = 0$, contrary to hypothesis. Consequently, $\widetilde{\rho}$ is a $*$-isomorphism. It carries $C(X)$ onto $\rho(C(X))$. Since it is weakly continuous, it carries $L^\infty(\mu)$ onto $\mathfrak{M}$.

Finally, put the two pieces together. The map $\pi\widetilde{\rho}$ is a $*$-isomorphism of $L^\infty(\mu)$ onto $L^\infty(\nu)$ which is the identity on $C(X)$. Hence by Lemma II.2.4, this is the identity map. Consequently, both $\widetilde{\rho}$ and $\pi = \widetilde{\rho}^{-1}$ are continuous between $L^\infty(\mu)$ endowed with the weak-$*$ topology and $\mathfrak{M}$ with the WOT topology. Therefore $\widetilde{\rho}$ is a homeomorphism. ∎

Applying this to a single normal operator yields the so called **L^∞ functional calculus**.

Corollary II.2.6 *If N is a normal operator on a separable Hilbert space, then there is a regular Borel measure μ on $\sigma(N)$ and a $*$-isomorphism φ_N of $L^\infty(\mu)$ onto $W^*(N)$ which is a homeomorphism from the weak-$*$ topology on $L^\infty(\mu)$ to the weak operator topology on $W^*(N)$.*

This yields the **spectral measure** of a normal operator as follows. Define a projection valued measure E_N on Borel subsets X of $\mathbb{C}$ by $E_N(X) := \varphi_N(\chi_X)$. This assigns a projection to each Borel subset of the plane. Since φ_N is a homomorphism, we obtain the identities

$$E_N(X \cap Y) = E_N(X)E_X(Y)$$

and
$$E_N(X \cup Y) = E_N(X) + E_N(Y) - E_N(X \cap Y).$$

Countable additivity is a consequence of the fact that φ_N is a homeomorphism from the weak-$*$ topology on $L^\infty(\mu)$ to the weak operator topology on $W^*(N)$. Indeed, suppose that X_n are pairwise disjoint Borel sets; and let $X = \bigcup_{n \geq 1} X_n$. Then $\sum_{k=1}^n \chi_{X_k}$ converges weak-$*$ to χ_X and thus $\sum_{k=1}^n E_N(X_k)$ converges in the weak operator topology to $E_N(X)$. As the limit is a projection, it follows that this converges in the strong operator topology as well. Hence we obtain that
$$\text{SOT-}\sum_{k \geq 1} E_N(X_k) = E_N(X).$$

We also observe that the measure μ is determined only up to its measure class of equivalent measures. Indeed, two measures have the same non-zero characteristic functions if and only if they have the same null sets which occurs if and only if they are equivalent measures. On the other hand, if μ and ν are equivalent measures, then the Radon–Nikodym Theorem says that $d\nu = h\,d\mu$ where $h > 0$ a.e. $d\mu$. Hence we may define a unitary operator from $L^2(\nu)$ onto $L^2(\mu)$ by $Uf = h^{1/2} f$. Indeed,
$$\|Uf\|^2_{L^2(\mu)} = \int |f|^2 h\,d\mu = \int |f|^2\,d\nu = \|f\|^2_{L^2(\nu)}.$$

Moreover,
$$M_g^\mu Uf = gfh^{1/2} = UM_g^\nu f;$$

whence $\operatorname{Ad} U$ is a unitary equivalence between the two cyclic representations. Therefore we have proven:

Corollary II.2.7 *Two cyclic representations of $C(X)$ are unitarily equivalent if and only if the corresponding measures are equivalent.*

A simple device allows us to obtain the same result for arbitrary separably acting abelian von Neumann algebras by showing that they are singly generated, and thus arise as a representation of some $C(X)$.

Lemma II.2.8 *Let $\mathfrak{M}$ be an abelian von Neumann algebra on a separable Hilbert space. Then there is a self-adjoint operator A such that $\mathfrak{M} = W^*(A)$. Moreover, if $\mathfrak{A}$ is a separable C^*-subalgebra of $\mathfrak{M}$, it can be arranged that $\mathfrak{A} \subset C^*(A)$.*

Proof. The unit ball of $\mathcal{B}(\mathcal{H})$ in the weak operator topology is metrizable by Proposition I.6.3 and compact. Hence there is a countable family of self-adjoint operators which are WOT-dense in the unit ball of $\mathfrak{M}_{sa}$. Include in this family a countable dense subset of $\mathfrak{A}_{sa}$. Each self-adjoint operator is in the norm-closed span of its spectral projections corresponding to diadic intervals. Consequently, there is a countable collection $\{E_n\}$ of projections in $\mathfrak{M}$ that spans a C^*-algebra $\mathfrak{E}$ which contains $\mathfrak{A}$ and is WOT-dense in $\mathfrak{M}$.

Let $A = \sum_{n \geq 1} 3^{-n} E_n$. We claim that $C^*(A) = \mathfrak{E}$. Since A belongs to $\mathfrak{E}$, it suffices to show that each E_n belongs to $C^*(A)$. To this end, note that

$$0 \leq E_1^\perp A = E_1^\perp \sum_{n \geq 2} 3^{-n} E_n \leq \tfrac{1}{6} E_1^\perp$$

and similarly

$$\tfrac{1}{3} E_1 \leq E_1 A \leq \tfrac{1}{2} E_1.$$

Thus it follows that the spectrum $\sigma(A)$ is contained in $[0, \tfrac{1}{6}] \cup [\tfrac{1}{3}, \tfrac{1}{2}]$. So $C^*(A)$ contains the projection $E_A([\tfrac{1}{3}, \tfrac{1}{2}]) = E_1$. After subtracting off $\tfrac{1}{3} E_1$ from A, we may likewise show that E_2 belongs to $C^*(A)$; and recursively we can establish that $C^*(A) = \mathfrak{E}$. Thus $W^*(A) = \overline{\mathfrak{E}}^{\text{WOT}} = \mathfrak{M}$. ∎

Corollary II.2.9 *Let $\mathfrak{M}$ be an abelian von Neumann algebra on a separable Hilbert space. Then there is a compact subset X of $\mathbb{R}$ and a regular Borel probability measure on X such that $\mathfrak{M}$ is $*$-isomorphic and* WOT-*homeomorphic to $L^\infty(\mu)$.*

Proof. By Lemma II.2.8, there is a self-adjoint operator A such that $\mathfrak{M} = W^*(A)$. Let $X = \sigma(A)$. Then the L^∞ functional calculus for A of Corollary II.2.6 yields the measure μ on X and the desired $*$-isomorphism. ∎

II.3 Multiplicity Theory

In the previous section, we identified representations which are multiplicity free, or multiplicity one; and showed that every representation always has a cyclic subspace such that the restriction to that subspace is a $*$-isomorphism. The question remains about how a general abelian von Neumann algebra is structured. If $\mathfrak{M}$ acts as $L^\infty(\mu)$ on $\mathcal{H} = L^2(\mu)$, form the algebra $\mathfrak{M}^{(n)}$ acting on the direct sum $\mathcal{H}^{(n)}$ of n copies of $\mathcal{H}$ by

$$M_f^{(n)}(h_1 \oplus \cdots \oplus h_n) = fh_1 \oplus \cdots \oplus fh_n;$$

where $M_f^{(n)}$ is the direct sum $M_f \oplus \cdots \oplus M_f$ of n copies of M_f. We justify saying that this algebra has **uniform multiplicity n** by the following lemma.

Lemma II.3.1 *Let $\mathfrak{M}$ act as $L^\infty(\mu)$ on $\mathcal{H} = L^2(\mu)$. The commutant of $\mathfrak{M}^{(n)}$ is $\mathcal{M}_n(\mathfrak{M})$, the algebra of $n \times n$ matrices with coefficients in $L^\infty(\mu)$ acting on $\mathcal{H}^{(n)}$.*

Proof. It is evident that every $n \times n$ matrix with coefficients in $L^\infty(\mu)$ commutes with $\mathfrak{M}^{(n)}$. Conversely, if A commutes with $\mathfrak{M}^{(n)}$, write it as an $n \times n$ matrix $[A_{ij}]$ where $A_{ij} = P_{H_i} A | \mathcal{H}_j$ maps the j-th copy of $\mathcal{H}$ into the i-th copy. A routine matrix calculation shows that each A_{ij} must commute with $\mathfrak{M}$; and hence belongs to $\mathfrak{M}$ because it is maximal abelian. ∎

In a more coordinate free way, we say that an abelian von Neumann algebra $\mathfrak{N}$ has **uniform multiplicity** n provided that $\mathfrak{N}'$ is unitarily equivalent to $\mathcal{M}_n(\mathfrak{M})$ for some maximal abelian algebra $\mathfrak{M}$. This means that $\mathfrak{N}'$ contains a system of matrix units $\{E_{ij} : 1 \leq i,j \leq n\}$ such that $E_{11}\mathfrak{N}'|E_{11}\mathcal{H}$ is maximal abelian. This makes sense even when $n = \aleph_0$ if we replace $\mathcal{M}_n(\mathfrak{M})$ by the algebra of all bounded operators on $\mathcal{H}^{(\infty)}$ such that the matrix coefficients A_{ij} lie in $\mathfrak{M}$. By Theorem II.2.2, this general situation reduces to the case we have just considered. However, there is one important subtlety:

Lemma II.3.2 *If $\mathfrak{N}$ has uniform multiplicity, then the multiplicity is well defined.*

Proof. Suppose that there are cardinal numbers $m \leq n$ such that $\mathfrak{N}$ has both uniform multiplicity m and n. Hence $\mathfrak{N}'$ is unitarily equivalent to $\mathcal{M}_m(\mathfrak{M})$ and to $\mathcal{M}_n(\widetilde{\mathfrak{M}})$ for two masas $\mathfrak{M}$ and $\widetilde{\mathfrak{M}}$. Thus there is a unital homomorphism of $\mathcal{M}_n$ into $\mathcal{M}_m$ defined as follows: take the unital imbedding j of $\mathcal{M}_n$ into $\mathcal{M}_n(\widetilde{\mathfrak{M}})$ by sending a matrix (or operator when $n = \infty$) to the corresponding matrix with scalar entries. Follow that with the identification of $\mathcal{M}_n(\widetilde{\mathfrak{M}})$ with $\mathfrak{N}'$ and hence with $\mathcal{M}_m(\mathfrak{M})$. Then follow this with evaluation at any multiplicative linear functional φ of $\mathfrak{M}$. This determines a unital $*$-homomorphism $\varphi^{(m)}$ of $\mathcal{M}_m(\mathfrak{M})$ onto $\mathcal{M}_m$. The composition is a unital $*$-homomorphism of $\mathcal{M}_n$ into $\mathcal{M}_m$. This is only possible when $n = m$. (Recall that $\mathcal{M}_n$ is a simple algebra, and $\mathcal{B}(\mathcal{H})$ has only the Calkin algebra as a possible quotient. So by dimension alone we may distinguish these algebras.) ∎

So now we may try to decompose an arbitrary abelian von Neumann algebra into parts of different multiplicities. To this end, say that a projection P in an abelian von Neumann algebra $\mathfrak{N}$ has uniform multiplicity n if $\mathfrak{N}|P\mathcal{H}$ has uniform multiplicity n.

Lemma II.3.3 *Let $\mathfrak{N}$ be an abelian von Neumann algebra acting on a separable Hilbert space. The supremum of all projections in $\mathfrak{N}$ of uniform multiplicity n also has uniform multiplicity n.*

Proof. By Corollary II.2.9, $\mathfrak{N}$ is $*$-isomorphic to $L^\infty(\mu)$. Thus every projection P in $\mathfrak{N}$ corresponds to multiplication by a characteristic function χ_X under this isomorphism. Thus if P has uniform multiplicity n, then $(\mathfrak{N}|P\mathcal{H})'$ is isomorphic to $\mathcal{M}_n(L^\infty(\mu|_X))$.

Now notice that if P has uniform multiplicity n, then so does every subprojection of P in $\mathfrak{N}$. For any $P' \leq P$ corresponds to $\chi_{X'}$ for some measurable subset $X' \subset X$; and thus $(\mathfrak{N}|P'\mathcal{H})'$ is isomorphic to $\mathcal{M}_n(L^\infty(\mu|_{X'}))$.

Suppose that P_α are projections of uniform multiplicity n corresponding to subsets X_α. Then the supremum of the P_α is the supremum of a countable subset. (Equivalently, there is a countable collection of the measurable sets X_α whose union contains all the others modulo null sets.) Denote this set by $\{P_k : k \geq 1\}$. Replace P_k by $P'_k = P_k(\bigvee_{1 \leq j < k} P_j)^\perp$ so that they are pairwise orthogonal with

II.3. Multiplicity Theory

the same supremum. By the previous paragraph, each P'_k has multiplicity n and corresponds to a characteristic function $\chi_{X'_k}$. The sets X'_k are essentially disjoint; so we may assume that they are actually disjoint.

Then let $P = \sum_{k \geq 1} P'_k$. This is the supremum of the P_α's, and corresponds to the characteristic function χ_X where $X = \bigcup_{k \geq 1} X'_k$. Because $\mathfrak{N}'$ commutes with each P'_k, the restriction $\mathfrak{N}'|P\mathcal{H}$ is unitarily equivalent to the direct sum

$$\sum_{k \geq 1} \oplus \mathfrak{N}'|P'_k\mathcal{H} = \mathcal{M}_n(\sum_{k \geq 1} \oplus L^\infty(\mu|_{X'_k})) \simeq \mathcal{M}_n(\mathcal{L}^\infty(\mu|_X)).$$

Hence P has uniform multiplicity n. ∎

Lemma II.3.4 *Let $\mathfrak{N}$ be an abelian von Neumann algebra acting on a separable Hilbert space. Then there is a non-zero projection P in $\mathfrak{N}$ of uniform multiplicity.*

Proof. Recursively apply Proposition II.2.1 to choose unit vectors x_n and projections Q_n in $\mathfrak{N}'$ onto the cyclic subspace span$\{Ax_n : A \in \mathfrak{N}\}$ such that x_n is a separating vector for $\mathfrak{N}|(\sum_{i=1}^{n-1} Q_i)^\perp \mathcal{H}$. Let $\overline{Q_n}$ denote the smallest projection in $\mathfrak{N}$ such that $Q_n \leq \overline{Q_n}$. This is known as the **central cover** of Q_n. Since x_n is always chosen to be a separating vector for the compression algebra, it follows that $\overline{Q_1} = I$ and $\overline{Q_n} \geq \overline{Q_{n+1}}$ for all $n \geq 1$.

There are two cases to deal with. Either (i) there is an integer n_0 such that $\overline{Q_{n_0}} = I > \overline{Q_{n_0+1}}$, or (ii) $\overline{Q_n} = I$ for all $n \geq 1$.

In the first case, let $P = \overline{Q_{n_0+1}}^\perp$; and consider the algebra $\mathfrak{N}|P\mathcal{H}$. Then $Px_{n_0+1} = 0$. Since x_{n_0+1} is a separating vector for $\mathfrak{N}|(\sum_{i=1}^n Q_i)^\perp \mathcal{H}$, it follows that $P(\sum_{i=1}^n Q_i)^\perp = 0$; whence $P \leq \sum_{i=1}^n Q_i$. Let $E_{ii} = PQ_i$ for $1 \leq i \leq n_0$. Then it is easy to see that $\overline{E_{ii}} = P\overline{Q_i} = P$. Moreover Px_i is a cyclic and separating vector for $\mathfrak{N}|E_{ii}\mathcal{H}$. If P corresponds to the characteristic function χ_X, then we see that $\mathfrak{N}|E_{ii}\mathcal{H}$ is unitarily equivalent to $\mathfrak{M} = L^\infty(\mu|_X)$ acting on $L^2(\mu|_X)$. Hence $\mathfrak{N}|P\mathcal{H}$ is unitarily equivalent to $\mathfrak{M}^{(n_0)}$. So by Lemma II.3.1, P has uniform multiplicity n_0.

The second case is similar, though a bit more complicated. We start over, and use Zorn's Lemma to construct a maximal family of separating vectors x_n for $\mathfrak{N}$ such that the cyclic subspaces Q_n onto span$\{Ax_n : A \in \mathfrak{N}\}$ are pairwise orthogonal. The central cover $\overline{R}$ of $R = (\sum_{n \geq 1} Q_n)^\perp$ must have $\overline{R} < I$. For otherwise by Proposition II.2.1, there is a separating vector y for $\mathfrak{N}|R\mathcal{H}$ which would be separating for $\mathfrak{N}$ contradicting the maximality of the family $\{x_n\}$. Let $P = \overline{R}^\perp$.

As before, let $E_{ii} = PQ_i$. The restriction $\mathfrak{N}|E_{ii}\mathcal{H}$ has a cyclic and separating vector Px_i, and thus is maximal abelian. As above, each is isomorphic to $\mathfrak{M} = L^\infty(\mu|_X)$ acting on $L^2(\mu|_X)$ for $i \geq 1$. Consequently, $\mathfrak{N}|P\mathcal{H}$ is unitarily equivalent to $\mathfrak{M}^{(\infty)}$. Hence by Lemma II.3.1, P has uniform multiplicity $\aleph_0$. ∎

We now have all the ingredients to describe the structure of a general abelian von Neumann algebra.

Theorem II.3.5 *Let $\mathfrak{N}$ be an abelian von Neumann algebra acting on a separable Hilbert space. Then there is a uniquely defined family $\{P_n : 1 \leq n \leq \aleph_0\}$ of projections in $\mathfrak{N}$ which are pairwise orthogonal, partition the identity (that is, $\sum_n P_n = I$) and such that P_n has uniform multiplicity n for each n.*

If we identify $\mathfrak{N}$ with $L^\infty(\mu)$ via the $*$-isomorphism of Corollary II.2.9, then there are mutually singular measures $\mu_n \ll \mu$ such that $\mathfrak{N}$ is unitarily equivalent to

$$\sum_{n=1}^{\infty} \oplus L^\infty(\mu_n)^{(n)} \oplus L^\infty(\mu_{\aleph_0})^{(\aleph_0)}$$

acting on

$$\mathcal{H} = \sum_{n=1}^{\infty} \oplus L^2(\mu_n)^{(n)} \oplus L^2(\mu_{\aleph_0})^{(\aleph_0)}.$$

Proof. For each cardinal n, $1 \leq n \leq \aleph_0$, let P_n be the supremum of all projections in $\mathfrak{N}$ of uniform multiplicity n. By Lemma II.3.3, each P_n has uniform multiplicity n and is the largest projection with this property.

For $m \neq n$, we must have P_m orthogonal to P_m because the projection $P_m P_n$ has uniform multiplicity both m and n, which would contradict Lemma II.3.2 unless $P_m P_n = 0$. Let $Q = (\sum_n P_n)^\perp$; and set $\mathfrak{N}_0 = \mathfrak{N}|Q\mathcal{H}$. If $Q \neq 0$, we may find a non-zero subprojection $Q' \leq Q$ of uniform multiplicity by Lemma II.3.4. This contradicts the maximality of one of the P_n. Hence $Q = 0$; and therefore $\{P_n : 1 \leq n \leq \aleph_0\}$ forms a partition of the identity.

Each P_n is identified with a characteristic function χ_{X_n} in $L^\infty(\mu)$. The orthogonality implies that X_n are essentially disjoint. Hence the measures $\mu_n = \mu|_{X_n}$ are mutually singular. And the fact that P_n partition the identity shows that the union $\cup_n X_n = X$ is the whole space up to a null set. This yields the desired decomposition of $\mathfrak{N}$. ∎

Applying this to the von Neumann algebra generated by a single normal operator yields a much stronger version of the Spectral Theorem II.1.3. The spectral measure E_N given by the L^∞ functional calculus is now seen to be determined by the algebra $L^\infty(\mu)$ and the **multiplicity function** m_N defined by $m_N(x) = k$ for $x \in X_k$, $1 \leq k \leq \aleph_0$. This function is determined almost everywhere $d\mu$ because of the uniqueness of the decomposition. The measure μ is determined up to its measure class of all equivalent measures by Corollary II.2.7.

This immediately yields complete unitary invariants for normal operators.

Corollary II.3.6 *Two normal operators M and N acting on separable Hilbert spaces are unitarily equivalent if and only if they have equivalent spectral measures and multiplicity functions.*

II.4 The Weyl–von Neumann–Berg Theorem

There are other weaker notions of equivalence that are important. Say that two operators A and B are **approximately unitarily equivalent** (write $A \sim_a B$) if there is a sequence of unitary operators U_n such that $B = \lim_{n\to\infty} U_n A U_n^*$. It is evident that two operators are approximately unitarily equivalent if and only if they have the same norm-closed unitary orbit

$$\overline{\mathcal{U}}(A) = \overline{\{UAU^* : U \text{ unitary}\}}.$$

Two operators with the same closed unitary orbit have the same *observable data* in the sense that no finite set of measurements determined by vectors can distinguish the two operators. It will be convenient to simultaneously consider several related notions. Two operators A and B are **approximately unitarily equivalent relative to $\mathfrak{K}$** (write $A \sim_\mathfrak{K} B$) if in addition to having $B = \lim_{n\to\infty} U_n A U_n^*$, one also has that $B - U_n A U_n^*$ belongs to the ideal $\mathfrak{K}$ of compact operators for all n. Evidently, this second notion is stronger. However, we shall see that the two relations are, in fact, equivalent.

Analogously, we say that two representations σ and ρ of a *separable* C*-algebra $\mathfrak{A}$ are **approximately unitarily equivalent (relative to $\mathfrak{K}$)** if there is a sequence of unitary operators such that

$$\rho(A) = \lim_{n\to\infty} \operatorname{Ad} U \, \sigma(A) \quad \text{for all} \quad A \in \mathfrak{A}$$

(and in addition the range of $\rho - \operatorname{Ad} U \sigma$ is contained in the ideal $\mathfrak{K}$ for all $n \geq 1$). Again we write $\rho \sim_a \sigma$ and $\rho \sim_\mathfrak{K} \sigma$ respectively.

There is another apparently much weaker relation in the same vein. we will say that two representations ρ and σ are **weak approximately unitarily equivalent** ($\rho \sim_{wa} \sigma$) if there there are sequences U_n and V_n of unitary operators such that

$$\sigma(A) = \operatorname*{WOT-lim}_{n\to\infty} U_n \rho(A) U_n^* \quad \text{and} \quad \rho(A) = \operatorname*{WOT-lim}_{n\to\infty} V_n \sigma(A) V_n^*$$

for all A in $\mathfrak{A}$. Both directions are needed to obtain an equivalence relation. This notion readily implies a stronger version of itself with convergence in the strong-$*$ topology. Indeed,

$$\begin{aligned}\left\|\big(\sigma(A) - U_n\rho(A)U_n^*\big)x\right\|^2 &= \Big(\big(\sigma(A) - U_n\rho(A)U_n^*\big)x, \sigma(A)x\Big) \\ &+ \Big(\big(\sigma(A^*) - U_n\rho(A^*)U_n^*\big)\sigma(A)x, x\Big) \\ &- \Big(\big(\sigma(A^*A) - U_n\rho(A^*A)U_n^*\big)x, x\Big).\end{aligned}$$

This demonstrates convergence in the strong operator topology for both A and A^*, which is strong-$*$ convergence.

The corresponding notion for a single operator using the weak operator topology does not yield a satisfactory relation (Exercise II.10). However, the strong-$*$ topology does yield a useful notion. So to have parallel notation, we say that two operators A and B are **weak approximately unitarily equivalent** if there there are sequences U_n and V_n of unitary operators such that

$$B = \text{SOT*-lim}\, U_n A U_n^* \quad \text{and} \quad A = \text{SOT*-lim}\, V_n B V_n^*.$$

This relation is sufficient to guarantee that there is a $*$-isomorphism ρ from $C^*(A)$ onto $C^*(B)$ such that $\rho(A) = B$ and $\rho \sim_{wa}$ id. To see this, first notice that if $U_n X_i U_n^*$ converges SOT to Y_i for $i = 1, 2$, then

$$(U_n X_1 X_2 U_n^* - Y_1 Y_2)x = (U_n X_1 U_n^* - Y_1)Y_2 x + Y_1(U_n X_2 U_n^* - Y_2)x$$
$$+ (U_n X_1 U_n^* - Y_1)(U_n X_2 U_n^* - Y_2)x.$$

Whence $U_n X_1 X_2 U_n^*$ converges SOT to $Y_1 Y_2$. Consequently, if p is any polynomial in non-commuting variables, it follows that

$$\text{SOT-lim}\, U_n p(A, A^*) U_n^* = p(B, B^*)$$

and

$$\text{SOT-lim}\, V_n p(B, B^*) V_n^* = p(A, A^*).$$

Thus the map $\rho(p(A, A^*)) = p(B, B^*)$ is an isometric $*$-isomorphism, and extends to their C*-algebras. Evidently, we have shown that $\rho \sim_{wa} \text{id}_{C^*(A)}$.

The following theorem shows that normal operators are *diagonal plus compact*. This will enable us to show that every normal operator is approximately unitarily equivalent to a diagonal operator, which will facilitate their classification.

Theorem II.4.1 *Suppose that $\mathfrak{A}$ is a separable abelian C*-subalgebra of $\mathcal{B}(\mathcal{H})$. Then there is an orthonormal basis $\{e_k : k \geq 1\}$ for $\mathcal{H}$ so that $\mathfrak{A}$ is contained in $\mathfrak{D} + \mathfrak{K}$, where $\mathfrak{D}$ is the algebra of diagonal operators with respect to this basis. In addition, one may arrange that any given finite collection of projections in $\mathfrak{A}''$ belongs to $\mathfrak{D}$.*

Proof. By Lemma II.2.8, $\mathfrak{A}$ is contained in a C*-algebra $\mathfrak{E} = \overline{\text{span}(\mathcal{E})}$ generated by a countable, commuting family $\mathcal{E} = \{E_n, n \geq 1\}$ of projections. If there are finitely many projections given, we may assume that they are the first N in the list.

Construct an approximate unit of projections for $\mathfrak{K}$ which is quasi-central for $\mathfrak{E}$ as follows. Fix an orthonormal basis $x_1, x_2, \ldots$ for $\mathcal{H}$. For each projection E_i, let us denote $E_i^{(-1)} := I - E_i$ and $E_i^{(1)} := E_i$. Then for $k \geq N$, set

$$\mathcal{L}_k = \text{span}\{\prod_{i=1}^k E_i^{(\epsilon_i)} x_j : 1 \leq j \leq k,\ \epsilon_i = \pm 1\}.$$

Then $\mathcal{L}_k$ is an increasing sequence of finite dimensional subspaces with dense union. So the orthogonal projections F_k onto $\mathcal{L}_k$ increase strongly to I; whence it is an approximate unit for $\mathfrak{K}$. Also notice that $\mathcal{L}_k$ is invariant for E_n when $n \leq k$, and thus F_k commutes with these projections. Consequently,

$$\lim_{k \to \infty} F_k E_n - E_n F_k = 0 \quad \text{for all} \quad n \geq 1.$$

II.4. The Weyl–von Neumann–Berg Theorem

This conclusion extends to the closed linear span.

Let $D_n = E_n$ for $1 \leq n \leq N$ and $D_n = E_n(I - F_n)$ for $n > N$. Then the projection D_n exactly commutes with every F_k for $k \geq N$ and with all the other D_m's. Let $\mathfrak{D}$ be the C*-algebra generated by $\{D_n : n \geq 1\}$. It is immediate that this is an abelian C*-algebra such that

$$\mathfrak{A} \subset \mathfrak{E} \subset \mathfrak{D} + \mathfrak{K}.$$

It remains to show that $\mathfrak{D}$ is diagonalizable. As $\mathfrak{D}$ commutes with each F_k for $k \geq N$, the finite dimensional subspaces $\mathcal{H}_N = \mathcal{L}_N$ and $\mathcal{H}_k = \mathcal{L}_k \ominus \mathcal{L}_{k-1}$ for $k > N$ are all invariant for $\mathfrak{D}$. The restriction of $\mathfrak{D}$ to $\mathcal{H}_k$ is a commuting family of normal matrices, which is diagonalizable by the finite dimensional spectral theorem. Choose such a basis for each $\mathcal{H}_k$. The basis for $\mathcal{H}$ obtained by combining these bases for each k diagonalizes $\mathfrak{D}$. ∎

The following corollary asserts that every normal operator is a *small* compact perturbation of a diagonalizable operator. It is known as the **Weyl–von Neumann–Berg Theorem**.

Corollary II.4.2 *Every normal operator N on a separable Hilbert space can be expressed as a sum $N = D + K$ of a diagonal normal operator D and a compact operator K. Moreover for any $\varepsilon > 0$ and any n commuting Hermitian operators $A_1, \ldots, A_n$, there are simultaneously diagonal Hermitian operators D_i and compact operators K_i such that $A_i = D_i + K_i$ and $\|K_i\| < \varepsilon$.*

Proof. We require an efficient way to construct each A_i from projections. For convenience, let us translate each A_i by a multiple of the identity and scale them so that $0 \leq A_i \leq I$. This may require redefining ε.

For each A_i, consider the spectral projections

$$E_k^{(i)} = E_{A_i}\Big(\bigcup_{i=1}^{2^{k-1}} (2^{-k}(2i-1), 2^{-k}(2i)] \Big)$$

for $k \geq 1$. Then

$$A_i = \sum_{k \geq 1} 2^{-k} E_k^{(i)}.$$

Choose N large enough that $2^{-N} < \varepsilon$. Then apply the previous proof to the family $\mathcal{E} = \{E_k^{(i)} : k \geq 1, 1 \leq i \leq n\}$. We obtain a diagonal algebra $\mathfrak{D}$ which contains $D_k^{(i)} = E_k^{(i)}$ for $1 \leq k \leq N$ and $1 \leq i \leq n$, and for each other $E_k^{(i)}$, $\mathfrak{D}$ contains $D_k^{(i)} = E_k^{(i)} - R_{ik}$ where R_{ik} is a finite rank projection.

Then let

$$B_i = \sum_{k \geq 1} 2^{-k} D_k^{(i)}.$$

These are positive diagonal contractions in $\mathfrak{D}$. Moreover,
$$K_i := A_i - B_i = \sum_{k>N} 2^{-k} R_{ik}.$$
It follows that K_i are compact and $\|K_i\| \leq 2^{-N} < \varepsilon$. To obtain the result for a normal operator N, apply this result to its real and imaginary parts. ∎

Pushing this a bit harder, we obtain our classification theorem of normal operators up to approximate unitary equivalence. For convenience, we isolate a combinatorial part of the argument for later use.

Lemma II.4.3 *Suppose that X is a compact metric space. Let $\{\xi_k : k \geq 1\}$ and $\{\zeta_k : k \geq 1\}$ be two countable dense subsets of X such that each isolated point of X is repeated the same number of times in each sequence. Then given $\varepsilon > 0$, there is a permutation π of $\mathbb{N}$ so that $\mathrm{dist}(\xi_k, \zeta_{\pi(k)}) < \varepsilon$ for all $k \geq 1$ and*
$$\lim_{k \to \infty} \mathrm{dist}(\xi_k, \zeta_{\pi(k)}) = 0.$$

Proof. Let X_e denote the cluster set of X together with all isolated points which are repeated infinitely often in the sequences. First pair off the (finitely many) terms in each sequence which are isolated points that are at a distance at least $\varepsilon/2$ from X_e. These pairings are at zero distance from each other, and so do not affect the final result. Then restricting our attention to the remainder, we may suppose that every point in X is within distance $\varepsilon/2$ of X_e.

We construct π recursively. At the k-th stage, we will arrange that $\pi(j)$ and $\pi^{-1}(j)$ are defined for $1 \leq j \leq k$ so that when $\pi(i)$ is defined,
$$|\xi_i - \zeta_{\pi(i)}| < \max\{\mathrm{dist}(\xi_i, X_e) + 2^{-i}\varepsilon, \ \mathrm{dist}(\zeta_{\pi(i)}, X_e) + 2^{-\pi(i)}\varepsilon\}.$$
Indeed, if $\pi(k)$ is as yet undefined, choose ℓ not yet in the range of the partially defined function π so that
$$|\xi_k - \zeta_\ell| < \mathrm{dist}(\xi_k, X_e) + 2^{-k}\varepsilon.$$
This is possible since $\{\zeta_\ell\}$ are dense in X_e and isolated points of X_e are repeated infinitely often. Then we set $\pi(k) = \ell$. Likewise, if k is not in the range of the partially defined function π, choose an integer ℓ which is not in the domain of π yet so that
$$|\xi_\ell - \zeta_k| < \mathrm{dist}(\zeta_k, X_e) + 2^{-k}\varepsilon.$$
Then we set $\pi(\ell) = k$. Proceeding in this way, the function π is eventually defined on all of $\mathbb{N}$ so that it is a bijection satisfying the desired estimates. ∎

Theorem II.4.4 *Suppose that M and N are normal operators on separable Hilbert spaces. Then M and N are approximately unitarily equivalent (relative to $\mathfrak{K}$) if and only if*
  (i) $\sigma_e(M) = \sigma_e(N)$, and
  (ii) $\mathrm{null}(M - \lambda I) = \mathrm{null}(N - \lambda I)$ *for all λ in $\mathbb{C} \setminus \sigma_e(M)$*

II.4. The Weyl–von Neumann–Berg Theorem

Proof. Suppose that M and N are approximately unitarily equivalent normal operators. If U_n are unitary operators such that $N = \lim_{n\to\infty} U_n M U_n^*$, then it is evident that $X := \sigma(N) = \sigma(M)$ and that

$$f(N) = \lim_{n\to\infty} U_n f(M) U_n^* \quad \text{for all} \quad f \in C(X).$$

In particular, if $f = \chi_{\{x\}}$ is the characteristic function of an isolated point of X, then

$$E_N(\{x\}) = \chi_{\{x\}}(N) = \lim_{n\to\infty} U_n \chi_{\{x\}}(M) U_n^* = \lim_{n\to\infty} U_n E_M(\{x\}) U_n^*.$$

Hence it follows that these two projections have the same rank. This establishes (ii), and shows that M and N have the same isolated points in their essential spectra. Since every cluster point of $\sigma(M) = \sigma(N)$ is in the essential spectrum, (i) follows.

For the converse, we first show that the result is true when both M and N are diagonalizable. In this case, we may write $M = \text{diag}(\mu_k)$ with respect to a basis $\{e_k\}$ and $N = \text{diag}(\nu_k)$ with respect to a basis $\{f_k\}$ where $\{\mu_k\}$ and $\{\nu_k\}$ are dense subsets of $X = \sigma(M) = \sigma(N)$. Moreover, any isolated point of X must be repeated according to the multiplicity of the eigenvalue, which is the same for M and N by (i) and (ii). Let $X_e = \sigma_e(M)$.

Given any $\varepsilon > 0$, Lemma II.4.3 produces a permutation π of positive integers so that $|\mu_k - \nu_{\pi(k)}| < \varepsilon$ for all k and

$$\lim_{k\to\infty} |\mu_k - \nu_{\pi(k)}| = 0.$$

Thus the unitary operator given by $Ue_k = f_{\pi(k)}$ satisfies

$$M - U^* N U = \text{diag}(\mu_k - \nu_{\pi(k)}),$$

which is compact and has norm $\|M - U^* N U\| < \varepsilon$. As ε was arbitrary, it follows that M and N are approximately unitarily equivalent relative to $\mathfrak{K}$.

Now turn to the general case. It suffices to show that every normal operator M is approximately unitarily equivalent to a diagonal operator relative to $\mathfrak{K}$. Now M may be decomposed as $M = M_0 \oplus M'$ where M_0 is the summand of M corresponding to all isolated eigenvalues of finite multiplicity. The remaining summand M' is normal and $\sigma(M')$ is contained (possibly properly) in $X_e = \sigma_e(M)$. By Corollary II.4.2, for each positive integer k, there is a diagonalizable operator D_k such that $M' - D_k$ is compact and $\|M' - D_k\| < 1/2k$. In particular, the eigenvalues of D_k are at a distance of at most $1/2k$ from X_e and asymptotically, they approach X_e. So there is a diagonal compact perturbation D_k' of D_k so that $\sigma(D_k') \subset X_e$ and $\|M' - D_k'\| < 1/k$. Let $E_k = M_0 \oplus D_k'$. Each E_k is diagonalizable with the same isolated eigenvalues with the same multiplicities as M and essential spectrum X_e. So by the diagonal case, $E_k \sim_{\mathfrak{K}} E_1$ for all $k \geq 1$. But $M - E_k$ is compact and $\lim_{k\to\infty} \|M - E_k\| = 0$. Hence it follows that $M \sim_{\mathfrak{K}} E_1$.

The same holds for N, and thus M and N are approximately unitarily equivalent relative to $\mathfrak{K}$. ∎

Looking at the Weyl–von Neumann–Berg Theorem in a slightly different way, we obtain the corresponding result for representations of $C(X)$. Say that a representation of $C(X)$ is a **diagonal representation** if the range is diagonalizable.

Corollary II.4.5 *Let X be a compact metric space. Then every representation of $C(X)$ on a separable Hilbert space is approximately unitarily equivalent to a diagonal representation relative to $\mathfrak{K}$.*

Proof. Let ρ be a representation of $C(X)$. Let P_n be an enumeration of all minimal finite rank projections in $\rho(C(X))''$; and let P be the sum of these projections. Then $\rho_0 = P\rho|P\mathcal{H}$ is easily seen to be diagonalizable. Thus we may restrict our attention to $\widetilde{\rho} = P^\perp \rho | P^\perp \mathcal{H}$. This has the property that $\widetilde{\rho}(f)$ is never a non-zero compact operator. Indeed, if $\widetilde{\rho}(f) = K \neq 0$ is compact, then $\widetilde{\rho}(C(X))$ contains $C^*(K)$ which contains a finite rank projection E of minimal rank. Since P belongs to $\rho(C(X))''$, it follows that $K' = P^\perp \rho(f) P^\perp$ is a non-zero compact operator in $\rho(C(X))''$ unitarily equivalent to $K \oplus 0$. Hence there is a finite rank projection E' in $\rho(C(X))''$ orthogonal to P, contrary to the definition of P.

By Lemma II.2.8, there is a positive operator A such that

$$\widetilde{\rho}(C(X)) \subset C^*(A) \subset \widetilde{\rho}(C(X))'' = W^*(A).$$

Note that $\sigma_e(A) = \sigma(A)$. Indeed, if λ were an isolated eigenvalue of A, then the corresponding spectral projection would be a finite rank projection in $C^*(A)$, which is contained in $\widetilde{\rho}(C(X))''$ contradicting the construction of P. By the Weyl–von Neumann–Berg Theorem, A is approximately unitarily equivalent relative to $\mathfrak{K}$ to a diagonalizable operator B with

$$\sigma(B) = \sigma_e(B) = \sigma_e(A) = \sigma(A).$$

Thus by Theorem II.4.4, the $*$-representation of $C^*(A)$ given by

$$\tau(f(A)) = f(B) \quad \text{for} \quad f \in C(\sigma(A))$$

is approximately unitarily equivalent to the identity representation relative to $\mathfrak{K}$. Therefore $\tau\widetilde{\rho}$ is approximately unitarily equivalent to $\widetilde{\rho}$ relative to $\mathfrak{K}$. So $\rho_0 \oplus \tau\widetilde{\rho}$ is a diagonal representation of $C(X)$ which is approximately unitarily equivalent relative to $\mathfrak{K}$ to ρ. ∎

Now we can complete the picture for approximate unitary equivalence of representations of $C(X)$. The last condition (iv) should be seen as measuring the support of the representation and the rank of (characteristic functions of) isolated points in this support.

II.4. The Weyl–von Neumann–Berg Theorem

Theorem II.4.6 *Let X be a compact metric space. Suppose that ρ and σ are separable representations of $C(X)$. Then the following are equivalent:*

(i) $\rho \sim_{\mathfrak{K}} \sigma$;
(ii) $\rho \sim_a \sigma$;
(iii) $\rho \sim_{wa} \sigma$;
(iv) $\operatorname{rank} \rho(f) = \operatorname{rank} \sigma(f)$ *for all* $f \in C(X)$.

Proof. The implications (i) implies (ii) implies (iii) are trivial, and (iii) implies (iv) follows because rank is lower semi-continuous in the WOT topology. (To see this, note that the set of operators of rank at most k is WOT-closed.) So assume that (iv) holds. By Corollary II.4.5, we may replace ρ and σ by diagonalizable representations which are approximately equivalent relative to $\mathfrak{K}$ to ρ and σ. Thus it suffices to prove (iv) implies (i) in this special case. Clearly condition (iv) implies that $\ker \rho = \ker \sigma$. By Exercise I.14, there is a closed subset E of X so that ideal $\ker \rho$ has the form

$$\mathfrak{I}(E) = \{f \in C(X) : f|_E = 0\}.$$

Thus both ρ and σ factor through the restriction to $C(E)$. Without loss of generality, we may suppose that $E = X$ and that ρ and σ are injective.

There are sequences $\{\xi_n : n \geq 1\}$ and $\{\zeta_n : n \geq 1\}$ in X such that

$$\rho(f) = \operatorname{diag}(f(\xi_n)) \quad \text{and} \quad \sigma(f) = \operatorname{diag}(f(\zeta_n))$$

with respect to orthonormal bases $\{e_n\}$ for $\mathcal{H}_\rho$ and $\{f_n\}$ for $\mathcal{H}_\sigma$. The condition $\ker \rho = \ker \sigma = \{0\}$ is equivalent to the condition

$$\overline{\{\xi_n : n \geq 1\}} = \overline{\{\zeta_n : n \geq 1\}} = X.$$

For each isolated point $\{\eta\}$ of X, consider the characteristic function $f = \chi_\eta$. Condition (iv) guarantees that the spectral projections $\rho(f)$ and $\sigma(f)$ have the same rank. Hence the isolated points of X are repeated the same number of times in the two sequences $\{\xi_n\}$ and $\{\zeta_n\}$. The remainder of the proof follows the proof of Theorem II.4.4. By Lemma II.4.3, given any positive integer n, there is a permutation π_n of $\mathbb{N}$ such that

$$\max_k \operatorname{dist}(\xi_k, \zeta_{\pi_n(k)}) < 1/n \quad \text{and} \quad \lim_{k \to \infty} \operatorname{dist}(\xi_k, \zeta_{\pi_n(k)}) = 0.$$

Let U_n be the unitary operator that implements this permutation between the bases for $\mathcal{H}_\rho$ and $\mathcal{H}_\sigma$. For any f in $C(X)$,

$$\rho(f) - U_n \sigma(f) U_n^* = \operatorname{diag}(f(\xi_k) - f(\zeta_{\pi_n(k)})).$$

The uniform continuity of f shows that this is a compact operator and that it has small norm for sufficiently large n. Thus U_n implement an approximate unitary equivalence relative to $\mathfrak{K}$. ∎

II.5 Voiculescu's Theorem

In this section, we will prove a generalization of the Weyl–von Neumann Theorem valid for arbitrary *separable* C*-algebras. While this has nothing directly to do with the title of this chapter, it is an important direct generalization of the results in the previous two sections on the Weyl–von Neumann–Berg Theorem.

The starting point is an approximation theorem for states known as **Glimm's Lemma**. Using nets rather than sequences, this lemma is valid for arbitrary C*-subalgebras of $\mathcal{B}(\mathcal{H})$.

Lemma II.5.1 *Let φ be a state on a separable C*-subalgebra $\mathfrak{A}$ of $\mathcal{B}(\mathcal{H})$ with the property that $\varphi(\mathfrak{A} \cap \mathfrak{K}) = 0$. Then there is a sequence of unit vectors x_n converging weakly to 0 such that the corresponding vector states $\psi_n(A) = (Ax_n, x_n)$ on $\mathfrak{A}$ converge weak-$*$ (pointwise) to φ.*

Proof. Without loss of generality, we may assume that $\mathfrak{A}$ contains $\mathfrak{K}$ by the simple expedient of replacing it with $\mathfrak{A} + \mathfrak{K}$ and defining $\varphi(A + K) = \varphi(A)$, which is well defined by Corollary I.5.6.

Consider the set $\mathcal{S}_e$ of all states ψ on $\mathfrak{A}$ for which there is an sequence of unit vectors x_n converging weakly to 0 in $\mathcal{H}$ such that $\psi(A) = \lim_{n \to \infty} (Ax_n, x_n)$ for all A in $\mathfrak{A}$. First note that $\mathcal{S}_e$ is non-empty. Indeed, take a countable dense subset $\{A_i\}$ of $\mathfrak{A}_{sa}$ and any orthonormal sequence y_k. Then a standard diagonalization argument yields a subsequence so that $\psi(A_i) = \lim_{k \to \infty} (A_i y_{n_k}, y_{n_k})$ exists for all $i \geq 1$. By uniform continuity, it converges on all of $\mathfrak{A}$. Clearly every state ψ in $\mathcal{S}_e$ annihilates the compact operators.

Secondly, we claim that $\mathcal{S}_e$ is convex. For suppose that ψ and φ are two states of this form corresponding to sequences x_k and y_n converging weakly to 0. Recursively choose unit vectors y'_k as follows. For each k, let P_k be the projection onto

$$\operatorname{span}\{A_i x_k : 1 \leq i \leq k\} \cup \{y'_j : 1 \leq j < k\}.$$

Since $\lim_{n \to \infty} \|P_k y_n\| = 0$, there is an integer n_k so large that $\|P_k y_{n_k}\| < 2^{-k}$. Now a small perturbation will yield a unit vector y'_k near y_{n_k} orthogonal to the range of P_k. Then $\{y'_k\}$ is orthonormal by construction. Since it is asymptotic to y_{n_k}, it determines the same state φ. Let $z_k = (x_k + y'_k)/\sqrt{2}$. This is a sequence of unit vectors converging weakly to 0. Moreover, when $k > i$,

$$(A_i z_k, z_k) = \tfrac{1}{2}((A_i x_k, x_k) + (A_i x_k, y'_k) + (y'_k, A_i x_k) + (A_i y'_k, y'_k))$$
$$= \tfrac{1}{2}(A_i x_k, x_k) + \tfrac{1}{2}(A_i y'_k, y'_k)$$

which converges to $\tfrac{1}{2}(\psi(A_i) + \varphi(A_i))$. Hence $\tfrac{1}{2}(\psi + \varphi)$ belongs to $\mathcal{S}_e$.

Another diagonalization argument shows that $\mathcal{S}_e$ is closed in the weak-$*$ topology. So suppose that φ is not in $\mathcal{S}_e$. Then by the Hahn–Banach Theorem, there is

II.5. Voiculescu's Theorem

a self-adjoint operator A in $\mathfrak{A}$ such that

$$\varphi(A) \notin W_e(A) := \{\psi(A) : \psi \in \mathcal{S}_e\}. \tag{1}$$

Because $W_e(A)$ is the continuous image of a compact convex set by a linear map, it is a closed convex set. Since A is Hermitian, $W_e(A)$ is a subset of $\mathbb{R}$.

Let the convex hull of $\sigma_e(A)$ be the interval $[a, b]$. It will be shown that $W_e(A)$ equals $[a, b]$ also. Replace A by $f(A)$ where f is the function $a \vee x \wedge b$. Since $\pi f(A) = f(\pi A) = \pi(A)$, it follows that $f(A) - A$ is compact. Therefore $\varphi(f(A)) = \varphi(A)$ and $W_e(f(A)) = W_e(A)$. We have arranged that $aI \leq A \leq bI$, so that $a \leq \varphi(A) \leq b$ and $a \leq (Ax, x) \leq b$ for every unit vector x. Hence $W_e(A)$ is contained in $[a, b]$.

The spectral projection $E_A(b - \frac{1}{n}, b]$ is infinite rank for all $n > 0$. If x_n are orthogonal unit vectors in $E_A(b - \frac{1}{n}, b]\mathcal{H}$, then $(Ax_n, x_n) > b - \frac{1}{n}$. Use a diagonalization argument to obtain a subsequence which determines an element ψ in $\mathcal{S}_e$. Then $\psi(A) \geq b$. Similarly, there is state ψ' in $\mathcal{S}_e$ so that $\psi'(A) \leq a$. Hence $W_e(A)$ contains $[a, b]$. This establishes the claim. However, $\varphi(A)$ belongs to $[a, b]$ as well, contradicting (1). ∎

This lemma can be strengthened to apply to certain finite dimensional maps. A **positive linear map** between C*-algebras is just a linear map which takes positive operators to positive operators. Hence it is a map which preserves the order structure. If φ is a map between C*-algebras $\mathfrak{A}$ and $\mathfrak{B}$, it induces a map $\varphi^{(n)}$ from $\mathcal{M}_n(\mathfrak{A})$ into $\mathcal{M}_n(\mathfrak{B})$ by

$$\varphi^{(n)}([A_{ij}]) = [\varphi(A_{ij})].$$

Say that φ is **n-positive** if $\varphi^{(n)}$ is positive and **completely positive** is it is n-positive for all $n \geq 1$. These notions will be explored in more detail in Chapter IX. For the moment, we notice that if ρ is a *-representation of $\mathfrak{A}$ into $\mathcal{B}(\mathcal{H})$ and X is a bounded operator from $\mathcal{K}$ into $\mathcal{H}$, then $\varphi(A) = X^*\rho(A)X$ is completely positive. Indeed, $\rho^{(n)}$ is a *-representation, and therefore is positive. Then a calculation shows that

$$\varphi^{(n)}([A_{ij}]) = X^{(n)*}\rho^{(n)}([A_{ij}])X^{(n)}$$

which is positive when $[A_{ij}]$ is positive.

Lemma II.5.2 *Let φ be a unital completely positive map of a separable C*-subalgebra $\mathfrak{A}$ of $\mathcal{B}(\mathcal{H})$ into $\mathcal{M}_n$ such that $\varphi(\mathfrak{A} \cap \mathfrak{K}) = 0$. Then there is a sequence V_k of isometries of $\mathbb{C}^n$ into $\mathcal{H}$ such that* WOT–$\lim_{k \to \infty} V_k = 0$ *and*

$$\lim_{k \to \infty} \|\varphi(A) - V_k^* A V_k\| = 0 \quad \text{for all} \quad A \in \mathcal{A}.$$

Proof. Notice that $\varphi^{(n)}$ is a positive map of $\mathcal{M}_n(\mathfrak{A})$ into $\mathcal{M}_n(\mathcal{M}_n)$. Fix an orthonormal basis y_i, $1 \leq i \leq n$, for $\mathbb{C}^n$ and consider the state

$$\Phi([A_{ij}]) = \frac{1}{n}\sum_{i,j=1}^{n}(\varphi(A_{ij})y_j, y_i) = (\varphi^{(n)}([A_{ij}])\mathbf{y}, \mathbf{y})$$

where $\mathbf{y} = (y_1, y_2, \ldots, y_n)^t/\sqrt{n}$. By the previous lemma, there are unit vectors $\mathbf{x}^k = (x_1^k, \ldots, x_n^k)$ in $\mathcal{H}^{(n)}$ converging weakly to zero such that

$$\Phi([A_{ij}]) = \lim_{k\to\infty}([A_{ij}]\mathbf{x}^k, \mathbf{x}^k).$$

Define $U_k y_i = \sqrt{n}\, x_i^k$. Then

$$\delta_{ij} = (y_j, y_i) = n\Phi(E_{ij}) = \lim_{k\to\infty} n(E_{ij}\mathbf{x}^k, \mathbf{x}^k) = \lim_{k\to\infty} n(x_j^k, x_i^k).$$

It follows that the set $\sqrt{n}\, x_i^k$ is almost orthonormal. A routine estimate shows that $U_k^* U_k - I$ has small norm. Therefore the partial isometry V_k in the polar decomposition of U_k is an isometry; and $\lim_{k\to\infty}\|V_k - U_k\| = 0$. Also

$$\lim_{k\to\infty}\left|\left((\varphi(A) - U_k^* AU_k)y_j, y_i\right)\right| = \lim_{k\to\infty}\left|n\Phi(AE_{ij}) - n(Ax_j^k, x_i^k)\right| = 0.$$

Hence

$$\lim_{k\to\infty}\|\varphi(A) - V_k^* AV_k\| = \lim_{k\to\infty}\|\varphi(A) - U_k^* AU_k\| = 0.$$

If P is any finite rank projection, then

$$0 = \varphi(P) = \lim_{k\to\infty}\|V_k^* PV_k\| = \lim_{k\to\infty}\|PV_k\|^2.$$

Thus V_k converges to 0 in the weak operator topology. ∎

We wish to note a minor variant of this lemma which will be useful. Suppose that $\mathcal{A}$ is a finite subset of $\mathfrak{A}$, $\varepsilon > 0$ and $\mathcal{N}$ is a finite dimensional subspace of $\mathcal{H}$. Then there is an isometry V from $\mathbb{C}^n$ into $\mathcal{N}^\perp$ such that $\|\varphi(A) - V^* AV\| < \varepsilon$ for each A in $\mathcal{A}$. Indeed, for k sufficiently large, one obtains

$$\|\varphi(A) - V_k^* AV_k\| < \varepsilon/2 \quad \text{for} \quad A \in \mathcal{A} \quad \text{and} \quad \|P_\mathcal{N} V_k\| < \varepsilon/4.$$

Let V be the isometry from the polar decomposition of $P_\mathcal{N}^\perp V_k$. A simple estimate yields the desired inequality.

This brings us to the key technical result of this section which is a generalization of the Weyl–von Neumann Theorem to arbitrary C*-algebras.

Theorem II.5.3 *Suppose that φ is a completely positive map from a separable C*-subalgebra $\mathfrak{A}$ of $\mathcal{B}(\mathcal{H})$ into $\mathcal{B}(\mathcal{K})$ such that $\varphi(\mathfrak{A} \cap \mathfrak{K}) = 0$. Then there is a sequence V_k of isometries of $\mathcal{K}$ into $\mathcal{H}$ such that $\varphi(A) - V_k^* AV_k$ is compact for all $k \geq 1$ and all A in $\mathfrak{A}$, and*

$$\lim_{k\to\infty}\|\varphi(A) - V_k^* AV_k\| = 0 \quad \text{for all} \quad A \in \mathfrak{A}.$$

II.5. Voiculescu's Theorem

Proof. By Theorem I.9.16 and Exercise I.34, there is a sequential quasicentral approximate unit for the finite rank operators on $\mathcal{K}$ relative to $\mathfrak{B} = C^*(\varphi(\mathfrak{A})) + \mathfrak{K}$. Thus there is an increasing sequence of finite rank positive contractions which tends strongly to the identity and is quasicentral for $\mathfrak{B}$.

By Exercise II.8, there is a sequence $\delta_n > 0$ with the property that for all $E \geq 0$ and $\|A\| \leq 1$ in any C*-algebra,

$$\|EA - AE\| < \delta_n \quad \text{implies} \quad \|E^{1/2}A - AE^{1/2}\| < 2^{-n}.$$

(In fact, $\delta_n = 2^{-2n-1}$ will suffice. See Exercise II.9.) Let $\{A_i, i \geq 1\}$ be a dense subset of the unit ball of $\mathfrak{A}_{sa}$. By dropping to a subsequence E_n of the quasicentral approximate unit, one may arrange that

$$\|\varphi(A_i)E_n - E_n\varphi(A_i)\| < \tfrac{1}{2}\delta_{n+1} \quad \text{for} \quad 1 \leq i \leq n+1.$$

Define $F_n = (E_n - E_{n-1})^{1/2}$ (using $E_0 = 0$). Then

$$\|\varphi(A_i)F_n - F_n\varphi(A_i)\| < 2^{-n} \quad \text{for} \quad 1 \leq i \leq n.$$

Thus it follows that

$$\sum_{n \geq 1} \|\varphi(A_i)F_n - F_n\varphi(A_i)\| < \infty \quad \text{for all} \quad i \geq 1.$$

Let P_n be the (increasing) sequence of finite rank projections onto the ranges of E_n. By the previous lemma and the subsequent remark, we may construct a sequence U_n of isometries from $P_n\mathcal{K}$ into $\mathcal{H}$ such that $\text{Ran}(U_n)$ is orthogonal to

$$\mathcal{N}_n = \text{span}\{\text{Ran}(U_i), A_j\text{Ran}(U_i), 1 \leq i \leq n-1, 1 \leq j \leq n\}$$

and

$$\|P_n\varphi(A_i)P_n - U_n^*A_iU_n\| < 2^{-n} \quad \text{for} \quad 1 \leq i \leq n.$$

Hence

$$\sum_{n \geq 1} \|P_n\varphi(A_i)P_n - U_n^*A_iU_n\| < \infty \quad \text{for all} \quad i \geq 1.$$

Define $V = \text{SOT-}\sum_{n \geq 1} U_nF_n$. This is an isometry because the U_n's have orthogonal ranges, whence

$$V^*V = \text{SOT-}\sum_{n=1}^{\infty} F_n^2 = I.$$

Notice that the orthogonality of $\mathcal{N}_n$ and $\text{Ran}\, U_n$ shows that

$$U_n^*A_iU_m = 0 \quad \text{for} \quad 1 \leq i \leq \max\{m, n\}.$$

Hence
$$\varphi(A_i) - V^*A_iV = \sum_{n\geq 1}\varphi(A_i)F_n^2 - \sum_{m\geq 1}\sum_{n\geq 1}F_mU_m^*A_iU_nF_n$$
$$= \sum_{n\geq 1}(\varphi(A_i)F_n - F_n\varphi(A_i))F_n + \sum_{m\neq n}F_mU_m^*A_iU_nF_n$$
$$+ \sum_{n\geq 1}F_n(P_n\varphi(A_i)P_n - U_n^*A_iU_n)F_n$$

The first and third sums are norm convergent sums of finite rank operators, and thus are compact for all $i \geq 1$; and the second sum is finite for each A_i. Thus this sum is compact. An arbitrary element A of $\mathfrak{A}$ can be approximated by a linear combination of the A_i's, from which we deduce that $\varphi(A) - V^*AV$ is compact for every A in $\mathfrak{A}$.

To get a sequence with the asymptotic norm property, we employ a simple trick. Apply the result just demonstrated to the map $\varphi^{(\infty)}(A) = \varphi(A)^{(\infty)}$ acting on $\mathcal{K}^{(\infty)}$. The resulting isometry V may be written as $[V_1 \ V_2 \ V_3 \ \cdots]$. The compactness of $\varphi(A)^{(\infty)} - V^*AV$ implies that

$$\lim_{k\to\infty}\|\varphi(A) - V_k^*AV_k\| = 0. \qquad\blacksquare$$

Corollary II.5.4 *Suppose that ρ is a non-degenerate representation of a separable C^*-subalgebra $\mathfrak{A}$ of $\mathcal{B}(\mathcal{H})$ into $\mathcal{B}(\mathcal{K})$ such that $\rho(\mathfrak{A}\cap\mathfrak{K}) = 0$. Then there is a sequence V_k of isometries of $\mathcal{K}$ into $\mathcal{H}$ such that $V_k\rho(A) - AV_k$ is compact for all $k\geq 1$ and all A in $\mathfrak{A}$, and*

$$\lim_{k\to\infty}\|V_k\rho(A) - AV_k\| = 0 \quad \text{for all} \quad A\in\mathfrak{A}.$$

Proof. Find isometries V_k using the theorem above. Then
$$(V_k\rho(A) - AV_k)^*(V_k\rho(A) - AV_k) =$$
$$\rho(A^*)(\rho(A) - V_k^*AV_k) + (\rho(A^*) - V_k^*A^*V_k)\rho(A) - (\rho(A^*A) - V_k^*A^*AV_k).$$
Hence this is compact, and the norms tend to 0 as k increases. $\qquad\blacksquare$

We now obtain the most important corollary known as **Voiculescu's Theorem**.

Corollary II.5.5 *Suppose that ρ is a non-degenerate representation of a separable C^*-subalgebra $\mathfrak{A}$ of $\mathcal{B}(\mathcal{H})$ into $\mathcal{B}(\mathcal{K})$ such that $\rho(\mathfrak{A}\cap\mathfrak{K}) = 0$. Then* id $\sim_{\mathfrak{K}}$ id $\oplus\rho$.

Proof. Apply the previous corollary to $\rho^{(\infty)}$ acting on $\mathcal{K}^{(\infty)}$. Let V be an isometry such that $AV - V\rho(A)^{(\infty)}$ is compact for all A in $\mathfrak{A}$. Then let J_n be the canonical injection of $\mathcal{K}$ onto the nth summand of $\mathcal{K}^{(\infty)}$; and let S_n be the infinite shift operator

$$S_n = \sum_{i=1}^{n-1}J_iJ_i^* + \text{SOT-}\sum_{i\geq n}J_{i+1}J_i^*.$$

II.5. Voiculescu's Theorem

Then define unitary operators W_n from $\mathcal{H} \oplus \mathcal{K}$ onto $\mathcal{H}$ by

$$W_n = \begin{bmatrix} I - VV^* + VS_nV^* & VJ_n \end{bmatrix}.$$

A simple calculation shows that this is unitary. It has the effect of imbedding $\rho(A)$ into the range of V_n, which is made available by shifting the ranges of $V_i\mathcal{H}$ onto $V_{i+1}\mathcal{H}$ for $i \geq n$.

One computes that $AW_n - W_n(A \oplus \rho(A))$ equals the 1×2 matrix

$$\begin{bmatrix} A(I - VV^*) - (I - VV^*)A + AVS_nV^* - VS_nV^*A & AVJ_n - VJ_n\rho(A) \end{bmatrix}.$$

Now

$$AVV^* - VV^*A = (AV - V\rho(A)^{(\infty)})V^* + V(A^*V - V\rho(A^*)^{(\infty)})^*;$$

and because $\rho^{(\infty)}$ commutes with S_n,

$$AVS_nV^* - VS_nV^*A =$$
$$(AV - V\rho(A)^{(\infty)})S_nV^* + VS_n(A^*V - V\rho(A^*)^{(\infty)})^*.$$

Let P_n denote the projection SOT-$\sum_{i \geq n} J_i J_i^*$. Since $S_n - I = P_n(S_n - I)$, we see that the 1,1 entry of $AW_n - W_n(A \oplus \rho(A))$ is

$$(AV - V\rho(A)^{(\infty)})P_n(S_n - I)V^* + V(S_n - I)P_n(A^*V - V\rho(A^*)^{(\infty)})^*$$

which is compact for all A in $\mathfrak{A}$. Similarly, the 1,2 entry is

$$AVJ_n - VJ_n\rho(A) = (AV - V\rho(A)^{(\infty)})J_n.$$

Moreover, since P_n and J_n tend strongly to 0, the norm of these terms converges to 0. Hence $\text{id} \sim_\mathfrak{K} \text{id} \oplus \rho$. ∎

Let π denote the quotient map of $\mathcal{B}(\mathcal{H})$ onto the Calkin algebra.

Corollary II.5.6 *Suppose that ρ_1 and ρ_2 are representations of a separable C*-algebra $\mathfrak{A}$ on separable spaces such that*

$$\ker \rho_1 = \ker \pi\rho_1 = \ker \rho_2 = \ker \pi\rho_2.$$

Then $\rho_1 \sim_\mathfrak{K} \rho_2$.

Proof. By hypothesis, both $\mathfrak{A}_i = \rho_i(\mathfrak{A})$ are separable C*-algebras containing no compact operators. There is a unique *-isomorphism σ of $\mathfrak{A}_1$ onto $\mathfrak{A}_2$ such that $\rho_2 = \sigma\rho_1$. Let id_i denote the identity representation of $\mathfrak{A}_i$. By the previous corollary,

$$\rho_1 = \text{id}_1 \rho_1 \sim_\mathfrak{K} (\text{id}_1 \oplus \sigma)\rho_1 = \rho_1 \oplus \rho_2.$$

Likewise, $\rho_1 \oplus \rho_2 \sim_\mathfrak{K} \rho_2$. ∎

We are now ready to classify all representations up to approximate unitary equivalence analogous to Theorem II.4.6. The remaining technicality is an easy result about C*-algebras of compact operators.

Lemma II.5.7 *Let $\mathfrak{A}$ be a C*-subalgebra of the compact operators, and let σ and ρ be two non-degenerate separable representations of $\mathfrak{A}$. Then σ and ρ are unitarily equivalent if and only if*

$$\mathrm{rank}(\sigma(A)) = \mathrm{rank}(\rho(A)) \quad \text{for all} \quad A \in \mathfrak{A}.$$

Proof. The necessity of the condition is clear. For the converse, let π_i denote the inequivalent irreducible representations of $\mathfrak{A}$. By Theorem I.10.7, there are cardinals n_i and m_i in $\{0, 1, \ldots, \aleph_0\}$ so that

$$\sigma \simeq \sum_i \oplus \pi_i^{(n_i)} \quad \text{and} \quad \rho \simeq \sum_i \oplus \pi_i^{(m_i)}.$$

Let E_i denote a rank one projection in $\mathfrak{K}(\mathcal{H}_i) = \pi_i(\mathfrak{A})$. Then

$$n_i = \mathrm{rank}(\sigma(E_i)) = \mathrm{rank}(\rho(E_i)) = m_i$$

for all i. Hence σ and ρ are unitarily equivalent. ∎

Theorem II.5.8 *Let $\mathfrak{A}$ be a separable C*-algebra, and let σ and ρ be non-degenerate representations of $\mathfrak{A}$ on separable Hilbert spaces. Then the following are equivalent:*

(i) $\sigma \sim_{\mathfrak{K}} \rho$,
(ii) $\sigma \sim_a \rho$,
(iii) $\sigma \sim_{wa} \rho$,
(iv) $\mathrm{rank}(\sigma(A)) = \mathrm{rank}(\rho(A))$ *for all* $A \in \mathfrak{A}$.

Proof. Clearly (i) to (iv) are successively weaker notions; where the implication (iii) implies (iv) follows from the fact that the set of operators of rank at most k is WOT-closed for every $k \geq 0$.

Condition (iv) applied to rank zero yields that $\ker \sigma = \ker \rho$. The spectral theory of compact Hermitian operators shows that every compact operator in any C*-algebra is the norm limit of finite rank operators in the C*-algebra. Thus if $\mathfrak{F}$ denotes the space of finite rank operators and π is the quotient of $\mathcal{B}(\mathcal{H})$ by the compact operators, we obtain

$$\ker \pi\sigma = \sigma^{-1}(\mathfrak{K}) = \overline{\sigma^{-1}(\mathfrak{F})}$$
$$= \overline{\rho^{-1}(\mathfrak{F})} = \rho^{-1}(\mathfrak{K}) = \ker \pi\rho.$$

Let $\mathfrak{B} = \sigma(\mathfrak{A})$; and define a representation of $\mathfrak{B}$ by

$$\tau(B) = \rho(\sigma^{-1}(B)).$$

II.5. Voiculescu's Theorem

This is well defined since σ and ρ have the same kernel; and condition (iv) shows that τ preserves rank. Let

$$\mathcal{H}_0 = \text{Ran}(\mathfrak{B} \cap \mathfrak{K}) \qquad \mathcal{H}_1 = \mathcal{H}_\sigma \ominus \mathcal{H}_0$$
$$\mathcal{K}_0 = \text{Ran}(\rho(\mathfrak{A}) \cap \mathfrak{K}) \quad \text{and} \quad \mathcal{K}_1 = \mathcal{H}_\rho \ominus \mathcal{K}_0.$$

Define γ to be the identity representation of $\mathfrak{B}$. Let

$$\gamma_i = \gamma|_{\mathcal{H}_i} \quad \text{and} \quad \tau_i = \tau|_{\mathcal{K}_i} \quad \text{for} \quad i = 0, 1.$$

We have observed that τ maps $\mathfrak{B} \cap \mathfrak{K}$ onto $\tau(\mathfrak{B}) \cap \mathfrak{K}$, and preserves rank. Hence by Lemma II.5.7, the restriction of τ_0 to this subalgebra is unitarily equivalent to the identity representation γ_0 on this subalgebra. By Lemma I.9.14, it follows that γ_0 and τ_0 are unitarily equivalent. Since τ_1 is a representation of $\mathfrak{B}$ which annihilates $\mathfrak{B} \cap \mathfrak{K}$, Corollary II.5.5 shows that

$$\gamma \sim_\mathfrak{K} \gamma \oplus \tau_1 = \gamma_0 \oplus \gamma_1 \oplus \tau_1 \simeq \tau_0 \oplus \tau_1 \oplus \gamma_1 = \tau \oplus \gamma_1.$$

By symmetry, consideration of $\rho(\mathfrak{A})$ shows that $\tau \sim_\mathfrak{K} \tau \oplus \gamma_1$ as well. Hence $\gamma \sim_\mathfrak{K} \tau$, which is equivalent to the statement $\sigma \sim_\mathfrak{K} \rho$. ∎

Finally we will obtain a corollary that generalizes the Weyl-von Neumann Theorem directly. Recall that the irreducible representations of $C(X)$ are one dimensional. Thus a representation of $C(X)$ is diagonalizable if and only if it is the direct sum of irreducible representations.

Corollary II.5.9 *Every representation of a separable C*-algebra $\mathfrak{A}$ on a separable Hilbert space is approximately unitarily equivalent to a direct sum of irreducible representations.*

Proof. As in the previous section, a representation of $\mathfrak{A}$ decomposes as a direct sum $\rho = \rho_e \oplus \widetilde{\rho}$ where ρ_e is the compression to the range $\mathcal{K}$ of $(\rho(\mathfrak{A}) \cap \mathfrak{K})\mathcal{H}$. Let σ_e be the compression representation of $\mathfrak{B} := \rho(\mathfrak{A})$ onto $\mathcal{K}$, and let σ' be the compression map to $\mathcal{K}^\perp$. By construction, $\widetilde{\rho}(A)$ factors as $\sigma'\rho$. Clearly, σ' annihilates the compact operators in $\mathfrak{B}$.

By Theorem I.10.7, the identity representation of $\mathfrak{B} \cap \mathfrak{K}$ is the sum of irreducible representations. By Lemmas I.9.15 and I.9.14, it follows that σ_e is the direct sum of irreducible representations.

From the GNS construction Theorem I.9.12, there is a direct sum σ of irreducible representations of $\mathfrak{B}$ such that $\ker \sigma = \ker \sigma'$. Now the range of $\tau = \sigma^{(\infty)}$ contains no compact operators either. So by the Corollary II.5.5,

$$\text{id}_\mathfrak{B} = \sigma_e \oplus \sigma' \sim_\mathfrak{K} \sigma_e \oplus \sigma' \oplus \tau \sim_\mathfrak{K} \sigma_e \oplus \tau.$$

Hence

$$\rho = \text{id}_\mathfrak{B} \rho \sim_\mathfrak{K} (\sigma_e \oplus \tau)\rho = \rho_e \oplus \tau\rho$$

which is the direct sum of irreducible representations. ∎

Exercises

II.1 *(Fuglede's Theorem)* If X commutes with a normal operator N, show that it also commutes with N^*.
HINT: Consider the *bounded* entire function
$$f(z) = e^{zN^*}Xe^{-zN^*} = e^{zN^*}e^{-\bar{z}N}Xe^{\bar{z}N}e^{-zN^*}.$$

II.2 Show that if M and N are normal and $MX = XN$, then $M^*X = XN^*$.
HINT: $\begin{bmatrix} M & 0 \\ 0 & N \end{bmatrix}$ commutes with $\begin{bmatrix} 0 & X \\ 0 & 0 \end{bmatrix}$.

II.3 Show that two similar normal operators are unitarily equivalent.
HINT: Take the unitary part of the polar decomposition of the similarity and apply Fuglede's Theorem.

II.4 (a) Suppose that N is normal. If $r > 0$, show that
$$\{x \in \mathcal{H} : \lim_{n\to\infty} r^{-n}\|(N - \lambda I)^n x\| = 0\}$$
is the spectral subspace of N corresponding to $\mathbb{D}_r(\lambda) = \{z : |z - \lambda| < r\}$.
(b) Show that if X commutes with N, it commutes with the spectral projections for every open disk. Hence deduce that X lies in $C^*(N)'$. (This is another proof of Fuglede's Theorem.)

II.5 (a) If $A = A^*$ has a cyclic vector x, show that the subspaces
$$\mathcal{M}_k = \text{span}\{\prod_{i=1}^{k} E_i^{(\epsilon_i)} x \quad \text{for} \quad \epsilon_i = \pm 1\}.$$
have dense union. With this choice, show that the perturbations R_k of Corollary II.4.2 have $\text{rank}(R_k) \leq 2^k$. Hence show that the compact operator K constructed lies in the Schatten $\mathcal{C}_p$ class, and has small $\mathcal{C}_p$ norm for any $p > 1$.
(b) Remove the condition on cyclicity by using the fact that A is the direct sum of cyclic operators.
(c) Generalize this to show that n commuting Hermitian operators may be diagonalized modulo the $\mathcal{C}_p$ class for any $p > n$.

II.6 Say that two operators A and B are **compalent** if there is a unitary operator U such that $B - UAU^*$ is compact. Prove that two normal operators are compalent if and only if they have the same essential spectrum.

II.7 Two operators A and B are **quasi-similar** if there are injective operators X and Y with dense range such that $AX = XB$ and $YA = BY$. Show that quasi-similar *normal* operators are unitarily equivalent.
NOTE: in general, quasi-similar operators need not even have the same spectrum.

Exercises

II.8 Show that if f is continuous on a compact subset $X \subset \mathbb{C}$, then for each $\varepsilon > 0$, there is a $\delta > 0$ so that whenever A and B are normal operators such that $\sigma(A) \cup \sigma(B) \subset X$ and $\|A - B\| < \delta$, then $\|f(A) - f(B)\| < \varepsilon$. HINT: Prove it first for polynomials in z and $\bar{z}$; then approximate f.

II.9 If A is positive, show that $\|A^{1/2}T - TA^{1/2}\| \leq 2\|T\|^{1/2}\|AT - TA\|^{1/2}$. HINT: Show that when $T = U$ is unitary, this is valid with constant 1 instead of 2. Then prove it for $T = T^*$ by applying the result to the unitary $U_s = (sT + iI)(sT - iI)^{-1}$ for $s > 0$.

II.10 Let A be a positive contraction such that both $0, 1 \in \sigma_e(A)$; and let P be a projection such that $\text{rank}(P) = \text{rank}(P^\perp) = \infty$. Show that there are unitary operators U_n and V_n such that

$$P = \text{WOT-lim}\, U_n A U_n^* \quad \text{and} \quad A = \text{WOT-lim}\, V_n P V_n^*.$$

This shows that the weak operator topology yields a very weak relation for single operators.

II.11 For nonseparable C*-algebras, the definition of approximate unitary equivalence requires nets of unitaries. Show that two representations of a C*-algebra $\mathfrak{A}$ are (weak) approximately unitarily equivalent (relative to $\mathfrak{K}$) if and only if this holds for the restriction to every separable subalgebra.

II.12 Let $\mathfrak{D}$ denote the diagonal algebra of ℓ^∞ acting on ℓ^2. Let τ denote a multiplicative linear functional on ℓ^∞ which annihilates c_0. Show that id and $\sigma = \text{id} \oplus \tau^{(\infty)}$ are approximately unitarily equivalent relative to $\mathfrak{K}$. However, show that there is no *sequence* of unitaries U_n such that $\text{WOT-lim}_{n \to \infty} U_n D U_n^* = \sigma(D)$ for all D in $\mathfrak{D}$. Nor is there a unitary U such that $UDU^* - \sigma(D)$ is compact for all D in $\mathfrak{D}$.

Notes and Remarks.

Spectral theory for self-adjoint operators goes back to Hilbert at the turn of the century. For normal operators, it is due to von Neumann[1929]. The L^∞ functional calculus is developed later in von Neumann [1931]. The multiplicity theory is due to Hellinger and Hahn. Weyl and von Neumann showed that Hermitian operators are diagonal plus compact. The normal case is due to Berg [1971]. Glimm's lemma was proven for a different context in Glimm [1960b]; and the generalization to finite dimensional maps is due to Bunce and Salinas [1976]. Voiculescu [1976] proved his non-commutative extension. Arveson [1977] put this all in an appropriate abstract framework using quasicentral approximate units. It is this approach that is used here even in the commutative case. His paper also shows the equivalence of weak approximate equivalence with the other stronger notions. Hadwin [1977] proves the even weaker version in terms of rank. In Hadwin [1980], he develops the appropriate analogue of these results for the non-separable case.

CHAPTER III

Approximately Finite Dimensional (AF) C*-algebras

The purpose of this chapter is to make a detailed study of a special class of C*-algebras which are built up in a natural way from finite dimensional algebras. This class has an interesting variety of examples but at the same time it admits a complete classification which will be dealt with in the next chapter. This classification is described in terms of a group called K_0 that reflects the structure of the projections in the algebra. It is significant that one can also give a complete description of the allowable groups. This will permit the construction of interesting algebras.

We begin with the building blocks, the finite dimensional C*-algebras. We may apply the structure theory for algebras of compact operators, Theorem I.10.8, to finite dimensional C*-algebras since every finite dimensional C*-algebra can be represented $*$-isomorphically as a C*-algebra acting on a finite dimensional Hilbert space. (See Exercise I.36.)

Theorem III.1.1 *Every finite dimensional C*-algebra $\mathfrak{A}$ is $*$-isomorphic to the direct sum of full matrix algebras*

$$\mathfrak{A} \simeq \mathcal{M}_{n_1} \oplus \cdots \oplus \mathcal{M}_{n_k}.$$

In particular, every finite dimensional C-algebra is unital.*

Proof. This is an immediate application of Theorem I.10.8. The number of summands must be finite and the dimensions n_i must also be finite for every i in order that the direct sum be finite dimensional. The direct sum of the unit on each summand is the identity element. ∎

We will make frequent use of the fact that $\mathcal{M}_n$ has a nice basis consisting of the **matrix units** $E_{ij} := e_i e_j^*$ for $1 \leq i, j \leq n$ associated to any orthonormal basis $e_1, \ldots, e_n$ for the underlying Hilbert space. Thus any finite dimensional C*-algebra $\mathfrak{A} \simeq \mathcal{M}_{n_1} \oplus \cdots \oplus \mathcal{M}_{n_k}$ has a system of matrix units which we will denote by

$$\{E_{ij}^{(s)} : 1 \leq s \leq k,\ 1 \leq i,j \leq n_s\}.$$

The representation theory of finite dimensional algebras is now a direct consequence of Theorem I.10.7.

Corollary III.1.2 *If π is a non-degenerate $*$-representation of a finite dimensional C$*$-algebra $\mathfrak{A} = \mathcal{M}_{n_1} \oplus \cdots \oplus \mathcal{M}_{n_k}$, then there are cardinal numbers $\alpha_1, \ldots, \alpha_k$ so that π is unitarily equivalent to $\mathrm{id}_1^{(\alpha_1)} \oplus \cdots \oplus \mathrm{id}_k^{(\alpha_k)}$ where id_j denotes the identity representation of $\mathcal{M}_{n_j}$.*

III.2 AF algebras

The next simplest C$*$-algebras may well be those algebras that can be built up from finite dimensional ones. While one might come up with several possible definitions of what this might mean, we shall see that several of the most natural choices turn out to define the same class.

A C$*$-algebra $\mathfrak{A}$ is **approximately finite dimensional** or **AF** if it is the closure of an increasing union of finite dimensional subalgebras $\mathfrak{A}_k$. When $\mathfrak{A}$ is unital, we further stipulate that $\mathfrak{A}_0$ consist of the scalar multiples of the identity element 1.

To understand how such a union might develop, we need to understand how one finite dimensional C$*$-algebra may fit into another.

Lemma III.2.1 *Suppose that φ is a unital homomorphism of a finite dimensional C$*$-algebra $\mathfrak{A}_1 \simeq \mathcal{M}_{n_1} \oplus \cdots \oplus \mathcal{M}_{n_k}$ into another finite dimensional C$*$-algebra $\mathfrak{A}_2 \simeq \mathcal{M}_{m_1} \oplus \cdots \oplus \mathcal{M}_{m_\ell}$. Then φ is determined up to unitary equivalence (in $\mathfrak{A}_2$) by an $\ell \times k$ matrix $A = [a_{ij}]$ in $\mathcal{M}_{\ell k}(\mathbb{N}_0)$ with non-negative integer entries such that*

$$A \begin{pmatrix} n_1 \\ n_2 \\ \vdots \\ n_k \end{pmatrix} = \begin{pmatrix} m_1 \\ m_2 \\ \vdots \\ m_\ell \end{pmatrix}.$$

Proof. Let ε_i, for $1 \leq i \leq \ell$, denote the canonical surjection of $\mathfrak{A}_2$ onto the summand $\mathcal{M}_{m_i}$. Then $\varphi_i := \varepsilon_i \varphi$ is a unital homomorphism of $\mathfrak{A}_1$ into $\mathcal{M}_{m_i}$. By Corollary III.1.2, φ_i is determined up to conjugation by a unitary in $\mathcal{M}_{m_i}$ by the multiplicities a_{ij} in $\mathbb{N}_0$, $1 \leq j \leq k$, so that

$$\varphi_i \simeq \mathrm{id}_{n_1}^{(a_{i1})} \oplus \cdots \oplus \mathrm{id}_{n_k}^{(a_{ik})}.$$

Since φ_i is unital, a comparison of dimensions yields

$$\sum_{i=1}^{k} a_{ij} n_j = m_i.$$

Since $\varphi \simeq \varphi_1 \oplus \cdots \oplus \varphi_\ell$, we see that φ is determined up to unitary equivalence by the non-negative integer valued $\ell \times k$ matrix $A = [a_{ij}]$ of **partial multiplicities**. The integer a_{ij} is the multiplicity of the imbedding of the summand $\mathcal{M}_{n_j}$ of $\mathfrak{A}_1$ into the summand $\mathcal{M}_{m_i}$ of $\mathfrak{A}_2$. ∎

Consider imbeddings which need not be unital. Then $\varphi(I)$ is a projection in $\mathfrak{A}_2$. By restricting to the range of this projection, we reclaim the unital case at the expense of reducing the size of the blocks in $\mathfrak{A}_2$. Hence we obtain:

Corollary III.2.2 *Suppose that φ is a homomorphism of a finite dimensional C*-algebra $\mathfrak{A}_1 \simeq \mathcal{M}_{n_1} \oplus \cdots \oplus \mathcal{M}_{n_k}$ into another finite dimensional C*-algebra $\mathfrak{A}_2 \simeq \mathcal{M}_{m_1} \oplus \cdots \oplus \mathcal{M}_{m_\ell}$. Then φ is determined up to unitary equivalence (in $\mathfrak{A}_2$) by an $\ell \times k$ matrix $A = [a_{ij}]$ in $\mathcal{M}_{\ell k}(\mathbb{N}_0)$ of partial multiplicities such that*

$$\sum_{i=1}^{k} a_{ij} n_j \leq m_i \quad \text{for} \quad 1 \leq i \leq \ell.$$

This allows us to describe the imbedding of $\mathfrak{A}_1$ into $\mathfrak{A}_2$ in a simple graphical way. Represent $\mathfrak{A}_1$ by the k-tuple $\{(1,1) = n_1, \ldots, (1,k) = n_k\}$ and $\mathfrak{A}_2$ by the ℓ-tuple $\{(2,1) = m_1, \ldots, (2,\ell) = m_\ell\}$. Denote the imbedding of $\mathfrak{A}_1$ into $\mathfrak{A}_2$ by drawing a_{ij} arrows from $(1,j)$ to $(2,i)$ to indicate the partial multiplicity of the imbedding. The sequence of these pictures for a sequence of imbeddings α_p of $\mathfrak{A}_p$ into $\mathfrak{A}_{p+1}$ is called the **Bratteli diagram** of the sequence. This is an infinite graph consisting of nodes $(p,i) = n$ representing the fact that the i-th summand of $\mathfrak{A}_p$ is isomorphic to $\mathcal{M}_n$. There will be $a_{ji}^{(p)}$ arrows from (p,i) to $(p+1,j)$ corresponding to the partial multiplicity of the mapping. This is also often called the Bratteli diagram of the *algebra*, but this is an abuse of terminology as the sequence is not unique. We will see though how to compare two diagrams.

Example III.2.3 Consider the C*-algebra $\mathbb{C}I + \mathfrak{K}$. Choose an increasing sequence P_n of projections with $\text{rank}(P_n) = n$ converging strongly to the identity. Let $\mathfrak{A}_n := \mathbb{C}P_n^\perp + P_n \mathfrak{K} P_n \simeq \mathbb{C} \oplus \mathcal{M}_n$. It is easy to see that $\cup_n \mathfrak{A}_n$ is dense in $\mathbb{C}I + \mathfrak{K}$. The imbedding of $\mathfrak{A}_n$ into $\mathfrak{A}_{n+1}$ imbeds $\mathcal{M}_n$ once into $\mathcal{M}_{n+1}$. Since $P_n^\perp = P_{n+1}^\perp + E_{n+1}$ where E_{n+1} is a rank 1 projection less than P_{n+1}, the scalars are imbedded once into each summand of $\mathfrak{A}_{n+1}$. We obtain the Bratteli diagram:

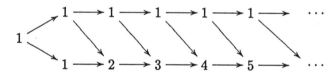

The ideal $\mathfrak{K}$ corresponds to the sequence

$$\mathfrak{B}_n := \mathfrak{A}_n \cap \mathfrak{K} = P_n \mathfrak{K} P_n \simeq \mathcal{M}_n.$$

So we obtain the (directed) subset of the diagram above to represent $\mathfrak{K}$:

$$1 \longrightarrow 2 \longrightarrow 3 \longrightarrow 4 \longrightarrow \quad \cdots$$

Example III.2.4 The **CAR algebra** $\mathfrak{A}$ is obtained as the union of subalgebras $\mathfrak{A}_k \simeq \mathcal{M}_{2^k}$ where the imbedding φ_k of $\mathcal{M}_{2^k}$ into $\mathcal{M}_{2^{k+1}}$ is given by the multiplicity 2 imbedding

$$\varphi_k(A) = \begin{bmatrix} A & 0 \\ 0 & A \end{bmatrix}.$$

III.2. AF algebras

This has the Bratteli diagram

$$1 \rightrightarrows 2 \rightrightarrows 4 \rightrightarrows 8 \rightrightarrows 16 \rightrightarrows 32 \rightrightarrows \cdots$$

The image of φ_k sits inside the smaller subalgebra $\mathfrak{B}_k \simeq \mathcal{M}_{2^k} \oplus \mathcal{M}_{2^k}$ of $\mathfrak{A}_{k+1}$. So $\mathfrak{A}$ is also the increasing union of the $\mathfrak{B}_k$. An inspection of the action of φ_{k+1} on $\mathfrak{B}_k$ shows that

$$\varphi_{k+1}\left(\begin{bmatrix} B_1 & 0 \\ 0 & B_2 \end{bmatrix}\right) = \left[\begin{array}{cc|cc} B_1 & 0 & 0 & 0 \\ 0 & B_2 & 0 & 0 \\ \hline 0 & 0 & B_1 & 0 \\ 0 & 0 & 0 & B_2 \end{array}\right].$$

Hence the partial multiplicities of the imbedding of $\mathfrak{B}_k$ into $\mathfrak{B}_{k+1}$ is given by $\begin{bmatrix} 1 & 1 \\ 1 & 1 \end{bmatrix}$. So we obtain an alternative diagram

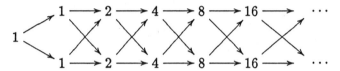

These two diagrams can be intertwined because of the relationship

$$\mathfrak{A}_1 \subset \mathfrak{B}_1 \subset \mathfrak{A}_2 \subset \mathfrak{B}_2 \subset \mathfrak{A}_3 \subset \mathfrak{B}_3 \subset \mathfrak{A}_4 \subset \mathfrak{B}_4 \subset \ldots$$

Except for dropping to subsequences and making small perturbations, this kind of intertwining must occur for two diagrams to represent the same algebra.

Example III.2.5 A finite dimensional commutative C*-algebra is spanned by finitely many pairwise orthogonal projections. Thus a separable commutative C*-algebra is AF precisely when it is spanned by its projections. This is equivalent to the spectrum being totally disconnected. A typical case of this phenomenon occurs for $C(X)$ where X is the Cantor set. Using the traditional 'middle thirds' construction, we obtain X as the intersection of a decreasing family X_n of closed sets where X_n consists of 2^n disjoint intervals, and each interval contains exactly two intervals of X_{n+1}. Let $\mathfrak{A}_n$ be the subalgebra of functions in $C(X)$ which are constant on the intervals of X_n.

Notice that $\mathfrak{A}_n$ is isomorphic to $\mathbb{C}^{2^n}$. The imbedding of $\mathfrak{A}_n$ into $\mathfrak{A}_{n+1}$ splits each minimal projection into the sum of two minimal projections. Thus we obtain the diagram

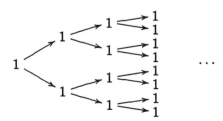

Example III.2.6 Finally we give an example which we won't be able to identify at the moment, but will come up later in our discussion. Start with a sequence $\mathfrak{A}_k = \mathcal{M}_{m_k} \oplus \mathcal{M}_{n_k}$ given recursively by $m_1 = n_1 = 1$ and a repeated use of the partial imbedding matrix $\begin{bmatrix} 1 & 1 \\ 1 & 0 \end{bmatrix}$. This yields the recursion relation

$$m_{k+1} = m_k + n_k \quad \text{and} \quad n_{k+1} = m_k$$

which is easily recognized as the Fibonacci sequence. We obtain the diagram

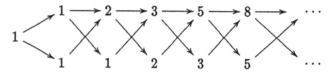

While it is premature to discuss when different diagrams yield the same algebra, let us at least observe that a diagram yields a unique algebra.

Proposition III.2.7 *If $\mathfrak{A} = \overline{\cup_{n \geq 1} \mathfrak{A}_n}$ and $\mathfrak{B} = \overline{\cup_{n \geq 1} \mathfrak{B}_n}$ have the same Bratteli diagram, then they are isomorphic.*

Proof. Let the injection of $\mathfrak{A}_n$ into $\mathfrak{A}_{n+1}$ be denoted by α_n, and similarly let β_n denote the imbedding of $\mathfrak{B}_n$ into $\mathfrak{B}_{n+1}$. At each stage, there is a natural isomorphism φ_n of $\mathfrak{A}_n$ onto $\mathfrak{B}_n$. Thus $\varphi_{n+1}\alpha_n$ and $\beta_n\varphi_n$ are injections of $\mathfrak{A}_n$ into $\mathfrak{B}_{n+1}$ with the same partial multiplicities (because the partial multiplicities of α_n and β_n agree). Hence by Corollary III.1.2 there is a unitary U_{n+1} in $\mathfrak{B}_{n+1}$ so that

$$\beta_n\varphi_n = \mathrm{Ad}(U_{n+1})\varphi_{n+1}\alpha_n.$$

Let $\psi_1 = \varphi_1$ and set $V_1 = I$. Recursively define

$$V_{n+1} = \beta_n(V_n)U_{n+1} \quad \text{and} \quad \psi_{n+1} = \mathrm{Ad}(V_{n+1})\varphi_{n+1} \quad \text{for} \quad n \geq 1.$$

Then a calculation shows that

$$\begin{aligned} \beta_n\psi_n &= \beta_n \mathrm{Ad}(V_n)\varphi_n = \mathrm{Ad}(\beta_n(V_n))\beta_n\varphi_n \\ &= \mathrm{Ad}(\beta_n(V_n))\mathrm{Ad}(U_{n+1})\varphi_{n+1}\alpha_n \\ &= \mathrm{Ad}(\beta_n(V_n)U_{n+1})\varphi_{n+1}\alpha_n = \psi_{n+1}\alpha_n. \end{aligned}$$

Hence the following diagram commutes.

$$\begin{array}{ccccccccc} \mathfrak{A}_1 & \xrightarrow{\alpha_1} & \mathfrak{A}_2 & \xrightarrow{\alpha_2} & \mathfrak{A}_3 & \xrightarrow{\alpha_3} & \mathfrak{A}_4 & \xrightarrow{\alpha_4} & \cdots \\ \psi_1 \downarrow & & \psi_2 \downarrow & & \psi_3 \downarrow & & \psi_4 \downarrow & & \\ \mathfrak{B}_1 & \xrightarrow{\beta_1} & \mathfrak{B}_2 & \xrightarrow{\beta_2} & \mathfrak{B}_3 & \xrightarrow{\beta_3} & \mathfrak{B}_4 & \xrightarrow{\beta_4} & \cdots \end{array}$$

From this it is evident that $\psi := \cup_n \psi_n$ is a $*$-isomorphism of $\cup_{n \geq 1} \mathfrak{A}_n$ onto $\cup_{n \geq 1} \mathfrak{B}_n$. As ψ is isometric, it extends by continuity to a $*$-isomorphism of $\mathfrak{A}$ onto $\mathfrak{B}$. ∎

III.3 Perturbations

In making approximations in AF algebras, any finite dimensional subalgebra $\mathfrak{B}$ of $\mathfrak{A} = \overline{\cup_{n\geq 1}\mathfrak{A}_n}$ will be 'almost contained' in $\mathfrak{A}_n$ for n sufficiently large. We will show that by *twisting* $\mathfrak{B}$ slightly by a unitary close to the identity, it can be moved inside $\mathfrak{A}_n$. This first lemma shows how to move projections.

Lemma III.3.1 *Given any $\varepsilon > 0$ and $n \in \mathbb{N}$, there exists a positive real number $\delta = \delta(\varepsilon, n)$ so that if $P_1, \ldots, P_n$ are pairwise orthogonal projections in a C*-algebra $\mathfrak{D}$ and $\mathfrak{B}$ is a C*-subalgebra of $\mathfrak{D}$ such that $\mathrm{dist}(P_i, \mathfrak{B}) < \delta$ for $1 \leq i \leq n$, then there are pairwise orthogonal projections Q_i in $\mathfrak{B}$ such that $\|P_i - Q_i\| < \varepsilon$ for $1 \leq i \leq n$. If $\sum_{i=1}^n P_i = 1$, then we may arrange that $\sum_{i=1}^n Q_i = 1$ also.*

Proof. First let $n = 1$ and set $\delta = \min\{\varepsilon/2, 1/3\}$; and suppose that P_1 is a projection in $\mathfrak{D}$ such that $\mathrm{dist}(P_1, \mathfrak{B}) < \delta$. Pick X in $\mathfrak{B}$ such that $\|P_1 - X\| < \delta$. Since $P_1 = P_1^*$, we may let $B = (X + X^*)/2$ and obtain $\|P_1 - B\| < \delta$. Because

$$\|(P - \lambda I)^{-1}\| = \mathrm{dist}(\lambda, \{0, 1\})^{-1},$$

it follows that $B - \lambda I$ is invertible if $|\lambda| > \delta$ and $|\lambda - 1| > \delta$. Hence the spectrum $\sigma(B)$ is contained in $[-\delta, \delta] \cup [1 - \delta, 1 + \delta]$. These intervals are disjoint because $\delta \leq 1/3$. Using the continuous functional calculus for normal operators, we obtain a projection $Q_1 := \chi_{[1-\delta, 1+\delta]}(B)$ in $\mathfrak{B}$. Moreover,

$$\|P_1 - Q_1\| \leq \|P_1 - B\| + \|z - \chi_{[1-\delta,1+\delta]}\|_{\sigma(B)} < \delta + \delta \leq \varepsilon.$$

For the general case, proceed by induction. Suppose that we have found projections Q_i in $\mathfrak{B}$ for $1 \leq i \leq n-1$ such that $\|P_i - Q_i\| < c_i \delta$. Set $P = \sum_{i=1}^{n-1} P_i$ and $Q = \sum_{i=1}^{n-1} Q_i$. Choose $B = B^*$ in $\mathfrak{B}$ as above with $\|P_n - B\| < \delta$. Then

$$\|P_n - (I-Q)B(I-Q)\| = \|(I-P)P_n(I-P) - (I-Q)B(I-Q)\|$$
$$\leq \|(Q-P)P_n(I-P)\| + \|(I-Q)(P_n - B)(I-P)\|$$
$$\quad + \|(I-Q)B(Q-P)\|$$
$$\leq (1 + \|B\|)\|Q - P\| + \|P_n - B\|$$
$$\leq (2 + \delta)\sum_{i=1}^{n-1} c_i \delta + \delta \quad < c'_n \delta$$

where $c'_n = 1 + 3\sum_{i=1}^{n-1} c_i$.

Proceeding as above, we obtain a projection Q_n in $(I-Q)\mathfrak{B}(I-Q)$ such that $\|(I-Q)B(I-Q) - Q_n\| < c'_n \delta$; whence $\|P_n - Q_n\| < 2c'_n \delta$. It remains to choose δ sufficiently small to obtain the desired estimate. When $\sum_{i=1}^n P_i = I$, we see that

$$\|I - \sum_{i=1}^n Q_i\| \leq \sum_{i=1}^n \|P_i - Q_i\| \leq (\sum_{i=1}^n c_i)\delta < 1$$

provided that δ is sufficiently small. But the only projection within distance 1 of the identity is I itself. Thus $\sum_{i=1}^{n} Q_i = I$. ∎

Now we are able to show that when a finite dimensional C*-algebra is almost contained in another C*-algebra, it can be twisted inside by conjugating by a unitary near the identity. We make no effort to make our estimates independent of the dimension, but this can be done with a lot of work. A natural measure of how close $\mathfrak{A}$ is to being included in $\mathfrak{B}$ is

$$\sup_{A \in \mathfrak{A}, \|A\| \leq 1} \mathrm{dist}(A, \mathfrak{B}).$$

However, as we will be working with matrix units, it will be convenient to take this supremum only over a standard basis. These two measures will be comparable within a constant factor that depends on the dimension.

Lemma III.3.2 *Given any $\varepsilon > 0$ and $n \in \mathbb{N}$, there exists a positive real number $\delta = \delta(\varepsilon, n)$ so that whenever $\mathfrak{A}$ and $\mathfrak{B}$ are C*-subalgebras of a unital C*-algebra $\mathfrak{D}$ with $\dim \mathfrak{A} \leq n$ and such that $\mathfrak{A}$ has a system of matrix units $\{E_{ij}^{(s)}\}$ satisfying $\mathrm{dist}(E_{ij}^{(k)}, \mathfrak{B}) < \delta$, then there is a unitary U in $C^*(\mathfrak{A}, \mathfrak{B})$ with $\|U - I\| < \varepsilon$ so that $U\mathfrak{A}U^*$ is contained in $\mathfrak{B}$.*

Proof. We may suppose that $\mathfrak{A}$ and $\mathfrak{B}$ are unital by adjoining the identity if necessary. It is easy to see that if E is a unit for $\mathfrak{A}$, then $\mathfrak{A}^{\sim} = \mathfrak{A} \oplus \mathbb{C}E^\perp$. So the set of matrix units for $\mathfrak{A}$ may be expanded to include $E^\perp$. An easy estimate which we omit shows that $\mathrm{dist}(E^\perp, \mathfrak{B}^{\sim}) < n\delta$.

First let us prove this theorem for $\mathfrak{A} \simeq \ell_n^\infty$. The matrix units for $\mathfrak{A}$ are the n pairwise orthogonal minimal projections $P_1, \ldots, P_n$. We may assume that $\varepsilon < 1$; set $\eta = (n+1)^{-1}\varepsilon$. Using $\delta = \delta(\eta, n)$ of the preceding lemma, we obtain pairwise orthogonal projections Q_i in $\mathfrak{B}$ such that $\|P_i - Q_i\| < \eta$. Define $X = \sum_{i=1}^{n} Q_i P_i$. Then a calculation shows that

$$X^*X = \sum_{i=1}^{n} P_i Q_i P_i \geq (1-\eta) \sum_{i=1}^{n} P_i = (1-\eta)I.$$

Similarly $XX^* \geq (1-\eta)I$. So X is invertible. Moreover, we have the identities $Q_i X = XP_i$ for $1 \leq i \leq n$.

Let U be the unitary in the polar decomposition of $X = U|X|$. As X is invertible, Theorem I.8.1 shows that U belongs to $C^*(X)$ which is contained in $C^*(\mathfrak{A}, \mathfrak{B})$. Clearly, each P_i commutes with $|X|$. So

$$UP_i = X|X|^{-1}P_i = XP_i|X|^{-1} = Q_i X|X|^{-1} = Q_i U.$$

III.3. Perturbations

Hence $UP_iU^* = Q_i$ for $1 \leq i \leq n$. It remains to make the norm estimate:

$$\|U - I\| \leq \|U - X\| + \|X - I\|$$

$$\leq \|X\| \, \||X|^{-1} - I\| + \sum_{i=1}^{n} \|(Q_i - P_i)P_i\|$$

$$\leq ((1-\eta)^{-1/2} - 1) + n\eta < (n+1)\eta = \varepsilon.$$

For general $\mathfrak{A}$ with matrix units $\{E_{ij}^{(s)} : 1 \leq s \leq k, 1 \leq i,j \leq n_s\}$, let $\mathfrak{M}$ be the masa spanned by the diagonal matrix units $\{E_{ii}^{(s)}\}$. Let $\delta = \delta(\eta', n)$ where $\eta' = (6n+6)^{-1}\varepsilon$. Using the unitary obtained above for $\mathfrak{M}$ and this choice of δ, we obtain a unitary U so that $U\mathfrak{M}U^*$ is contained in $\mathfrak{B}$ and $\|U - I\| < \varepsilon/6$. The matrix units for $U\mathfrak{A}U^*$, say $F_{ij}^{(s)} := UE_{ij}^{(s)}U^*$, satisfy the norm estimate

$$\mathrm{dist}(F_{ij}^{(s)}, \mathfrak{B}) < \delta + \|E_{ij}^{(s)} - F_{ij}^{(s)}\|$$

$$\leq \delta + 2\|U - I\| < 3\varepsilon/6 = \varepsilon/2.$$

Choose $X_{1j}^{(s)}$ in $\mathfrak{B}$ so that $\|F_{1j}^{(s)} - X_{1j}^{(s)}\| < \varepsilon/2$. As $F_{jj}^{(s)}$ belong to $\mathfrak{B}$, we may replace $X_{1j}^{(s)}$ with $F_{11}^{(s)} X_{1j}^{(s)} F_{jj}^{(s)}$ without increasing this norm estimate. Assume that this has been done.

Notice that $X_{1j}^{(s)}$ is bounded below by $1 - \varepsilon/2$ on the range of $F_{jj}^{(s)}$ and has range equal to the range of $F_{11}^{(s)}$. Thus the polar decomposition for $X_{1j}^{(s)}$ yields a partial isometry $G_{1j}^{(s)}$ in $\mathfrak{B}$ with the same co-domain and range. As in the commutative case, we find that

$$\|X_{1j}^{(s)} - G_{1j}^{(s)}\| < (1 - \varepsilon/2)^{-1/2} - 1 < \varepsilon/3.$$

Extend this to a full system by defining $G_{ij}^{(s)} = G_{1i}^{(s)*} G_{1j}^{(s)}$. Then define a unitary operator

$$W := \sum_{s=1}^{k} \sum_{j=1}^{n_s} G_{1j}^{(s)*} F_{1j}^{(s)}.$$

An easy computation confirms that $G_{ij}^{(s)} W = W F_{ij}^{(s)}$, whence

$$WF_{ij}^{(s)} W^* = G_{ij}^{(s)} \quad \text{for all} \quad 1 \leq s \leq k, \quad 1 \leq i,j \leq n_s.$$

Finally we obtain the norm estimate (using the fact that a sum of terms with orthogonal domains and ranges has norm equal to the maximum of the norm of each term)

$$\|W - I\| = \max \|G_{1j}^{(s)*} F_{1j}^{(s)} - F_{1j}^{(s)*} F_{1j}^{(s)}\|$$

$$= \max \|G_{1j}^{(s)} - F_{1j}^{(s)}\| < 5\varepsilon/6.$$

The desired unitary WU thus satisfies
$$\|WU - I\| \leq \|W - I\|\|U\| + \|U - I\| < 5\varepsilon/6 + \varepsilon/6 = \varepsilon. \qquad \blacksquare$$

For ease of our constructions, we improve this result in the case that $\mathfrak{A}$ and $\mathfrak{B}$ contain a common subalgebra.

Corollary III.3.3 *If in addition to the hypotheses of the preceding lemma, we have an algebra $\mathfrak{A}_1$ contained in $\mathfrak{A} \cap \mathfrak{B}$, then we may choose the unitary U to lie in the commutant of $\mathfrak{A}_1$.*

Proof. This can be accomplished with superior constants by doing the construction over again with $\mathfrak{A}_1$ in mind. However, we choose to give a more transparent argument that has the minor flaw of requiring a smaller choice of δ.

Let $\delta = \delta(\varepsilon/3, \dim \mathfrak{A})$ from the last lemma. Obtain a unitary operator U in $C^*(\mathfrak{A}, \mathfrak{B})$ such that $U\mathfrak{A}U^* \subset \mathfrak{B}$ and $\|U - I\| < \varepsilon/3$. Fix a set $\{F_{ij}^{(s)}\}$ of matrix units for $\mathfrak{A}_1$; and let $G_{ij}^{(s)} = UF_{ij}^{(s)}U^*$ be the corresponding system for $U\mathfrak{A}_1U^*$. Define

$$W = \sum_{s=1}^{k} \sum_{j=1}^{n_s} G_{1j}^{(s)*} F_{1j}^{(s)}$$

as in the proof above. This lies in $\mathfrak{B}$ and satisfies $WF_{ij}^{(s)}W^* = G_{ij}^{(s)}$ for all i, j, s and

$$\|W - I\| \leq \max \|G_{1j}^{(s)} - F_{1j}^{(s)}\|$$
$$= \max \|(U - I)F_{1j}^{(s)} - F_{1j}^{(s)}(U - I)\| \leq 2\|U - I\|.$$

Thus the unitary $V = W^*U$ commutes with $\mathfrak{A}_1$, satisfies $V\mathfrak{A}V^* \subset \mathfrak{B}$ and

$$\|V - I\| \leq \|W - I\|\|U\| + \|U - 1\| \leq 3\|U - I\| < \varepsilon. \qquad \blacksquare$$

Now we can obtain a characterization of AF algebras which is not dependent on choosing a sequence of subalgebras.

Theorem III.3.4 *A C^*-algebra $\mathfrak{A}$ is AF if and only if it is separable and:*

(∗) *for all $\varepsilon > 0$ and $A_1, \ldots, A_n$ in $\mathfrak{A}$, there exists a finite dimensional C^*-subalgebra $\mathfrak{B}$ of $\mathfrak{A}$ such that $\mathrm{dist}(A_i, \mathfrak{B}) < \varepsilon$ for $1 \leq i \leq n$.*

Moreover, if $\mathfrak{A}_1$ is a finite dimensional subalgebra of $\mathfrak{A}$, then we may choose $\mathfrak{B}$ so that it contains $\mathfrak{A}_1$.

Proof. If $\mathfrak{A}$ is AF, the other properties are easily established. So assume that $\mathfrak{A}$ is separable and (∗) holds. Fix a countable dense subset $\{A_i : i \geq 1\}$ of the unit ball of $\mathfrak{A}$ with $A_1 = 0$ and a sequence ε_i decreasing monotonely to 0. Proceed recursively starting with $\mathfrak{A}_1$ if it is given, and with the identity adjoined if $\mathfrak{A}$ is unital. (Otherwise start with the subalgebra of scalars.)

For the purpose of proceeding by induction, assume that a finite dimensional subalgebra $\mathfrak{A}_k$ has been found so that $\mathrm{dist}(A_i, \mathfrak{A}_k) < \varepsilon_k$ for all $1 \leq i \leq k$.

III.3. Perturbations

Let $\delta = \delta(\varepsilon_{k+1}/3, \dim \mathfrak{A}_k)$ as in Lemma III.3.2; and fix a set of matrix units $\{E_{ij}^{(s)}\}$ for $\mathfrak{A}_k$. Using property (*), find a finite dimensional subalgebra $\mathfrak{B}$ of $\mathfrak{A}$ so that $\mathrm{dist}(E_{ij}^{(s)}, \mathfrak{B}) < \delta$ for all the matrix units and $\mathrm{dist}(A_i, \mathfrak{B}) < \varepsilon_{k+1}/3$ for $1 \leq i \leq k+1$. Then by Lemma III.3.2, there is a unitary U in $\mathfrak{A}$ so that $U\mathfrak{A}_k U^* \subset \mathfrak{B}$ and $\|U - I\| < \varepsilon_{k+1}/3$. Let $\mathfrak{A}_{k+1} := U^* \mathfrak{B} U$. Clearly this is a finite dimensional algebra containing $\mathfrak{A}_k$. Moreover,

$$\mathrm{dist}(A_i, \mathfrak{A}_{k+1}) = \mathrm{dist}(UA_i U^*, \mathfrak{B}) \leq 2\|U - I\| + \varepsilon_{k+1}/3 < \varepsilon_{k+1}.$$

Proceeding recursively, we construct an increasing sequence of finite dimensional algebras with dense union. So $\mathfrak{A}$ is AF. ∎

It turns out that there is quite a strong uniqueness of the chain of subalgebras of an AF algebra. Of course, one may always drop to a subsequence, but this will not change the union. The union is unique up to conjugation by a unitary. This allows us to (almost) intertwine two chains determining the algebra.

Theorem III.3.5 *Suppose that $\mathfrak{A}$ is an AF algebra which is the closure of the increasing union of two chains $\mathfrak{A} = \overline{\cup_{n \geq 1} \mathfrak{A}_n} = \overline{\cup_{n \geq 1} \mathfrak{B}_n}$. Then for any $\varepsilon > 0$, there is a unitary operator W in $\mathfrak{A}^\sim$ with $\|W - I\| < \varepsilon$ so that*

$$\cup_{n \geq 1} \mathfrak{A}_n = W \cup_{n \geq 1} \mathfrak{B}_n W^*.$$

In particular, there are subsequences m_i and n_i of $\mathbb{N}$ so that

$$\mathfrak{A}_{m_i} \subset W \mathfrak{B}_{n_i} W^* \subset \mathfrak{A}_{m_{i+1}} \text{ for all } i \geq 1.$$

Proof. Choose positive numbers ε_i so that $\prod_{i \geq 1}(1 + \varepsilon_i)^2 < 1 + \varepsilon$. Let $m_1 = 1$ and $\delta_1 := \delta(\varepsilon_1, \dim \mathfrak{A}_1)$ as in Lemma III.3.2. As the matrix units of $\mathfrak{A}_1$ are almost contained in the dense union of the $\mathfrak{B}_n$'s, there is an integer n_1 such that $\mathfrak{A}_1$ is within δ_1 of being contained in $\mathfrak{B}_{n_1}$. Hence there is a unitary U_1 so that

$$U_1 \mathfrak{A}_1 U_1^* \subset \mathfrak{B}_{n_1} \quad \text{and} \quad \|U_1 - I\| < \varepsilon_1.$$

Now let $\eta_1 := \delta(\varepsilon_1, \dim \mathfrak{B}_{n_1})$. Proceeding in the same way, find an integer $m_2 > 1$ so that $U_1^* \mathfrak{B}_{n_1} U_1$ is η_1-contained in $\mathfrak{A}_{m_2}$. By Corollary III.3.3, there is a unitary V_1 in $\mathfrak{A}$ commuting with $\mathfrak{A}_1$ so that

$$V_1 U_1^* \mathfrak{B}_{n_1} U_1 V_1^* \subset \mathfrak{A}_{m_2} \quad \text{and} \quad \|V_1 - I\| < \varepsilon_1.$$

We will proceed recursively. Suppose that at the k-th stage, we have found integers $m_1 < \cdots < m_{k+1}$ and $n_1 < \cdots < n_k$ and unitary operators U_i and V_i for $1 \leq i \leq k$ satisfying $\|U_i - I\| < \varepsilon_i$ and $\|V_i - I\| < \varepsilon_i$ such that

$$\mathfrak{A}_{m_i} \subset \widetilde{\mathfrak{B}}_{n_i} := V_i U_i^* \ldots V_1 U_1^* \mathfrak{B}_{n_i} U_1 V_1^* \ldots U_i V_i^* \subset \mathfrak{A}_{m_{i+1}}$$

and so that V_i commutes with $\mathfrak{A}_i$ and U_{i+1} commutes with $\widetilde{\mathfrak{B}}_{n_i}$ for $1 \leq i \leq k$. To obtain the next stage, repeat the argument of the second paragraph.

The condition on ε_i guarantees that the infinite product

$$W := \lim_{i \to \infty} V_i U_i^* \ldots V_1 U_1^*$$

exists and satisfies $\|W - I\| < \varepsilon$. Moreover, if we set
$$W_k := V_k U_k^* \ldots V_1 U_1^* \quad \text{and} \quad W^{(k)} = \lim_{i \to \infty} V_i U_i^* \ldots V_{k+1} U_{k+1}^*,$$
we see that $W^{(k)}$ commutes with $\widetilde{\mathfrak{B}}_{n_k}$; whence
$$W \mathfrak{B}_{n_k} W^* = W^{(k)} \widetilde{\mathfrak{B}}_{n_k} W^{(k)*} = \widetilde{\mathfrak{B}}_{n_k}. \qquad \blacksquare$$

We obtain the following immediate consequence.

Corollary III.3.6 *If $\mathfrak{A} = \overline{\cup_{n \geq 1} \mathfrak{A}_n}$ and $\mathfrak{B} = \overline{\cup_{n \geq 1} \mathfrak{B}_n}$ are $*$-isomorphic AF algebras, then $\cup_{n \geq 1} \mathfrak{A}_n$ and $\cup_{n \geq 1} \mathfrak{B}_n$ are $*$-isomorphic.*

Example III.3.7 The purpose of this example is to show that even if $\mathfrak{A}_n$ is isomorphic to $\mathfrak{B}_n$ for all $n \geq 1$, the algebras need not be isomorphic. The nature of the imbeddings is critical.

Consider the algebra $\mathfrak{A} = C(X)$ where X is the Cantor set with a chain of subalgebras $\mathfrak{A}_n$ isomorphic to $\ell_{2^n}^\infty$ as in Example III.2.5. For the second algebra, let $\mathfrak{B} = C(Y)$ where $Y = \{0\} \cup \{n^{-1} : n \geq 1\}$. Let $\mathfrak{B}_n$ be the subalgebra of functions in $C(Y)$ which are constant on $[0, 2^{-n}]$. It is readily apparent that $\mathfrak{B}_n$ is also isomorphic to $\ell_{2^n}^\infty$. But as the spectrum of isomorphic $C(K)$ spaces are homeomorphic, we deduce that $\mathfrak{A}$ and $\mathfrak{B}$ are not isomorphic.

The Bratteli diagram of $\mathfrak{B}$ is quite different from that of $\mathfrak{A}$:

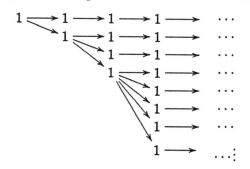

III.4 Ideals and Quotients

We will now see that the ideal structure of an AF algebra can be read off of its Bratteli diagram. First we show that ideals are **inductive**, meaning that they come from ideals of the defining sequence. Notice that the finite dimensionality of the subalgebras $\mathfrak{A}_n$ is not used. So this result may be used later for more general inductive limits.

Lemma III.4.1 *Let $\mathfrak{J}$ be an ideal of a C^*-algebra $\mathfrak{A} = \overline{\cup_{n \geq 1} \mathfrak{A}_n}$, where $\mathfrak{A}_n$ is an increasing sequence of C^*-subalgebras of $\mathfrak{A}$. Then*
$$\mathfrak{J} = \overline{\cup_{n \geq 1} (\mathfrak{J} \cap \mathfrak{A}_n)} = \overline{\mathfrak{J} \cap \cup_{n \geq 1} \mathfrak{A}_n}.$$
Consequently, when $\mathfrak{A}$ is AF, so is $\mathfrak{J}$.

III.4. Ideals and Quotients

Proof. Consider the commutative diagram

$$\begin{array}{ccccc}
\mathfrak{J} \cap \mathfrak{A}_n & \xrightarrow{\iota} & \mathfrak{A}_n & \xrightarrow{\pi} & \mathfrak{A}_n/(\mathfrak{J} \cap \mathfrak{A}_n) \\
\downarrow\iota & & \downarrow\iota & & \downarrow\alpha \\
\mathfrak{J} & \xrightarrow{\iota} & \mathfrak{A}_n + \mathfrak{J} & \xrightarrow{\pi} & \mathfrak{A}_n + \mathfrak{J}/\mathfrak{J}
\end{array}$$

Here ι denotes the canonical injections, and π denotes the canonical quotient maps. The map α is an isomorphism by Corollary I.5.6. In particular, for A in $\mathfrak{A}_n$,

$$\mathrm{dist}(A, \mathfrak{A}_n \cap \mathfrak{J}) = \mathrm{dist}(A, \mathfrak{J}).$$

Hence if J belongs to $\mathfrak{J}$ and $\varepsilon > 0$, there is an element A in $\mathfrak{A}_n$ (for n large enough) so that $\|J - A\| < \varepsilon$. Thus there is an element J' in $\mathfrak{J} \cap \mathfrak{A}_n$ so that $\|A - J'\| < \varepsilon$. Hence $\|J - J'\| < 2\varepsilon$. It follows that $\mathfrak{J} = \cup_{n \geq 1}(\mathfrak{J} \cap \mathfrak{A}_n)$.

If $\mathfrak{A}_n$ are all finite dimensional, then so are $\mathfrak{J} \cap \mathfrak{A}_n$; and hence $\mathfrak{J}$ is AF. ∎

Now we can describe all the ideals. A subset S of a Bratteli diagram $\mathcal{D}$ is **directed** if whenever some (p, i) belongs to S and $(p, i) \to (p+1, j)$ in $\mathcal{D}$, then $(p+1, j)$ also belongs to S. The subset S is **hereditary** if whenever every image of a vertex at the next level belongs to S, then the vertex also belongs to S. That is, whenever for some (p, i) in $\mathcal{D}$ all the $(p+1, j)$ belong to S for which there is an arrow $(p, i) \to (p+1, j)$ in $\mathcal{D}$, then (p, i) also belongs to S.

Theorem III.4.2 *Let $\mathfrak{A}$ be an AF algebra with a Bratteli diagram $\mathcal{D}$. Then the ideals of $\mathfrak{A}$ are in a one to one correspondence with directed hereditary subsets S of $\mathcal{D}$.*

Proof. An ideal $\mathfrak{J}_p$ of $\mathfrak{A}_p \simeq \mathcal{M}_{(p,1)} \oplus \cdots \oplus \mathcal{M}_{(p,k)}$ corresponds to the sum of a subset of these summands because $\mathcal{M}_n$ is simple. So given an ideal $\mathfrak{J}$ of $\mathfrak{A}$, let S denote the subset of $\mathcal{D}$ corresponding to the union of the summands corresponding to each $\mathfrak{J}_p := \mathfrak{J} \cap \mathfrak{A}_p$. Since Lemma III.4.1 shows that the sequence $\mathfrak{J}_p$ uniquely determines $\mathfrak{J}$, one sees that $\mathfrak{J}$ can be recovered from knowledge of S.

Let α_p denote the injection of $\mathfrak{A}_p$ into $\mathfrak{A}_{p+1}$. To see that S is directed, note that if $\mathcal{M}_{(p,i)}$ belongs to $\mathfrak{J}_p$, then whenever $(p, i) \to (p+1, j)$, there is a non-empty intersection

$$\mathfrak{J} \cap \mathcal{M}_{(p+1,j)} \supset \alpha_p(\mathcal{M}_{(p,i)}) \cap \mathcal{M}_{(p+1,j)} \neq \{0\}.$$

Hence $\mathfrak{J}$ contains $\mathcal{M}_{p+1,j}$ and so $(p+1, j)$ belongs to S.

To see that S is hereditary, suppose that $(p+1, j)$ belongs to S for all j in

$$J := \{j : (p, i) \to (p+1, j) \text{ in } \mathcal{D}\}.$$

Then since

$$\alpha_p(\mathcal{M}_{(p,i)}) \subset \sum_{j \in J} \oplus \mathcal{M}_{(p+1,j)} \subset \mathfrak{J},$$

it follows that (p, i) belongs to S as well.

Conversely, if $\mathcal{S}$ is a subset of $\mathcal{D}$ which is directed and hereditary, let $\mathfrak{J}_p$ denote the ideal of $\mathfrak{A}_p$ corresponding to the set of vertices (p,i) in $\mathcal{S}$. Since $\mathcal{S}$ is directed, this is an increasing net of ideals. The closure of the union is an ideal $\mathfrak{J}$ of $\mathfrak{A}$. The hereditary property ensures that $\mathfrak{J}_p = \mathfrak{A}_p \cap \mathfrak{J}_{p+1}$ from which we deduce that $\mathfrak{J}_p = \mathfrak{A}_p \cap \mathfrak{J}$. Hence this is the sequence we associated to $\mathfrak{J}$ as in the first paragraph. So the subset $\mathcal{S}$ is the canonical choice associated to $\mathfrak{J}$. ∎

Corollary III.4.3 *A unital AF algebra $\mathfrak{A}$ with Bratteli diagram $\mathcal{D}$ is simple if and only if for each (p,i) in $\mathcal{D}$, there is an integer $q > p$ so that $(p,i) \to (q,j)$ for all j.*

Proof. If $\mathcal{D}$ has this property, then as every non-zero ideal $\mathfrak{J}$ will have non-zero intersection with some $\mathfrak{A}_p$, we find a (p,i) in the associated subset $\mathcal{S}$ of $\mathcal{D}$. Hence by the directed property and the hypothesis of the corollary, $\mathcal{S}$ contains all (q,j) at some higher lever. But this means that $\mathfrak{J}$ contains $\mathfrak{A}_q$, and so it contains the identity. Therefore $\mathfrak{A}$ is simple.

Conversely, suppose that the condition fails for a certain (p,i). Then the directed subset $\mathcal{S}'$ of $\mathcal{D}$ generated by (p,i) is a proper subset at every level. Increase this to the smallest directed and hereditary subset $\mathcal{S}$ containing (p,i) by adding in any vertex that maps completely into $\mathcal{S}'$ at some higher level. This is still proper at every level because each $\mathfrak{A}_n$ contains the identity of $\mathfrak{A}$. Consequently, it corresponds to a proper ideal and $\mathfrak{A}$ is not simple. ∎

Now we can proceed to compute the quotient algebra.

Theorem III.4.4 *If $\mathfrak{J}$ is an ideal of an AF algebra $\mathfrak{A}$ corresponding to a subset $\mathcal{S}$ of the Bratteli diagram $\mathcal{D}$, then $\mathfrak{A}/\mathfrak{J}$ is AF and corresponds to the diagram $\mathcal{D} \setminus \mathcal{S}$.*

Proof. First notice that $\mathfrak{A}/\mathfrak{J}$ is the increasing union of subalgebras $\mathfrak{A}_n + \mathfrak{J}/\mathfrak{J}$, which are finite dimensional, and thus the quotient is AF. By the proof of Theorem III.4.2, we have that $\mathfrak{A}_n + \mathfrak{J}/\mathfrak{J} \simeq \mathfrak{A}_n/\mathfrak{J}_n$. This algebra is isomorphic to

$$\sum_{(n,i) \notin \mathcal{S}} \oplus \mathcal{M}_{(n,i)}$$

which corresponds to the n-th level of $\mathcal{D} \setminus \mathcal{S}$. Moreover the partial imbeddings of (n,i) into $(n+1,j)$ are unchanged if neither belongs to $\mathcal{S}$. Hence the Bratteli diagram of $\mathfrak{A}/\mathfrak{J}$ is $\mathcal{D} \setminus \mathcal{S}$. ∎

III.5 Examples

Example III.5.1 UHF Algebras. A C*-algebra is called **uniformly hyperfinite** or **UHF** if it is the increasing union of unital subalgebras isomorphic to full matrix algebras $\mathcal{M}_{k_n}$. Since a unital imbedding of $\mathcal{M}_m$ into $\mathcal{M}_n$ requires $m|n$, we have an increasing sequence $k_1 | k_2 | k_3 | \ldots$. Thus for each prime integer p, there is a unique ε_p in $\mathbb{N} \cup \{\infty\}$ which is the supremum of the exponents of powers of p which divide k_n as n tends to infinity. We define the **supernatural number** associated

III.5. Examples

to the sequence $\mathfrak{A}_n$ to be the formal product $\delta(\mathfrak{A}) := \prod_{p \text{ prime}} p^{\epsilon_p}$. The following theorem of **Glimm** classifies UHF algebras.

Theorem III.5.2 *Two UHF algebras $\mathfrak{A}$ and $\mathfrak{B}$ are isomorphic if and only if*

$$\delta(\mathfrak{A}) = \delta(\mathfrak{B}).$$

Proof. Let $\mathfrak{A}$ and $\mathfrak{B}$ correspond to sequences k_m and ℓ_n respectively. If $\mathfrak{A}$ and $\mathfrak{B}$ are isomorphic, Corollary III.3.6 shows that $\cup_{m \geq 1} \mathfrak{A}_m$ and $\cup_{n \geq 1} \mathfrak{B}_n$ are isomorphic. Thus as in Theorem III.3.5, there are subsequences m_i and n_i so that $\mathfrak{A}_{m_i} \simeq \mathcal{M}_{k_{m_i}}$ is isomorphic to a subalgebra of $\mathfrak{B}_{n_i} \simeq \mathcal{M}_{\ell_{n_i}}$ and thus $k_{m_i} | \ell_{n_i}$; and $\mathfrak{B}_{n_i}$ is isomorphic to a subalgebra of $\mathfrak{A}_{m_{i+1}}$, whence $\ell_{n_i} | k_{m_{i+1}}$. Hence $\mathfrak{A}$ and $\mathfrak{B}$ determine the same supernatural number.

Conversely, suppose that $\delta(\mathfrak{A}) = \delta(\mathfrak{B})$. Then there are subsequences m_i and n_i such that $k_{m_i} | \ell_{n_i} | k_{n_{i+1}}$. For convenience, let us renumber so that $m_i = n_i = i$; and let α_n denote the imbedding of $\mathfrak{A}_n$ into $\mathfrak{A}_{n+1}$; and similarly let β_n denote the imbedding of $\mathfrak{B}_n$ into $\mathfrak{B}_{n+1}$. Define a *-isomorphism from $\cup_{n \geq 1} \mathfrak{A}_n$ onto $\cup_{n \geq 1} \mathfrak{B}_n$ as follows. There is a unique imbedding (up to unitary equivalence) φ_1 of $\mathfrak{A}_1$ into $\mathfrak{B}_1$ by Lemma III.2.1. Likewise there is an imbedding ψ_1' of $\mathfrak{B}_1$ into $\mathfrak{A}_2$. The composition $\psi_1' \varphi_1$ is equivalent to the imbedding α_1 of $\mathfrak{A}_1$ into $\mathfrak{A}_2$. Thus by Lemma III.2.1, there is a unitary operator W_1 in $\mathfrak{A}_2$ so that $\mathrm{Ad}(W_1) \psi_1' \varphi_1 = \alpha_1$. Let $\psi_1 = \mathrm{Ad}(W_1) \psi_1'$.

Continue in this way to construct injections φ_n of $\mathfrak{A}_n$ into $\mathfrak{B}_n$ and ψ_n of $\mathfrak{B}_n$ into $\mathfrak{A}_{n+1}$ so that the diagram

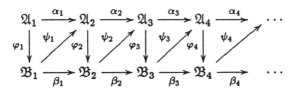

commutes. Then by construction, $\varphi = \overline{\cup_{n \geq 1} \varphi_n}$ is the desired isomorphism; and $\psi = \overline{\cup_{n \geq 1} \psi_n}$ is its inverse. ∎

Corollary III.5.3 *UHF algebras are simple.*

Proof. Since the Bratteli diagram consists of a single node at each level, it is evident that the only directed hereditary and non-empty subset is the whole diagram. Thus UHF algebras are simple by Corollary III.4.3. ∎

Example III.5.4 The CAR algebra. The acronym CAR abbreviates **canonical anticommutation relations**, a notion which comes from quantum mechanics. Consider a linear map α of a Hilbert space $\mathcal{H}$ into $\mathcal{B}(\mathcal{H})$ such that for all f, g in $\mathcal{H}$

$$\alpha(f)\alpha(g) + \alpha(g)\alpha(f) = 0 \tag{1}$$

and
$$\alpha(f)^*\alpha(g) + \alpha(g)\alpha(f)^* = (g,f)I. \tag{2}$$

Let $\mathfrak{A}$ denote the C*-algebra generated by $\{\alpha(f) : f \in \mathcal{H}\}$. We will establish that this algebra is independent of the choice of the function α, and that it is in fact the 2^∞ UHF algebra.

First, notice that if f is a unit vector, then using $g = f$ yields
$$\alpha(f)^2 = 0 \quad \text{and} \quad \alpha(f)^*\alpha(f) + \alpha(f)\alpha(f)^* = I.$$

Multiplying the second equation by $\alpha(f)^*\alpha(f)$ yields
$$(\alpha(f)^*\alpha(f))^2 = \alpha(f)^*\alpha(f) =: E(f)$$

is a projection. Hence $\alpha(f)\alpha(f)^* = E(f)^\perp$. The operator $\alpha(f)$ is a partial isometry with domain $E(f)$ and range $E(f)^\perp$. Hence
$$C^*(\alpha(f)) = \text{span}\{\alpha(f), \alpha(f)^*, E(f), E(f)^\perp\} \simeq \mathcal{M}_2.$$

Let $E_{21}^{(1)} = \alpha(f)$, $E_{12}^{(1)} = \alpha(f)^*$, $E_{11}^{(1)} = E(f)$, and $E_{22}^{(1)} = E(f)^\perp$.

Next note that if f and g are orthogonal unit vectors, then
$$[\alpha(g), E(f)] = \alpha(g)\alpha(f)^*\alpha(f) - \alpha(f)^*\alpha(f)\alpha(g)$$
$$= \alpha(g)\alpha(f)^*\alpha(f) + \alpha(f)^*\alpha(g)\alpha(f)$$
$$= (g,f)\alpha(f) = 0.$$

Thus $\alpha(g)$ commutes with $E(f)$, and consequently so does $E(g)$.

Let $V_1 := I - 2E(f) = E(f)^\perp - E(f)$. Then a calculation using (1) shows that
$$V_1\alpha(g)\alpha(f) = -V_1\alpha(f)\alpha(g) = \alpha(f)V_1\alpha(g)$$

and
$$V_1\alpha(g)\alpha(f)^* = -V_1\alpha(f)^*\alpha(g) = \alpha(f)^*V_1\alpha(g).$$

Hence $V_1\alpha(g)$ commutes with $C^*(\alpha(f))$.

Note that $C^*(V_1\alpha(g)) = \text{span}\{V_1\alpha(g), V_1\alpha(g)^*, E(g), E(g)^\perp\} \simeq \mathcal{M}_2$ and commutes with $C^*(\alpha(f))$. Set
$$E_{21}^{(2)} = V_1\alpha(g), \quad E_{12}^{(2)} = V_1\alpha(g)^*, \quad E_{11}^{(2)} = E(g), \text{ and } E_{22}^{(2)} = E(g)^\perp.$$

It is now evident that $C^*(\alpha(f), \alpha(g))$ is isomorphic to $\mathcal{M}_4$ with a standard basis $E_{ij}^{(1)}E_{k\ell}^{(2)}$ for $1 \leq i,j,k,\ell \leq 2$.

Now fix an orthonormal basis $\{f_n : n \geq 1\}$ for $\mathcal{H}$. Set
$$V_0 = I \quad \text{and} \quad V_n = \prod_{i=1}^{n-1}(I - 2E(f_i)) \quad \text{for} \quad n \geq 2.$$

III.5. Examples

Then we may define matrix units for a copy of $\mathcal{M}_2$ by

$$E_{11}^{(n)} = E(f_n), \quad E_{21}^{(n)} = V_n\alpha(f_n), \quad E_{12}^{(n)} = V_n\alpha(f_n)^*, \quad \text{and} \quad E_{22}^{(n)} = E(f_n)^\perp.$$

It follows by extending the analysis above that these copies of $\mathcal{M}_2$ commute with each other, and hence $\mathfrak{A}_n := C^*(\{\alpha(f_i) : 1 \leq i \leq n\})$ is isomorphic to $\mathcal{M}_{2^n}$ with a standard basis consisting of the matrix units $E_{\varphi\psi} := \prod_{k=1}^n E_{\varphi(k)\psi(k)}^{(k)}$ for all functions φ and ψ of $\{1, 2, \ldots, n\}$ into $\{1, 2\}$.

Thus $\mathfrak{A} = \overline{\bigcup_{n=1}^\infty \mathfrak{A}_n}$ is the 2^∞ UHF algebra of Example III.2.4.

Example III.5.5 The GICAR algebra. GICAR stands for **gauge invariant** CAR and is also called the **current algebra**. If U is a unitary operator on $\mathcal{H}$, define an automorphism of the CAR algebra $\mathfrak{A}$ by

$$\varphi_U(\alpha(f)) = \alpha(Uf) \quad \text{for all} \quad f \in \mathcal{H}.$$

It is easy to see that this map preserves the relations (1) and (2) above, and thus φ_U extends to $*$-automorphism of $\mathfrak{A}$. In particular, consider the scalar unitaries λI for λ in the unit circle $\mathbb{T}$. These give rise to the so called **gauge automorphisms** $\chi_\lambda := \varphi_{\lambda I}$ for λ in $\mathbb{T}$ determined by the identities

$$\chi_\lambda(\alpha(f)) = \alpha(\lambda f) = \lambda\alpha(f) \quad \text{and} \quad \chi_\lambda(\alpha(f)^*) = \bar{\lambda}\alpha(f)$$

for all f in $\mathcal{H}$.

Define the subalgebra $\mathfrak{A}^0$ of the CAR algebra which is invariant for all the gauge automorphisms:

$$\mathfrak{A}^0 := \{A \in \mathfrak{A} : \chi_\lambda(A) = A \text{ for all } \lambda \in \mathbb{T}\}.$$

Note that if $f_1, \ldots, f_n, g_1, \ldots, g_m$ are vectors in $\mathcal{H}$, then

$$A := \alpha(f_1)^* \ldots \alpha(f_n)^* \alpha(g_1) \ldots \alpha(g_m)$$

satisfies $\chi_\lambda(A) = \lambda^{m-n} A$. In particular, A belongs to $\mathfrak{A}^0$ when $m = n$. It will turn out that $\mathfrak{A}^0$ is spanned by the terms of this type. In particular, note that $\chi_\lambda(E(f)) = E(f)$ for all f in $\mathcal{H}$; and hence we also have $\chi_\lambda(V_n) = V_n$ for the symmetries V_n described in the previous section. Therefore the matrix unit $E_{ij}^{(n)}$ is transformed by the rule

$$\chi_\lambda(E_{ij}^{(n)}) = \lambda^{i-j} E_{ij}^{(n)} \quad \text{for} \quad 1 \leq i, j \leq 2.$$

Define $\mathfrak{A}_n^0 := \mathfrak{A}^0 \cap \mathfrak{A}_n$. Let us show that $\mathfrak{A}_n^0$ is isomorphic to $\sum_{k=0}^n \oplus \mathcal{M}_{\binom{n}{k}}$. Consider the action of χ_λ on a standard basis element of $\mathfrak{A}_n$

$$\chi_\lambda\left(\prod_{k=1}^n E_{i_k j_k}^{(k)}\right) = \lambda^{(\sum_{k=1}^n i_k - \sum_{k=1}^n j_k)} \prod_{k=1}^n E_{i_k j_k}^{(k)}.$$

So with respect to this basis, χ_λ acts as a diagonal matrix on $\mathfrak{A}_n$. So we see immediately that the fixed point algebra is spanned by those matrix units satisfying

$\sum_{k=1}^{n} i_k = \sum_{k=1}^{n} j_k$. As these are sums of 1's and 2's, the possible sums are $n, n+1, \ldots, 2n$.

For each integer $0 \le s \le n$, the span
$$\mathcal{F}_s^n = \text{span}\{E_{\varphi\psi} : \sum_{k=1}^{n} \varphi(k) = \sum_{k=1}^{n} \psi(k) = n + s\}$$
is an algebra. Indeed, a product $E_{\varphi\psi} E_{\varphi'\psi'} = \delta_{\psi\varphi'} E_{\varphi\psi'}$ is non-zero if and only if $\varphi' = \psi$, which implies that $\sum_{k=1}^{n} \varphi(k) = n + s = \sum_{k=1}^{n} \psi'(k)$. In fact this formula shows that this basis is a set of matrix units for a full matrix algebra. The dimension is computed from the fact that there must be exactly s 2's and $n - s$ 1's in each term. This means that there are $\binom{n}{s}$ possible functions φ of $\{1, 2, \ldots, n\}$ into $\{1, 2\}$ with this sum. So $\mathcal{F}_s^n \simeq \mathcal{M}_{\binom{n}{s}}$. Moreover our product identity shows that $\mathcal{F}_s^n \mathcal{F}_t^n = 0$ for $s \ne t$. So this decomposition is a direct sum as claimed.

Next consider the imbedding of $\mathfrak{A}_n^0$ into $\mathfrak{A}_{n+1}^0$. For $j = 1, 2$, let φj denote the function of $\{1, 2, \ldots, n+1\}$ into $\{1, 2\}$ given by $\varphi j(k) = \varphi(k)$ for $1 \le k \le n$ and $\varphi j(n+1) = j$. In the imbedding of $\mathfrak{A}_n$ into $\mathfrak{A}_{n+1}$, the matrix unit $E_{\varphi\psi}$ is sent to $E_{\varphi 1 \psi 1} + E_{\varphi 2 \psi 2}$. If $E_{\varphi\psi}$ lies in $\mathcal{F}_s^n$, then
$$\sum_{k=1}^{n+1} \varphi 1(k) = \sum_{k=1}^{n+1} \psi 1(k) = (n+1) + s$$
and
$$\sum_{k=1}^{n+1} \varphi 2(k) = \sum_{k=1}^{n+1} \psi 2(k) = (n+1) + s + 1.$$
Hence $E_{\varphi 1 \psi 1}$ lies in $\mathcal{F}_s^{n+1}$ and $E_{\varphi 2 \psi 2}$ lies in $\mathcal{F}_{s+1}^{n+1}$. This yields a Bratteli diagram for the sequence $\mathfrak{A}_n^0$ which looks like Pascal's triangle:

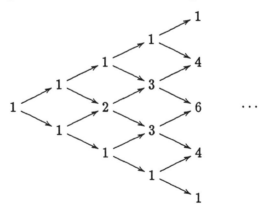

Next we show that the map $\lambda \to \chi_\lambda(A)$ is continuous for each A in $\mathfrak{A}$. This is evident for the matrix units $E_{\varphi\psi}$ from the explicit formula established above. Since each $\mathfrak{A}_n$ is finite dimensional, we see that $\lambda \to \chi_\lambda(A)$ is continuous for A

III.6. Extensions

in $\mathfrak{A}_n$. Given A in $\mathfrak{A}$ and $\varepsilon > 0$, choose an approximant B in $\mathfrak{A}_n$ for some n so that $\|A - B\| < \varepsilon$. Then choose a $\delta > 0$ so that $|\lambda - \lambda'| < \delta$ implies that $\|\chi_\lambda(B) - \chi_{\lambda'}(B)\| < \varepsilon$. Then for $|\lambda - \lambda'| < \delta$,

$$\|\chi_\lambda(A) - \chi_{\lambda'}(A)\|$$
$$\leq \|\chi_\lambda(A) - \chi_\lambda(B)\| + \|\chi_\lambda(B) - \chi_{\lambda'}(B)\| + \|\chi_{\lambda'}(B) - \chi_{\lambda'}(A)\|$$
$$< \varepsilon + \varepsilon + \varepsilon = 3\varepsilon.$$

This establishes continuity at A.

We wish to establish that $\mathfrak{A}^0 = \overline{\cup_{n \geq 1} \mathfrak{A}_n^0}$. Let A belong to $\mathfrak{A}^0$ and let $\varepsilon > 0$. Choose an element B in $\mathfrak{A}_n$ for some n so that $\|A - B\| < \varepsilon$. Then define

$$B_0 := \int_{\mathbb{T}} \chi_\lambda(B)\, d\lambda$$

where $d\lambda$ is normalized Lebesgue measure on the circle. This integral makes sense in the Riemann sense because the integrand is norm continuous. I claim that B_0 lies in $\mathfrak{A}_n^0$. It lies in $\mathfrak{A}_n$ because $\chi_\lambda(\mathfrak{A}_n) = \mathfrak{A}_n$. But it is also gauge invariant because

$$\chi_{\lambda'}(B_0) = \int_{\mathbb{T}} \chi_{\lambda'} \chi_\lambda(B)\, d\lambda = \int_{\mathbb{T}} \chi_{\lambda + \lambda'}(B)\, d\lambda$$
$$= \int_{\mathbb{T}} \chi_\mu(B)\, d\mu = B_0$$

Moreover this is a good approximant to A because

$$\|A - B_0\| = \left\|\int_{\mathbb{T}} \chi_\lambda(A - B)\, d\lambda\right\|$$
$$\leq \int_{\mathbb{T}} \|\chi_\lambda(A - B)\|\, d\lambda = \|A - B\| < \varepsilon$$

Hence $\cup_{n \geq 1} \mathfrak{A}_n^0$ is dense in $\mathfrak{A}^0$.

This algebra has a rich ideal structure. The reader can verify that the ideals correspond to the directed sets generated by any single vertex of the diagram.

III.6 Extensions

A C*-algebra $\mathfrak{E}$ is an **extension** of a C*-algebra $\mathfrak{A}$ by a C*-algebra $\mathfrak{B}$ if there is a monomorphism j of $\mathfrak{A}$ onto an ideal of $\mathfrak{E}$ such that $\mathfrak{E}/j(\mathfrak{A}) \simeq \mathfrak{B}$. In other words, there is a short exact sequence

$$0 \longrightarrow \mathfrak{A} \xrightarrow{j} \mathfrak{E} \xrightarrow{\tau} \mathfrak{B} \longrightarrow 0.$$

This is the inverse operation of taking quotients. However, there is in general more than one extension of $\mathfrak{A}$ by $\mathfrak{B}$. Later, we will see some examples of how such extensions are classified. In this section, we will show that extensions of AF algebras by AF algebras are AF.

The key device is to lift a projection in the quotient algebra to a projection in $\mathfrak{A}$. After that, the result follows using straightforward arguments.

Lemma III.6.1 *Suppose that $\mathfrak{J}$ is an AF ideal of a C*-algebra $\mathfrak{A}$. Then for each projection p in $\mathfrak{A}/\mathfrak{J}$, there is a projection P in $\mathfrak{A}$ such that $P + \mathfrak{J} = p$.*

Proof. Let the quotient map be denoted by τ. First lift p to a positive contraction in $\mathfrak{A}$. Indeed, if $\tau(X) = p$, then $H = f(\operatorname{Re} X)$ suffices, where $f(x) = x \wedge 1 \vee 0$. So

$$\tau(H) = f(\tau(\operatorname{Re} X)) = f(\operatorname{Re} \tau(X)) = f(p) = p.$$

Then $\tau(e^{2\pi i H}) = e^{2\pi i p} = \dot{I}$ because $\sigma(p) = \{0, 1\}$; and hence $W = e^{2\pi i H}$ is a unitary element in $\mathfrak{J} + \mathbb{C}I$.

Since $\mathfrak{J}$ is AF, we may write it as an increasing union of finite dimensional subalgebras $\mathfrak{J}_n$. For any $\delta > 0$, there is an integer n and a unitary U in $\mathfrak{J}_n + \mathbb{C}I$ such that $\|W - U\| < \delta$. As $\tau(U) = \alpha I$ for some scalar α, we have

$$|\alpha - 1| = \|\tau(U - W)\| < \delta.$$

Replace U by $V = \alpha^{-1}U$. Then $\tau(V) = \dot{I}$ and $\|W - V\| < 2\delta$.

Since V lies in a finite dimensional algebra, it has finite spectrum. Thus there is a positive operator A in $\mathfrak{J}_n + \mathbb{C}I$ such that $V = e^{2\pi i A}$ and $0 \le A < I$. Because

$$\dot{I} = \tau(V) = e^{2\pi i \tau(A)},$$

it follows that $\tau(A)$ is a projection. And as $\|A\| < 1$, this means that $\tau(A) = 0$; so that A belongs to $\mathfrak{J}_n$.

Since the spectrum of A is finite, $E := \chi_{(\frac{1}{3}, \frac{2}{3})}(A)$ is a projection in $\mathfrak{J}_n$. Define $K = (I - E)H(I - E)$. We claim that for δ sufficiently small, $\frac{1}{2}$ is not in $\sigma(K)$. Once this claim is established, we may set $P = \chi_{(\frac{1}{2}, 1]}(K)$. This is a projection in $\mathfrak{A}$. Moreover, since E belongs to $\mathfrak{J}$,

$$\tau(P) = \chi_{(\frac{1}{2}, 1]}(\tau(K)) = \chi_{(\frac{1}{2}, 1]}(\tau(H)) = p.$$

Consider the functions $f(e^{2\pi i t}) = t - t^2$ and $g(e^{2\pi i t}) = t^2 - t^3$ for $0 \le t \le 1$, which are defined and continuous on the unit circle $\mathbb{T}$. By continuity, for each $\varepsilon > 0$, there is a $\delta > 0$ so that if U and V are unitary such that $\|U - V\| < \delta$, then $\|f(U) - f(V)\| < \varepsilon$ and $\|g(U) - g(V)\| < \varepsilon$ (see Exercise II.8). Thus with this choice of δ for any given ε, we obtain

$$\|(H - H^2) - (A - A^2)\| = \|f(W) - f(V)\| < \varepsilon$$

and

$$\|(H^2 - H^3) - (A^2 - A^3)\| = \|g(W) - g(V)\| < \varepsilon.$$

III.6. Extensions

Notice that
$$\sup_{0\leq a\leq \frac{1}{3}} (a-a^2) = \sup_{\frac{2}{3}\leq a\leq 1} (a-a^2) = \tfrac{2}{9}$$
$$= \inf_{\frac{1}{3}\leq a\leq \frac{2}{3}} (a-a^2) < \sup_{\frac{1}{3}\leq a\leq \frac{2}{3}} (a-a^2) = \tfrac{1}{4}.$$

Hence
$$\tfrac{2}{9}E \leq A - A^2 \leq \tfrac{1}{4}E + \tfrac{2}{9}E^\perp.$$

Therefore
$$E^\perp H^2 E^\perp - K^2 = E^\perp H^2 E^\perp - E^\perp H E^\perp H E^\perp$$
$$= E^\perp H E H E^\perp = E^\perp (H-A) E (H-A) E^\perp$$
$$\leq \tfrac{9}{2} E^\perp (H-A)(A-A^2)(H-A) E^\perp.$$

Thus
$$\|E^\perp H^2 E^\perp - K^2\| \leq \tfrac{9}{2}(\|H(A-A^2) - (A^2-A^3)\|)\|H-A\|$$
$$\leq \tfrac{9}{2}\left(\|H\|\|(A-A^2)-(H-H^2)\| + \|(H^2-H^3)-(A^2-A^3)\|\right)\|H-A\|$$
$$< \tfrac{9}{2}(2\varepsilon) = 9\varepsilon.$$

We used the fact that $\|H-A\| \leq 1$ because both are positive contractions. Finally we compute
$$K - K^2 = E^\perp H E^\perp - K^2 = E^\perp (H-H^2) E^\perp + E^\perp H^2 E^\perp - K^2.$$

Thus
$$\|K - K^2\| \leq \|E^\perp \left((H-H^2) - (A-A^2)\right) E^\perp\|$$
$$+ \|E^\perp (A-A^2) E^\perp\| + \|E^\perp H^2 E^\perp - K^2\|$$
$$< \varepsilon + \tfrac{2}{9} + 9\varepsilon < \tfrac{1}{4}$$

provided that $\varepsilon < \tfrac{1}{360}$. Since $\tfrac{1}{2} - \left(\tfrac{1}{2}\right)^2 = \tfrac{1}{4}$, this implies that $\tfrac{1}{2}$ does not belong to $\sigma(K)$, completing the proof. ∎

The next lemma manipulates matrix units much like we did earlier in this chapter.

Lemma III.6.2 *Suppose that $\mathfrak{J}$ is an AF ideal of a C*-algebra $\mathfrak{A}$ and that $\mathfrak{B}$ is a finite dimensional subalgebra of $\mathfrak{A}/\mathfrak{J}$. Then there is a (not necessarily unital) *-monomorphism ρ of $\mathfrak{B}$ into $\mathfrak{A}$ such that $\tau\rho = \mathrm{id}_\mathfrak{B}$.*

Proof. Use Lemma III.6.1 to lift the minimal central projections of $\mathfrak{B}$ to pairwise orthogonal projections in $\mathfrak{A}$. To do this, suppose that the minimal central projections are $p_1, \ldots, p_k$. First lift p_1 to a projection P_1. Then consider the ideal $P_1^\perp \mathfrak{J} P_1^\perp$ in $P_1^\perp \mathfrak{A} P_1^\perp$. This is also an AF ideal, so the procedure may be repeated. This reduces the problem to lifting $\mathcal{M}_n$ from $p_i \mathfrak{B} p_i$ to $P_i \mathfrak{A} P_i$.

Let e_{ij} for $1 \leq i,j \leq n$ be a set of matrix units for $\mathcal{M}_n$. As above, find pairwise orthogonal projections $E_1, \ldots, E_n$ such that $\tau(E_j) = e_{jj}$ for $1 \leq j \leq n$. For each j, let V_j be chosen so that $\tau(V_j) = e_{j1}$. Then define $Q_j = E_1 V_j^* E_j V_j E_1$. Since $\tau(Q_j) = e_{11}$, we see that $Q_j = E_1 - J_j$ where J_j is a positive contraction in $\mathfrak{J}$.

By Exercise III.1, there is an increasing sequence of projections which form an approximate identity for $E_1 \mathfrak{J} E_1$. Choose a projection F in $E_1 \mathfrak{J} E_1$ such that

$$\|(E_1 - F) J_j (E_1 - F)\| < \tfrac{1}{3} \quad \text{for} \quad 1 \leq j \leq n.$$

By replacing E_1 with $E_1' = E_1 - F$ in the definition of Q_j above, we obtain a perturbation $Q_j' = E_1' - J_j'$ with $\|J_j'\| < \tfrac{1}{3}$. So the polar decomposition yields the factorizations $E_j V_j E_1' = E_{j1} Q_j'^{1/2}$ where E_{j1} is a partial isometry with initial space E_1' and $\tau(E_{j1}) = e_{j1}$. Thus the range space is $E_j' := E_{j1} E_{j1}^* \leq E_j$ still satisfies $\tau(E_j') = e_{jj}$.

Set $E_{ij} := E_{i1} E_{j1}^*$. It is readily verified that this forms a set of matrix units in $\mathfrak{A}$ which maps onto e_{ij} in the quotient. ∎

Now we have the necessary ingredients to complete the proof of the main result.

Theorem III.6.3 *Suppose that*

$$0 \longrightarrow \mathfrak{J} \xrightarrow{j} \mathfrak{A} \xrightarrow{\tau} \mathfrak{B} \longrightarrow 0$$

is an exact sequence of C*-algebras and that $\mathfrak{J}$ and $\mathfrak{B}$ are AF. Then $\mathfrak{A}$ is AF.

Proof. Using Theorem III.3.4, we take $A_1, \ldots, A_n$ in $\mathfrak{A}$ and $\varepsilon > 0$ and try to find a finite dimensional subalgebra that almost contains them. Since $\tau(A_1), \ldots, \tau(A_n)$ lie in the AF algebra $\mathfrak{B}$, there is a finite dimensional subalgebra $\mathfrak{B}_0$ of $\mathfrak{B}$ and elements b_i in $\mathfrak{B}_0$ such that $\|\tau(A_i) - b_i\| < \varepsilon/3$ for $1 \leq i \leq n$.

Lemma III.6.2 provides a $*$-monomorphism ρ of $\mathfrak{B}_0$ onto a subalgebra $\mathfrak{A}_0$ of $\mathfrak{A}$ such that $\tau\rho = \mathrm{id}_{\mathfrak{B}_0}$. Choose J_i in $\mathfrak{J}$ so that $\|A_i - \rho(b_i) - J_i\| < \varepsilon/3$. Then it suffices to approximate $B_i := \rho(b_i) + J_i$ within $2\varepsilon/3$ by elements of a finite dimensional subalgebra. Note that the B_i lie in $\mathfrak{A}_0 + \mathfrak{J}$.

The proof is completed by constructing an approximate identity of projections for $\mathfrak{J}$ which commutes with $\mathfrak{A}_0$. To this end, let $E_{ij}^{(s)}$ for $1 \leq s \leq k$, $1 \leq i,j \leq n_s$ be a set of matrix units for $\mathfrak{A}_0$; and let P be the unit of $\mathfrak{A}_0$. For each s, $E_{11}^{(s)} \mathfrak{J} E_{11}^{(s)}$ is AF, and thus has a sequential approximate identity of projections, say $F_n^{(s)}$. Similarly, $P^\perp \mathfrak{J} P^\perp$ has an approximate identity $F_n^{(0)}$ of projections. Define

$$F_n = F_n^{(0)} + \sum_{s=1}^{k} \sum_{i=1}^{n_s} E_{i1}^{(s)} F_n^{(s)} E_{1i}^{(s)}.$$

This is readily seen to be an approximate identity for $\mathfrak{J}$. In addition,

$$\begin{aligned}
E_{ij}^{(s)} F_n &= E_{ij}^{(s)} E_{j1}^{(s)} F_n^{(s)} E_{1j}^{(s)} \\
&= E_{i1}^{(s)} F_n^{(s)} E_{1j}^{(s)} \\
&= E_{i1}^{(s)} F_n^{(s)} E_{1i}^{(s)} E_{ij}^{(s)} = F_n E_{ij}^{(s)}.
\end{aligned}$$

Hence each F_n commutes with $\mathfrak{A}_0$.

For each A in $\mathfrak{A}_0$ and J in $\mathfrak{J}$, we have

$$\lim_{n\to\infty} F_n(A+J)F_n + F_n^\perp A F_n^\perp = A + \lim_{n\to\infty} F_n J F_n = A + J.$$

So there is an integer n sufficiently large so that

$$\|B_i - (F_n B_i F_n + F_n^\perp \rho(b_i) F_n^\perp)\| < \varepsilon/3.$$

These approximants all lie in the algebra $F_n \mathfrak{J} F_n + F_n^\perp \mathfrak{A}_0 F_n^\perp$. Since this is a direct sum, and $F_n \mathfrak{J} F_n$ is AF and $F_n^\perp \mathfrak{A}_0 F_n^\perp$ is finite dimensional, this algebra is AF. So there are elements C_i lying in a finite dimensional subalgebra such that

$$\|F_n B_i F_n + F_n^\perp \rho(b_i) F_n^\perp - C_i\| < \varepsilon/3.$$

This is the desired approximating set. So $\mathfrak{A}$ is AF. ∎

Exercises

III.1 If $\mathfrak{A}$ is AF, show that it has an approximate identity consisting of an increasing sequence of projections.
HINT: Use the units of the subalgebras $\mathfrak{A}_n$.

III.2 Show that if P is a projection in an AF algebra $\mathfrak{A}$, then $P\mathfrak{A}P$ is AF.
NOTE: This result was used several times in the last section.

III.3 Show that a direct limit of finite dimensional C*-algebras can be replaced by a direct limit with injective connecting maps with the same limit.

III.4 Suppose that $\mathfrak{A}$ is a C*-subalgebra of a finite dimensional C*-algebra $\mathfrak{B}$. If a system of matrix units for $\mathfrak{A}$ is given, show that there is a system of matrix units for $\mathfrak{B}$ such that each matrix unit of $\mathfrak{A}$ is a sum of matrix units of $\mathfrak{B}$.

III.5 Let $\mathfrak{A}$ be a UHF algebra with unique trace τ, and let π_τ be the representation of $\mathfrak{A}$ associated to τ by the GNS construction. Prove that τ extends to a faithful WOT-continuous trace on $\pi_\tau(\mathfrak{A})''$.

III.6 Show that a C*-subalgebra of an AF algebra need not be AF.
HINT: Show that $C[0,1]$ can be imbedded as a subalgebra of $C(X)$ where X is the Cantor set.

III.7 Show that $\Phi(A) = \int_{\mathbb{T}} \chi_\lambda(A) \, d\lambda$ is a contractive projection of the CAR algebra onto the GICAR algebra. Moreover, show that Φ is positive and faithful.

III.8 Find all the ideals of the GICAR algebra.

III.9 A **derivation** of a C*-algebra $\mathfrak{A}$ is a (bounded) linear map δ of $\mathfrak{A}$ into itself such that
$$\delta(AB) = \delta(A)B + A\delta(B).$$
A derivation δ is called **inner** if there is an element X in $\mathfrak{A}$ such that $\delta(A) = \delta_X(A) = XA - AX$. It is **approximately inner** if it is the pointwise limit of inner derivations.

(a) Show that every derivation of a finite dimensional C*-algebra is inner.
hint: Choose a finite group $\mathcal{G}$ of unitaries which span $\mathfrak{A}$. Define
$$X = |\mathcal{G}|^{-1} \sum_{U \in \mathcal{G}} \delta(U)U.$$

(b) Show that every derivation of an AF algebra is approximately inner.
HINT: Show that $\|X\| \leq \|\delta\|$ in part (a).

(c) Show that every derivation of the compact operators has the form δ_X for some X in $\mathcal{B}(\mathcal{H})$.

III.10 A derivation is self-adjoint or a $*$-derivation if $\delta(A^*) = \delta(A)^*$. In this case, show that $\alpha_t = e^{t\delta} := \sum_{n \geq 0} \frac{1}{n!}(t\delta)^n$ is a $*$-automorphism for all real values t.
HINT: Show that $\delta^n(AB) = \sum_{k=0}^{n} \binom{n}{k} \delta^k(A) \delta^{n-k}(B)$.

Notes and Remarks.

Almost all of the material on AF algebras in this chapter is taken from the seminal paper of Bratteli [1972]. The exceptions are the classification of UHF algebras which is taken from Glimm [1960]; and the last section. The original proof of the result on extensions due to Brown [1981] used K-theory. The proof given here is due to Choi [1983].

CHAPTER IV

K-theory for AF C*-algebras

K_0 is a functor that assigns an ordered abelian group to each ring based on the structure of idempotents in the matrix algebras over the ring. It turns out to be a useful and frequently computable algebraic invariant. For C*-algebras, it often carries a lot of important information. In particular, it turns out to be a complete invariant for AF algebras. Moreover, there is a complete description of all possible K_0 groups of AF algebras. This has proven to be a useful tool for constructing C*-algebras with particular properties.

In a ring $\mathcal{R}$, say that two idempotents P and Q are (von Neumann) **equivalent** if there are elements X, Y in $\mathcal{R}$ such that

$$P = XY \quad \text{and} \quad Q = YX.$$

In a C*-algebra $\mathfrak{A}$, two projections P and Q are $*$-**equivalent** if there is an element X in $\mathfrak{A}$ (necessarily a partial isometry) such that

$$P = X^*X \quad \text{and} \quad Q = XX^*.$$

It is evident that these are equivalence relations. In a C*-algebra, it is more natural and more convenient to consider only projections rather than arbitrary idempotents. The first easy proposition shows that this choice does not affect things. Moreover, since equivalence and $*$-equivalence are the same relation on projections, we will use the term equivalence only after this proposition.

Proposition IV.1.1 *In a C*-algebra, every idempotent is equivalent to a projection; and equivalent projections are $*$-equivalent.*

Proof. Suppose that $E = E^2$ is an idempotent in a C*-algebra $\mathfrak{A}$. Set

$$X = I + (E - E^*)^*(E - E^*) = I - E - E^* + EE^* + E^*E.$$

This is a strictly positive (hence invertible) element of $\mathfrak{A}^\sim$. A calculation yields $EX = EE^*E = XE$. Thus E and E^* both commute with X^{-1}. Therefore, we may define a self-adjoint element $P = EE^*X^{-1}$. Furthermore,

$$P^2 = X^{-1}(EE^*E)E^*X^{-1} = X^{-1}XEE^*X^{-1} = P.$$

So P is a projection; and it satisfies $EP = P$ and $PE = X^{-1}EE^*E = E$. Thus E and P are equivalent.

Now suppose that P and Q are equivalent projections, and let X and Y be given so that $P = XY$ and $Q = YX$. Let $X_0 = PXQ$ and $Y_0 = QYP$. Then

$$X_0 Y_0 = PXQYP = P(XY)^2 P = P$$

and

$$Y_0 X_0 = QYPXQ = Q(YX)^2 Q = Q.$$

Notice that since the range of X_0 equals the range of P and the range of X_0^* equals the range of Q, the partial isometry U in the polar decomposition $X_0 = U|X_0|$ satisfies $UU^* = P$ and $U^*U = Q$. It must be shown that U belongs to $\mathfrak{A}$. Now

$$P = P^*P = X_0^* Y_0^* Y_0 X_0 \leq \|Y_0\|^2 X_0^* X_0 \leq \|Y_0\|^2 \|X_0\|^2 P.$$

Thus $\|Y_0\|^{-2} P \leq X_0^* X_0 \leq \|X_0\|^2 P$. So $\sigma(X_0^* X_0)$ is contained in the interval $\{0\} \cup [\|Y_0\|^{-2}, \|X_0\|^2]$; and $\chi(X_0^* X_0) = P$, where χ is the function such that $\chi(0) = 0$ and $\chi(x) = 1$ on $\sigma(X_0^* X_0) \setminus \{0\}$. Therefore $U = X_0 g(X_0^* X_0)$ where g is the continuous function on $\sigma(X_0^* X_0)$ such that $g(0) = 0$ and $g(x) = x^{-1/2}$ for $x \geq \|Y_0\|^{-2}$. In particular, U belongs to $\mathfrak{A}$ and thus P and Q are $*$-equivalent projections in $\mathfrak{A}$. ∎

The first clue that K_0 has a topological nature comes from the fact that it is a **homotopy invariant**. Equivalence satisfies homotopy invariance if whenever $p(t) = P_t$, for $0 \leq t \leq 1$, is a continuous path of projections in $\mathfrak{A}$, then P_0 and P_1 are equivalent. This will then imply the homotopy invariance of K_0.

Proposition IV.1.2 *If P and Q are projections such that $\|P - Q\| < 1$, then they are equivalent. Thus equivalence is a homotopy invariant.*

Proof. For the first part, suppose that $\|P - Q\| = 1 - \varepsilon < 1$ and consider the operator $X = PQ$. Then

$$P \geq XX^* = PQP = P - P(P - Q)P \geq \varepsilon P.$$

Similarly, $\varepsilon Q \leq X^*X \leq Q$. As in the previous proposition, the partial isometry U in the polar decomposition of X belongs to $\mathfrak{A}$ and yields $UU^* = P$ and $U^*U = Q$.

If $p(t) = P_t$ is a continuous path of projections, continuity implies that there are real numbers $0 = t_0 < t_1 < \cdots < t_n = 1$ so that $\|P_{t_i} - P_{t_{i-1}}\| < 1/2$ for $1 \leq i \leq n$. Hence by the first part of this proposition, each $P_{t_{i-1}}$ is equivalent to P_{t_i}. Thus P_0 and P_1 are equivalent. ∎

Corollary IV.1.3 *If $\mathfrak{A} = \overline{\cup_{n \geq 1} \mathfrak{A}_n}$ is an AF algebra, then every projection in $\mathfrak{A}$ is equivalent to a projection in $\cup_{n \geq 1} \mathfrak{A}_n$. Moreover, if P and Q in $\cup_{n \geq 1} \mathfrak{A}_n$ are equivalent in $\mathfrak{A}$, they are equivalent in $\cup_{n \geq 1} \mathfrak{A}_n$.*

Proof. By the density of $\cup_{n \geq 1} \mathfrak{A}_n$ in $\mathfrak{A}$, each projection P in $\mathfrak{A}$ is close to $\mathfrak{A}_n$ for n sufficiently large. So Lemma III.3.1 shows that there is a projection Q in $\mathfrak{A}_n$ with $\|P - Q\| < 1/2$. Therefore P and Q are equivalent by Proposition IV.1.2.

IV. K-theory for AF C*-algebras

If P and Q in $\cup_{n\geq 1}\mathfrak{A}_n$ are equivalent in $\mathfrak{A}$ by a partial isometry U, choose an element $X = PXQ$ in $\mathfrak{A}_n$ for some n sufficiently large so that $\|U - X\| < 1/2$. As above, the partial isometry V in the polar decomposition of X belongs to $\mathfrak{A}_n$ and implements the equivalence. ∎

Now we extend our definitions to deal with matrix algebras.

Let $\mathcal{P}(\mathfrak{A})$ denote the collection of all projections in $\cup_{n\geq 1}\mathcal{M}_n(\mathfrak{A})$. This is a semigroup under the operation of direct sum. Say that two projections in $\mathcal{P}(\mathfrak{A})$, say P in $\mathcal{M}_m(\mathfrak{A})$ and Q in $\mathcal{M}_n(\mathfrak{A})$ with $m \leq n$, are **equivalent** (write $P \sim Q$) if $P \oplus 0_{n-m}$ is equivalent to Q in $\mathcal{M}_n(\mathfrak{A})$. Say that P and Q are **stably equivalent** (write $P \approx Q$) if there is a projection R in $\mathcal{P}(\mathfrak{A})$ so that $P \oplus R \sim Q \oplus R$. Stable equivalence is evidently an equivalence relation. Let $K_0^+(\mathfrak{A})$ denote the collection of stable equivalence classes, denoted by $[P]$ for P in $\mathcal{P}(\mathfrak{A})$.

We need the following routine results about $K_0(\mathfrak{A})^+$. In particular, part (a) shows that direct sum induces a well defined operation on $K_0(\mathfrak{A})^+$.

Lemma IV.1.4 *Let P, P', Q, Q' and R be projections in $\mathcal{P}(\mathfrak{A})$.*
  (a) *If $P \approx P'$ and $Q \approx Q'$, then $P \oplus Q \approx P' \oplus Q'$.*
  (b) $P \oplus Q \sim Q \oplus P$.
  (c) *If $PQ = 0$ in $\mathcal{M}_n(\mathfrak{A})$, then $P + Q \sim P \oplus Q$.*
  (d) *If $P \oplus R \approx Q \oplus R$, then $P \approx Q$.*

Proof. All parts are easy. We will prove (c) as an example. Suppose that P, Q are mutually orthogonal projections in $\mathcal{M}_n(\mathfrak{A})$. Let $X = \begin{bmatrix} P & 0 \\ Q & 0 \end{bmatrix}$. Then

$$X^*X = \begin{bmatrix} P+Q & 0 \\ 0 & 0 \end{bmatrix} \quad \text{and} \quad XX^* = \begin{bmatrix} P & 0 \\ 0 & Q \end{bmatrix}. \qquad \blacksquare$$

Corollary IV.1.5 $K_0^+(\mathfrak{A})$ *is an abelian cancellation semigroup with the operation $[P] + [Q] := [P \oplus Q]$ and with zero element $[0]$.*

Proof. By Lemma IV.1.4 (a), the sum operation on $K_0^+(\mathfrak{A})$ is well defined. By part (b), it is commutative. By (d), $[P] + [R] = [Q] + [R]$ implies that $[P] = [Q]$, so $K_0^+(\mathfrak{A})$ has the cancellation property. Associativity follows because direct sum is an associative operation. Finally, it is clear that $[0]$ is a zero element. ∎

Say that a C*-algebra $\mathfrak{A}$ satisfies **cancellation** if $\sim$ and $\approx$ are the same equivalence relation on $\mathcal{P}(\mathfrak{A})$. This isn't always the case, but it simplifies things when it is true.

Theorem IV.1.6 *If $\mathfrak{A}$ is an AF algebra, then $\mathfrak{A}$ satisfies cancellation, and $K_0^+(\mathfrak{A})$ is generated by the projections in $\mathfrak{A}$.*

Proof. Suppose that $\mathfrak{A} = \overline{\cup_{n\geq 1}\mathfrak{A}_n}$. By Corollary IV.1.3, every projection in $\mathcal{M}_p(\mathfrak{A})$ is equivalent to a projection P in $\mathcal{M}_p(\mathfrak{A}_n)$ for some n sufficiently large. Suppose that $\mathfrak{A}_n \simeq \mathcal{M}_{k_1} \oplus \cdots \oplus \mathcal{M}_{k_s}$. Then $\mathcal{M}_p(\mathfrak{A}_n) \simeq \mathcal{M}_{pk_1} \oplus \cdots \oplus \mathcal{M}_{pk_s}$. So we may write $P \simeq P_1 \oplus \cdots \oplus P_s$, and set $\pi_i = \text{rank}(P_i)$.

Since two projections in $\mathcal{M}_n$ are equivalent if and only if they have the same rank, two projections $P \simeq P_1 \oplus \cdots \oplus P_s$ and $Q \simeq Q_1 \oplus \cdots \oplus Q_s$ in $\mathfrak{A}_n$ are equivalent in $\mathfrak{A}_n$ exactly when $\operatorname{rank}(P_i) = \operatorname{rank}(Q_i)$ for $1 \le i \le s$.

Suppose that $P \approx Q$ in $\mathfrak{A}$, and let R be chosen so that $P \oplus R \sim Q \oplus R$. By Corollary IV.1.3, $P \oplus R$ and $Q \oplus R$ are equivalent in $\mathfrak{A}_m$ for some sufficiently large m. We may suppose without loss of generality that P and Q lie in $\mathcal{M}_p(\mathfrak{A}_m)$ and R lies in $\mathcal{M}_q(\mathfrak{A}_m)$. Using the notation above, write

$$P \simeq P_1 \oplus \cdots \oplus P_s, \quad Q \simeq Q_1 \oplus \cdots \oplus Q_s, \quad \text{and} \quad R \simeq R_1 \oplus \cdots \oplus R_s.$$

Then in $\mathfrak{A}_m$, $P \oplus R \sim Q \oplus R$ if and only if $\operatorname{rank}(P_i \oplus R_i) = \operatorname{rank}(Q_i \oplus R_i)$, which occurs if and only if $\operatorname{rank}(P_i) = \operatorname{rank}(Q_i)$ which is equivalent to $P \sim Q$.

Let $E_{11}^{(i)}$ denote the rank 1 projection in the ith summand of $\mathfrak{A}_m$. Then

$$[P] = \sum_{i=1}^{s} \pi_i [E_{11}^{(i)}].$$

So $K_0^+(\mathfrak{A})$ is generated by the projections in $\mathfrak{A}$. ∎

IV.2 K_0

An abelian cancellation semigroup S generates an abelian group G, called the **Grothendieck group** of S, consisting of all 'differences' of elements of S modulo the natural equivalence. That is, consider the collection of all formal differences $s - t$ for s, t in S and identify $s_1 - t_1$ with $s_2 - t_2$ if and only if $s_1 + t_2 = s_2 + t_1$ in S. The cancellation property is exactly what is needed to show that this is an equivalence relation. Indeed, suppose that $s_1 - t_1 = s_2 - t_2$ and $s_2 - t_2 = s_3 - t_3$. By definition this means that

$$s_1 + t_2 = s_2 + t_1 \quad \text{and} \quad s_2 + t_3 = s_3 + t_2.$$

Adding these two equations and using commutativity, we obtain

$$s_1 + t_3 + (s_2 + t_2) = s_3 + t_1 + (s_2 + t_2).$$

So by cancellation, $s_1 + t_3 = s_3 + t_1$ and thus $s_1 - t_1 = s_3 - t_3$.

Define addition by the rule $(s_1 - t_1) + (s_2 - t_2) = (s_1 + s_2) - (t_1 + t_2)$. It is routine to verify that addition is well defined, commutative and associative. The inverse of $s - t$ is $t - s$, and the zero is $0 - 0$. So G is an abelian group.

There is a natural imbedding of S into G given by the map taking s to $s - 0$. This is injective because of the cancellation argument above. So with this identification, we have $G = S - S$. Under good conditions, S determines a cone in G which yields a useful partial order.

If $\mathfrak{A}$ is a C*-algebra, define $K_0(\mathfrak{A})$ to be the Grothendieck group of $K_0^+(\mathfrak{A})$.

Example IV.2.1 For the moment, we content ourselves with a very elementary example. Let $\mathfrak{A}$ be a finite dimensional C*-algebra, say that $\mathfrak{A} \simeq \mathcal{M}_{n_1} \oplus \cdots \oplus \mathcal{M}_{n_k}$. As we have seen, two projections $P = P_1 \oplus \cdots \oplus P_k$ and $Q = Q_1 \oplus \cdots \oplus Q_k$

IV.2. K_0

in matrix algebras over $\mathfrak{A}$ are equivalent if and only if $\text{rank}(P_i) = \text{rank}(Q_i)$ for all $1 \leq i \leq k$. As all ranks are possible, $K_0^+(\mathfrak{A})$ equals $\mathbb{N}_0^k$, the semigroup of non-negative integer k-tuples. Hence that the enveloping group is $K_0(\mathfrak{A}) = \mathbb{Z}^k$.

Define a relation on $K_0(\mathfrak{A})$ by setting $x \leq y$ if $y - x \in K_0^+(\mathfrak{A})$. Say that u is an **order unit** of $K_0(\mathfrak{A})$ if for every x in $K_0(\mathfrak{A})$ there is a positive integer n such that $-nu \leq x \leq nu$.

The following easy proposition establishes that $\leq$ is a preorder (a reflexive and transitive relation that is not necessarily symmetric). Then we establish a sufficient condition to imply that it is a partial order.

Proposition IV.2.2 *Let $\mathfrak{A}$ be a C*-algebra. Then*

(i) $\leq$ *is a preorder on $K_0(\mathfrak{A})$.*
(ii) *if $x \leq y$, then $x + z \leq y + z$ for all $z \in K_0(\mathfrak{A})$.*
(iii) *if $\mathfrak{A}$ is unital, then $[I]$ is an order unit for $K_0(\mathfrak{A})$.*

Proof. Since 0 belongs to $K_0^+(\mathfrak{A})$, $x \leq x$ for all x in $K_0(\mathfrak{A})$. Also since $K_0^+(\mathfrak{A})$ contains $K_0^+(\mathfrak{A}) + K_0^+(\mathfrak{A})$, we see that $x \leq y$ and $y \leq z$ implies that $x \leq z$. So $\leq$ is a preorder.

Part (ii) is immediate from the definition.

When $\mathfrak{A}$ is unital, any projection P in $\mathcal{M}_n(\mathfrak{A})$ satisfies $0 \leq P \leq I_n$. Hence $0 \leq [P] \leq n[I]$. So if Q is a projection in $\mathcal{M}_m(\mathfrak{A})$, we obtain

$$-m[I] \leq [P] - [Q] \leq n[I]. \qquad \blacksquare$$

A subset S of an abelian group G is a **cone** if $S \cap (-S) = \{0\}$ and $G = S - S$. In this case, the preorder $\leq$ defined on G by S is a partial order. For if $x \leq y \leq x$, then $x - y$ belongs to both S and $-S$, whence $x = y$. The property $G = S - S$ implies, among other things, that given x and y in G, there is an element larger than both of them. To see this, write $x = a - b$ and $y = c - d$ for elements a, b, c, d in S. Then both x and y are dominated by $a + c$. Note that this does not mean that there is necessarily a least element greater than x and y.

A C*-algebra $\mathfrak{A}$ is called **finite** if whenever P and Q are equivalent projections in $\mathfrak{A}$ such that $P \leq Q$, then $P = Q$. The algebra is **stably finite** if $\mathcal{M}_n(\mathfrak{A})$ is finite for all $n \geq 1$. Since equivalence in a matrix algebra $\mathcal{M}_n$ is determined by rank, it is evident that finite dimensional C*-algebras are stably finite.

When $\mathfrak{A}$ is unital, finiteness is equivalent to the statement that $\mathfrak{A}$ contains no proper isometries. Indeed, if S is an isometry in $\mathfrak{A}$, then $SS^* \leq S^*S = I$. So finiteness implies that $SS^* = I$; whence S is unitary. Conversely, suppose that $\mathfrak{A}$ is a unital C*-algebra with no proper isometries, and let P and Q be equivalent projections in $\mathfrak{A}$ such that $P \leq Q$. If X were a partial isometry in $\mathfrak{A}$ such that $P = XX^*$ and $Q = X^*X$, then $S = X + Q^\perp$ would be an isometry in $\mathfrak{A}$ with range projection $P + Q^\perp$. Since X must be unitary, it follows that $P = Q$.

In fact, the term finite comes from the well known property of matrix algebras that left invertibility implies invertibility (see Exercise IV.6).

Theorem IV.2.3 *AF algebras are stably finite.*

Proof. Let $\mathfrak{A} = \overline{\cup_{n\geq 1}\mathfrak{A}_n}$. Since $\mathcal{M}_k(\mathfrak{A}) = \overline{\cup_{n\geq 1}\mathcal{M}_k(\mathfrak{A}_n)}$, it follows that $\mathcal{M}_k(\mathfrak{A})$ is also AF. So it suffices to show that AF algebras are finite.

Suppose that P and Q are equivalent projections in $\mathfrak{A}$ with $P \leq Q$. Use Lemma III.3.1 to obtain projections P' and Q' in some $\mathfrak{A}_n$ (for n sufficiently large) so that $P' \leq Q'$ and $\|P' - P\| < 1/2$ and $\|Q' - Q\| < 1/2$. By Proposition IV.1.2,
$$[P'] = [P] = [Q] = [Q'].$$
By the finiteness of $\mathfrak{A}_n$, $P' = Q'$. Hence the projection $Q - P$ has norm less than 1, which implies that $P = Q$. ∎

Theorem IV.2.4 *If $\mathfrak{A}$ is stably finite, then $K_0^+(\mathfrak{A})$ is a cone.*

Proof. Suppose that P in $\mathcal{M}_k(\mathfrak{A})$ and Q in $\mathcal{M}_\ell(\mathfrak{A})$ are projections in matrix algebras over $\mathfrak{A}$ such that $[P] = -[Q]$ is an element of $K_0^+(\mathfrak{A}) \cap (-K_0^+(\mathfrak{A}))$. Then
$$[0] = [P] + [Q] = [P \oplus Q].$$
Since $0 \leq P \oplus Q$ in $\mathcal{M}_{k+\ell}(\mathfrak{A})$, finiteness shows that $P \oplus Q = 0$; and therefore $[P] = [Q] = [0]$. The requirement that $K_0(\mathfrak{A}) = K_0^+(\mathfrak{A}) - K_0^+(\mathfrak{A})$ follows from the definition of $K_0(\mathfrak{A})$. ∎

An immediate consequence of our results is the following.

Corollary IV.2.5 *If $\mathfrak{A}$ is stably finite, then $K_0(\mathfrak{A})$ is a partially ordered group. When $\mathfrak{A}$ is unital, it also has an order unit.*

IV.3 Dimension Groups

For the rest of this chapter, we restrict our attention to AF algebras. Call the ordered group $(K_0(\mathfrak{A}), K_0^+(\mathfrak{A}))$ associated to an AF algebra a **dimension group**. Notice that $(K_0(\mathcal{M}_n), K_0^+(\mathcal{M}_n)) = (\mathbb{Z}, \mathbb{Z}_+)$ is the same for all the full matrix algebras $\mathcal{M}_n$, $n \geq 1$. To distinguish between dimensions, we introduce another notion. A **scale** on a dimension group (G, G^+) is a subset S of G^+ which is

- **hereditary**: if $0 \leq g \leq s$ and $s \in S$, then $g \in S$.
- **directed**: if $s_1, s_2 \in S$ then there is an $s \in S$ so that $s_1 \leq s$ and $s_2 \leq s$.
- **generating**: every $g \in G^+$ is the sum of finitely many elements of S.

For an AF algebra, we define the scale
$$\Gamma(\mathfrak{A}) := \{[P] : P \text{ a projection in } \mathfrak{A}\}.$$
This is a generating set by Theorem IV.1.6. Suppose that Q is a projection in $\mathfrak{A}$ and P is a projection in $\mathcal{M}_n(\mathfrak{A})$ such that $[P] \leq [Q]$. Then there is a partial isometry X in $\mathcal{M}_n(\mathfrak{A})$ such that $XX^* = P$ and $X^*X \leq Q \oplus O_{n-1}$. Therefore there is a projection P' in $\mathfrak{A}$ such that $X^*X = P' \oplus O_{n-1}$. Hence $[P] = [P']$ belongs to $\Gamma(\mathfrak{A})$. This shows that $\Gamma(\mathfrak{A})$ is hereditary. To verify that $\Gamma(\mathfrak{A})$ is directed, let P and Q be projections in $\mathfrak{A}$. By Corollary IV.1.3, there is an integer n and projections

IV.3. Dimension Groups

P' and Q' in $\mathfrak{A}_n$ equivalent to P and Q respectively. Write $\mathfrak{A}_n = \sum_{i=1}^{k} \oplus \mathcal{M}_{n_i}$, $P = \sum_{i=1}^{k} \oplus P_i$ and $Q = \sum_{i=1}^{k} \oplus Q_i$. By Example IV.2.1, the order in $K_0(\mathfrak{A}_n)$ is determined by the rank of each summand. Choose a projection R_i in $\mathcal{M}_{n_i}$ with

$$\text{rank}(R_i) = \max\{\text{rank}(P_i), \text{rank}(Q_i)\}.$$

Then $R = \sum_{i=1}^{k} \oplus R_i$ is a projection in $\mathfrak{A}_n$ with $[P] \leq [R]$ and $[Q] \leq [R]$. So $\Gamma(\mathfrak{A})$ is directed. This shows that $\Gamma(\mathfrak{A})$ is a scale.

In particular, if $\mathfrak{A}$ is unital, $[I]$ is an order unit for $K_0(\mathfrak{A})$; so

$$\Gamma(\mathfrak{A}) = \{g \in K_0^+(\mathfrak{A}) : 0 \leq g \leq [I]\}.$$

In the non-unital case, there need not be an order unit. For example, the abelian C*-algebra c_0 of all sequences converging to 0 has countably many pairwise orthogonal minimal projections E_i. Together they generate $K_0^+(c_0) \simeq \sum_{i=1}^{\infty} \mathbb{Z}_+$ which is a cone in $K_0(c_0) \simeq \sum_{i=1}^{\infty} \mathbb{Z}$. The scale is

$$\Gamma(c_0) = \{(n_i) : n_i \in \{0,1\}, n_i = 0 \text{ except finitely often}\}.$$

It is easy to see that there is no order unit in this case.

A **positive homomorphism** of a dimension group $(G, G^+, \Gamma(G))$ into another dimension group $(H, H^+, \Gamma(H))$ is a homomorphism ψ of G into H such that $\psi(G^+) \subset H^+$. The map ψ is called **contractive** if $\psi(\Gamma(G)) \subset \Gamma(H)$. Also ψ is said to be **unital** if G and H have order units u_G and u_H determining $\Gamma(G)$ and $\Gamma(H)$ respectively, and $\psi(u_G) = u_H$. A positive unital homomorphism is always contractive.

If φ is a *-homomorphism of a C*-algebra $\mathfrak{A}$ into another C*-algebra $\mathfrak{B}$, then there are *-homomorphisms $\varphi^{(n)}$ of $\mathcal{M}_n(\mathfrak{A})$ into $\mathcal{M}_n(\mathfrak{B})$ obtained by applying φ to each matrix entry. So we may define a natural map φ_* from $K_0(\mathfrak{A})$ into $K_0(\mathfrak{B})$ by

$$\varphi_*([P]) := [\varphi^{(n)}(P)] \quad \text{for all} \quad P \in \mathcal{P}(\mathfrak{A}).$$

In order to show that φ_* is well defined, one must verify that $P \sim Q$ implies $\varphi(P) \sim \varphi(Q)$, and that $P \approx Q$ implies $\varphi(P) \approx \varphi(Q)$. Both of these facts follow trivially from the definitions. Also since

$$\varphi^{(m+n)}(P \oplus Q) = \varphi^{(m)}(P) \oplus \varphi^{(n)}(Q),$$

it follows readily that φ_* is a homomorphism of $K_0^+(\mathfrak{A})$ into $K_0^+(\mathfrak{B})$. This extends uniquely to a positive homomorphism of $K_0(\mathfrak{A})$ into $K_0(\mathfrak{B})$. Moreover, since projections in $\mathfrak{A}$ are taken to projections in $\mathfrak{B}$, φ_* is contractive. When φ is a unital map of unital algebras, φ_* is also unital.

Example IV.3.1 Consider the case of a homomorphism $\varphi : \mathfrak{A} \to \mathfrak{B}$ between finite dimensional C*-algebras. Let $\mathfrak{A} \simeq \mathcal{M}_{n_1} \oplus \cdots \oplus \mathcal{M}_{n_k}$ and $\mathfrak{B} \simeq \mathcal{M}_{m_1} \oplus \cdots \oplus \mathcal{M}_{m_\ell}$.

By Corollary III.2.2, φ is determined up to unitary equivalence in $\mathfrak{B}$ by an $\ell \times k$ matrix $A = [a_{ij}]$ in $\mathcal{M}_{\ell k}(\mathbb{N}_0)$ such that

$$\sum_{i=1}^{k} a_{ij} n_j \leq m_i \quad \text{for} \quad 1 \leq i \leq \ell.$$

In the unital case, these are all equalities.

$K_0(\mathfrak{A})$ is the scaled dimension group $(\mathbb{Z}^k, \mathbb{Z}_+^k, [0, [I_\mathfrak{A}]])$, and $K_0(\mathfrak{B})$ equals $(\mathbb{Z}^\ell, \mathbb{Z}_+^\ell, [0, [I_\mathfrak{B}]])$, where $[I_\mathfrak{A}] = (n_1, \ldots, n_k)^t$ and $[I_\mathfrak{B}] = (m_1, \ldots, m_\ell)^t$. Let E_j denote a minimal projection in the j-th summand of $\mathfrak{A}$. Then $[E_j] = e_j$, the vector with j-th coordinate 1 and all others equal to 0. Now $\varphi(E_j) = F_1 \oplus \cdots \oplus F_\ell$, where F_i is a projection of rank a_{ij} in $\mathcal{M}_{n_i}$. Consequently,

$$[\varphi(E_j)] = (a_{1j}, \ldots, a_{nj})^t = Ae_i.$$

Hence $\varphi_* = A$.

We need to recall some properties of direct limits of groups, suitably enhanced for dimension groups. Suppose that $(G_m, G_m^+, \Gamma(G_m))$ are dimension groups and $\gamma_{mn} : G_m \to G_n$ are contractive homomorphisms for all $m \leq n$. The direct limit $G = \varinjlim G_n$ is the group $(G, G^+, \Gamma(G))$ with the following properties:

(i) there are homomorphisms $\gamma_m : G_m \to G$ so that $\gamma_n \gamma_{mn} = \gamma_m$ for all $m \leq n$.
(ii) $G^+ := \cup_{m \geq 1} \gamma_m(G_m^+)$ and $\Gamma(G) := \cup_{m \geq 1} \gamma_m(\Gamma(G_m))$.
(iii) if H is any (ordered) group and $\rho_n : G_n \to H$ are homomorphisms such that $\rho_m = \rho_n \gamma_{mn}$ for all $m \leq n$, then there is a homomorphism $\rho : G \to H$ so that $\rho_m = \rho \gamma_m$ for all $m \geq 1$. If ρ_m are all positive or contractive, then so is ρ.

Condition (iii) is the universal property that distinguishes the direct limit. For detailed information about direct limits, we recommend consulting a book on algebra. However, we will prove one basic result that will be used repeatedly.

Lemma IV.3.2 *Let $G = \varinjlim G_n$ with connecting maps γ_{mn}. Suppose that φ is a homomorphism of a finitely generated group H into G_m such that $\gamma_m \varphi = 0$. Then there is an integer $n \geq m$ such that $\gamma_{mn} \varphi = 0$.*

Proof. The key is the fact that if g is in G_m and $\gamma_{mn}(g) \neq 0$ for all $n \geq m$, then $\gamma_m(g) \neq 0$. To see this, consider the group $K := \prod_{n \geq 1} G_n / \sum_{n \geq 1} G_n$, the group of all sequences (g_n) of elements g_n in G_n modulo the relation that identifies sequences which eventually agree. Define homomorphisms Γ_n of G_n into K by $\Gamma_n(g) = [(g_k)]$ where

$$g_k = \begin{cases} 0 & k < n \\ \gamma_{nk}(g) & k \geq n \end{cases}.$$

IV.3. Dimension Groups

It is readily verified that $\Gamma_n \gamma_{mn} = \Gamma_m$ for all $m < n$. Therefore by the universal property (iii) of the direct limit, there is a homomorphism Γ of G into K so that $\Gamma \gamma_m = \Gamma_m$ for all $m \geq 1$. If g belongs to G_m and $\gamma_{mn}(g) \neq 0$ for all $n \geq m$, then $\Gamma \gamma_m(g) = \Gamma_m(g) \neq 0$. Hence $\gamma_m(g) \neq 0$.

Suppose that $\varphi : H \to G_m$ satisfies the hypotheses of the lemma. Applying the key fact established above to each generator h_i of H, we find integers $n_i > m$ so that $\gamma_{mn_i} \varphi(h_i) = 0$. Thus with $n = \max\{n_i\}$, we obtain $\gamma_{mn} \varphi = 0$. ∎

The following result will allow us to compute the K_0 group of an AF algebra.

Theorem IV.3.3 *If $\mathfrak{A} = \overline{\cup_{n \geq 1} \mathfrak{A}_n}$ is an AF algebra, then*

$$K_0(\mathfrak{A}) = \varinjlim K_0(\mathfrak{A}_n)$$

as a scaled dimension group. If all imbeddings are unital, then the order unit is the direct limit of the order units.

Proof. We are given $*$-homomorphisms $\varphi_{mn} : \mathfrak{A}_m \to \mathfrak{A}_n$ and $\varphi_n : \mathfrak{A}_n \to \mathfrak{A}$ so that $\varphi_m = \varphi_n \varphi_{mn}$ for all $m \leq n$. Therefore we obtain contractive homomorphisms $\varphi_{mn*} : K_0(\mathfrak{A}_m) \to K_0(\mathfrak{A}_n)$ and $\varphi_{n*} : K_0(\mathfrak{A}_n) \to K_0(\mathfrak{A})$ so that $\varphi_{m*} = \varphi_{n*} \varphi_{mn*}$ for all $m \leq n$.

Let $(G, G^+, \Gamma(G)) := \varinjlim K_0(\mathfrak{A}_n)$. By the definition of direct limits, there are homomorphisms $\psi_n : K_0(\mathfrak{A}_n) \to G$ so that $\psi_m = \psi_n \varphi_{mn*}$ for all $m \leq n$. Applying the universal property (iii) of direct limits of scaled dimension groups, we obtain a contractive homomorphism $\psi : G \to K_0(\mathfrak{A})$ so that the following diagram commutes:

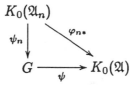

We will complete the proof by showing that ψ is an isomorphism.

Let $[P]$ be an element of $\Gamma(K_0(\mathfrak{A}))$. By Corollary IV.1.3, there is an integer n and a projection Q in $\mathfrak{A}_n$ so that $\varphi_{n*}([Q]) = [P]$. Let $g = \psi_n([Q])$. This lies in $\Gamma(G)$ by definition. Now $\psi(g) = [P]$. Hence ψ maps $\Gamma(G)$ onto $\Gamma(K_0(\mathfrak{A}))$. Since $\Gamma(G)$ generates G^+ and G, and similarly $\Gamma(K_0(\mathfrak{A}))$ generates $K_0^+(\mathfrak{A})$ and $K_0(\mathfrak{A})$, it follows that ψ is surjective.

Suppose that $\psi(g) = 0$. Choose an element x in $K_0(\mathfrak{A}_m)$, for m sufficiently large, such that $\psi_m(x) = g$. Then choose projections P and Q in $\mathcal{P}(\mathfrak{A}_m)$ so that $x = [P]_m - [Q]_m$ (where the notation $[P]_m$ denotes the equivalence class of P in $K_0(\mathfrak{A}_m)$). Since

$$0 = \psi(\psi_m(x)) = \varphi_{m*}(x) = [P] - [Q],$$

we see that $P \approx Q$ in $\mathfrak{A}$. By Corollary IV.1.3, there is an integer $n \geq m$ so that $[P]_n = [Q]_n$. Thus $\varphi_{mn*}(x) = 0$. So $g = \psi_n(\varphi_{mn*}(x)) = 0$. It follows that ψ is injective. ∎

Example IV.3.4 Consider the CAR algebra $\mathfrak{A}$ (see Examples III.2.4 and III.5.4) which is a direct limit of the full matrix algebras $\mathfrak{A}_n = \mathcal{M}_{2^n}$ with connecting maps $\varphi_{mn}(A) = A^{(2^{n-m})}$, where $A^{(k)}$ denotes the direct sum of k copies of A. We have seen that $K_0(\mathcal{M}_{2^n}) = (\mathbb{Z}, \mathbb{Z}_+)$. Since these algebras and the imbeddings are all unital, we need to keep track of the order unit. The map of $K_0^+(\mathcal{M}_{2^n})$ into $\mathbb{Z}_+$ is given by rank. So the order unit is $u_n = 2^n$. Thus the dimension group is $K_0(\mathcal{M}_{2^n}) = (\mathbb{Z}, \mathbb{Z}_+, [0, 2^n])$. By computing rank again, one sees that the intertwining maps are given by $\varphi_{mn*}(k) = 2^{n-m}k$. It is convenient to normalize so that the order unit is always 1. Then $K_0(\mathfrak{A}_n) = (2^{-n}\mathbb{Z}, 2^{-n}\mathbb{Z}_+, [0, 1])$ and the maps φ_{mn} are the natural inclusions. It follows that

$$K_0(\mathfrak{A}) = (\mathbb{Z}\left[\tfrac{1}{2}\right], \mathbb{Z}_+\left[\tfrac{1}{2}\right], [0,1])$$

where $\mathbb{Z}\left[\tfrac{1}{2}\right]$ is the group of diadic rationals.

Example IV.3.5 A much easier example to analyze is the algebra of compact operators $\mathfrak{K}$. Think of $\mathfrak{K}$ as the direct limit of matrix algebras $\mathcal{M}_n$ with imbeddings $\varphi_{mn} : \mathcal{M}_m \to \mathcal{M}_n$ given by $\varphi_{mn}(A) = A \oplus O_{n-m}$. Then the corresponding imbeddings φ_{mn*} of $K_0(\mathcal{M}_m) = (\mathbb{Z}, \mathbb{Z}_+, [0, m])$ into $K_0(\mathcal{M}_n) = (\mathbb{Z}, \mathbb{Z}_+, [0, n])$ is the identity map. Thus $K_0(\mathfrak{K}) = (\mathbb{Z}, \mathbb{Z}_+, \mathbb{Z}_+)$.

Example IV.3.6 Consider the Fibonacci algebra of Example III.2.6. Recall that $\mathfrak{A}_k = \mathcal{M}_{m_k} \oplus \mathcal{M}_{n_k}$, where m_k and n_k are given recursively by $m_1 = n_1 = 1$ and a repeated use of the partial imbedding matrix $A = \left[\begin{smallmatrix}1 & 1\\1 & 0\end{smallmatrix}\right]$, which yields the recursion relation

$$m_{k+1} = m_k + n_k \quad \text{and} \quad n_{k+1} = m_k.$$

It is easy to see that

$$K_0(\mathfrak{A}_k) = (\mathbb{Z}^2, \mathbb{Z}_+^2, [\left(\begin{smallmatrix}0\\0\end{smallmatrix}\right), \left(\begin{smallmatrix}m_k\\n_k\end{smallmatrix}\right)]).$$

Since a projection $P \oplus Q$ is sent to $(P \oplus Q) \oplus P$, the map from $K_0(\mathfrak{A}_k)$ to $K_0(\mathfrak{A}_{k+1})$ is given by the matrix A:

$$\varphi_{k,k+1}(\left(\begin{smallmatrix}m\\n\end{smallmatrix}\right)) = A\left(\begin{smallmatrix}m\\n\end{smallmatrix}\right) = \left(\begin{smallmatrix}m+n\\m\end{smallmatrix}\right).$$

IV.3. Dimension Groups

Since $\varphi_{k,k+1}$ is an isomorphism for each $k \geq 1$, the limit group $K_0(\mathfrak{A})$ is isomorphic to $\mathbb{Z}^2$. Indeed, map $K_0(\mathfrak{A}_n)$ onto $K_0(\mathfrak{A})$ by the map A^{-n}. So

$$K_0(\mathfrak{A}_m) = \mathbb{Z}^2 \xrightarrow{A^{n-m}} K_0(\mathfrak{A}_n) = \mathbb{Z}^2$$
$$\searrow_{A^{-m}} \qquad \downarrow_{A^{-n}}$$
$$K_0(\mathfrak{A}) = \mathbb{Z}^2$$

is a commutative diagram for all $m \leq n$.

The order unit for $K_0(\mathfrak{A}_k)$ is $\binom{m_k}{n_k} = A^k \binom{1}{0}$. Thus the order unit for $K_0(\mathfrak{A})$ is $\binom{1}{0}$. The positive cone is obtained as

$$K_0^+(\mathfrak{A}) = \bigcup_{n \geq 1} A^{-n} \mathbb{Z}_+^2.$$

Setting $n_0 = 0$, we obtain the Fibonacci sequence $n_k = n_{k-2} + n_{k-1}$ for all $k \geq 2$. An easy induction argument shows that

$$A^{-k} = \begin{bmatrix} (-1)^k n_{k-1} & (-1)^{k-1} n_k \\ (-1)^{k-1} n_k & (-1)^k n_{k+1} \end{bmatrix}$$

These are unimodular matrices, so they map $\mathbb{Z}^2$ onto itself. The image of $\mathbb{Z}_+^2$ is the cone generated by

$$\alpha_k = \begin{pmatrix} (-1)^k n_{k-1} \\ (-1)^{k-1} n_k \end{pmatrix} \quad \text{and} \quad \beta_k = \begin{pmatrix} (-1)^{k-1} n_k \\ (-1)^k n_{k+1} \end{pmatrix}.$$

From the well known formula

$$n_k = \frac{\tau^{k+1} + (-\tau)^{1-k}}{\tau^2 + 1}$$

where $\tau := \frac{1+\sqrt{5}}{2}$, we obtain

$$\lim_{k \to \infty} \frac{n_{2k+1}}{n_{2k}} = \tau_+ \quad \text{and} \quad \lim_{k \to \infty} \frac{n_{2k}}{n_{2k-1}} = \tau_-.$$

Thus $\alpha_{2k} = \beta_{2k-1}$ is approaching the line segment $y = \tau x$, $x > 0$; and similarly, $\alpha_{2k+1} = \beta_{2k}$ is approaching the line segment $y = \tau x$, $x < 0$; and $\tau x + y > 0$ for all of these vectors. Thus the union of these cones is

$$K_0^+(\mathfrak{A}) = \{\binom{m}{n} : \tau m + n \geq 0\}.$$

The map from $\mathbb{Z}^2$ into $(\mathbb{R}, \mathbb{R}_+, 1)$ given by $f(\binom{m}{n}) = m + \tau^{-1} n$ is a positive unital homomorphism. Thus we see that the order on $K_0(\mathfrak{A})$ is a total order. Indeed, we obtain the isomorphism

$$(K_0(\mathfrak{A}), K_0^+(\mathfrak{A}), \binom{1}{0}) \simeq (\mathbb{Z} + \tau^{-1}\mathbb{Z}, (\mathbb{Z} + \tau^{-1}\mathbb{Z}) \cap \mathbb{R}_+, 1).$$

Example IV.3.7 In this last example, we will compute the K_0 group of the GICAR algebra of Example III.5.5. From the Bratteli diagram, which is like Pascal's triangle, we obtain that $K_0(\mathfrak{A}_n) \simeq (\mathbb{Z}^{n+1}, \mathbb{Z}_+^{n+1})$. The map α_{n*} from $K_0(\mathfrak{A}_n)$ into $K_0(\mathfrak{A}_{n+1})$ is given by

$$\alpha_{n*}(a_0, a_1, \ldots, a_n) = (a_0, a_0 + a_1, a_1 + a_2, \ldots, a_{n-1} + a_n, a_n).$$

It is helpful to represent the group $K_0(\mathfrak{A}_n)$ as $\mathcal{P}_n = \{p \in \mathbb{Z}[x] : \deg p \leq n\}$ and $\mathcal{P}_n^+ = \mathbb{Z}_+[x] \cap \mathcal{P}_n$. Then the map α_{n*} becomes multiplication by $x + 1$. As we have seen before, the natural choice of maps from $\mathcal{P}_n$ into the limit group (G, G^+) is given by multiplication by $(x + 1)^{-n}$. The limit group then consists of

$$G = \{(x+1)^{-n} p(x) : p \in \mathbb{Z}[x], \deg p \leq n, n \geq 1\}.$$

The positive cone consists of those elements for which p has all positive coefficients. Thus in order to determine if $(x + 1)^{-n} p(x)$ belongs to G^+, one must check the other representatives $(x + 1)^{-n-N}((x + 1)^N p(x))$. So this element belongs to G^+ if there is an integer N for which $(x + 1)^N p(x)$ has all positive coefficients.

Lemma IV.3.8 *For each pair (a, b) in $\mathbb{R}^2$ with $b \neq 0$, there is an integer $N \geq 0$ such that $(x + 1)^N (x^2 - 2ax + a^2 + b^2)$ has all positive coefficients.*

Proof. Rewrite the quadratic as $x^2 - \alpha x + \beta$; and notice that $4\beta > \alpha^2$. We may assume that $\alpha > 0$, for otherwise $N = 0$ suffices. Compute

$$(x+1)^N (x^2 - \alpha x + \beta) = (x^2 - \alpha x + \beta) \sum_{k=0}^{N} \binom{N}{k} x^k$$

$$= \sum_{k=0}^{N+2} \left(\beta \binom{N}{k} - \alpha \binom{N}{k-1} + \binom{N}{k-2}\right) x^k$$

$$= \sum_{k=0}^{N+2} \frac{N!}{k!(N+2-k)!} a_{N,k} x^k$$

where

$$a_{N,k} = \beta(N+2-k)(N+1-k) - \alpha k(N+2-k) + k(k-1)$$
$$= (1 + \alpha + \beta)k^2 - (2\beta + \alpha)Nk - (3\beta + 2\alpha + 1)k + \beta(N^2 + 3N + 2)$$
$$= (1 + \alpha + \beta)(k - \frac{\beta + \alpha/2}{1 + \alpha + \beta} N)^2 + \beta(3N - 3k + 2) - (2\alpha + 1)k$$
$$\quad + (1 + \alpha + \beta)^{-1} N^2 \left((1 + \alpha + \beta)\beta - (\beta + \alpha/2)^2\right)$$
$$\geq (1 + \alpha + \beta)^{-1} N^2 (\beta - \frac{\alpha^2}{4}) - 4\beta - (2\alpha + 1)(N + 2)$$

This is positive for large N independent of k. ∎

Proposition IV.3.9 *Given a non-zero polynomial p in $\mathbb{R}[x]$, there is an integer N so that $(x + 1)^N p(x)$ has all non-negative coefficients if and only if $p(x) > 0$ on $(0, \infty)$.*

Proof. Clearly every non-zero polynomial with non-negative coefficients is strictly positive on $(0, \infty)$. So this condition is necessary. Conversely, a polynomial is strictly positive on the positive real line precisely when it has no roots there and its leading coefficient is positive. So it factors as a product

$$p(x) = c \prod_i (x + \lambda_i) \prod_j (x^2 - 2a_j x + a_j^2 + b_j^2)$$

where $\lambda_i \geq 0$ and $b_j > 0$. By the previous lemma, there are integers $N_j \geq 0$ so that $(x + 1)^{N_j}(x^2 - 2a_j x + a_j^2 + b_j^2)$ each has non-negative coefficients. Thus $N = \sum_j N_j$ will suffice. ∎

Hence G^+ consists of those elements $(x + 1)^{-n} p(x)$ such that $\deg p \leq n$ and $p(x) > 0$ on $(0, \infty)$. We can obtain a less unwieldy formulation by making a change of variables. Substitute $y = (x + 1)^{-1}$. The function $(x + 1)^{-n} p(x)$ for a polynomial $\sum_{j=0}^n a_j x^j$ is transformed into

$$y^n p(\tfrac{1}{y} - 1) = \sum_{j=0}^n a_j (1 - y)^j y^{n-j}.$$

Thus this map carries G onto $Z[y]$. This transformation is a conformal map that carries the positive real line onto $(0, 1)$. Thus a function is positive on the positive real line if and only if the transformed function is positive on $(0, 1)$. So G^+ consists of those polynomials in $\mathbb{Z}[y]$ which are positive on $(0, 1)$. The order unit is the constant polynomial 1.

IV.4 Elliott's Theorem

In this section, we will prove that the dimension group of a C*-algebra is a complete invariant for AF algebras. As well as classifying AF algebras in a useful algebraic way, this result has proven to be the prototype for a much larger classification scheme for more general C*-algebras.

In order to recover an AF algebra from its K-theory, we need to be able to recover maps from their induced homomorphisms. Example IV.3.1 essentially indicates how this can be done in the finite dimensional case.

Lemma IV.4.1 *Suppose that $\mathfrak{A}$ and $\mathfrak{B}$ are finite dimensional C*-algebras and that ψ is a contractive homomorphism of $K_0(\mathfrak{A})$ into $K_0(\mathfrak{B})$. Then there is a *-homomorphism φ of $\mathfrak{A}$ into $\mathfrak{B}$ such that $\varphi_* = \psi$, and φ is unique up to unitary equivalence in $\mathfrak{B}$. Moreover, if ψ is unital, then so is φ.*

Proof. Let $\mathfrak{A} \simeq \mathcal{M}_{n_1} \oplus \cdots \oplus \mathcal{M}_{n_k}$ and $\mathfrak{B} \simeq \mathcal{M}_{m_1} \oplus \cdots \oplus \mathcal{M}_{m_\ell}$. $K_0(\mathfrak{A})$ is the scaled dimension group $(\mathbb{Z}^k, \mathbb{Z}_+^k, [0, [I_\mathfrak{A}]])$ where $[I_\mathfrak{A}] = (n_1, \ldots, n_k)^t$. Similarly,

$K_0(\mathfrak{B})$ equals $(\mathbb{Z}^\ell, \mathbb{Z}_+^\ell, [0, [I_\mathfrak{B}]])$ where $[I_\mathfrak{B}] = (m_1, \ldots, m_\ell)^t$. The homomorphism ψ of $\mathbb{Z}^k$ into $\mathbb{Z}^\ell$ is given by an $\ell \times k$ integer matrix $A = [a_{ij}]$ so that $\psi(z) = Az$ for each vector z in $\mathbb{Z}^k$. Since ψ is contractive, it follows that $a_{ij} \geq 0$ and $\psi([I_\mathfrak{A}]) \leq [I_\mathfrak{B}]$. This latter condition translates into the inequalities

$$\sum_{j=1}^k a_{ij} n_j \leq m_i \quad \text{for} \quad 1 \leq i \leq \ell.$$

When ψ is unital, these are equalities.

From Corollary III.2.2, we see that these are precisely the conditions required on a set of partial multiplicities of an imbedding of $\mathfrak{A}$ into $\mathfrak{B}$. By that corollary, there is a $*$-homomorphism φ of $\mathfrak{A}$ into $\mathfrak{B}$ with these partial multiplicities. It is unique up to unitary equivalence in $\mathfrak{B}$, and is unital when all the inequalities are equalities. By Example IV.3.1, it follows that $\varphi_* = \psi$. ∎

We next improve this result by replacing the image algebra $\mathfrak{B}$ by an AF algebra.

Lemma IV.4.2 *Let $\mathfrak{A}$ be a finite dimensional C*-algebra; and let $\mathfrak{B} = \overline{\cup_{n \geq 1} \mathfrak{B}_n}$ be an AF algebra with the imbeddings of $\mathfrak{B}_m$ into $\mathfrak{B}$ be denoted by β_m. Suppose that ψ is a contractive homomorphism of $K_0(\mathfrak{A})$ into $K_0(\mathfrak{B})$. Then there is an integer m and a $*$-homomorphism φ of $\mathfrak{A}$ into $\mathfrak{B}_m$ such that $\beta_{m*} \varphi_* = \psi$. Moreover, φ is unique up to unitary equivalence in $\mathfrak{B}$; and if ψ is unital, then so is φ.*

Proof. Let $\mathfrak{A}$ be isomorphic to $\mathcal{M}_{n_1} \oplus \cdots \oplus \mathcal{M}_{n_k}$. Let $I_\mathfrak{A}$ be the unit for $\mathfrak{A}$, and let E_j, for $1 \leq j \leq k$, be the minimal projections of $\mathfrak{A}$. Since ψ maps projections of $\mathfrak{A}$ into $\Gamma(\mathfrak{B}) = \cup_{n \geq 1} \beta_{n*} \Gamma(\mathfrak{B}_n)$, there is an integer n sufficiently large so that $\psi([I_\mathfrak{A}])$ and $\psi([E_j])$, for $1 \leq j \leq k$, all belong to $\beta_{n*} \Gamma(\mathfrak{B}_n)$. Choose projections J and F_j in $\mathfrak{B}_n$ so that $\beta_{n*}([J]) = \psi([I_\mathfrak{A}])$ and $\beta_{n*}([F_j]) = \psi([E_j])$ for $1 \leq j \leq k$. In the unital case, take $J = I_{\mathfrak{B}_n}$. Since

$$\beta_{n*}([J] - \sum_{j=1}^k n_j [F_j]) = 0,$$

Lemma IV.3.2 shows that there is an integer $m \geq n$ so that

$$\beta_{nm*}([J] - \sum_{j=1}^k n_j [F_j]) = 0.$$

Let $F_j' := \beta_{nm}(F_j)$ and $J' = \beta_{nm}(J)$. Define a map ρ from $K_0(\mathfrak{A})$ into $K_0(\mathfrak{B}_m)$ by setting $\rho([E_j]) = [F_j']$ and extending it to a homomorphism. By construction, ρ is positive. Since

$$\rho(I_\mathfrak{A}) = \sum_{j=1}^k n_j [F_j'] = [J'] \in \Gamma(\mathfrak{B}_m),$$

IV.4. Elliott's Theorem

it follows that ρ is contractive. By construction, we also have $\beta_{m*}\rho = \psi$.

Apply the previous lemma to ρ to obtain a homomorphism φ of $\mathfrak{A}$ into $\mathfrak{B}_m$ such that $\varphi_* = \rho$ (and is unital in the unital case). This is the desired map.

Suppose that $\varphi' : \mathfrak{A} \to \mathfrak{B}_{m'}$ is another map with these properties. Choose an integer $\ell \geq \max\{m, m'\}$ and replace φ and φ' by $\beta_{m\ell}\varphi$ and $\beta_{m'\ell}\varphi'$ respectively, so that they map into the same subalgebra. Now $\beta_{\ell*}(\varphi_* - \varphi'_*) = 0$. Again by Lemma IV.3.2 there is an integer $p \geq \ell$ so that $\beta_{\ell p*}(\varphi_* - \varphi'_*) = 0$. Hence by Corollary III.2.2, $\beta_{\ell p}\varphi$ and $\beta_{\ell p}\varphi'$ are unitarily equivalent in $\mathfrak{B}_p$. ∎

We are now able to prove the main classification theorem for AF algebras.

Theorem IV.4.3 *Two AF algebras $\mathfrak{A}$ and $\mathfrak{B}$ are $*$-isomorphic if and only if their scaled dimension groups are isomorphic. Moreover, given a scaled dimension group isomorphism $\rho : K_0(\mathfrak{A}) \to K_0(\mathfrak{B})$, there is a $*$-isomorphism φ of $\mathfrak{A}$ onto $\mathfrak{B}$ such that $\varphi_* = \rho$.*

Proof. As the second statement implies the non-trivial part of the first, we will only prove the second statement. Let $\mathfrak{A} = \overline{\cup_{m\geq 1}\mathfrak{A}_m}$ and $\mathfrak{B} = \overline{\cup_{n\geq 1}\mathfrak{B}_n}$ with connecting maps α_m and β_n respectively. By Theorem IV.3.3,

$$K_0(\mathfrak{A}) = \varinjlim K_0(\mathfrak{A}_m) \quad \text{and} \quad K_0(\mathfrak{B}) = \varinjlim K_0(\mathfrak{B}_n).$$

The plan is to construct maps to make the following diagram commutative:

$$\begin{array}{ccccccccc}
\mathfrak{A}_{m_1} & \xrightarrow{\alpha_{m_1 m_2}} & \mathfrak{A}_{m_2} & \xrightarrow{\alpha_{m_2 m_3}} & \mathfrak{A}_{m_3} & \xrightarrow{\alpha_{m_3 m_4}} & \cdots & \longrightarrow & \mathfrak{A} \\
\varphi_1 \downarrow & \psi_1 \nearrow & \varphi_2 \downarrow & \psi_2 \nearrow & \varphi_2 \downarrow & \psi_3 \nearrow & & \varphi \downarrow \uparrow \psi \\
\mathfrak{B}_{n_1} & \xrightarrow[\beta_{n_1 n_2}]{} & \mathfrak{B}_{n_2} & \xrightarrow[\beta_{n_2 n_3}]{} & \mathfrak{B}_{n_3} & \xrightarrow[\beta_{n_3 n_4}]{} & \cdots & \longrightarrow & \mathfrak{B}
\end{array}$$

and so that $\varphi_* = \rho$ and $\psi_* = \rho^{-1}$.

Choose two increasing sequences of positive integers $m_1 < m_2 < \ldots$ and $n_1 < n_2 < \ldots$ as follows. Let $m_1 = 1$. Apply Lemma IV.4.2 to the map $\rho\alpha_{m_1*}$ of $K_0(\mathfrak{A}_{m_1})$ into $K_0(\mathfrak{B})$ to obtain an integer $n_1 \geq 1$ and a $*$-homomorphism φ_1 of $\mathfrak{A}_{m_1}$ into $\mathfrak{B}_{n_1}$ such that $\beta_{n_1*}\varphi_{1*} = \rho\alpha_{m_1*}$.

Now apply Lemma IV.4.2 to the map $\rho^{-1}\beta_{n_1*}$ of $K_0(\mathfrak{B}_{n_1})$ into $K_0(\mathfrak{A})$ to obtain a positive integer m and a $*$-homomorphism ψ of $\mathfrak{B}_{n_1}$ into $\mathfrak{A}_m$ such that $\beta_{m*}\psi_* = \rho^{-1}\beta_{m_1*}$. Then notice that

$$\alpha_{m*}(\psi_*\varphi_{1*} - \alpha_{m_1 m*}) = 0.$$

Thus by Lemma IV.3.2 there is an integer $m_2 \geq m$ such that

$$\alpha_{mm_2*}(\psi_*\varphi_{1*} - \alpha_{m_1 m*}) = 0.$$

Hence by Corollary III.2.2 there is a unitary U in $\mathfrak{A}_{m_2}$ so that

$$\mathrm{Ad}(U)\alpha_{mm_2}\psi\varphi_1 = \alpha_{m_1 m_2}.$$

Set $\psi_1 := \mathrm{Ad}(U)\alpha_{mm_2}\psi$.

Now proceed recursively. If $m_1, \ldots, m_k$, $n_1, \ldots, n_{k-1}$, $\varphi_1, \ldots, \varphi_{k-1}$ and $\psi_1, \ldots, \psi_{k-1}$ are all defined so that the diagram above commutes, follow the procedure of the previous paragraph to obtain $n_k > n_{k-1}$ and a map φ_k of $\mathfrak{A}_{m_k}$ into $\mathfrak{B}_{n_k}$ so that

$$\beta_{n_k*}\varphi_{k*} = \rho\alpha_{m_k*} \quad \text{and} \quad \varphi_k\psi_{k-1} = \beta_{n_{k-1}n_k}.$$

Then choose $m_{k+1} > m_k$ and a homomorphism ψ_k from $\mathfrak{B}_{n_k}$ into $\mathfrak{A}_{m_{k+1}}$ so that

$$\alpha_{m_{k+1}*}\psi_{k*} = \rho^{-1}\beta_{n_k*} \quad \text{and} \quad \psi_k\varphi_k = \alpha_{m_k m_{k+1}}.$$

Let $\varphi : \mathfrak{A} \to \mathfrak{B}$ and $\psi : \mathfrak{B} \to \mathfrak{A}$ be the maps determined by the maps φ_k, ψ_k and the properties of direct limits so that the diagrams

$$\begin{array}{ccc} \mathfrak{A}_{m_k} & \xrightarrow{\alpha_{m_k}} & \mathfrak{A} \\ \varphi_k \downarrow & & \downarrow \varphi \\ \mathfrak{B}_{n_k} & \xrightarrow{\beta_{n_k}} & \mathfrak{B} \end{array} \quad \text{and} \quad \begin{array}{ccc} \mathfrak{A}_{m_{k+1}} & \xrightarrow{\alpha_{m_{k+1}}} & \mathfrak{A} \\ \psi_k \uparrow & & \uparrow \psi \\ \mathfrak{B}_{n_k} & \xrightarrow{\beta_{n_k}} & \mathfrak{B} \end{array}$$

commute for all $k \geq 1$. As the restriction of $\varphi|\alpha_{m_k}(\mathfrak{A}_{m_k}) = \varphi_k$ is injective for all $k \geq 1$ and $\psi\varphi|\mathfrak{A}_{m_k} = \alpha_{m_k}$, it follows that $\psi\varphi = \mathrm{id}_\mathfrak{A}$. Similarly, $\varphi\psi = \mathrm{id}_\mathfrak{B}$. So $\mathfrak{A}$ and $\mathfrak{B}$ are isomorphic. Moreover, at the level of K-theory, we have the commutative diagrams

$$\begin{array}{ccc} K_0(\mathfrak{A}_{m_k}) & \xrightarrow{\alpha_{m_k*}} & K_0(\mathfrak{A}) \\ \varphi_{k*} \downarrow & & \downarrow \varphi_* \\ K_0(\mathfrak{B}_{n_k}) & \xrightarrow{\beta_{n_k*}} & K_0(\mathfrak{B}) \end{array} \quad \text{and} \quad \begin{array}{ccc} K_0(\mathfrak{A}_{m_k}) & \xrightarrow{\alpha_{m_k*}} & K_0(\mathfrak{A}) \\ \varphi_{k*} \downarrow & & \downarrow \rho \\ K_0(\mathfrak{B}_{n_k}) & \xrightarrow{\beta_{n_k*}} & K_0(\mathfrak{B}) \end{array}$$

By the uniqueness of the direct limit of the maps φ_{k*}, we obtain $\varphi_* = \rho$. Similarly, $\psi_* = \rho^{-1}$. ∎

IV.5 Applications

Since $K_0(\mathfrak{A})$ is a complete invariant, it must be possible to read off various algebraic information about $\mathfrak{A}$ from $K_0(\mathfrak{A})$. In this section, we will see some examples of this.

Ideals. First consider the ideals of $\mathfrak{A}$, which were classified in terms of subsets of the Bratteli diagram in Theorem III.4.2. A subgroup H of a partially ordered group (G, G^+) is called an **order ideal** provided that $H^+ := H \cap G^+$ is hereditary (meaning that if $0 \leq g \leq h$ for some g in G and h in H^+, then g belongs to H) and $H = H^+ - H^+$.

IV.5. Applications

Proposition IV.5.1 *The ideals of an AF algebra $\mathfrak{A}$ are in one-to-one correspondence with the order ideals of $K_0(\mathfrak{A})$ via the map which takes each ideal $\mathfrak{J}$ to $K_0(\mathfrak{J})$.*

Proof. Suppose that $\mathfrak{J}$ is an ideal of $\mathfrak{A}$. Then $K_0(\mathfrak{J})$ is the subgroup of $K_0(\mathfrak{A})$ generated by $\Gamma(\mathfrak{J}) = \{[P] : P = P^2 \in \mathfrak{J}\}$. In particular,

$$K_0^+(\mathfrak{J}) = K_0^+(\mathfrak{A}) \cap K_0(\mathfrak{J}).$$

To see that $K_0^+(\mathfrak{J})$ is hereditary, suppose that $0 \leq [Q] \leq [P]$, where $[Q]$ is in $K_0^+(\mathfrak{A})$ and $[P]$ belongs to $K_0^+(\mathfrak{J})$. We may suppose that there is a common integer n so that Q is a projection in $\mathcal{M}_n(\mathfrak{A})$ and P is a projection in $\mathcal{M}_n(\mathfrak{J})$. By Theorem IV.1.6, Q is unitarily equivalent to a projection $Q' \leq P$. Thus Q' belongs to $\mathcal{M}_n(\mathfrak{J})$ by Theorem I.5.3, and so Q also lies in $\mathcal{M}_n(\mathfrak{J})$. Since $K_0(\mathfrak{J}) = K_0^+(\mathfrak{J}) - K_0^+(\mathfrak{J})$, we see that $K_0(\mathfrak{J})$ is an order ideal of $K_0(\mathfrak{A})$.

Conversely, suppose that H is an order ideal of $K_0(\mathfrak{A})$. Let $\mathfrak{J}$ be the ideal generated by the set $\mathcal{S}$ consisting of those projections P in $\mathfrak{A}$ such that $[P]$ belongs to $\Gamma(H) := H \cap \Gamma(K_0(\mathfrak{A}))$. We claim that

$$\mathfrak{J} = \{X \in \mathfrak{A} : \text{there exist } P_i \in \mathcal{S} \text{ so that } X^*X \leq \sum P_i\}.$$

Indeed, the right hand side, say $\mathfrak{X}$, is contained in the ideal $\mathfrak{J}$ by Theorem I.5.3. One shows that $\mathfrak{X}$ is closed under sums by using a clever identity. Suppose that

$$X^*X \leq \sum_{i=1}^{m} P_i \quad \text{and} \quad Y^*Y \leq \sum_{j=1}^{n} Q_j$$

for projections P_i and Q_j in S. Then

$$(X \pm Y)^*(X \pm Y) \leq (X+Y)^*(X+Y) + (X-Y)^*(X-Y)$$
$$= 2X^*X + 2Y^*Y \leq 2\sum_{i=1}^{m} P_i + 2\sum_{j=1}^{n} Q_j$$

and so $X \pm Y$ belong to $\mathfrak{X}$. To show that $\mathfrak{X}$ is an ideal, it suffices by Theorem I.8.4 to show that UX and XU belong to $\mathfrak{X}$ for every X in $\mathfrak{X}$ and every unitary U in $\mathfrak{A}^\sim$. Now UX belongs to $\mathfrak{X}$ because

$$(UX)^*UX = X^*X \leq \sum_{i=1}^{m} P_i.$$

Also

$$(XU)^*XU = U^*X^*XU \leq \sum_{i=1}^{m} U^*P_iU.$$

Since $[U^*P_iU] = [P_i]$ belongs to $\Gamma(H)$, we see that U^*P_iU belongs to $\mathcal{S}$ and so UX is in $\mathfrak{X}$. This establishes our claim.

If Q is a projection in $\mathfrak{J}$, the claim shows that there are projections P_i in $\mathcal{S}$ so that $Q \leq \sum_{i=1}^m P_i$ and therefore $[Q] \leq \sum_{i=1}^m [P_i]$. As H^+ is hereditary, Q also belongs to $\mathcal{S}$. Therefore $\Gamma(K_0(\mathfrak{J})) = \Gamma(H)$; and hence $K_0(\mathfrak{J}) = H$. This shows that the map from ideals $\mathfrak{J}$ to order ideals $K_0(\mathfrak{J})$ is surjective. On the other hand, suppose that $K_0(\mathfrak{J}_1) = K_0(\mathfrak{J}_2) = H$ for two ideals $\mathfrak{J}_1$ and $\mathfrak{J}_2$. As $\mathfrak{J}_1$ and $\mathfrak{J}_2$ are AF, they are generated by their projections. If $\mathfrak{J}_2$ contains a projection P, then $[P]$ belongs to $\Gamma(K_0(\mathfrak{J}_1))$ and thus P also belongs to $\mathfrak{J}_1$. Consequently $\mathfrak{J}_1$ and $\mathfrak{J}_2$ have the same projections and are therefore equal. ∎

A dimension group is called **simple** if it has no proper order ideals. We obtain the following immediate corollary.

Corollary IV.5.2 *An AF algebra $\mathfrak{A}$ is simple if and only if $(K_0(\mathfrak{A}), K_0^+(\mathfrak{A}))$ is a simple dimension group.*

Traces. Generalizing the notion of the trace of a matrix, a (normalized) **trace** on a C*-algebra $\mathfrak{A}$ is a state τ such that $\tau(XY) = \tau(YX)$ for all X and Y in $\mathfrak{A}$. Using Theorem I.8.4, we can easily see that to verify that τ is a trace, it suffices to show that $\tau(UXU^*) = \tau(X)$ for every X in $\mathfrak{A}$ and every unitary element U in $\mathfrak{A}^\sim$. Also, because the projections span an AF algebra, it suffices to verify that $\tau(UPU^*) = \tau(P)$ for every projection P and unitary U in $\mathfrak{A}^\sim$. In particular, τ induces a well defined map τ_* on $K_0(\mathfrak{A})$ by $\tau_*([P]) = \tau^{(n)}(P)$ for projections P in $\mathcal{M}_n(\mathfrak{A})$. Hence we can reasonably expect to find a K-theoretic characterization of traces.

The key observation is that τ_* is a contractive homomorphism of $K_0(\mathfrak{A})$ into $(\mathbb{R}, \mathbb{R}_+, [0,1])$. Indeed, for every projection P in $\mathfrak{A}$, $\tau(P)$ belongs to $[0,1]$ because τ is positive and contractive. If P and Q are projections in $\mathcal{M}_m(\mathfrak{A})$ and $\mathcal{M}_n(\mathfrak{A})$ respectively, then

$$\tau^{(m+n)}(P \oplus Q) = \tau^{(m)}(P) + \tau^{(n)}(Q);$$

from which we deduce that τ_* is a homomorphism. So we define a **state** φ on a dimension group $(G, G^+, \Gamma(G))$ to be a positive homomorphism into $(\mathbb{R}, \mathbb{R}_+, [0,1])$ such that $\varphi(\Gamma(G)) = \varphi(G^+) \cap [0,1]$. In the unital case, this latter condition is merely that $\varphi([I]) = 1$.

Theorem IV.5.3 *The traces on an AF algebra $\mathfrak{A}$ are in a natural one-to-one correspondence with states on $K_0(\mathfrak{A})$ via the map which takes a trace τ to τ_*.*

Proof. Any trace on $\mathfrak{A}$ extends to a trace on the unitization $\mathfrak{A}^\sim$ so that $\tau(I) = 1$. So for convenience, we suppose that our algebras are unital. We have already observed that if τ is a trace, then τ_* is a state on $K_0(\mathfrak{A})$.

Conversely, suppose that ρ is a state on $K_0(\mathfrak{A})$. Define a map τ on the *rational* span of the projections in $\mathfrak{A}$ by

$$\tau\left(\sum_{i=1}^m r_i P_i\right) := \sum_{i=1}^m r_i \rho([P_i]).$$

IV.5. Applications

To see that this is well defined and contractive, suppose that

$$\left\|\sum_{i=1}^{m} r_i P_i - \sum_{j=1}^{n} s_j Q_j\right\| < 1/k.$$

Choose an integer N such that Nr_i, Ns_j and N/k are all integers for $1 \le i \le m$ and $1 \le j \le n$. Then

$$-\frac{N}{k}I \le \sum_{i=1}^{m} Nr_i P_i - \sum_{j=1}^{n} Ns_j Q_j \le \frac{N}{k}I.$$

It follows (see Exercise IV.12) that

$$-\frac{N}{k}[I] \le \sum_{i=1}^{m} Nr_i [P_i] - \sum_{j=1}^{n} Ns_j [Q_j] \le \frac{N}{k}[I].$$

Hence applying ρ and dividing by N yields

$$-\frac{1}{k} \le \sum_{i=1}^{m} r_i \rho([P_i]) - \sum_{j=1}^{n} s_j \rho([Q_j]) \le \frac{1}{k}.$$

Thus τ extends to a well defined state on $\mathfrak{A}_{sa}$ by continuity and hence by complexification to all of $\mathfrak{A}$. Moreover, since

$$\tau(UPU^*) = \rho([UPU^*]) = \rho([P]) = \tau(P)$$

for every projection P and unitary U in $\mathfrak{A}^\sim$, the remarks preceding the proof imply that τ is a trace. As this definition of τ is the only possible one with $\tau_* = \rho$, there is a unique trace for each state on $K_0(\mathfrak{A})$. ∎

Example IV.5.4 Consider the algebra $\mathfrak{K}$ of compact operators. Then

$$K_0(\mathfrak{K}) = (\mathbb{Z}, \mathbb{Z}_+, \mathbb{Z}_+).$$

The only homomorphism taking $\mathbb{Z}_+$ into $[0,1]$ is the zero map. So there are no (finite) traces.

On a finite dimensional C*-algebra $\mathfrak{A} = \mathcal{M}_{n_1} \oplus \cdots \oplus \mathcal{M}_{n_k}$ we have

$$K_0(\mathfrak{A}) = (\mathbb{Z}^k, \mathbb{Z}_+^k, [0,[I]]) \quad \text{where} \quad [I] = \begin{pmatrix} n_1 \\ \vdots \\ n_k \end{pmatrix}.$$

Let E_j denote a rank one projection in the j-th summand, and let I_j denote the unit of the j-th summand. If τ is a trace on $\mathfrak{A}$, let $t_j := \tau(I_j)$. Since $[I_j] = n_j [E_j]$ and $[I] = \sum_{j=1}^k [I_j]$, we must have

$$\sum_{j=1}^{k} t_j = 1 \quad \text{and} \quad \tau(E_j) = n_j^{-1} t_j.$$

Thus τ is a convex combination of the unique normalized trace on each summand.

Example IV.5.5 Consider the Fibonacci algebra $\mathfrak{A}$ from Example IV.3.6. Recall that
$$K_0(\mathfrak{A}) = (\mathbb{Z}^2, \{\binom{m}{n} : \tau m + n \geq 0\}, \binom{1}{0}))$$
which is isomorphic to $(\mathbb{Z} + \tau^{-1}\mathbb{Z}, (\mathbb{Z} + \tau^{-1}\mathbb{Z}) \cap \mathbb{R}_+, 1)$. Since this is a total order, there is a unique unital homomorphism of $K_0(\mathfrak{A})$ into $(\mathbb{R}, \mathbb{R}_+, 1)$, namely $\tau_*(\binom{m}{n}) = m + \tau^{-1}n$. So there is a unique trace on this algebra.

Automorphisms. The automorphism group of a C*-algebra is often difficult to deal with. In the case of AF algebras, certain simplifications occur.

An automorphism α of a C*-algebra $\mathfrak{A}$ is called **inner** if there is a unitary operator U in $\tilde{\mathfrak{A}}$ such that $\alpha = \text{Ad}\,U$, which means that $\alpha(A) = UAU^*$. The set of all inner automorphisms forms a normal subgroup $\text{Inn}(\mathfrak{A})$. An automorphism α is **approximately inner** if it is the pointwise limit of inner automorphisms. That is, there is a net α_k of inner automorphisms such that $\alpha(A) = \lim_k \alpha_k(A)$ for every A in $\mathfrak{A}$. When $\mathfrak{A}$ is separable, a sequence will suffice. The subgroup of approximately inner automorphisms is denoted by $\overline{\text{Inn}}(\mathfrak{A})$.

We also define several other normal subgroups of $\text{Aut}(\mathfrak{A})$. Let $\text{Aut}_0(\mathfrak{A})$ denote the connected component of the identity map id in the topology of pointwise convergence; and let $\text{Aut}_p(\mathfrak{A})$ denote the path connected component of the identity. Finally, let $\text{Id}(\mathfrak{A})$ denote the set of automorphisms α such that $\alpha_* = \text{id}_*$ on $K_0(\mathfrak{A})$.

Remark IV.5.6 (a) $\text{Aut}_p(\mathfrak{A}) \subset \text{Aut}_0(\mathfrak{A}) \subset \text{Id}(\mathfrak{A})$ for any C*-algebra. The first inclusion is true for all topological spaces. For each projection P in $\mathcal{M}_n(\mathfrak{A})$, let
$$\mathcal{U}_P = \{\alpha \in \text{Aut}(\mathfrak{A}) : \alpha^{(n)}(P) \sim P\}$$
$$= \{\alpha \in \text{Aut}(\mathfrak{A}) : \alpha_*([P]) = [P]\}.$$
Suppose that P and Q are projections in $\mathfrak{A}$ such that $\|P - Q\| < 1$. Then by Proposition IV.1.2, P is equivalent to Q; whence $[P] = [Q]$. Thus $\mathcal{U}_P$ is both open and closed, and so contains $\text{Aut}_0(\mathfrak{A})$. Thus
$$\text{Aut}_0(\mathfrak{A}) \subset \bigcap_{P \in \mathcal{P}(\mathfrak{A})} \mathcal{U}_P = \text{Id}(\mathfrak{A}).$$

(b) $\overline{\text{Inn}}(\mathfrak{A}) \subset \text{Id}(\mathfrak{A})$. If $\alpha = \text{Ad}\,U$, then $\alpha_*([P]) = [U^{(n)}PU^{(n)*}] = [P]$. Thus $\text{Inn}(\mathfrak{A}) \subset \text{Id}(\mathfrak{A})$. By part (a), $\text{Id}(\mathfrak{A})$ is closed. So it contains $\overline{\text{Inn}}(\mathfrak{A})$ too.

Theorem IV.5.7 *If $\mathfrak{A}$ is an AF algebra, then*
$$\overline{\text{Inn}}(\mathfrak{A}) = \text{Aut}_p(\mathfrak{A}) = \text{Aut}_0(\mathfrak{A}) = \text{Id}(\mathfrak{A}).$$

Proof. Using the remarks above, it suffices to show that
$$\text{Id}(\mathfrak{A}) \subset \overline{\text{Inn}}(\mathfrak{A}) \cap \text{Aut}_p(\mathfrak{A}).$$

IV.5. Applications

Write $\mathfrak{A} = \overline{\cup_{n \geq 1} \mathfrak{A}_n}$ as an increasing union of finite dimensional subalgebras. Fix an α in $\mathrm{Id}(\mathfrak{A})$. Since $\mathfrak{A} = \overline{\cup_{n \geq 1} \alpha(\mathfrak{A}_n)}$, Theorem III.3.5 shows that there is a unitary W in $\mathfrak{A}^\sim$ such that

$$W \cup_{n \geq 1} \alpha(\mathfrak{A}_n) W^* = \cup_{n \geq 1} \mathfrak{A}_n.$$

Furthermore, we may assume that $\|W - I\| < 1/2$ so that W is in the connected component of I, and thus $\mathrm{Ad}\, W$ belongs to $\mathrm{Aut}_p(\mathfrak{A}) \cap \mathrm{Inn}(\mathfrak{A})$. Hence we may replace α by $(\mathrm{Ad}\, W)\alpha$.

So for every integer n, there is an integer m such that $\alpha(\mathfrak{A}_n)$ is contained in $\mathfrak{A}_m$. Let E_i, $1 \leq i \leq k$, be a set of minimal projections in $\mathfrak{A}_n$. Since $\alpha_* = \mathrm{id}_*$, we have $\alpha(E_i) \sim E_i$ in $\mathfrak{A}$. By Corollary IV.1.3, these projections are equivalent in some $\mathfrak{A}_{m_i}$. Taking $m' = \max\{m, m_i\}$, Corollary III.2.2 provides a unitary U_n in $\mathfrak{A}_{m'}$ such that $\alpha|\mathfrak{A}_n = \mathrm{Ad}\, U_n|\mathfrak{A}_n$. By dropping to a subsequence, we may suppose that $\alpha(\mathfrak{A}_n)$ is contained in $\mathfrak{A}_{n+1}$ and that U_n belongs to $\mathfrak{A}_{n+1}$. Clearly, $\alpha(A) = \lim_{n \to \infty} \mathrm{Ad}\, U_n(A)$ for all A in $\cup_{n \geq 1} \mathfrak{A}_n$. Since all automorphisms are contractive, this extends to the closure; whence α belongs to $\overline{\mathrm{Inn}(\mathfrak{A})}$.

Next, notice that since $\mathrm{Ad}\, U_{n+1}|\mathfrak{A}_n = \mathrm{Ad}\, U_n|\mathfrak{A}_n$, the unitary element $U_{n+1} U_n^*$ belongs to the finite dimensional C*-algebra $\mathfrak{A}_n' \cap \mathfrak{A}_{n+2}$. The unitary group is connected in $\mathcal{M}_n$ and thus in any finite dimensional C*-algebra. Therefore, we may choose a continuous path of unitaries $W_s^{(n)}$ in $\mathfrak{A}_n' \cap \mathfrak{A}_{n+2}$ for $0 \leq s \leq 1$ so that $W_0^{(n)} = I$ and $W_1^{(n)} = U_{n+1} U_n^*$. Then define a continuous path of unitaries in $\mathfrak{A}$ by

$$U_t = W_{\sigma_n(t)}^{(n)} U_n \quad \text{for} \quad 1 - \tfrac{1}{n} \leq t \leq 1 - \tfrac{1}{n+1}, \; n \geq 1$$

where $\sigma_n(t) = (n^2 + n)t - n^2 + 1$ maps $[1 - \tfrac{1}{n}, 1 - \tfrac{1}{n+1}]$ linearly onto $[0, 1]$. By construction, $\mathrm{Ad}\, U_t|\mathfrak{A}_n = \alpha|\mathfrak{A}_n$ for all $1 - \tfrac{1}{n} \leq t < 1$. Consequently, $\lim_{t \to 1} \mathrm{Ad}\, U_t(A) = \alpha(A)$ for all A in $\mathfrak{A}$. Thus α is the endpoint of the path $\mathrm{Ad}\, U_t$, and therefore belongs to $\mathrm{Aut}_p(\mathfrak{A}) \cap \overline{\mathrm{Inn}(\mathfrak{A})}$. ∎

Corollary IV.5.8 *If $\mathfrak{A}$ is a unital AF algebra such that $K_0(\mathfrak{A})$ is totally ordered and Archimedean, then $\mathrm{Aut}(\mathfrak{A}) = \overline{\mathrm{Inn}(\mathfrak{A})}$. In particular, this is the case for UHF algebras.*

Proof. Since $K_0(\mathfrak{A})$ is totally ordered and Archimedean, there is a unique unital homomorphism into $(\mathbb{R}, \mathbb{R}_+, 1)$. Hence there is a unique unital automorphism of $K_0(\mathfrak{A})$, namely id_*. But for any automorphism α, the map α_* is a unital automorphism of $K_0(\mathfrak{A})$. Hence $\mathrm{Aut}(\mathfrak{A}) = \mathrm{Id}(\mathfrak{A})$ which, by Theorem IV.5.7, equals $\overline{\mathrm{Inn}(\mathfrak{A})}$.

A UHF algebra $\mathfrak{A} = \varinjlim \mathcal{M}_{n_k}$ has the K_0 group $K_0(\mathfrak{A}) = \cup_{k \geq 1} n_k^{-1} \mathbb{Z}$ with order structure inherited from the reals and order unit 1. Thus the hypotheses hold in this case. ∎

IV.6 Riesz Groups

In this section, we develop some properties of a class of partially ordered groups that will turn out to determine the possible K_0 groups of an AF algebra. The proof of this important fact will come in the following section.

A partially ordered group (G, G^+) is called **unperforated** if whenever $g \in G$, $k \in \mathbb{N}$ and $kg \in G^+$, then g belongs to G^+.

Proposition IV.6.1 *If $\mathfrak{A}$ is an AF algebra, the group $K_0(\mathfrak{A})$ is unperforated.*

Proof. When $\mathfrak{A}$ is finite dimensional, this is clear since G is isomorphic to $(\mathbb{Z}^n, \mathbb{Z}_+^n)$ in this case. Suppose that $\mathfrak{A} = \overline{\cup_{n \geq 1} \mathfrak{A}_n}$ with connecting maps α_{mn}. Then $K_0(\mathfrak{A}) = \varinjlim K_0(\mathfrak{A}_n)$ by Theorem IV.3.3.

Suppose that g is in $K_0(\mathfrak{A})$ and $k \geq 1$ such that kg belongs to $K_0^+(\mathfrak{A})$. Then there is an integer m sufficiently large so that there are elements x in $K_0(\mathfrak{A}_m)$ and y in $K_0^+(\mathfrak{A}_m)$ for which $\alpha_{m*}(x) = g$ and $\alpha_{m*}(y) = kg$. Then $\alpha_{m*}(y - kx) = 0$. Hence there is an integer $n \geq m$ so that $\alpha_{mn*}(y - kx) = 0$. Let $x' = \alpha_{mn*}(x)$ and $y' = \alpha_{mn*}(y)$. Then $y' = kx'$ is positive in $K_0(\mathfrak{A}_n)$, which is unperforated. Therefore x' is positive. So $g = \alpha_{n*}(x')$ is also positive. ∎

The following proposition establishes the equivalence of several decomposition properties that are somewhat weaker than a lattice property. This property will turn out to be the key notion. A partially ordered group (G, G^+) which is unperforated and satisfies the Riesz interpolation property of the following theorem will be called a **Riesz group**.

Theorem IV.6.2 *For a partially ordered group (G, G^+) the following are equivalent:*

(i) *Riesz interpolation property: For every x_1, x_2, y_1, y_2 in G such that $x_i \leq y_j$ for $i, j \in \{1, 2\}$, there is an element z in G such that $x_i \leq z \leq y_j$ for $i, j \in \{1, 2\}$.*

(ii) *Riesz decomposition property: For every x, y_1, y_2 belonging to G^+ such that $x \leq y_1 + y_2$, there are elements x_i in G^+ such that $x = x_1 + x_2$ and $x_i \leq y_i$ for $i = 1, 2$.*

(iii) *another Riesz decomposition property: For every x_1, x_2, y_1, y_2 in G^+ such that $x_1 + x_2 = y_1 + y_2$, there are elements z_{ij} in G^+ such that $x_i = z_{i1} + z_{i2}$ and $y_j = z_{1j} + z_{2j}$ for $i, j \in \{1, 2\}$.*

Proof. (i)⇒(ii). Given the hypotheses of (ii), notice that 0 and $x - y_1$ are both less than x and y_2. Hence by (i), there is an interpolating element z so that

$$\{0, x - y_1\} \leq z \leq \{x, y_2\}.$$

Then let $x_1 = x - z$ and $x_2 = z$. Clearly $x = x_1 + x_2$. We already have $0 \leq z \leq y_2$. Since $x - y_1 \leq z \leq x$, we have $0 \leq x - z \leq y_1$.

(ii)⇒(iii). Given the hypotheses of (iii), apply (ii) to $x_1 \leq y_1 + y_2$ to obtain positive elements $z_{11} \leq y_1$ and $z_{12} \leq y_2$ such that $x_1 = z_{11} + z_{12}$. Then define

IV.6. Riesz Groups

$z_{21} = y_1 - z_{11}$ and $z_{22} = y_2 - z_{12}$. These are positive and are designed to satisfy the necessary identities for y_1 and y_2. It remains to verify

$$z_{21} + z_{22} = y_1 - z_{11} + y_2 - z_{12}$$
$$= (y_1 + y_2) - (z_{11} + z_{12}) = x_1 + x_2 - x_1 = x_2.$$

(iii)⇒(i). Given the hypotheses of (i), apply (iii) to the identity

$$(y_1 - x_1) + (y_2 - x_2) = (y_1 - x_2) + (y_2 - x_1).$$

This yields elements z_{ij} in G^+ such that

$$y_1 - x_1 = z_{11} + z_{12} \qquad y_2 - x_2 = z_{21} + z_{22}$$
$$y_1 - x_2 = z_{11} + z_{21} \qquad y_2 - x_1 = z_{12} + z_{22}$$

Set $z = x_1 + z_{12} = y_1 - z_{11} = x_2 + z_{21} = y_2 - z_{22}$. From this, it is evident that $x_i \leq z \leq y_j$ for $i, j \in \{1, 2\}$. ∎

An ordered group G is **lattice ordered** if for each x, y in G, there is a *least upper bound* $z = x \vee y$ in G; that is, $x \leq z$, $y \leq z$, and if $x \leq a$ and $y \leq a$, then $z \leq a$. The greatest lower bound of x and y is $x \wedge y = -\bigl((-x) \vee (-y)\bigr)$. Lattice ordered groups satisfy the Riesz interpolation property; for if $x_i \leq y_j$ for $i, j \in \{1, 2\}$, then $z = x_1 \vee x_2$ is an interpolant.

A routine induction argument yields the following generalizations.

Corollary IV.6.3 *If G is a Riesz group, then*
  (i) *For all x_i, y_j in G such that $x_i \leq y_j$ for $1 \leq i \leq k$ and $1 \leq j \leq \ell$, there is an element z in G such that $x_i \leq z \leq y_j$ for $1 \leq i \leq k$ and $1 \leq j \leq \ell$.*
  (ii) *For every x, y_j in G^+ such that $x \leq \sum_{j=1}^{\ell} y_j$, there are elements x_j in G^+ such that $x = \sum_{j=1}^{\ell} x_j$ and $x_j \leq y_j$ for $1 \leq j \leq \ell$.*
  (iii) *For all x_i, y_j in G^+ such that $\sum_{i=1}^{k} x_i = \sum_{j=1}^{\ell} y_j$, there are elements z_{ij} in G^+ for $1 \leq i \leq k$ and $1 \leq j \leq \ell$ such that*

  $$x_i = \sum_{j=1}^{\ell} z_{ij} \quad \text{and} \quad y_j = \sum_{i=1}^{k} z_{ij}.$$

To bring this back to C*-algebras, we need the following easy result.

Proposition IV.6.4 *If $\mathfrak{A}$ is an AF algebra, then $K_0(\mathfrak{A})$ is a Riesz group.*

Proof. If $\mathfrak{A}_n$ is a finite dimensional C*-algebra, $K_0(\mathfrak{A}_n) \simeq (\mathbb{Z}^k, \mathbb{Z}_+^k)$ which is a lattice ordered group. Hence it satisfies the Riesz interpolation property.

In the general case, $K_0(\mathfrak{A}) = \varinjlim K_0(\mathfrak{A}_n)$. Suppose that $x_i \leq y_j$ in $K_0(\mathfrak{A})$ for $i, j \in \{1, 2\}$. Proceeding as in the proof of Proposition IV.6.1, we can find $a_i \leq b_j$ in $K_0(\mathfrak{A}_n)$ such that $\alpha_{n*}(a_i) = x_i$ and $\alpha_{n*}(b_j) = y_j$. Let c be an interpolating point in $K_0(\mathfrak{A}_n)$ for a_i and b_j. Then $z = \alpha_{n*}(c)$ is the desired interpolant for x_i's and y_j's. ∎

IV.7 The Effros–Handelman–Shen Theorem

Our goal is to show that every Riesz group is the K_0 group of an AF algebra. Since these K_0 groups are all direct limits of copies of $(\mathbb{Z}^k, \mathbb{Z}^k_+)$, we first obtain a technical result about homomorphisms from $(\mathbb{Z}^k, \mathbb{Z}^k_+)$ into a Riesz group that will be the key building block in the proof of the main result.

Lemma IV.7.1 *Let (G, G^+) be a Riesz group. Suppose that φ is a positive homomorphism of $(\mathbb{Z}^k, \mathbb{Z}^k_+)$ into (G, G^+). Then there is an integer n and positive homomorphisms $\sigma : \mathbb{Z}^k \to \mathbb{Z}^n$ and $\psi : \mathbb{Z}^n \to G$ such that $\varphi = \psi\sigma$ and $\ker \sigma = \ker \varphi$.*

Proof. As every subgroup of $\mathbb{Z}^k$ is a finitely generated free group, there is a basis a_i, $1 \leq i \leq m$, for $\ker \varphi$ so that $\ker \varphi = \sum_{i=1}^m \mathbb{Z}a_i$. It suffices to find positive homomorphisms $\sigma_1 : \mathbb{Z}^k \to \mathbb{Z}^{n_1}$ and $\varphi_1 : \mathbb{Z}^{n_1} \to G$ so that $\varphi = \varphi_1\sigma_1$ and $\ker \sigma_1$ contains a_1. For then, m repetitions will construct σ with all m generators of $\ker \varphi$ in its kernel.

Let $a = a_1$. Rename the standard basis of $\mathbb{Z}^k$ according to the sign of the coefficients in the expansion of a as $e_1, \ldots, e_r$ for those with positive coefficients, $f_1, \ldots, f_s$ for negative coefficients, and $g_1, \ldots, g_t$ for zero coefficients. Thus there are positive integers m_i and n_j such that

$$a = \sum_{i=1}^r m_i e_i - \sum_{j=1}^s n_j f_j.$$

Let $p = p(a) = \max\{m_i, n_j\}$ and $d(a) = |\{i : m_i = p\} \cup \{j : n_j = p\}|$. Replace a by $-a$ if necessary and re-order the basis so that $m_1 = p$. We further reduce our task to the construction of σ such that

$$(p(\sigma(a)), d(\sigma(a))) < (p(a), d(a))$$

in the lexicographic order. That is, we decrease d keeping p fixed or we decrease p. Again, a repeated application of this procedure will eventually produce a homomorphism σ_1 so that $\bigl(p(\sigma_1(a)), d(\sigma_1(a))\bigr) = (0, 0)$, meaning that $\sigma_1(a) = 0$.

Let $\varepsilon_i := \varphi(e_i)$ and $\eta_j := \varphi(f_j)$. Then

$$0 = \varphi(a) = \sum_{i=1}^r m_i \varepsilon_i - \sum_{j=1}^s n_j \eta_j.$$

Hence

$$0 \leq p\varepsilon_1 \leq \sum_{i=1}^r m_i \varepsilon_i = \sum_{j=1}^s n_j \eta_j \leq p\sum_{j=1}^s \eta_j.$$

Since G is unperforated, we obtain $0 \leq \varepsilon_1 \leq \sum_{j=1}^s \eta_j$. By the Riesz interpolation property, there are elements θ_j in G^+ with $\theta_j \leq \eta_j$ such that $\varepsilon_1 = \sum_{j=1}^s \theta_j$.

IV.7. The Effros–Handelman–Shen Theorem

Set $n = r - 1 + 2s + t$ and define maps $\sigma : \mathbb{Z}^k \to \mathbb{Z}^n$ and $\psi : \mathbb{Z}^n \to G$ as follows. Denote a basis for $\mathbb{Z}^n$ by $h'_1, \ldots, h'_s, e'_2, \ldots, e'_r, f'_1, \ldots, f'_s, g'_1, \ldots, g'_t$. Then set

$$\sigma(e_1) = \sum_{j=1}^{s} h'_j \qquad \psi(h'_j) = \theta_j \qquad \text{for } 1 \le j \le s$$
$$\sigma(e_i) = e'_i \qquad \psi(e'_i) = \varepsilon_i \qquad \text{for } 2 \le i \le r$$
$$\sigma(f_j) = f'_j + h'_j \qquad \psi(f'_j) = \eta_j - \theta_j \qquad \text{for } 1 \le j \le s$$
$$\sigma(g_k) = g'_k \qquad \psi(g'_k) = \varphi(g_k) \qquad \text{for } 1 \le k \le t$$

Clearly, both σ and ψ are positive. Compute

$$\psi\sigma(e_1) = \psi(\sum_{j=1}^{s} h'_j) = \sum_{j=1}^{s} \theta_j = \varepsilon_1 = \varphi(e_1)$$
$$\psi\sigma(e_i) = \psi(e'_i) = \varepsilon_i = \varphi(e_i) \qquad \text{for } 2 \le i \le r$$
$$\psi\sigma(f_j) = \psi(f'_j + h'_j) = (\eta_j - \theta_j) + \theta_j = \eta_j = \varphi(f_j) \qquad \text{for } 1 \le j \le s$$
$$\psi\sigma(g_k) = \psi(g'_k) = \varphi(g_k) \qquad \text{for } 1 \le k \le t$$

Thus $\psi\sigma = \varphi$.

Finally, we compute

$$\sigma(a) = \sum_{i=1}^{r} m_i \sigma(e_i) - \sum_{j=1}^{s} n_j \sigma(f_j)$$
$$= p \sum_{j=1}^{s} h'_j + \sum_{i=2}^{r} m_i e'_i - \sum_{j=1}^{s} n_j (f'_j + h'_j)$$
$$= \sum_{j=1}^{s} (p - n_j) h'_j + \sum_{i=2}^{r} m_i e'_i - \sum_{j=1}^{s} n_j f'_j$$

Since $p - n_j < p$ for all j, the total number of coefficients equal to p has been decreased by one.

Thus the first stage of the construction has been successfully completed. The lemma follows from repeated application. ∎

The next theorem shows that Riesz groups are precisely the groups which are direct limits of groups isomorphic to $(\mathbb{Z}^k, \mathbb{Z}^k_+)$. Together with Proposition IV.6.4, this is most of the way to our main result.

Theorem IV.7.2 *If G is a countable Riesz group, then G is the direct limit of a sequence $G_k \simeq (\mathbb{Z}^{n_k}, \mathbb{Z}^{n_k}_+)$.*

Proof. We may assume that $G \ne \{0\}$. Let $g_1, g_2, \ldots$ be an enumeration of $G^+ \setminus \{0\}$. Let $G_0 = \{0\}$, and let $\varphi_0 : G_0 \to G$ be the zero map. For the purpose of induction, suppose that $G_k \simeq (\mathbb{Z}^{n_k}, \mathbb{Z}_+^{n_k})$ and $\varphi_k : G_k \to G$ is a positive homomorphism such that $g_1, \ldots, g_k$ belong to $\varphi(G_k^+)$. We will construct an ordered group $G_{k+1} \simeq (\mathbb{Z}^{n_{k+1}}, \mathbb{Z}_+^{n_{k+1}})$ and positive homomorphisms $\sigma_k : G_k \to G_{k+1}$ and $\varphi_{k+1} : G_{k+1} \to G$ so that $\varphi_{k+1}\sigma_k = \varphi_k$, g_{k+1} belongs to $\varphi_{k+1}(G_{k+1}^+)$ and $\ker \sigma_k = \ker \varphi_k$.

Let $K = G_k \oplus \mathbb{Z}$ and define positive homomorphisms $\sigma : G_k \to K$ and $\varphi : K \to G$ by $\sigma(g) = (g, 0)$ and $\varphi(g, n) = \varphi_k(g) + n g_{k+1}$. By Lemma IV.7.1, there is a group $G_{k+1} \simeq (\mathbb{Z}^{n_{k+1}}, \mathbb{Z}_+^{n_{k+1}})$ and there are positive homomorphisms $\sigma' : K \to G_{k+1}$ and $\varphi_{k+1} : G_{k+1} \to G$ so that $\varphi_{k+1}\sigma' = \varphi$ and $\ker \sigma' = \ker \varphi$. We have the commutative diagram

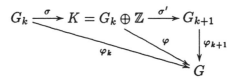

Set $\sigma_k := \sigma'\sigma$. Then

$$\varphi_{k+1}\sigma_k = (\varphi_{k+1}\sigma')\sigma = \varphi\sigma = \varphi_k.$$

and since

$$\sigma(\ker \varphi_k) = \ker \varphi_k \oplus 0 \subset \ker \varphi = \ker \sigma',$$

it follows that $\ker \varphi_k \subset \ker \sigma_k$. Hence $\ker \varphi_k = \ker \sigma_k$.

Let $H = \varinjlim G_k$ and let the maps from G_k to H be denoted by ψ_k. Let ρ be the positive homomorphism from H to G obtained from the universal property of direct limits so that the following diagram commutes.

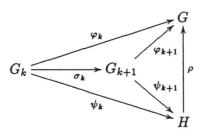

We will verify that ρ is an isomorphism. Notice that ρ is surjective because for each g_k, there is an element a_k in G_k^+ so that $\varphi_k(a_k) = g_k$. Hence $h_k := \psi_k(a_k)$ belongs to H^+ and

$$\rho(h_k) = \rho\psi_k(a_k) = \varphi_k(a_k) = g_k.$$

Thus $\rho(H^+) = G^+$ (as containment follows from the positivity of ρ). Hence

$$\rho(H) = \rho(H^+) - \rho(H^+) = G^+ - G^+ = G.$$

IV.7. The Effros–Handelman–Shen Theorem

Now suppose that $\rho(h) = 0$. Then $h = \psi_k(a)$ for some element a in G_k for k sufficiently large. Therefore
$$0 = \rho(h) = \rho\psi_k(a) = \varphi_k(a).$$
Hence a lies in $\ker \varphi_k = \ker \sigma_k$. Thus $h = \psi_{k+1}\sigma_k(a) = 0$. So ρ is an isomorphism. ∎

Now we are prepared to collect on the hard work. Since it is often easier to verify that a group is Riesz than to build an AF algebra, this will supply us with a potent tool for finding AF algebras with desirable properties.

Theorem IV.7.3 *If G is a countable Riesz group with scale $\Gamma(G)$, then there is a unital AF algebra $\mathfrak{A}$ such that*
$$(K_0(\mathfrak{A}), K_0^+(\mathfrak{A}), \Gamma(\mathfrak{A})) \simeq (G, G^+, \Gamma(G)).$$

Proof. Using Theorem IV.7.2, write G as a direct limit of groups H_k isomorphic to $(\mathbb{Z}^{n_k}, \mathbb{Z}_+^{n_k})$ with connecting maps $\varphi_{k\ell}$ such that $\ker \varphi_k = \ker \varphi_{k,k+1}$.

First suppose that G has an order unit u_G such that $\Gamma(G) = [0, u_G]$. We may suppose that u_G lies in the image of H_1, say $u_G = \varphi_1(u_1)$ for a certain element u_1. Define $u_k = \varphi_{1k}(u_1)$. Let G_k denote the subgroup of elements of H_k given by
$$G_k = \bigcup_{n \geq 1} \{h \in H_k : -nu_k \leq h \leq nu_k\}.$$
Then $G_k \simeq (\mathbb{Z}^{m_k}, \mathbb{Z}_+^{m_k})$, where $m_k \leq n_k$ is the number of non-zero coefficients of u_k in the standard basis for H_k. Moreover, u_k is an order unit for G_k.

Since any g in G_k satisfies $-nu_k \leq g \leq nu_k$, it follows that
$$-nu_\ell \leq \varphi_{k\ell}(g) \leq nu_\ell.$$
So $\varphi_{k\ell}(G_k)$ is contained in G_ℓ. Thus $\varinjlim G_k$ is the subgroup $\bigcup_{k \geq 1} \varphi_k(G_k)$ of G. However, if g belongs to G, there is an integer n so that $-nu_G \leq g \leq nu_G$. Since $nu_G - g$ and $g + nu_G$ are positive, there is an integer k and positive elements x, y in H_k^+ so that $\varphi_k(x) = nu_G - g$ and $\varphi_k(y) = g + nu_G$. Thus $a := nu_k - x$ and $b := y - nu_k$ satisfy $\varphi_k(a) = \varphi_k(b) = g$. Let $z = \varphi_{k,k+1}(a)$. Then $z = \varphi_{k,k+1}(b)$ as well because $a - b$ belongs to $\ker \varphi_k = \ker \varphi_{k,k+1}$. Hence
$$z = nu_{k+1} - \varphi_{k,k+1}(x) \leq nu_{k+1}$$
and
$$z = -nu_{k+1} + \varphi_{k,k+1}(y) \geq -nu_{k+1}.$$
Therefore z belongs to G_{k+1}; and consequently $g = \varphi_{k+1}(z)$ belongs to $\varinjlim G_k$. Thus $G = \varinjlim G_k$ as desired.

Let $u_k = (n_{k1}, \ldots, n_{km_k})^t$. Define
$$\mathfrak{A}_k = \mathcal{M}_{n_{k1}} \oplus \cdots \oplus \mathcal{M}_{n_{km_k}}.$$

Then $K_0(\mathfrak{A}_k) = G_k$ as a scaled dimension group. By Lemma IV.4.1, there is a homomorphism $\alpha_k : \mathfrak{A}_k \to \mathfrak{A}_{k+1}$ so that $\alpha_{k*} = \varphi_k$. Let $\mathfrak{A} = \varinjlim \mathfrak{A}_n$. By Theorem IV.3.3,

$$K_0(\mathfrak{A}) = \varinjlim K_0(\mathfrak{A}_k) = \varinjlim G_k = G$$

as scaled dimension groups.

In the general case, enumerate $\Gamma(G) = \{g_i : i \geq 1\}$. Since this is a directed set, choose elements v_n in $\Gamma(G)$ such that $g_i \leq v_n$ for $1 \leq i \leq n$. From the arguments above, we can find an integer k_n and element u_n in $H_{k_n}^+$ so that $\varphi_{k_n}(u_n) = v_n$ and so that $\varphi_{n_{k-1}n_k}(u_{k-1}) \leq u_k$. Then proceed as in the unital case to construct an AF algebra $\mathfrak{A} = \varinjlim \mathfrak{A}_n$ such that $K_0(\mathfrak{A}_n) \simeq (G_{k_n}, G_{k_n}^+, [0, u_n])$. The only additional thing to notice is that in the limit,

$$\bigcup_{n \geq 1} \varphi_{k_n}([0, u_k]) = \bigcup_{n \geq 1} [0, v_n] = \Gamma(G).$$

This shows that $\Gamma(\mathfrak{A}) = \Gamma(G)$ as desired. ∎

IV.8 Blackadar's Simple Unital Projectionless C*-algebra

In this section, we will provide an application of the ideas presented about AF algebras to construct a simple C*-algebra that has no projections except for zero and the identity. This example, constructed by Blackadar, was the first C*-algebra known to have this property. Of course, it is not an AF algebra, but it is built up from one in a clever way.

The 2-adic numbers. Let $\overline{\mathbb{Q}}$ denote the group of 2-adic numbers consisting of elements which are formal sums $q = \sum_{k=m}^{\infty} \varepsilon_k 2^k$ where $m \in \mathbb{Z}$ and $\varepsilon_k \in \{0, 1\}$. Addition and multiplication are defined in the usual way. This makes $\overline{\mathbb{Q}}$ into a ring. The subring $\overline{\mathbb{Z}}$ of 2-adic integers consists of those elements with $m = 0$. This subring contains a copy of the integers corresponding to sequences (ε_k) which are eventually constant. For example, the formal sum $\sum_{k \geq 0} 2^k$ represents -1. In the same way, $\overline{\mathbb{Q}}$ contains the diadic rational numbers. Because 2 is a prime, it can be verified that $\overline{\mathbb{Q}}$ is a field. (We will not need this fact.)

There is a metric on $\overline{\mathbb{Q}}$ given by the norm $\|q\| = 2^{-m}$, where m is the least integer such that $\varepsilon_m = 1$ (and $\|0\| = 0$). It is easy to verify that

$$\|q_1 - q_3\| \leq \max\{\|q_1 - q_2\|, \|q_2 - q_3\|\}.$$

This is the topology inherited by $\overline{\mathbb{Q}}$ as a subset of the Cantor set $\prod_{n=-\infty}^{\infty} \{0,1\}_n$. From this, it is evident that $\overline{\mathbb{Z}}$ is compact; and so $\overline{\mathbb{Q}} = \cup_{n \geq 1} 2^{-n} \overline{\mathbb{Z}}$ is σ-compact. It is also evident from the definition of the metric that $\mathbb{Z}$ is a dense subset of $\overline{\mathbb{Z}}$.

Suppose that G is a compact open subgroup of $(\overline{\mathbb{Q}}, +)$. Then it is contained in $2^{-m}\overline{\mathbb{Z}}$ for some m and it contains a basic open neighbourhood of 0, say $2^k\overline{\mathbb{Z}}$.

IV.8. Blackadar's Simple Unital Projectionless C*-algebra

It is not difficult to see that $2^{-m}\overline{\mathbb{Z}}/2^k\overline{\mathbb{Z}}$ is the cyclic group on 2^{m+k} elements. So we deduce that the compact open subgroups have the form $2^n\overline{\mathbb{Z}}$ for n in $\mathbb{Z}$. Every compact open set will be the union of cosets of these subgroups. Indeed, each point a will have a neighbourhood $a + 2^k\overline{\mathbb{Z}}$ contained in the clopen set. A simple argument allows one to reduce to a finite union of disjoint cosets.

As $\overline{\mathbb{Q}}$ is a locally compact topological group (under addition), there is a Haar measure μ on $\overline{\mathbb{Q}}$ which is unique if we normalize it so that $\mu(\overline{\mathbb{Z}}) = 1$. We summarize these facts in the following theorem. Details may be found in Hewitt and Ross, *Abstract Harmonic Analysis*, §II.10.

Theorem IV.8.1 $\overline{\mathbb{Q}}$ *is a locally compact, σ-compact, totally disconnected topological field. $\overline{\mathbb{Z}}$ is a compact open subring which contains $\mathbb{Z}$ as a dense subset. The compact open subgroups of $(\overline{\mathbb{Q}}, +)$ are $2^n\overline{\mathbb{Z}}$ for $n \in \mathbb{Z}$. Every clopen subset of $\overline{\mathbb{Q}}$ is a finite disjoint union of cosets of these subgroups. There is a (unique) Haar measure μ on $\overline{\mathbb{Q}}$ such that $\mu(\overline{\mathbb{Z}}) = 1$.*

An AF algebra. Let $G = (C_c(\overline{\mathbb{Q}}, \mathbb{Z}), +)$ be the additive group of continuous, integer valued, compactly supported functions on $\overline{\mathbb{Q}}$. Define a state θ on G by

$$\theta(g) := \int g \, d\mu.$$

It is easy to verify that θ takes values in the diadic rational numbers. Define a cone on G by

$$G^+ := \{g \in G : \theta(g) > 0\} \cup \{0\}.$$

Clearly G is a countable torsion free group. Let $\chi_{\overline{\mathbb{Z}}}$ denote the characteristic function of $\overline{\mathbb{Z}}$; and define this to be the order unit of G.

Proposition IV.8.2 $(G, G^+, \chi_{\overline{\mathbb{Z}}})$ *is a simple dimension group with unique state θ.*

Proof. That $G^+ + G^+$ is contained in G^+ is clear. And if g is in G with $|\theta(g)| < n$, then $\theta(g + n\chi_{\overline{\mathbb{Z}}}) = \theta(g) + n > 0$. Hence $g = (g + n\chi_{\overline{\mathbb{Z}}}) - n\chi_{\overline{\mathbb{Z}}}$ belongs to $G^+ - G^+$. It also follows similarly that

$$-n\chi_{\overline{\mathbb{Z}}} < g < n\chi_{\overline{\mathbb{Z}}}$$

so that $\chi_{\overline{\mathbb{Z}}}$ is an order unit.

To see that G is unperforated, suppose that $ng > 0$, meaning that

$$0 < \theta(ng) = n\theta(g).$$

Then $\theta(g) > 0$, and hence $g > 0$. Also G is a lattice ordered group with the operations of min and max. So it is also a Riesz group.

Suppose that H is a non-zero order ideal of G. Then it contains a positive element h. So for some integer n, $\theta(h) > 2^{-n}$. Hence $\theta(2^n h - \chi_{\overline{\mathbb{Z}}}) > 0$. That is, $0 \leq \chi_{\overline{\mathbb{Z}}} \leq 2^n h$. By the hereditary property of order ideals, $\chi_{\overline{\mathbb{Z}}}$ belongs to H and thus $H = G$. So G is simple.

If τ is a state on G and $\theta(g) = 0$, then for every n in $\mathbb{Z}$, $\theta(\chi_{\overline{\mathbb{Z}}} + ng) = 1 > 0$ and hence $\chi_{\overline{\mathbb{Z}}} + ng > 0$. Thus

$$0 \leq \tau(\chi_{\overline{\mathbb{Z}}} + ng) = 1 + n\tau(g) \quad \text{for all} \quad n \in \mathbb{Z}.$$

Hence $\tau(g) = 0$.

Consider $\chi_{2^n\overline{\mathbb{Z}}}$ for $n \geq 0$. For each integer $0 \leq a < 2^n$,

$$\theta(\chi_{2^n\overline{\mathbb{Z}}} - \chi_{2^n\overline{\mathbb{Z}}+a}) = 2^{-n} - 2^{-n} = 0.$$

Hence $\tau(\chi_{2^n\overline{\mathbb{Z}}+a}) = \tau(\chi_{2^n\overline{\mathbb{Z}}})$. And since

$$1 = \tau(\chi_{\overline{\mathbb{Z}}}) = \tau\Big(\sum_{a=0}^{2^n-1} \chi_{2^n\overline{\mathbb{Z}}+a}\Big) = 2^n\tau(\chi_{\overline{\mathbb{Z}}}),$$

we deduce that $\tau(\chi_{2^n\overline{\mathbb{Z}}+a}) = 2^{-n} = \theta(\chi_{2^n\overline{\mathbb{Z}}+a})$ for all $n \geq 0$ and $0 \leq a < 2^n$. Since these characteristic functions span G, it follows that $\tau = \theta$. ∎

By the Effros–Handelman–Shen Theorem IV.7.3, there is an AF algebra $\mathfrak{B}$ such that $(K_0(\mathfrak{B}), K_0^+(\mathfrak{B}), [I])$ is order isomorphic to $(G, G^+, \chi_{\overline{\mathbb{Z}}})$. As G is simple, Corollary IV.5.2 implies that $\mathfrak{B}$ is also simple. By Theorem IV.5.3, $\mathfrak{B}$ has a unique trace associated to the unique state θ on G.

Define two automorphisms of G by

$$\sigma_* g(x) = g(x-1) \quad \text{and} \quad \alpha_* g(x) = g(x/2).$$

It is easy to see that both σ_* and α_* are positive since Haar measure is translation invariant and is scaled by 2 on multiplying by $1/2$. Set $u = \chi_{\overline{\mathbb{Z}}}$ and $v = \chi_{2\overline{\mathbb{Z}}}$. Then we have,

$$\sigma_*(u) = u \quad \alpha_*(u) = v \quad \text{and} \quad \sigma_*(v) = \chi_{1+2\overline{\mathbb{Z}}} = u - v.$$

In addition, σ_* and α_* are related by $\sigma_*^2 \alpha_* = \alpha_* \sigma_*$. Indeed,

$$\sigma_*^2 \alpha_* g(x) = \alpha_* g(x-2) = g(\tfrac{x}{2} - 1)$$
$$= \sigma_* g(x/2) = \alpha_* \sigma_* g(x).$$

We also point out a fact that will be needed later: u is a minimal fixed point of σ_* in G^+. Indeed, if $\sigma_*(g) = g$, then g is constant on cosets of $\mathbb{Z}$, and hence by continuity, it is constant on cosets of $\overline{\mathbb{Z}}$. Hence $\theta(g)$ is a positive integer. As $\theta(u) = 1$, this is a minimal fixed point.

By Elliott's Theorem IV.4.3, there is an automorphism σ of $\mathfrak{B}$ which induces σ_*. There is a projection P in $\mathfrak{B}$ with $[P] = v$. Since

$$\sigma_*([P]) = [I] - [P] = [I - P],$$

there is a partial isometry U in $\mathfrak{B}$ such that

$$U\sigma(P)U^* = I - P.$$

IV.8. Blackadar's Simple Unital Projectionless C*-algebra

Likewise, there is a partial isometry V so that $V\sigma(I-P)V^* = P$. Let
$$W = (I-P)U + PV.$$
This is a unitary operator such that $W\sigma(P)W^* = I - P$. Replace σ by $\operatorname{Ad} W \sigma$. Then $\sigma(P) = I - P$. Because $(\operatorname{Ad} W)_* = \operatorname{id}_*$, this doesn't affect σ_*.

Since α_* is not unital, it does not come from an automorphism of $\mathfrak{B}$. Define $\mathfrak{B}_0 = P\mathfrak{B}P$. Then using Exercise IV.13, we may compute $K_0(\mathfrak{B}_0)$. It has order unit $[P] = v = \chi_{2\overline{\mathbb{Z}}}$. However, this is an order unit for all of G by the same argument that shows that $\chi_{\overline{\mathbb{Z}}}$ is an order unit. Consequently,
$$(K_0(\mathfrak{B}_0), K_0^+(\mathfrak{B}_0), [I_{\mathfrak{B}_0}]) \simeq (G, G^+, v).$$

It follows that α_* is an order isomorphism of $K_0(\mathfrak{B})$ onto $K_0(\mathfrak{B}_0)$. By Elliott's Theorem IV.4.3, there is an isomorphism α of $\mathfrak{B}$ onto $\mathfrak{B}_0$ which induces α_* on the K_0 groups. Since $\sigma^2(P) = P$, we have $\sigma^2(\mathfrak{B}_0) = \mathfrak{B}_0$. Let $\rho = \sigma^2|_{\mathfrak{B}_0}$ belong to $\operatorname{Aut}(\mathfrak{B}_0)$. Because $K_0(\mathfrak{B}_0) = G$, it follows that $\rho_* = \sigma_*^2$. Consider $\beta = \alpha^{-1}\rho\alpha\sigma^{-1}$. This is an automorphism of $\mathfrak{B}$ such that
$$\beta_* = \alpha_*^{-1}\sigma_*^2\alpha_*\sigma_*^{-1} = \operatorname{id}_*.$$

By Theorem IV.5.7, β lies in the path component of the identity automorphism. Let β_t be a continuous path of automorphisms such that $\beta_0 = \operatorname{id}$ and $\beta_1 = \beta$. Then define the path $\alpha_t := \alpha\beta_t$ of isomorphisms of $\mathfrak{B}$ onto $\mathfrak{B}_0$ from $\alpha_0 = \alpha$ to $\alpha_1 = \rho\alpha\sigma^{-1}$.

The Mapping Cone. Now define a C*-algebra Γ known as the **mapping cone** of σ by
$$\Gamma = \{f \in C([0,1], \mathfrak{B}) : f(1) = \sigma(f(0))\}.$$
This is a unital C*-algebra that is neither simple nor AF. It will be used to build our example. Let us identify the space obtained from $[0, 1]$ by identifying 0 and 1 with the unit circle $\mathbb{T}$. This is necessarily in the spectrum of Γ because the value of $f(1)$ is determined by $f(0)$.

Proposition IV.8.3 Γ *is a unital C*-algebra with no proper projections. The ideals of Γ are precisely $\mathfrak{J}_A := \{f \in \Gamma : f|_A = 0\}$ for closed subsets A of $\mathbb{T} = [0,1]$ mod 1.*

Proof. If $f = f^* = f^2$, then $f(t)$ is a projection for every t. By the homotopy property of K_0 (Proposition IV.1.2), we have that $[f(t)] = [f(0)]$ for all t in $[0,1]$. Thus
$$[f(0)] = [f(1)] = [\sigma(f(0))] = \sigma_*([f(0)]).$$
Hence $0 \le [f(0)] \le [I]$ is a fixed point of σ_*. By our earlier comments, this implies that $f(0)$ is scalar; and thus f is scalar.

Suppose that $\mathfrak{J}$ is an ideal of Γ. Let
$$A = \ker \mathfrak{J} = \{t \in \mathbb{T} : f(t) = 0 \text{ for all } f \in \mathfrak{J}\}.$$

If $t \notin A$, choose f in $\mathfrak{J}$ such that $f(t) = B \neq 0$. Then since $\mathfrak{B}$ is simple, there are X_i and Y_i in $\mathfrak{B}$ such that $\sum_{i=1}^n X_i B Y_i = I$. Pick any elements x_i and y_i in Γ such that $x_i(t) = X_i$ and $y_i(t) = Y_i$. (When $t = 0$, the corresponding value at $t = 1$ is determined.) Then $\mathfrak{J}$ contains $h = \sum_{i=1}^n x_i f y_i$, which satisfies $h(t) = I$. Set $k_t = F(\operatorname{Re} h)$ where $F(x) = x \wedge 1 \vee 0$. It follows that $0 \leq k_t \leq 1$, k_t belongs to $\mathfrak{J}$ and $k_t(t) = I$.

The proof is completed by a partition of unity argument. Fix g in $\mathfrak{J}_A$ and $\varepsilon > 0$. The set $B = \{t \in \mathbb{T} : \|g(t)\| \geq \varepsilon\}$ is compact and disjoint from A. For t in B, there is a neighbourhood

$$\mathcal{O}_t = \{s \in \mathbb{T} : k_t(s) > (1-\varepsilon)I\}.$$

By compactness, there is a finite open cover $\mathcal{O}_{t_i}$ of B for certain points t_i in $\mathbb{T}$. Then

$$k = F\left((1-\varepsilon)^{-1} \sum_{i=1}^n k_{t_i}\right)$$

belongs to $\mathfrak{J}$, $0 \leq k \leq 1$ and $k(t) = I$ for all t in B. Thus

$$\operatorname{dist}(g, \mathfrak{J}) \leq \|g - gk\| \leq \sup_{t \in \mathbb{T} \setminus B} \|g(t)\| \|I - k(t)\| \leq \varepsilon.$$

Hence $\mathfrak{J} = \mathfrak{J}_A$. ∎

Define an endomorphism ψ of Γ by

$$\psi f(t) = \alpha_t(f(\tfrac{t}{2})) + \sigma^{-1} \alpha_1(f(\tfrac{t+1}{2})).$$

Note that α_t maps $\mathfrak{B}$ into $\mathfrak{B}_0 = P\mathfrak{B}P$; while $\sigma^{-1}\alpha_1 = \sigma\alpha\sigma^{-1}$ maps $\mathfrak{B}$ into $P^\perp \mathfrak{B} P^\perp$. Consequently, this is a direct sum. It follows that ψ is an endomorphism. Moreover, it is readily seen to be injective. To verify that ψf actually lies in Γ, compute

$$\sigma \psi f(0) = \sigma\left(\alpha(f(0)) + \sigma^{-1}\alpha_1(f(\tfrac{1}{2}))\right)$$
$$= \alpha_1(f(\tfrac{1}{2})) + \sigma\alpha(\sigma^{-1}f(1))$$
$$= \alpha_1(f(\tfrac{1}{2})) + \sigma^{-1}\alpha_1(f(1)) = \psi f(1).$$

The map ψ is unital since $\psi 1(t) = P + P^\perp = I$.

Let $\Gamma_n = \Gamma$ for $n \geq 1$, and set $\psi_n = \psi : \Gamma_n \to \Gamma_{n+1}$. Then define a C*-algebra $\mathfrak{A} = \varinjlim(\Gamma_n, \psi_n)$. We identify Γ_n with its image in $\mathfrak{A}$. This map is injective because each ψ_n is injective. So $\mathfrak{A} = \overline{\cup_{n \geq 1} \Gamma_n}$.

Theorem IV.8.4 *The C*-algebra $\mathfrak{A}$ is a simple unital projectionless C*-algebra.*

Proof. If E is a projection in $\mathfrak{A}$, then there is an integer n sufficiently large that $\operatorname{dist}(E, \Gamma_n) < 1/4$. Hence by Lemma III.3.1, there is a projection F in Γ_n such that $\|E - F\| < 1/2$. By Proposition IV.8.3, F is 0 or 1. Thus E is also scalar. So $\mathfrak{A}$ has no non-trivial projections.

Suppose that $\mathfrak{J}$ is an ideal of $\mathfrak{A}$. By Lemma III.4.1,
$$\mathfrak{J} = \overline{\bigcup_{n \geq 1} \mathfrak{J} \cap \Gamma_n}.$$
As each $\mathfrak{J}_n = \mathfrak{J} \cap \Gamma_n$ is an ideal of Γ_n, it has the form $\mathfrak{J}_{A_n}$ for a certain closed subset of $\mathbb{T}$ by Proposition IV.8.3. Moreover, the nature of the imbedding of Γ_n into Γ_{n+1} yields
$$\psi(\mathfrak{J}_{A_n}) = \mathfrak{J}_{A_{n+1}} \cap \psi(\Gamma_n).$$
This implies that
$$A_n = \tfrac{1}{2} A_{n+1} + (\tfrac{1}{2} A_{n+1} + \tfrac{1}{2}).$$
So A_n is invariant under translation by $\tfrac{1}{2}$. Inductively, we deduce that A_n is invariant under translation by 2^{-k} for all $k \geq 1$. Since A_n is closed, this implies that it is empty or equal to all of $\mathbb{T}$. If A_n is empty, then $\mathfrak{J}_n$ contains the identity and hence $\mathfrak{J} = \mathfrak{A}$. On the other hand, if $A_n = \mathbb{T}$ for every $n \geq 1$, then $\mathfrak{J}_n = \{0\}$ for all n; whence $\mathfrak{J} = \{0\}$. Thus $\mathfrak{A}$ is simple. ∎

Exercises

IV.1   Show that if two projections are homotopic in the set of idempotents, then they are homotopic in the set of projections.

IV.2   Show that equivalent projections need not be homotopic.
HINT: Consider the projections
$$P = \begin{bmatrix} 1 & 0 \\ 0 & 0 \end{bmatrix} \quad \text{and} \quad Q = \begin{bmatrix} t & \sqrt{t(1-t)} \\ \sqrt{t(1-t)} & 1-t \end{bmatrix}$$
in $\mathcal{M}_2(C[0,1])$. Show that they are contained in a common C*-subalgebra in which they are equivalent but not homotopic.

IV.3   Suppose that $\mathfrak{A}$ is contained in $\mathcal{B}(\mathcal{H})$. Show that if E is an idempotent, then the equivalent projection computed in Proposition IV.1.1 is the orthogonal projection onto the range of E.

IV.4   Two homomorphisms of $\mathfrak{A}$ into $\mathfrak{B}$ are homotopic is there is a pointwise continuous path of homomorphisms connecting them. Show that if α is homotopic to β, then $\alpha_* = \beta_*$.

IV.5   A C*-algebra is **contractible** if the identity automorphism is homotopic to the zero map. Show that the **cone** of $\mathfrak{A}$,
$$C\mathfrak{A} = \{f \in C([0,1], \mathfrak{A}) : f(0) = 0\}$$
is contractible.

IV.6   Show that if $\mathfrak{A}$ is finite and unital, then left invertible elements are invertible.

IV.7 Show that a UHF algebra $\mathfrak{A}$ has a unique trace τ. Then show that τ_* is an isomorphism of $K_0(\mathfrak{A})$ into $\mathbb{R}$.

IV.8 Find all the order ideals of K_0 of the GICAR algebra. Compare this description of the ideals of the algebra with the Bratteli diagram approach.

IV.9 Show that if $\mathfrak{A}$ has a faithful trace, then it is stably finite.

IV.10 Show that $\text{Inn}(\mathfrak{A})$, $\overline{\text{Inn}(\mathfrak{A})}$, $\text{Aut}_0(\mathfrak{A})$ and $\text{Id}(\mathfrak{A})$ are all normal subgroups of $\text{Aut}(\mathfrak{A})$.

IV.11 In a finite dimensional C*-algebra $\mathfrak{A}$, show that $[P] \leq [Q]$ if and only if $\tau(P) \leq \tau(Q)$ for every trace on $\mathfrak{A}$.

IV.12 Suppose that P_i and Q_j are projections in an AF algebra such that
$$\sum_{i=1}^{m} P_i \leq \sum_{j=1}^{n} Q_j.$$
Show that $\sum_{i=1}^{m}[P_i] \leq \sum_{j=1}^{n}[Q_j]$.
HINT: Reduce the question to a finite dimensional subalgebra, and use the previous exercise.

IV.13 If $\mathfrak{A}$ is AF and P is a projection in $\mathfrak{A}$, show that
$$K_0(P\mathfrak{A}P) = \bigcup_{n \geq 1}[-n[P], n[P]], \quad K_0^+(P\mathfrak{A}P) = K_0(P\mathfrak{A}P) \cap K_0^+(\mathfrak{A})$$
and the order unit is $[P]$.

IV.14 Show that every countable torsion free abelian group G is the direct limit of a sequence of $\mathbb{Z}^{n_i}$. Show that G can be given an order to make it into a dimension group.

IV.15 If $0 \to \mathfrak{J} \to \mathfrak{E} \to \mathfrak{A} \to 0$ is an exact sequence of AF algebras, show that $0 \to K_0(\mathfrak{J}) \to K_0(\mathfrak{E}) \to K_0(\mathfrak{A}) \to 0$ is an exact sequence of ordered groups.

IV.16 Show that $G = \mathbb{Z}^2$ with the order
$$G^+ = \{(m,n) : m = n = 0 \text{ or } m > 0, n > 0\}$$
is an ordered group that fails to have the Riesz interpolation property. Show that $0 \to (\mathbb{Z}, \mathbb{Z}_+) \to (G, G^+) \to (\mathbb{Z}, \mathbb{Z}_+) \to 0$ is an exact sequence of ordered groups. Thus extensions of Riesz groups need not be Riesz groups.

IV.17 (a) If $\mathfrak{A}$ is AF, consider the imbeddings of $\mathcal{M}_n(\mathfrak{A})$ into $\mathcal{M}_{n+1}(\mathfrak{A})$ given by $\alpha_n(A) = \begin{bmatrix} A & 0 \\ 0 & 0 \end{bmatrix}$. Show that $\mathfrak{A} \otimes \mathfrak{K} := \varinjlim \mathcal{M}_n(\mathfrak{A})$ is AF; and show that
$$(K_0(\mathfrak{A} \otimes \mathfrak{K}), K_0^+(\mathfrak{A} \otimes \mathfrak{K}), \Gamma(\mathfrak{A} \otimes \mathfrak{K})) \simeq (K_0(\mathfrak{A}), K_0^+(\mathfrak{A}), K_0^+(\mathfrak{A})).$$

(b) If $\mathfrak{A}$ and $\mathfrak{B}$ are AF, show that $\mathfrak{A} \otimes \mathfrak{K}$ and $\mathfrak{B} \otimes \mathfrak{K}$ are isomorphic if and only if $(K_0(\mathfrak{A}), K_0^+(\mathfrak{A}))$ and $(K_0(\mathfrak{B}), K_0^+(\mathfrak{B}))$ are isomorphic.

IV.18 Let $\mathfrak{A}$ be Blackadar's projectionless C*-algebra.

(a) Show that every trace on Γ has the form $\int_0^1 \tau(f(t))\, d\mu$ for some probability measure μ on $[0, 1]$, where τ is the unique trace on $\mathfrak{B}$.

(b) Define traces on each Γ_n by $\tau_n(f) = \int_0^1 \tau(f(t))\, dt$. Then verify that $\tau_{n+1}\psi = \tau_n$. Thus deduce that $\overline{\tau} = \varinjlim \tau_n$ is a trace on $\mathfrak{A}$.

(c) Show that $\mathfrak{A}$ has a unique trace.

HINT: Show that the restriction to each Γ_n corresponds to a measure μ_n which is invariant under diadic translations; and thus is Lebesgue measure.

Notes and Remarks.

Equivalence of projections was first introduced for von Neumann algebras by Murray and von Neumann [1936]. K-theory comes from algebraic topology. A classical introduction to topological K-theory is Atiyah [1967]. It was first introduced to C*-algebras by Elliott [1978], although his paper Elliott [1976] classifying AF algebras by their equivalence classes of projections makes no reference to the enveloping group. The basic reference for Riesz groups is Fuchs [1965]. Elliott [1978] pointed out that dimension groups have the Riesz property. Effros, Handelman and Shen [1980] proved Theorem IV.7.3. Blackadar [1980] constructed this first known example of a simple projectionless C*-algebra. Blackadar's paper also contains Theorem IV.5.7.

CHAPTER V

C*-algebras of Isometries

In this chapter, we will study several important classes of C*-algebras built from isometries and weighted shifts.

V.1 Toeplitz Operators

The most important non-normal operator is the **unilateral shift**. It is much studied because it displays many interesting phenomena that occur in infinite dimensions but not in finite dimensional spaces. This operator is defined on an orthonormal basis $\{e_n : n \geq 0\}$ by

$$S \sum_{n \geq 0} a_n e_n = \sum_{n \geq 0} a_n e_{n+1}.$$

It is immediately evident that S is an isometry with range equal to the closed span of $\{e_n : n \geq 1\}$. This is a proper subspace of codimension 1. Thus S is a **proper isometry**, meaning that it is isometric but not unitary. Moreover, it is Fredholm with index

$$\text{ind}(S) = \text{null}(S) - \text{null}(S^*) = -1.$$

There is a convenient representation of S which allows us to utilize an important connection with function theory. Let $L^2(\mathbb{T})$ denote the square integrable functions on the unit circle with respect to Lebesgue measure. This has an orthonormal basis $e_n = z^n$ for $n \in \mathbb{Z}$; where z is the identity function on $\mathbb{T}$. Let H^2 or $H^2(\mathbb{T})$ denote the subspace of $L^2(\mathbb{T})$ spanned by $\{e_n(z) : n \geq 0\}$. This is the subspace of L^2 functions of analytic type. Indeed, the expansion $h = \sum_{n \geq 0} a_n e_n$ is the Fourier series of h. The Poisson extension of h to the unit disk is $\sum_{n \geq 0} a_n z^n$, which converges uniformly on compact subsets of the unit disk $\mathbb{D}$; and therefore determines an analytic function $\hat{h}(z)$ on $\mathbb{D}$.

It is an easy exercise to verify that

$$\lim_{r \to 1^-} \|\hat{h}(re^{i\theta})\|_2 = \|h\|_2.$$

It also follows from Fatou's theorem that the radial limits exist almost everywhere

$$\lim_{r \to 1^-} \hat{h}(re^{i\theta}) = h(e^{i\theta}) \text{ a.e.}$$

V.1. Toeplitz Operators

While we will not make use of this fact here, this shows that there is an intimate connection with analytic function theory.

If g is a bounded measurable function on the circle, $g \in L^\infty(\mathbb{T})$, define a multiplication operator M_g on $L^2(\mathbb{T})$ by $M_g f = gf$. This is easily seen to be a normal operator with norm $\|M_g\| = \|g\|_\infty$. Define the **Toeplitz operator** T_g on H^2 by

$$T_g h = P_{H^2} g h.$$

This is just the compression of M_g to the subspace H^2. In particular, the operator T_z acts on the basis e_n by $T_z e_n = e_{n+1}$. Thus T_z is unitarily equivalent to the unilateral shift.

Proposition V.1.1 *For g in L^∞, $T_g^* = T_{\bar g}$ and $\|T_g\| = \|T_g\|_e = \|g\|_\infty$.*

Proof. The first statement follows from a simple calculation:

$$(T_g^* f, h) = (f, P_{H^2} g h) = (f, gh) = (\bar g f, h) = (T_{\bar g} f, h)$$

for all f, h in H^2. Now

$$\|T_g\|_e \leq \|T_g\| \leq \|M_g\| = \|g\|_\infty.$$

Let $\varepsilon > 0$ be given. Since the trigonometric polynomials (polynomials in z and z^{-1}) are dense in $L^2(\mathbb{T})$, there is a polynomial $p = \sum_{k=-N}^{N} a_k e_k$ with $\|p\|_2 = 1$ such that $\|gp\|_2 > \|g\|_\infty - \varepsilon$. Then for all integers $n > N$, $z^n p$ belongs to H^2. Suppose that $gp = \sum_{k=-\infty}^{\infty} b_k e_k$. Then

$$T_g z^n p = P_{H^2} z^n g p = \sum_{k=-n}^{\infty} b_k e_{k+n}.$$

Hence

$$\lim_{n \to \infty} \|T_g z^n p\|_2 = \|gp\|_2 > \|g\|_\infty - \varepsilon.$$

Moreover, the sequence $z^n p$ tends weakly to zero (and in fact, $z^{3Ns} p$ are pairwise orthogonal for $s \geq 1$). Hence we deduce that $\|T_g\|_e = \|g\|_\infty$. ∎

Let H^∞ denote the subalgebra of L^∞ of functions of analytic type. (Consequently, $H^\infty = H^2 \cap L^\infty$.) An operator is **subnormal** if it is the restriction of a normal operator to an invariant subspace.

Proposition V.1.2 *For h in H^∞, the space H^2 is invariant for M_h; and thus analytic Toeplitz operators are subnormal, $T_h = M_h|_{H^2}$. For every g in L^∞ and h in H^∞, one has*

$$T_g T_h = T_{gh} \quad \text{and} \quad T_{\bar h} T_g = T_{\bar h g}.$$

Proof. Let $h = \sum_{n \geq 0} h_n z^n$. Then $M_h z^k = \sum_{n \geq 0} h_n z^{n+k}$ belongs to H^2 for all $k \geq 0$. Thus H^2 is invariant for M_h, and so $T_h f = hf$ for all f in H^2. Consequently,

$$T_g T_h f = T_g h f = P_{H^2} g h f = T_{gh} f.$$

Finally,

$$T_{\bar{h}} T_{\bar{g}} = (T_{\bar{g}} T_h)^* = T_{\bar{g}h}^* = T_{\bar{h}g}. \qquad \blacksquare$$

A key fact is that the shift T_z commutes with all Toeplitz operators modulo the compact operators.

Proposition V.1.3 *For every g in L^∞, the commutator $T_z T_g - T_g T_z$ has rank at most one.*

Proof. Notice that $T_g T_z = T_{gz}$ by the previous proposition. So working in L^2,

$$T_z T_g - T_{gz} = \left(P_{H^2} M_z P_{H^2} M_g P_{H^2} - P_{H^2} M_z M_g P_{H^2} \right) | H^2$$
$$= P_{H^2} M_z P_{H^2}^\perp M_g P_{H^2} | H^2.$$

Since $P_{H^2} M_z P_{H^2}^\perp = e_0 e_{-1}^*$ is rank one, the commutator has rank at most one. $\blacksquare$

Corollary V.1.4 *For all g in L^∞ and f in $C(\mathbb{T})$, the semi-commutators $T_f T_g - T_{fg}$ and $T_g T_f - T_{gf}$ are compact.*

Proof. Given $\varepsilon > 0$, there is a trigonometric polynomial $p = \sum_{k=-N}^{N} a_k z^k$ such that $\|f - p\|_\infty < \varepsilon$. As in the proof of the previous proposition, we obtain

$$T_g T_f - T_{gf} = P_{H^2} M_g P_{H^2}^\perp M_f P_{H^2}.$$

So it suffices to show that $P_{H^2}^\perp M_f P_{H^2}$ is compact. The range of $P_{H^2}^\perp M_p P_{H^2}$ is contained in $\operatorname{span}\{z^{-k} : 1 \leq k \leq N\}$; and thus it is finite rank. Since

$$\|P_{H^2}^\perp (M_f - M_p) P_{H^2}\| \leq \|f - p\|_\infty < \varepsilon,$$

it follows that $P_{H^2}^\perp M_f P_{H^2}$ is the norm limit of finite rank operators and thus is compact. The other term is handled by taking adjoints. $\blacksquare$

Now we are prepared to compute the C*-algebra generated by T_z.

Theorem V.1.5 *The C*-algebra generated by T_z has the form*

$$\mathcal{T}(C(\mathbb{T})) := \{T_f + K : f \in C(\mathbb{T}) \text{ and } K \in \mathfrak{K}\},$$

where $C(\mathbb{T})$ is the space of continuous functions on $\mathbb{T}$ and $\mathfrak{K}$ is the ideal of compact operators $\mathfrak{K}$. This algebra is irreducible and contains $\mathfrak{K}$ as its unique minimal ideal. The map $s(f) = T_f$ is a continuous section of the exact sequence

$$0 \longrightarrow \mathfrak{K} \xrightarrow{\iota} C^*(T_z) \underset{s}{\overset{\pi}{\rightleftarrows}} C(\mathbb{T}) \longrightarrow 0$$

V.1. Toeplitz Operators

Proof. Suppose that P is a projection commuting with T_z. Then P commutes with $I - T_z T_z^* = e_0 e_0^*$. So $Pe_0 = e_0$ or $Pe_0 = 0$. In the former case,
$$Pe_n = PT_z^n e_0 = T_z^n Pe_0 = e_n$$
for all $n \geq 0$. Thus $P = I$. While in the second case, we similarly find $P = 0$. Hence $C^*(T_z)$ is irreducible.

Since $C^*(T_z)$ contains a non-zero compact operator, it must contain all of $\mathfrak{K}$ by Corollary I.10.4. It also clearly contains T_p for every trigonometric polynomial p (as these are polynomials in T_z and $T_{z^{-1}} = T_z^*$). As the trigonometric polynomials are dense in $C(\mathbb{T})$, $C^*(T_z)$ contains all T_f for f in $C(\mathbb{T})$.

Any non-zero ideal containing a non-zero operator X will contain non-zero compact operators of the form XF for some finite rank operator F. Therefore, by Lemma I.9.15, this ideal is irreducible; and hence contains all the compact operators by Corollary I.10.4. So $\mathfrak{K}$ is the unique minimal ideal.

On the other hand, Corollary V.1.4 shows that products of Toeplitz operators with continuous symbol are Toeplitz operators (with continuous symbol) plus some compact operator. Thus $\mathcal{T}(C(\mathbb{T}))$ is a $*$-algebra. To see that it is norm closed, suppose that $T_{f_n} + K_n$ converges to an operator X, where f_n belong to $C(\mathbb{T})$ and K_n are compact. Then
$$\|f_n - f_m\|_\infty = \|T_{f_n} - T_{f_m}\|_e \leq \|T_{f_n} + K_n - (T_{f_m} + K_m)\|.$$
So f_n is a Cauchy sequence with uniform limit f in $C(\mathbb{T})$. It follows that K_n is also Cauchy with compact limit K. Hence $X = T_f + K$. Therefore $\mathcal{T}(C(\mathbb{T}))$ is closed. So $\mathcal{T}(C(\mathbb{T}))$ is a C*-algebra containing T_z and contained in $C^*(T_z)$; whence they are equal.

Let π be the quotient map of $C^*(T_z)$ onto $C^*(T_z)/\mathfrak{K}$. By Corollary V.1.4, this quotient algebra is abelian. Moreover, the map $\pi s : C(\mathbb{T}) \to C^*(T_z)/\mathfrak{K}$ is isometric by Proposition V.1.1, and an algebra $*$-homomorphism by Corollary V.1.4. Hence it is a $*$-isomorphism. With this identification, it is clear that s is a continuous section of the quotient map. ∎

Since an operator is invertible modulo the compact operators precisely when it is Fredholm, we obtain the following sharp spectral theorem for Toeplitz operators with continuous symbol.

Theorem V.1.6 *For f in $C(\mathbb{T})$, T_f is Fredholm if and only if 0 is not in the range of f. In this case, $\mathrm{ind}(T_f) = -\mathrm{wind}(f)$, where $\mathrm{wind}(f)$ is the winding number of the oriented curve $f(\mathbb{T})$ about 0.*

Proof. Since $C^*(T_z)/\mathfrak{K} \simeq C(\mathbb{T})$, it follows that
$$\sigma_e(T_f) = \sigma_{C(\mathbb{T})}(f) = f(\mathbb{T}).$$
So if f_t is a homotopy of non-vanishing functions in $C(\mathbb{T})$, the continuity of the Fredholm index and winding number shows that
$$\mathrm{ind}(T_{f_0}) = \mathrm{ind}(T_{f_1}) \quad \text{and} \quad \mathrm{wind}(f_0) = \mathrm{wind}(f_1).$$

Every non-vanishing function f in $C(\mathbb{T})$ is homotopic to z^n where n is the winding number of f about 0. Thus

$$\text{ind}(T_f) = \text{ind}(T_{z^n}) = -n = -\text{wind}(f). \qquad \blacksquare$$

Lemma V.1.7 *Either* $\ker T_g = 0$ *or* $\ker T_g^* = 0$ *for every non-zero g in L^∞. So if T_g is Fredholm of index 0, then it is invertible.*

Proof. Suppose, to the contrary, that h and k are non-zero functions in H^2 such that $T_g h = 0 = T_{\bar{g}} k$. Then $\overline{gh}$ and $g\bar{k}$ lie in $\overline{(H^2)^\perp} = H_0^2$, the space of H^2 functions such that $\hat{h}(0) = 0$. Now $H_0^2 H^2$ is contained in H_0^1, the space of L^1 functions on $\mathbb{T}$ such that $\hat{f}(n) = 0$ for $n \le 0$. Thus $f = (\overline{gh})k$ and $\bar{f} = (g\bar{k})h$ lie in H_0^1. But then $\hat{f}(n) = 0 = \hat{\bar{f}}(n) = 0$ for $n \le 0$; whence $\hat{f}(n) = 0$ for all n in $\mathbb{Z}$. It follows that $f = 0$, since the Fourier transform is injective on L^1.

To complete the proof, we need to quote a result of F. and M. Riesz from function theory that asserts that a non-zero function h in H^2 does not vanish on a subset of $\mathbb{T}$ of positive measure (see Exercise V.4). Thus $h\bar{k} \ne 0$ a.e., and consequently $g = 0$ a.e., contrary to fact. Hence at least one of $\ker T_g$ or $\ker T_g^*$ must be zero. A Fredholm operator T_g of index 0 satisfies $\text{null}(T_g) = \text{null}(T_g^*)$. Thus both kernels are 0, and therefore T_g is invertible. $\blacksquare$

The following complete spectral picture for a Toeplitz operator with continuous symbol is now immediate.

Corollary V.1.8 *If f belongs to $C(\mathbb{T})$, then $\sigma_e(T_f) = f(\mathbb{T})$ and*

$$\sigma(T_f) = f(\mathbb{T}) \cup \{\lambda : \text{wind}(f - \lambda) \ne 0\}.$$

V.2 Isometries

In this section, we consider the C*-algebra of a general proper isometry. The first step is the **Wold decomposition** that provides a structure theory for isometries.

Theorem V.2.1 *If S is an isometry on a Hilbert space $\mathcal{H}$, then there is a cardinal number α and a unitary operator U (possibly vacuous) such that S is unitarily equivalent to $T_z^{(\alpha)} \oplus U$.*

Proof. Let $\mathcal{M} = \ker S^* = (\text{Ran } S)^\perp$; and let $\alpha = \dim \mathcal{M}$. Notice that $S^n \mathcal{M}$ is orthogonal to $S^m \mathcal{M}$ for $n > m \ge 0$. Indeed,

$$(S^n x, S^m y) = (S^{n-m} x, y) = 0 \quad \text{for all} \quad x, y \in \mathcal{M}.$$

Fix an orthonormal basis $\{e_0^i : 0 \le i < \alpha\}$ for $\mathcal{M}$. Define $e_k^i = S^k e_0^i$ for $k \ge 1$ and $0 \le i < \alpha$. For fixed k, this forms an orthonormal basis for $S^k \mathcal{M}$. Together, they form a basis for $\sum \oplus_{k \ge 0} S^k \mathcal{M} =: \mathcal{K}$. It is evident that the restriction of S to

V.3. Bunce–Deddens Algebras

$\mathcal{N}_i := \text{span}\{e_k^i : k \geq 0\}$ is unitarily equivalent to the unilateral shift T_z for each $0 \leq i < \alpha$. Hence the restriction to $\mathcal{K}$ is unitarily equivalent to $T_z^{(\alpha)}$. In particular,

$$S\mathcal{K} = \sum_{k \geq 1} \oplus S^k \mathcal{M} = \mathcal{K} \ominus \mathcal{M} = \mathcal{K} \cap \text{Ran } S \subset \mathcal{K}.$$

Now $S\mathcal{K}^\perp$ is orthogonal to $S\mathcal{K} + (\text{Ran } S)^\perp = \mathcal{K}$. So $S\mathcal{K}^\perp$ is contained in $\mathcal{K}^\perp$. On the other hand, $\mathcal{K}^\perp$ is contained in Ran S and is orthogonal to $S\mathcal{K}$. Hence $S\mathcal{K}^\perp$ must contain $\mathcal{K}^\perp$. So $S\mathcal{K}^\perp = \mathcal{K}^\perp$. This shows that the restriction $U = S|\mathcal{K}^\perp$ is unitary. Thus S is unitarily equivalent to $T_z^{(\alpha)} \oplus U$ as desired. ∎

Note that S is a proper isometry exactly when $\alpha \geq 1$. We are now prepared to prove **Coburn's Theorem**.

Theorem V.2.2 *If S is a proper isometry, then there is a unique $*$-isomorphism φ of $\mathrm{C}^*(T_z)$ onto $\mathrm{C}^*(S)$ such that $\varphi(T_z) = S$.*

Proof. By the Wold decomposition, S is unitarily equivalent to $T_z^{(\alpha)} \oplus U$, where U is unitary and $\alpha \geq 1$ since S is proper. Using the notation of the previous proof, define a subspace $\mathcal{N}_0 = \text{span}\{e_k^0 : k \geq 0\}$. Then $\mathcal{N}_0$ is a reducing subspace for S and $S|\mathcal{N}_0$ is unitarily equivalent to T_z. Let ψ be the $*$-homomorphism of $\mathrm{C}^*(S)$ into $\mathrm{C}^*(T_z)$ given by restriction to $\mathcal{N}_0$; that is $\psi(A) = A|\mathcal{N}_0$. Since $\psi(S) = T_z$ and the range of ψ is a C*-algebra by Theorem I.5.5, this map is surjective.

Define a map φ of $\mathrm{C}^*(T_z) = \mathcal{T}(\mathrm{C}(\mathbb{T}))$ into $\mathrm{C}^*(S)$ by

$$\varphi(T_f + K) = (T_f + K)^{(\alpha)} \oplus f(U) \quad \text{for } f \in \mathrm{C}(\mathbb{T}), \, K \in \mathfrak{K}.$$

Since the map taking $T_f + K$ in $\mathcal{T}(\mathrm{C}(\mathbb{T}))$ to f in $\mathrm{C}(\mathbb{T})$ is a $*$-homomorphism, and the normal functional calculus shows that the map taking f in $\mathrm{C}(\mathbb{T})$ to $f(U)$ is a $*$-homomorphism, the composition is a $*$-homomorphism from $\mathcal{T}(\mathrm{C}(\mathbb{T}))$ onto $\mathrm{C}^*(U)$. Thus φ is readily seen to be a $*$-monomorphism such that $\varphi(T_z) = S$. As the image is a C*-algebra, this map is also surjective. So it is a $*$-isomorphism. Moreover, $\varphi = \psi^{-1}$. ∎

V.3 Bunce–Deddens Algebras

A **weighted unilateral shift** is an operator T for which there is an orthonormal basis $\{e_k : k \geq 1\}$ and weights a_n such that $Te_n = a_n e_{n+1}$ for all $n \geq 1$. T is a **periodic** weighted shift if there is an integer n such that $a_{k+n} = a_k$ for all $k \geq 1$. We will study a class of C*-algebras which are the direct limit of certain C*-algebras of these weighted shifts.

Proposition V.3.1 *The C*-algebra $\mathfrak{W}(n)$ of all weighted shifts of period n with respect to a fixed basis $\{e_k : k \geq 1\}$ is unitarily equivalent to $\mathcal{M}_n(\mathcal{T}(\mathrm{C}(\mathbb{T})))$; and it is singly generated.*

Proof. Consider the shift T with weights $a_i = 1$ when i is a multiple of n and $a_i = 1/2$ otherwise. Decompose the Hilbert space as $\mathcal{H} = \mathcal{H}_1 \oplus \cdots \oplus \mathcal{H}_n$ where $\mathcal{H}_i = \text{span}\{e_{kn+i} : k \geq 0\}$ for $1 \leq i \leq n$. With respect to this decomposition, the operators T and T^*T have the matrix forms

$$T = \begin{bmatrix} 0 & 0 & \cdots & 0 & T_z \\ \frac{1}{2}I & 0 & \cdots & 0 & 0 \\ 0 & \frac{1}{2}I & \cdots & 0 & 0 \\ \vdots & \vdots & \ddots & \vdots & 0 \\ 0 & 0 & \cdots & \frac{1}{2}I & 0 \end{bmatrix} \quad \text{and} \quad T^*T = \begin{bmatrix} \frac{1}{4}I & 0 & \cdots & 0 & 0 \\ 0 & \frac{1}{4}I & \cdots & 0 & 0 \\ \vdots & \vdots & \ddots & \vdots & \vdots \\ 0 & 0 & \cdots & \frac{1}{4}I & 0 \\ 0 & 0 & \cdots & 0 & I \end{bmatrix}.$$

Since T^*T belongs to $\mathrm{C}^*(T)$, so does $E_{nn} = f(T^*T) = \text{diag}(0,\ldots,0,I)$, where $f(1) = 1$ and $f(\frac{1}{4}) = 0$. It follows that $T^{*(n-i)}E_{nn}T^{n-j} = 2^{-i-j}E_{ij}$ belongs to $\mathrm{C}^*(T)$, where E_{ij} denotes the $n \times n$ matrix with all zero entries except for an I in the i,j entry. These form a set of matrix units for $\mathcal{M}_n$. Now $\mathrm{C}^*(T)$ also contains $\mathrm{C}^*(TE_{n1})$:

$$TE_{n1} = \begin{bmatrix} T_z & 0 & \cdots & 0 & 0 \\ 0 & 0 & \cdots & 0 & 0 \\ \vdots & \vdots & \ddots & \vdots & \vdots \\ 0 & 0 & \cdots & 0 & 0 \\ 0 & 0 & \cdots & 0 & 0 \end{bmatrix} \quad \text{and} \quad \mathrm{C}^*(TE_{n1}) = \begin{bmatrix} \mathcal{T}(C(\mathbb{T})) & 0 & \cdots & 0 & 0 \\ 0 & 0 & \cdots & 0 & 0 \\ \vdots & \vdots & \ddots & \vdots & \vdots \\ 0 & 0 & \cdots & 0 & 0 \\ 0 & 0 & \cdots & 0 & 0 \end{bmatrix}.$$

Moving this around by the matrix units shows that $\mathrm{C}^*(T)$ contains the C*-algebra $\mathcal{M}_n(\mathcal{T}(C(\mathbb{T})))$. As T clearly belongs to this algebra, this is an equality.

Finally, any n-periodic weighted shift with weights $a_{1 \bmod n}, \ldots, a_{n \bmod n}$ has the form

$$\begin{bmatrix} 0 & 0 & \cdots & 0 & a_n T_z \\ a_1 I & 0 & \cdots & 0 & 0 \\ 0 & a_2 I & \cdots & 0 & 0 \\ \vdots & \vdots & \ddots & \vdots & 0 \\ 0 & 0 & \cdots & a_{n-1}I & 0 \end{bmatrix}.$$

Clearly this lies in $\mathcal{M}_n(\mathcal{T}(C(\mathbb{T})))$. Hence $\mathfrak{W}(n) = \mathrm{C}^*(T) = \mathcal{M}_n(\mathcal{T}(C(\mathbb{T})))$. ∎

The following corollary is immediate from Theorem V.1.5.

Corollary V.3.2 *The sequence* $0 \longrightarrow \mathfrak{K} \longrightarrow \mathfrak{W}(n) \longrightarrow \mathcal{M}_n(C(\mathbb{T})) \longrightarrow 0$ *is exact.*

V.3. Bunce–Deddens Algebras

It also follows that the C*-algebra $\mathcal{M}_n(C(\mathbb{T}))$ is generated by the operator

$$\widetilde{T} = \begin{bmatrix} 0 & 0 & \cdots & 0 & z \\ \frac{1}{2} & 0 & \cdots & 0 & 0 \\ 0 & \frac{1}{2} & \cdots & 0 & 0 \\ \vdots & \vdots & \ddots & \vdots & 0 \\ 0 & 0 & \cdots & \frac{1}{2} & 0 \end{bmatrix}.$$

Suppose that n_k is an increasing sequence such that n_k divides n_{k+1} for $k \geq 1$. The **Bunce–Deddens algebra** $\mathfrak{B}(\{n_k\})$ is the quotient by the compact operators $\mathfrak{K}$ of the C*-algebra generated by all weighted shifts (with respect to a fixed basis) of period n_k for $k \geq 1$. Let

$$\mathfrak{B}(n_k) = \mathfrak{W}(n_k)/\mathfrak{K} \simeq \mathcal{M}_{n_k}(C(\mathbb{T})).$$

Then, since n_k-periodic weighted shifts are also n_{k+1}-periodic, there is a natural injection β_k of $\mathfrak{B}(n_k)$ into $\mathfrak{B}(n_{k+1})$. It follows that $\mathfrak{B}(\{n_k\})$ is the direct limit of $\mathcal{M}_{n_k}(C(\mathbb{T}))$ with respect to the imbeddings β_k.

To understand the limit algebra, we need to understand the imbeddings β_k. So consider the imbedding β of $\mathfrak{B}(n)$ into $\mathfrak{B}(nm)$. Notice that the generator $\widetilde{T}_n$ of $\mathfrak{B}(n)$ from the corollary above is mapped to the $mn \times mn$ matrix $\widetilde{A}_n$ (which should be considered as a $m \times m$ matrix with $n \times n$ matrices as coefficients) in Figure V.1 below. An easy computation shows that $\beta(\widetilde{T}_n^* \widetilde{T}_n) = \widetilde{A}_n^* \widetilde{A}_n$ is the

$$\widetilde{A}_n = \begin{bmatrix} \begin{array}{ccccc|ccccc|c|ccccc} 0 & 0 & \cdots & 0 & 0 & 0 & 0 & \cdots & 0 & 0 & & 0 & 0 & \cdots & 0 & z \\ \frac{1}{2} & 0 & \cdots & 0 & 0 & 0 & 0 & \cdots & 0 & 0 & & 0 & 0 & \cdots & 0 & 0 \\ 0 & \frac{1}{2} & \cdots & 0 & 0 & 0 & 0 & \cdots & 0 & 0 & \cdots & 0 & 0 & \cdots & 0 & 0 \\ \vdots & \vdots & \ddots & \vdots & \vdots & \vdots & \vdots & \ddots & \vdots & \vdots & & \vdots & \vdots & \ddots & \vdots & \vdots \\ 0 & 0 & \cdots & \frac{1}{2} & 0 & 0 & 0 & \cdots & 0 & 0 & & 0 & 0 & \cdots & 0 & 0 \\ \hline 0 & 0 & \cdots & 0 & 1 & 0 & 0 & \cdots & 0 & 0 & & 0 & 0 & \cdots & 0 & 0 \\ 0 & 0 & \cdots & 0 & 0 & \frac{1}{2} & 0 & \cdots & 0 & 0 & & 0 & 0 & \cdots & 0 & 0 \\ 0 & 0 & \cdots & 0 & 0 & 0 & \frac{1}{2} & \cdots & 0 & 0 & \cdots & 0 & 0 & \cdots & 0 & 0 \\ \vdots & \vdots & \ddots & \vdots & \vdots & \vdots & \vdots & \ddots & \vdots & \vdots & & \vdots & \vdots & \ddots & \vdots & \vdots \\ 0 & 0 & \cdots & 0 & 0 & 0 & 0 & \cdots & \frac{1}{2} & 0 & & 0 & 0 & \cdots & 0 & 0 \\ \hline & & \vdots & & & & & \vdots & & & \ddots & & & \vdots & & \\ \hline 0 & 0 & \cdots & 0 & 0 & 0 & 0 & \cdots & 0 & 0 & & 0 & 0 & \cdots & 0 & 0 \\ 0 & 0 & \cdots & 0 & 0 & 0 & 0 & \cdots & 0 & 0 & & \frac{1}{2} & 0 & \cdots & 0 & 0 \\ 0 & 0 & \cdots & 0 & 0 & 0 & 0 & \cdots & 0 & 0 & \cdots & 0 & \frac{1}{2} & \cdots & 0 & 0 \\ \vdots & \vdots & \ddots & \vdots & \vdots & \vdots & \vdots & \ddots & \vdots & \vdots & & \vdots & \vdots & \ddots & \vdots & \vdots \\ 0 & 0 & \cdots & 0 & 0 & 0 & 0 & \cdots & 0 & 0 & & 0 & 0 & \cdots & \frac{1}{2} & 0 \end{array} \end{bmatrix}.$$

FIGURE V.1

$nm \times nm$ diagonal matrix $\widetilde{D}_n = \text{diag}(d_i)$ where $d_{kn} = 1$ for $1 \leq k \leq m$, and $d_i = \frac{1}{4}$ otherwise.

Let $\widetilde{E}_{ij}^{(n)}$ be the canonical matrix units for $\mathfrak{B}(n)$. Since $\widetilde{E}_{nn}^{(n)} = f(\widetilde{T}_n^* \widetilde{T}_n)$ when $f(1) = 1$ and $f(\frac{1}{4}) = 0$, we obtain that

$$\beta(\widetilde{E}_{nn}^{(n)}) = f(\widetilde{D}_n) = \sum_{k=1}^{m} \widetilde{E}_{kn,kn}^{(mn)}.$$

From the relation

$$\widetilde{E}_{ij}^{(n)} = 2^{i+j} \widetilde{T}^{*(n-i)} \widetilde{E}_{nn}^{(n)} \widetilde{T}^{n-j}$$

it follows that for $1 \leq i, j \leq n$

$$\beta(\widetilde{E}_{ij}^{(n)}) = \sum_{k=0}^{m-1} \widetilde{E}_{km+i,km+j}^{(mn)} = \text{diag}(\widetilde{E}_{ij}^{(n)}, \ldots, \widetilde{E}_{ij}^{(n)}) \simeq \widetilde{E}_{ij}^{(n)} \otimes I_m.$$

Thus a scalar matrix X in $\mathcal{M}_n$ is sent to $\beta(X) = \text{diag}(X, \ldots, X) \simeq X \otimes I_m$.

Consider $\beta(z\widetilde{E}_{11}^{(n)}) = \beta(\widetilde{E}_{11}^{(n)} \widetilde{T} \widetilde{E}_{n1}^{(n)}) = (\widetilde{E}_{11}^{(n)} \otimes I_m) \widetilde{A} (\widetilde{E}_{n1}^{(n)} \otimes I_m)$ which has the matrix form

$$\begin{bmatrix} 0 & 0 & \cdots & 0 & z\widetilde{E}_{11}^{(n)} \\ \widetilde{E}_{11}^{(n)} & 0 & \cdots & 0 & 0 \\ 0 & \widetilde{E}_{11}^{(n)} & \cdots & 0 & 0 \\ \vdots & \vdots & \ddots & \vdots & 0 \\ 0 & 0 & \cdots & \widetilde{E}_{11}^{(n)} & 0 \end{bmatrix}.$$

Perform the **canonical shuffle** by permuting the basis vectors from $f_1, \ldots, f_{mn}$ to $f_1, f_{n+1}, \ldots, f_{(m-1)n+1}, f_2, f_{n+2}, \ldots, f_{(m-1)n}, f_{mn}$. This converts $\beta(\widetilde{E}_{ij}^{(n)})$ to the $n \times n$ matrix with i,j entry equal to I_m and all other entries equal to 0. Thus a scalar matrix $X = [x_{ij}]$ in $\mathcal{M}_n$ is sent to the matrix $[x_{ij} I_m]$; and the matrix $z\widetilde{E}_{11}^{(n)}$ is sent to

$$\begin{bmatrix} Z_m & 0 & \cdots & 0 \\ 0 & 0 & \cdots & 0 \\ \vdots & \vdots & \ddots & \vdots \\ 0 & 0 & \cdots & 0 \end{bmatrix}_{n \times n} \quad \text{where} \quad Z_m = \begin{bmatrix} 0 & 0 & \cdots & 0 & z \\ 1 & 0 & \cdots & 0 & 0 \\ 0 & 1 & \cdots & 0 & 0 \\ \vdots & \vdots & \ddots & \vdots & 0 \\ 0 & 0 & \cdots & 1 & 0 \end{bmatrix}_{m \times m}.$$

Thus β is unitarily equivalent to $I_n \otimes \alpha_m$, where α_m is the monomorphism of $C(\mathbb{T})$ into $\mathcal{M}_m(C(\mathbb{T}))$ given by $\alpha_m(z) = Z_m$. So now we concentrate on the analysis of α_m.

Identify $\mathbb{T}$ with $\mathbb{R}/\mathbb{Z}$ via the function $z(t) = e^{2\pi i t}$, $t \in \mathbb{R}$. The unitary matrix $Z_m(t)$ has the characteristic polynomial $p_t(\lambda) = \lambda^m - z(t)$, which has roots

$$\{z(\tfrac{t}{m}), z(\tfrac{t+1}{m}), \ldots, z(\tfrac{t+m-1}{m})\} = \{z(t') : mt' \equiv t \mod 1\}.$$

V.3. Bunce–Deddens Algebras

Let $t_i = \frac{t+i}{m}$, $0 \leq i < m$ be the solutions of $mt' \equiv t \mod 1$. The corresponding eigenvalues are

$$\xi_i(t) = \frac{1}{\sqrt{m}} \begin{pmatrix} z((m-1)t_i) \\ z((m-2)t_i) \\ \vdots \\ z(t_i) \\ 1 \end{pmatrix} \qquad Z_m \xi_i(t) = \frac{1}{\sqrt{m}} \begin{pmatrix} z(t) \\ z((m-1)t_i) \\ \vdots \\ z(2t_i) \\ z(t_i) \end{pmatrix} = z(t_i) \xi_i(t).$$

Consider the unitary matrices

$$U_t = \begin{bmatrix} \xi_0(t) & \xi_1(t) & \cdots & \xi_{m-1}(t) \end{bmatrix} = \left[\frac{1}{\sqrt{m}} z\left(\frac{(m-i)(t+j-1)}{m} \right) \right].$$

This is clearly continuous as a function on $\mathbb{R}$. And since $\xi_i(1) = \xi_{i+1}(0)$, we see that

$$U_0 = \begin{bmatrix} \xi_0(0) & \xi_1(0) & \cdots & \xi_{m-2}(0) & \xi_{m-1}(0) \end{bmatrix}$$

and

$$U_1 = \begin{bmatrix} \xi_1(0) & \xi_2(0) & \cdots & \xi_{m-1}(0) & \xi_0(0) \end{bmatrix}.$$

Thus $U_1 = U_0 V$ (or equivalently $U_0^* U_1 = V$), where V is the permutation shift

$$V = \begin{bmatrix} 0 & 0 & \cdots & 0 & 1 \\ 1 & 0 & \cdots & 0 & 0 \\ 0 & 1 & \cdots & 0 & 0 \\ \vdots & \vdots & \ddots & \vdots & 0 \\ 0 & 0 & \cdots & 1 & 0 \end{bmatrix}.$$

Let $D_t = \mathrm{diag}(z(\frac{t}{m}), z(\frac{t+1}{m}), \ldots, z(\frac{t+m-1}{m}))$. It follows that

$$Z_m(t) = U_t D_t U_t^* = U_0 (U_0^* U_t) D_t (U_0^* U_t)^* U_0^*.$$

Notice that $U_0^* U_t$ is a continuous path from I to V on $[0,1]$ and $VD_1V^* = D_0$. So the diagonal entries of D_t sweep out the m disjoint arcs of the circle between the m-th roots of unity. The twist obtained from V matches the endpoint of each arc to the initial point of the next. Hence, in effect, the values on the circle $\mathbb{T}$ are being wrapped around m times, with the gradual shift from I to V implementing the continuous matching of the *cuts* at the m-th roots of unity. For this reason, this imbedding of $C(\mathbb{T})$ into $\mathcal{M}_m(C(\mathbb{T}))$ is called an **m-times around imbedding**. Since $z(t)$ is the generator for $C(\mathbb{T})$, we obtain the formula

$$\alpha_m(f)(t) = U_t \, \mathrm{diag}\left(f(\tfrac{t}{m}), f(\tfrac{t+1}{m}), \ldots, f(\tfrac{t+m-1}{m}) \right) U_t^* = U_t f(D_t) U_t^*. \tag{1}$$

With this preparation, we are now ready to analyze the Bunce–Deddens algebras.

Theorem V.3.3 *The Bunce–Deddens algebra $\mathfrak{B}(\{n_k\})$ is simple.*

Proof. Let $\mathfrak{J}$ be a non-zero ideal of $\mathfrak{B}$. By Lemma III.4.1,

$$\mathfrak{J} = \overline{\bigcup_{k \geq 1} \mathfrak{J} \cap \mathfrak{B}(n_k)}.$$

Let $\mathfrak{J}_k := \mathfrak{J} \cap \mathfrak{B}(n_k)$. The ideals of $\mathfrak{B}(n_k) \simeq \mathcal{M}_{n_k}(C(\mathbb{T}))$ have the form

$$\mathfrak{J}_E^{(n_k)} = \{ f \in C(\mathbb{T}, \mathcal{M}_{n_k}) : f(t) = 0 \text{ for all } t \in E \}$$

for some closed set $E \subset \mathbb{T}$. Let E_k be the closed sets such that $\mathfrak{J}_k = \mathfrak{J}_{E_k}^{(n_k)}$.

Let $m_k := n_{k+1} n_k^{-1}$. Then the imbedding β_k of $\mathfrak{B}_k$ into $\mathfrak{B}_{k+1}$ is unitarily equivalent to $I_{n_k} \otimes \alpha_{m_k}$. Since $\beta_k(\mathfrak{J}_k) = \beta_k(\mathfrak{B}_{n_k}) \cap \mathfrak{J}_{k+1}$, we see from (1) that

$$E_k = \{\tfrac{t+i}{m_k} : t \in E_{k+1} \text{ and } 0 \leq i < m_k\}.$$

Repeated use of this observation (or consideration of the imbedding of $\mathfrak{B}_k$ into $\mathfrak{B}_\ell$ for large ℓ) shows that E_k is invariant under translation by $n_k n_\ell^{-1}$ for every $\ell > k$. So E_k is either empty or dense. In the first case, $\mathfrak{J}_k$ contains the identity and $\mathfrak{J} = \mathfrak{B}$. In the second case, $\mathfrak{J}_k = \{0\}$. Since this holds for all k, it follows that $\mathfrak{J} = \{0\}$. Therefore $\mathfrak{B}$ is simple. ∎

Theorem V.3.4 *Bunce–Deddens algebras are not AF.*

Proof. The way to distinguish Bunce–Deddens algebras from AF algebras is to use another topological invariant—this one associated to the invertible elements. In any finite dimensional C*-algebra, every invertible operator is triangularizable with invertible diagonal entries. Thus it can be connected to the identity by a continuous path of invertible elements in the algebra. This property readily extends to an AF algebra because every invertible element T in the algebra is close (within $\|T^{-1}\|^{-1}/2$) to an element A in a finite dimensional subalgebra. Every element this close to T is invertible; so the straight line from T to A lies in the set of invertible elements. Since A is connected to the identity in the invertibles, so is T.

On the other hand, consider the Bunce–Deddens algebra as a subalgebra of the Calkin algebra $\mathcal{Q}(\mathcal{H}) = \mathcal{B}(\mathcal{H})/\mathfrak{K}$. It always contains the image $\widetilde{T}_z$ of the unilateral shift T_z. This is an element of Fredholm index -1. The set of invertible elements of the Calkin algebra of fixed index form a component of the set of invertibles. In particular, index is constant on any continuous path. So $\widetilde{T}_z$ is not connected to the identity even in $\mathcal{Q}(\mathcal{H})$, and hence is in a separate component in the Bunce–Deddens algebra. It follows that Bunce–Deddens algebras are not AF. ∎

It turns out that the distinguishing invariant between Bunce–Deddens algebras is the supernatural number we used earlier to classify UHF algebras (Example III.5.1 and Theorem III.5.2).

Theorem V.3.5 *Two Bunce–Deddens algebras $\mathfrak{B}(\{m_k\})$ and $\mathfrak{B}(\{n_k\})$ are *-isomorphic if and only if their supernatural numbers $\delta(\{n_k\})$ and $\delta(\{m_k\})$ are equal.*

V.3. Bunce–Deddens Algebras

Proof. If the two sequences represent the same supernatural number, then one may drop to subsequences so that $m_k|n_k$ and $n_k|m_{k+1}$. Then the imbedding of $\mathfrak{B}(m_k)$ into $\mathfrak{B}(m_{k+1})$ factors through $\mathfrak{B}(n_k)$; and vice versa. So it is apparent that the limits are equivalent.

Conversely, notice that $\mathcal{M}_{m_k}$ is imbedded unitally into $\mathfrak{B}(m_k)$. Therefore if $\mathfrak{B}(\{m_k\})$ is $*$-isomorphic to $\mathfrak{B}(\{n_k\})$, then $\mathcal{M}_{m_k}$ is imbedded unitally into $\mathfrak{B}(\{n_k\})$. By Lemma III.3.2, there is an integer ℓ such that this imbedding can be conjugated by a unitary to an imbedding φ into $\mathfrak{B}(n_\ell)$ which is isomorphic to $\mathcal{M}_{n_\ell}(C(\mathbb{T}))$. Compose φ with the point evaluation δ_1 of $\mathcal{M}_{n_\ell}(C(\mathbb{T}))$ into $\mathcal{M}_{n_\ell}$ given by $\delta_1(f) = f(1)$. This map is a unital homomorphism, and thus $\delta_1\varphi$ is a unital homomorphism of $\mathcal{M}_{m_k}$ into $\mathcal{M}_{n_\ell}$. As $\mathcal{M}_{m_k}$ is simple, this is an isomorphism. This can only occur when $m_k|n_\ell$. Likewise, $n_\ell|m_{k'}$ for some k' sufficiently large. So these two algebras have the same supernatural number. ∎

Theorem V.3.6 *A Bunce–Deddens algebra has a unique trace, which is faithful and unital.*

Proof. Let τ_k be the trace on $\mathfrak{B}(n_k) \simeq \mathcal{M}_{n_k}(C(\mathbb{T}))$ given by

$$\tau_k(A) = \int_0^1 n_k^{-1} \operatorname{Tr} A(t)\, dt.$$

Clearly this is a faithful unital trace on $\mathfrak{B}(n_k)$. Moreover, by (1)

$$\tau_{k+1}\beta_k(A) = \int_0^1 n_{k+1}^{-1} \operatorname{Tr} \beta_k A(t)\, dt$$

$$= \int_0^1 n_{k+1}^{-1} \operatorname{Tr} U(t) \operatorname{diag}\left(A(\tfrac{t}{m}), A(\tfrac{t+1}{m}), \ldots, A(\tfrac{t+m-1}{m})\right) U(t)^*\, dt$$

$$= \int_0^1 n_{k+1}^{-1} \sum_{j=0}^{m_k-1} \operatorname{Tr} A(\tfrac{t+j}{m})\, dt = \tau_k(A).$$

Thus $\tau = \varinjlim \tau_k$ exists, and is a faithful unital trace on $\mathfrak{B}(\{n_k\})$.

Conversely, suppose that σ is a trace on $\mathfrak{B}(\{n_k\})$. Let σ_k denote the restriction of σ to $\mathfrak{B}(n_k)$. Since there is a unique trace on $\mathcal{M}_{n_k}$, there is a probability measure μ_k on $[0,1)$ such that

$$\sigma_k(A) = \int_0^1 n_k^{-1} \operatorname{Tr} A(t)\, d\mu_k.$$

(See Exercise IV.18(a).) But since $\sigma_k = \sigma_{k+1}\beta_k$, it follows as in the proof of simplicity that μ_k is invariant under translation by $n_k n_\ell^{-1}$ for every $\ell > k$. Thus μ_k equals Lebesgue measure for all k; whence $\sigma = \tau$. ∎

V.4 Cuntz Algebras

In this section, we examine another class of simple infinite C*-algebras generated by isometries. This time, the isometries will have orthogonal ranges. For $n \geq 2$, the **Cuntz algebra** $\mathcal{O}_n$ is the universal C*-algebra generated by isometries $S_1, \ldots, S_n$ such that

$$\sum_{i=1}^{n} S_i S_i^* = I \qquad (\ddagger)$$

It is clear that such isometries exist.

Construct a universal C*-algebra $\mathcal{O}_n$ with generators $S_1, \ldots, S_n$ satisfying ($\ddagger$) with the property that whenever $T_1, \ldots, T_n$ form another set of isometries satisfying ($\ddagger$), there is a (unique) homomorphism ρ of $\mathcal{O}_n$ onto $\mathrm{C}^*(\{T_1, \ldots, T_n\})$ such that $\rho(S_i) = T_i$ for $1 \leq i \leq n$. This can be done by forming a maximal collection $\{\pi_\alpha\}$ of inequivalent irreducible representations of the relation ($\ddagger$). These representations are necessarily on a separable Hilbert space. So we can form the representation $\pi = \sum \oplus \pi_\alpha$. From the GNS construction, it follows that every representation is *-equivalent to the direct sum of irreducible ones. So it follows that $\mathcal{O}_n = \mathrm{C}^*(\pi(S_i), 1 \leq i \leq n)$ has the desired universal property.

To calculate in $\mathcal{O}_n$, we work with **words** in the generators. Let **n** denote the set $\{1, \ldots, n\}$. For a sequence $\mu = (i_1, i_2, \ldots, i_k)$ in $\mathbf{n}^k$, set

$$S_\mu = S_{i_1} S_{i_2} \cdots S_{i_k}.$$

Also let $|\mu| = k$ denote the length of the word.

Notice that $S_i^* S_i = I$ and $S_j^* S_i = 0$ when $i \neq j$. Let $\nu = (j_1, \ldots, j_\ell)$. The product $S_\mu^* S_\nu$ will be non-zero only if there is perfect cancellation. That is, if $|\mu| = k$ and $|\nu| = \ell$, then

$$0 \neq S_{i_k}^* \cdots S_{i_2}^* S_{i_1}^* S_{j_1} S_{j_2} \cdots S_{j_\ell}$$

implies that $i_s = j_s$ for $1 \leq s \leq \min\{k, \ell\}$. We record this as a lemma for future reference:

Lemma V.4.1 *Suppose that $S_\mu^* S_\nu \neq 0$ with $k = |\mu|$ and $\ell = |\nu|$.*

(i) *If $|\mu| = |\nu|$, then $\mu = \nu$ and $S_\mu^* S_\nu = I$.*
(ii) *If $|\mu| < |\nu|$, there is a word ν' in $\mathbf{n}^{\ell-k}$ such that $\nu = \mu\nu'$ and $S_\mu^* S_\nu = S_{\nu'}$.*
(iii) *If $|\mu| > |\nu|$, there is a word μ' in $\mathbf{n}^{k-\ell}$ such that $\mu = \nu\mu'$ and $S_\mu^* S_\nu = S_{\mu'}^*$.*

Thus any word in the S_i's and S_j^*'s has a reduced expression with no S_i's to the right of any S_j^*'s. Hence *every non-zero word in the set $\{S_i, S_i^* : 1 \leq i \leq n\}$ has a unique reduced expression of the form $S_\mu S_\nu^*$.*

Let $\mathcal{W}_k^n$ denote the set of words in $\mathbf{n}^k$; and let $\mathcal{W}^n = \bigcup_{k \geq 0} \mathcal{W}_k^n$. Define

$$\mathfrak{F}_k^n = \mathrm{span}\{S_\mu S_\nu^* : \mu, \nu \in \mathcal{W}_k^n\} \quad \text{and} \quad \mathfrak{F}^n = \overline{\bigcup_{k \geq 1} \mathfrak{F}_k^n}.$$

V.4. Cuntz Algebras

Proposition V.4.2 $\mathfrak{F}_k^n$ *is isomorphic to* $\mathcal{M}_{n^k}$ *and* $\mathfrak{F}^n$ *is the UHF algebra of type* n^∞.

Proof. $\mathfrak{F}_k^n$ is spanned by $\{S_\mu S_\nu^* : \mu, \nu \in W_k^n\}$ which form a set of matrix units for $\mathcal{M}_{n^k}$ because of the identity

$$(S_\mu S_\nu^*)(S_{\mu'} S_{\nu'}^*) = S_\mu (S_\nu^* S_{\mu'}) S_{\nu'}^* = \delta_{\nu\mu'} S_\mu S_{\nu'}^*$$

which follows from Lemma V.4.1. The imbedding of $\mathfrak{F}_k^n$ into $\mathfrak{F}_{k+1}^n$ is unital, and thus has multiplicity n. Moreover,

$$\sum_{i=1}^n S_{\mu i} S_{\nu i}^* = S_\mu \Big(\sum_{i=1}^n S_i S_i^*\Big) S_\nu^* = S_\mu S_\nu^*.$$

This shows that the matrix units for $\mathfrak{F}_{k+1}^n$ are compatible with those of $\mathfrak{F}_k^n$.

It follows that $\mathfrak{F}^n$ is a direct limit of full matrix algebras of size n^k under unital imbeddings. Hence it is the UHF algebra of type n^∞. ∎

This subalgebra $\mathfrak{F}^n$ plays the role of a coefficient space for a Fourier type series for elements of $\mathcal{O}_n$. The first step is to identify the 0-th Fourier coefficient. An **expectation** of a C*-algebra onto a subalgebra is a positive, unital idempotent map. Expectations occur frequently in the study of operator algebras, and have many nice general properties.

Theorem V.4.3 *There is a faithful expectation* Φ_0 *of* $\mathcal{O}_n$ *onto* $\mathfrak{F}^n$.

Proof. For any complex number λ in $\mathbb{T}$, the isometries λS_i for $1 \leq i \leq n$ satisfy (‡) and generate $\mathcal{O}_n$. Thus there is an automorphism ρ_λ of $\mathcal{O}_n$ such that $\rho_\lambda(S_i) = \lambda S_i$ for $1 \leq i \leq n$. Note that $\rho_\lambda(S_i^*) = \overline{\lambda} S_i^*$ for each i.

A simple calculation shows that $\rho_\lambda(S_\mu S_\nu^*) = \lambda^{|\mu|-|\nu|} S_\mu S_\nu^*$. Thus it is apparent that the function $f_X(\lambda) = \rho_\lambda(X)$ is continuous when X is in the algebraic span of the words in the S_i's and S_j^*'s. Since these words are dense in $\mathcal{O}_n$ and automorphisms have norm 1, an easy estimate shows that f_X is continuous for each X in $\mathcal{O}_n$.

Define

$$\Phi_0(X) = \int_0^1 f_X(e^{2\pi i t})\, dt = \int_0^1 \rho_{e^{2\pi i t}}(X)\, dt$$

which makes sense as a Riemann integral. This is easily seen to be a unital positive contractive map on $\mathcal{O}_n$. Now

$$\Phi_0(S_\mu S_\nu^*) = \int_0^1 e^{2\pi i t(|\mu|-|\nu|)}\, dt\, S_\mu S_\nu^* = \begin{cases} 0 & \text{if } |\mu| \neq |\nu| \\ S_\mu S_\nu^* & \text{if } |\mu| = |\nu| \end{cases}.$$

So, on the algebraic span of words, it is apparent that Φ_0 maps into $\mathfrak{F}^n$ and is the identity map on a dense subset of $\mathfrak{F}^n$. So Φ_0 is a contractive projection of $\mathcal{O}_n$ onto $\mathfrak{F}^n$; in other words, it is an expectation.

If X is positive and non-zero, then $\rho_{e^{2\pi it}}(X)$ is positive and non-zero for all t. Thus the integral $\Phi_0(X)$ is also positive and non-zero; whence Φ_0 is faithful. ∎

A more algebraic way to compute this expectation is needed. The following simple calculation will help.

Lemma V.4.4 *Suppose that μ and ν are words in $\mathcal{W}^n$ such that $|\mu| \neq |\nu|$ and $\max\{|\mu|, |\nu|\} \leq k$. Set $S_\gamma = S_1^k S_2$. Then $S_\gamma^*(S_\mu S_\nu^*) S_\gamma = 0$.*

Proof. It follows from Lemma V.4.1 that $S_\gamma^* S_\mu = 0$ unless $S_\mu = S_1^{|\mu|}$, in which case $S_\gamma^* S_\mu = S_2^* S_1^{*(k-|\mu|)}$. Similarly, $S_\nu^* S_\gamma = 0$ unless $S_\nu = S_1^{|\nu|}$, in which case $S_\nu^* S_\gamma = S_1^{k-|\nu|} S_2$. So, even when both are non-zero, their product is

$$S_\gamma^* S_\mu S_\nu^* S_\gamma = S_2^* S_1^{*k-|\mu|} S_1^{k-|\nu|} S_2 = S_2^* S_1^{|\mu|-|\nu|} S_2 = 0,$$

where we make sense of S_1^m as $S_1^{*|m|}$ when $m < 0$. ∎

Lemma V.4.5 *For each positive integer m, there is an isometry W in $\mathcal{O}_n$ commuting with $\mathfrak{F}_m^n$ such that $\Phi_0(Y) = W^* Y W$ for every Y in*

$$\mathrm{span}\{S_\mu S_\nu^* : \max\{|\mu|, |\nu|\} \leq m\}.$$

Proof. Let $S_\gamma = S_1^{2m} S_2$ and define $W = \sum_{|\delta|=m} S_\delta S_\gamma S_\delta^*$. Then

$$W^*W = \sum_{|\delta|=m} \sum_{|\varepsilon|=m} S_\delta S_\gamma^* (S_\delta^* S_\varepsilon) S_\gamma S_\varepsilon^*$$

$$= \sum_{|\delta|=m} S_\delta S_\gamma^* S_\gamma S_\delta^* = \sum_{|\delta|=m} S_\delta S_\delta^* = I.$$

So W is an isometry. Moreover, if $|\mu| = m = |\nu|$, then

$$W S_\mu = \sum_{|\delta|=m} S_\delta S_\gamma (S_\delta^* S_\mu) = S_\mu S_\gamma$$

and similarly, $S_\nu^* W = S_\gamma S_\nu^*$. Hence the matrix unit $S_\mu S_\nu^*$ of $\mathfrak{F}_m^n$ satisfies

$$W S_\mu S_\nu^* = S_\mu S_\gamma S_\nu^* = S_\mu S_\nu^* W.$$

So W commutes with all Y in $\mathfrak{F}_m^n$. Consequently, $W^* X W = X$ for all X in $\mathfrak{F}_m^n$.

On the other hand, when $|\mu| \neq |\nu|$ and $\max\{|\mu|, |\nu|\} \leq m$, Lemma V.4.4 shows that

$$S_\gamma(S_\delta^* S_\mu S_\nu^* S_\delta) S_\gamma^* = 0 \quad \text{for all} \quad |\delta| = m;$$

so $W^* S_\mu S_\nu^* W = 0$. Consequently, $W^* Y W = \Phi_0(Y)$ for Y in the span of all words $S_\mu S_\nu^*$ such that $\max\{|\mu|, |\nu|\} \leq m$. ∎

Theorem V.4.6 *If $X \neq 0$ belongs to $\mathcal{O}_n$, then there are elements A, B in $\mathcal{O}_n$ such that $AXB = I$.*

Proof. Since $X \neq 0$ and the expectation Φ_0 onto $\mathfrak{F}^n$ is faithful, it follows that $\Phi_0(X^*X) \neq 0$. Multiply X by a scalar so that $\|\Phi_0(X^*X)\| = 1$. Then pick a self-adjoint element $Y = Y^*$ in the algebraic span of the words $S_\mu S_\nu^*$ so that $\|X^*X - Y\| < 1/4$. Then $\|\Phi_0(Y)\| > 3/4$. Let m be the maximum length of the words involved in the expansion of Y.

By Lemma V.4.5, there is an isometry W in $\mathcal{O}_n$ such that $W^*YW = \Phi_0(Y)$, which lies in $\mathfrak{F}_m^n$. By Proposition V.4.2, $\mathfrak{F}_m^n$ is a full matrix algebra. So there is a minimal idempotent E (from the diagonalization of $\Phi_0(Y)$) such that

$$E\Phi_0(Y) = \Phi_0(Y)E = \|\Phi_0(Y)\|E > \tfrac{3}{4}E.$$

Choose a unitary U in $\mathfrak{F}_m^n$ such that $UEU^* = E_{11} = S_1^m S_1^{*m}$; and set

$$Z = \|\Phi_0(Y)\|^{-1/2} S_1^{*m} U E W^*.$$

We have $\|Z\| \leq 2/\sqrt{3}$ and

$$ZYZ^* = \|\Phi_0(Y)\|^{-1} S_1^{*m} U E (W^*YW) E U^* S_1^m$$
$$= \|\Phi_0(Y)\|^{-1} S_1^{*m} U \|\Phi_0(Y)\| E U^* S_1^m = S_1^{m*} E_{11} S_1^m = I.$$

A simple calculation now shows that

$$\|I - ZX^*XZ^*\| \leq \|Z\|^2 \|Y - X^*X\| < \tfrac{4}{3}\tfrac{1}{4} = \tfrac{1}{3}.$$

Thus ZX^*XZ^* is invertible. Let $B = Z^*(ZX^*XZ^*)^{-1/2}$. Then

$$(B^*X^*)XB = I. \qquad \blacksquare$$

The following important corollary is immediate.

Corollary V.4.7 *$\mathcal{O}_n$ is simple. Thus if $T_1, \ldots, T_n$ are any n isometries such that $\sum_{i=1}^n T_i T_i^* = I$, then $C^*(T_1, \ldots, T_n)$ is isomorphic to $\mathcal{O}_n$.*

V.5 Simple Infinite C*-Algebras

A projection in a C*-algebra $\mathfrak{A}$ is called **infinite** if it is equivalent to a proper subprojection of itself. A projection P is **properly infinite** if there are orthogonal projections Q_1 and Q_2 such that $P \sim Q_1 \sim Q_2$ such that $Q_1 + Q_2 \leq P$. A C*-algebra is **(properly) infinite** if it contains a (properly) infinite projection. This definition is consistent with the definition of finite C*-algebra given in section IV.2.

We will show that infinite simple C*-algebras are properly infinite, and in fact always have a subalgebra which has $\mathcal{O}_n$ as a quotient.

Theorem V.5.1 *If $\mathfrak{A}$ is a simple infinite C*-algebra, then $\mathfrak{A}$ contains a projection Q and partial isometries T_i, $i \geq 1$, such that $T_i^*T_i = Q > \sum_{i=1}^n T_i T_i^*$ for all $n \geq 1$. In particular, $\mathfrak{A}$ is properly infinite.*

Proof. Let S be a partial isometry in $\mathfrak{A}$ such that $P = SS^* < Q = S^*S$. By working in $\mathfrak{B} = Q\mathfrak{A}Q$, we may arrange that $\mathfrak{B}$ is unital and $Q = I$. It is easy to verify that $\mathfrak{B}$ is simple; and it is infinite because Q is infinite.

Since $\mathfrak{B}$ is simple and $I - P \ne 0$, there are elements X_i in $\mathfrak{A}$ such that

$$\sum_{i=1}^{k} X_i^*(I - P)X_i = I.$$

(Prove it! See Exercise V.9.) Let $T_1 = \sum_{i=1}^{k} S^{i-1}(I-P)X_i$. Notice that $S^i(I-P)$ have pairwise orthogonal ranges for $i \geq 0$. Therefore

$$T_1^* T_1 = \sum_{i=1}^{k} \sum_{j=1}^{k} X_j^*(I - P)S^{*j-1}S^{i-1}(I - P)X_i$$

$$= \sum_{i=1}^{k} X_i^*(I - P)X_i = I$$

and

$$T_1 T_1^* \leq \sum_{i=1}^{k} S^{i-1}(I - P)S^{*i-1} = I - S^k S^{*k}.$$

Set $T_i = S^{k(i-1)}T_1$ for $i \geq 2$. Then

$$T_i T_i^* = S^{k(i-1)} T_1 T_1^* S^{*k(i-1)}$$

$$\leq S^{k(i-1)}(I - S^k S^{*k})S^{*k(i-1)}$$

$$= S^{k(i-1)} S^{*k(i-1)} - S^{ki} S^{*ki}.$$

Hence $T_i T_i^*$ are pairwise orthogonal projections. As each projection is equivalent to the identity, it follows that $\mathfrak{B}$ and $\mathfrak{A}$ are properly infinite. ∎

Lemma V.5.2 *Let $\mathfrak{E}_n$ be a C*-algebra generated by n isometries $S_1, \ldots, S_n$ such that $\sum_{i=1}^{n} S_i S_i^* = P < I$. Then the ideal $\langle P^\perp \rangle$ generated by $P^\perp$ is isomorphic to the compact operators and $\mathfrak{E}_n / \mathfrak{K} \simeq \mathcal{O}_n$.*

Proof. Since $S_i^* P^\perp = 0 = P^\perp S_i$, it is easy to verify using Lemma V.4.1 that $\langle P^\perp \rangle$ is spanned by

$$\{S_\mu P^\perp S_\nu^* : |\mu| < \infty, \, |\nu| < \infty\}.$$

Moreover it also follows from this that

$$(S_\mu P^\perp S_\nu^*)(S_\alpha P^\perp S_\beta^*) = \delta_{\nu\alpha} S_\mu P^\perp S_\beta^*.$$

Thus this set forms a set of matrix units for an algebra isomorphic to $\mathfrak{K}$. In particular, this shows that $P^\perp$ is a minimal projection in $\mathfrak{E}_n$.

In the quotient algebra $\mathfrak{E}_n / \mathfrak{K}$, the images $\widetilde{S}_i$ of the S_i's are isometries such that $\sum_{i=1}^{n} \widetilde{S}_i \widetilde{S}_i^* = \widetilde{I}$. Thus by Corollary V.4.7, this quotient is isomorphic to $\mathcal{O}_n$. ∎

Combining the last two results, we obtain

Corollary V.5.3 *If $\mathfrak{A}$ is a simple infinite C*-algebra, then $\mathcal{O}_n$ is the quotient of a subalgebra of $\mathfrak{A}$ for all $n \geq 1$.*

A minor modification of the argument yields the following useful lemma.

Lemma V.5.4 *If P and Q are projections in a simple C*-algebra $\mathfrak{A}$, and P is infinite, then Q is equivalent to a subprojection of P.*

Proof. First we show that there are elements Z_i such that $Q = \sum_{i=1}^n Z_i P Z_i^*$ (which is not immediate in the non-unital case). As $\mathfrak{A}$ is simple, there are elements X_i and Y_i in $\mathfrak{A}$ so that $\|Q - \sum_{i=1}^m X_i P Y_i\| < 1/2$. Hence

$$Q \leq \sum_{i=1}^m Q X_i P Y_i Q + Q Y_i^* P X_i^* Q$$

$$\leq \sum_{i=1}^m Q X_i P X_i^* Q + Q Y_i^* P Y_i Q =: A \leq cQ$$

where $c = \sum_{i=1}^m \|X_i\|^2 + \|Y_i\|^2$. By the functional calculus, there is a function f in $C_0(\sigma(A))$ such that $f(x) = x^{-1/2}$ on $[1, c]$. Hence

$$Q = f(A) A f(A) = \sum_{i=1}^m f(A) Q X_i P X_i^* Q f(A) + f(A) Q Y_i^* P Y_i Q f(A).$$

By Theorem V.5.1, there are partial isometries S_i in $\mathfrak{A}$ for $1 \leq i \leq n$ so that $\sum_{i=1}^n S_i S_i^* \leq P = S_i^* S_i$. Let $T = \sum_{i=1}^n Z_i P S_i^*$. Then

$$TT^* = \sum_{i=1}^n \sum_{j=1}^n Z_i P S_i^* S_j P Z_i^* = \sum_{i=1}^n Z_i P Z_i^* = Q.$$

In particular, T is a partial isometry and therefore $T^*T = PT^*TP$ is a subprojection of P. ∎

Now consider an even stronger notion. A C*-algebra is **purely infinite** if every hereditary subalgebra contains an infinite projection. The following result together with Theorem V.4.6 implies that $\mathcal{O}_n$ is purely infinite.

Theorem V.5.5 *If $\mathfrak{A}$ is a simple unital C*-algebra of dimension at least 2, then the following are equivalent:*

   (i) *$\mathfrak{A}$ is purely infinite.*
  (ii) *for every non-zero element A in $\mathfrak{A}$, there are elements X and Y such that $XAY = I$.*
 (iii) *for every non-zero positive element A in $\mathfrak{A}$ and $\varepsilon > 0$, there is an element X in $\mathfrak{A}$ with $\|X\| < \|A\|^{-1/2} + \varepsilon$ such that $XAX^* = I$.*

Proof. Suppose that (iii) holds and $A \neq 0$. Then there is an element X in $\mathfrak{A}$ such that $X(A^*A)X = I$ which proves (ii).

If (ii) holds and $A \neq 0$ is positive, find X and Y so that $XA^{1/2}Y = I$. Then

$$I = XA^{1/2}YY^*A^{1/2}X^* \leq \|Y\|^2 XAX^*.$$

Thus $Z = XAX^*$ is invertible, whence $I = (Z^{-1/2}X)A(X^*Z^{-1/2})$. This proves (iii) without the norm estimate. Suppose that $\mathfrak{B}$ is a hereditary subalgebra of $\mathfrak{A}$ and B is a non-scalar positive element of $\mathfrak{B}$ which is not invertible. Then there is an element X in $\mathfrak{A}$ so that $XBX^* = I$. Let $S = B^{1/2}X^*$. Then $S^*S = I$ and S is not invertible; so S is a proper isometry. Moreover,

$$P = SS^* = B^{1/2}X^*XB^{1/2}$$

belongs to $\mathfrak{B}$. This is an infinite projection in $\mathfrak{B}$ since SP belongs to $\mathfrak{B}$ and $(SP)^*SP = P$; while SPS^* is a subprojection of P orthogonal to $S(I - P)S^*$. This shows that $\mathfrak{B}$ is infinite, proving (i).

Finally suppose that (i) holds. Let A in $\mathfrak{A}$ be a positive element of norm 1, and let $0 < \varepsilon < 1/2$. Let $c = 1 - \varepsilon$. Define a function

$$f(t) = \begin{cases} 0 & 0 \leq t \leq 1 - \varepsilon \\ 1 - \varepsilon^{-1}(1 - t) & 1 - \varepsilon \leq t \leq 1 \end{cases}.$$

Let $\mathfrak{B}$ be the hereditary subalgebra $\overline{f(A)\mathfrak{A}f(A)}$. By (i), it contains an infinite projection P. From the definition of $\mathfrak{B}$, one obtains $P \leq E_A([1-\varepsilon, 1])$; and hence $PAP \geq (1 - \varepsilon)P$. By Lemma V.5.4, the identity is equivalent to a subprojection of P. Thus there is an isometry S in $\mathfrak{A}$ such that $SS^* \leq P$. So calculate

$$B := S^*AS = S^*PAPS \geq 1 - \varepsilon S^*PS = (1 - \varepsilon)I.$$

Therefore,

$$(B^{-1/2}S^*)A(SB^{-1/2}) = I.$$

Finally, $\|SB^{-1/2}\| \leq (1 - \varepsilon)^{-1/2} < 1 + \varepsilon$; which establishes (iii). ∎

Corollary V.5.6 *The Cuntz algebras $\mathcal{O}_n$ for $2 \leq n < \infty$ are purely infinite.*

V.6 Classification of Cuntz Algebras

Lemma V.5.2 suggests studying the possible *unital* C*-algebras $\mathfrak{E}$ such that

$$0 \longrightarrow \mathfrak{K} \overset{i}{\longrightarrow} \mathfrak{E} \overset{\rho}{\longrightarrow} \mathcal{O}_n \longrightarrow 0$$

is exact. These algebras are **extensions** of the compact operators by $\mathcal{O}_n$. Two extensions $\mathfrak{E}$ and $\mathfrak{E}'$ are called **strongly equivalent** if there is a $*$-isomorphism ψ

V.6. Classification of Cuntz Algebras

of $\mathfrak{E}$ onto $\mathfrak{E}'$ such that

$$\begin{array}{ccccccccc}
0 & \longrightarrow & \mathfrak{K} & \overset{\iota}{\longrightarrow} & \mathfrak{E} & \overset{\rho}{\longrightarrow} & \mathcal{O}_n & \longrightarrow & 0 \\
 & & \parallel & & \downarrow \psi & & \parallel & & \\
0 & \longrightarrow & \mathfrak{K} & \underset{\iota'}{\longrightarrow} & \mathfrak{E}' & \underset{\rho'}{\longrightarrow} & \mathcal{O}_n & \longrightarrow & 0
\end{array}$$

commutes. This turns out to be too fine an invariant. We will say that two extensions are **equivalent** if there is a $*$-isomorphism ψ of $\mathfrak{E}$ onto $\mathfrak{E}'$ such that $\psi(\mathfrak{K}) = \mathfrak{K}$ and

$$\begin{array}{ccccccccc}
0 & \longrightarrow & \mathfrak{K} & \overset{\iota}{\longrightarrow} & \mathfrak{E} & \overset{\rho}{\longrightarrow} & \mathcal{O}_n & \longrightarrow & 0 \\
 & & \downarrow \psi|\mathfrak{K} & & \downarrow \psi & & \parallel & & \\
0 & \longrightarrow & \mathfrak{K} & \underset{\iota'}{\longrightarrow} & \mathfrak{E}' & \underset{\rho'}{\longrightarrow} & \mathcal{O}_n & \longrightarrow & 0
\end{array}$$

commutes. The collection of equivalence classes of extensions will be denoted by $\text{Ext}(\mathcal{O}_n)$.

An important observation is that we may represent $\mathfrak{E}$ as a subalgebra of $\mathcal{B}(\mathcal{H})$ containing $\mathfrak{K}$. This will allow us to reformulate the equivalence question. Recall that the Calkin algebra is the quotient $\mathcal{Q}(\mathcal{H})$ of $\mathcal{B}(\mathcal{H})$ by the compact operators; and let the quotient map be denoted by π.

Lemma V.6.1 *Every automorphism φ of the C*-algebra of compact operators $\mathfrak{K}$ has the form $\text{Ad}\, U$ for some unitary operator in $\mathcal{B}(\mathcal{H})$.*

Proof. Fix a unit vector e in $\mathcal{H}$. Since ee^* is a minimal projection in $\mathfrak{K}$, so is $\varphi(ee^*)$ which therefore has the form ff^* for some unit vector f in $\mathcal{H}$. Define

$$Ux := \varphi(xe^*)f.$$

Then

$$\|Ux\|^2 = (\varphi(xe^*)f, \varphi(xe^*)f) = (\varphi(ex^*xe^*)f, f)$$
$$= (\|x\|^2 \varphi(ee^*)f, f) = \|x\|^2 (ff^*f, f) = \|x\|^2.$$

Thus U is isometric.

If y belongs to $\mathcal{H}$, choose a compact operator K such that $\varphi(K) = yf^*$. Then

$$U(Ke) = \varphi(Kee^*)f = \varphi(K)\varphi(ee^*)f = (yf^*)(ff^*)f = y.$$

Hence U is unitary.

Finally, for each K in $\mathfrak{K}$ and y in $\mathcal{H}$,

$$UKU^*y = \varphi((KU^*y)e^*)f = \varphi(K)\varphi((U^*y)e^*)f$$
$$= \varphi(K)U(U^*y) = \varphi(K)y.$$

Therefore $\varphi = \text{Ad}\, U$. ∎

Theorem V.6.2 *Let $\mathfrak{E}$ be an extension of $\mathfrak{K}$ by $\mathcal{O}_n$. Then the identity representation on $\mathfrak{K}$ extends to a faithful representation on $\mathcal{H}$ of the extension $\mathfrak{E}$ of $\mathfrak{K}$ by $\mathcal{O}_n$. This induces a $*$-monomorphism τ of $\mathcal{O}_n$ into the Calkin algebra. Two extensions $\mathfrak{E}$ and $\mathfrak{E}'$ are equivalent if and only if the associated monomorphisms τ and τ' are unitarily equivalent in the sense that there is a unitary U in $\mathcal{B}(\mathcal{H})$ such that $\tau' = \operatorname{Ad} \pi(U)\, \tau$.*

Proof. Let id denote the identity representation on $\mathfrak{K}$. By Lemma I.9.14, this extends uniquely to a representation σ of $\mathfrak{E}$ on $\mathcal{H}$. It is unital because $\sigma(I)K = K$ for every compact K. Define the map τ from $\mathcal{O}_n$ into $\mathcal{Q}(\mathcal{H})$ by $\tau(A) = \pi\sigma(\rho^{-1}(A))$. This is easily verified to be a well defined unital $*$-homomorphism. Since $\mathcal{O}_n$ is simple, τ is injective.

To see that σ is injective, suppose that $\sigma(E) = 0$. Then
$$\tau(\rho(E)) = \pi\sigma(E) = 0;$$
whence $\rho(E) = 0$. But then E is compact, so $E = \sigma(E) = 0$.

Let τ and τ' be two monomorphisms of $\mathcal{O}_n$ into the Calkin algebra, and let $\mathfrak{E} = \pi^{-1}\tau(\mathcal{O}_n)$ and $\mathfrak{E}' = \pi^{-1}\tau'(\mathcal{O}_n)$ be the corresponding extensions. Suppose there is a unitary U in $\mathcal{B}(\mathcal{H})$ such that $\tau' = (\operatorname{Ad} \pi U)\tau$. Then let $\psi = \operatorname{Ad} U$. This is a $*$-isomorphism of $\mathfrak{E}$ which takes $\mathfrak{K}$ onto itself and such that
$$\operatorname{Ad} U(\mathfrak{E}) = (\operatorname{Ad} U)\pi^{-1}(\tau(\mathcal{O}_n)) = \pi^{-1}(\operatorname{Ad} \pi U)(\tau(\mathcal{O}_n))$$
$$= \pi^{-1}(\tau'(\mathcal{O}_n)) = \mathfrak{E}'.$$

Conversely, suppose that $\mathfrak{E}$ and $\mathfrak{E}'$ are equivalent via an isomorphism ψ. By Lemma V.6.1, there is a unitary U in $\mathcal{B}(\mathcal{H})$ such that $\psi|\mathfrak{K} = \operatorname{Ad} U|\mathfrak{K}$. For any E in $\mathfrak{E}$ and K in $\mathfrak{K}$,
$$\psi(E)(UKU^*) = \psi(E)\psi(K) = \psi(EK)$$
$$= UEKU^* = (UEU^*)(UKU^*).$$
As the ranges of $U\mathfrak{K}U^*$ are dense in $\mathcal{H}$, we deduce that $\psi(E) = UEU^*$. Hence $\tau' = (\operatorname{Ad} \pi U)\tau$. ∎

This theorem suggests another notion of equivalence using unitaries in the Calkin algebra instead of unitaries in $\mathcal{B}(\mathcal{H})$. Say that two monomorphisms τ and τ' of $\mathcal{O}_n$ into the Calkin algebra are **weakly equivalent** if there is a unitary element u in $\mathcal{Q}(\mathcal{H})$ such that $\tau' = (\operatorname{Ad} u)\tau$. The collection of all weak equivalence classes will be denoted by $\operatorname{Ext}_w(\mathcal{O}_n)$.

We define an addition on $\operatorname{Ext}(\mathcal{O}_n)$ by setting $[\sigma] + [\tau] := [\sigma \oplus \tau]$ where $[\sigma]$ and $[\tau]$ are two equivalence classes of monomorphisms of $\mathcal{O}_n$ into the Calkin algebra. This defines a monomorphism of $\mathcal{O}_n$ into $\mathcal{Q}(\mathcal{H}) \oplus \mathcal{Q}(\mathcal{H})$ which is contained in $\mathcal{M}_2(\mathcal{Q}(\mathcal{H}))$. But, since $\mathcal{M}_2(\mathcal{B}(\mathcal{H})) \simeq \mathcal{B}(\mathcal{H})$ and the restriction to the compacts identifies $\mathcal{M}_2(\mathfrak{K})$ with $\mathfrak{K}$, this induces an isomorphism of $\mathcal{M}_2(\mathcal{Q}(\mathcal{H}))$ onto $\mathcal{Q}(\mathcal{H})$. To see that this addition is well defined, suppose that $\sigma' = (\operatorname{Ad} \pi U)\sigma$ and

V.6. Classification of Cuntz Algebras

$\tau' = (\mathrm{Ad}\,\pi V)\tau$. Then $W = U \oplus V$ defines a unitary in $\mathcal{M}_2(\mathcal{B}(\mathcal{H}))$ such that $(\mathrm{Ad}\,\pi W)(\sigma \oplus \tau) = \sigma' \oplus \tau'$. Because of the properties of direct sum, it is immediately evident that this operation is abelian and associative. Hence $\mathrm{Ext}(\mathcal{O}_n)$ becomes an abelian semigroup.

An extension τ is called **trivial** if there is a $*$-monomorphism σ of $\mathcal{O}_n$ into $\mathcal{B}(\mathcal{H})$ such that $\tau = \pi\sigma$. This means that the exact sequence for the extension splits:

$$0 \longrightarrow \mathfrak{K} \longrightarrow \mathfrak{E} \underset{\sigma}{\overset{\pi}{\rightleftarrows}} \mathcal{O}_n \longrightarrow 0$$

We will need the following result of **Voiculescu**.

Theorem V.6.3 *If $\mathfrak{A}$ is a separable C*-algebra, then all trivial extensions of $\mathfrak{K}$ by $\mathfrak{A}$ are equivalent; and they form a zero element for $\mathrm{Ext}(\mathfrak{A})$.*

Proof. Let τ_i be trivial extensions of $\mathfrak{A}$ for $i = 1, 2$; and let σ_i be $*$-monomorphisms of $\mathfrak{A}$ into $\mathcal{B}(\mathcal{H})$ such that $\pi\sigma_i = \tau_i$. Then

$$\ker \sigma_1 = \ker \pi\sigma_1 = \ker \sigma_2 = \ker \pi\sigma_2 = \{0\}.$$

Therefore by Corollary II.5.6 of Voiculescu's Weyl–von Neumann Theorem, σ_1 and σ_2 are approximately unitarily equivalent. Thus there is a unitary operator U such that $\sigma_2 - \mathrm{Ad}\,U\,\sigma_1$ has compact range; whence $\tau_2 = (\mathrm{Ad}\,\pi U)\tau_1$. Thus the trivial elements belong to a single equivalence class.

Now suppose that τ is an extension of $\mathfrak{A}$; and let $\mathfrak{E} = \pi^{-1}\tau(\mathfrak{A})$. If $\pi\sigma$ is a trivial extension of $\mathfrak{A}$, then the representation of $\mathfrak{E}$ given by $\rho = \sigma\tau^{-1}\pi$ annihilates the compact operators. Therefore by another Corollary II.5.5 of Voiculescu's Theorem, $\mathrm{id}_\mathfrak{E} \sim_\mathfrak{K} \mathrm{id}_\mathfrak{E} \oplus \rho$. Equivalently, τ and $\tau \oplus \pi\sigma$ are unitarily equivalent in $\mathcal{B}(\mathcal{H})/\mathfrak{K}$. Hence $[\tau] + [\pi\sigma] = [\tau]$. So $[\pi\sigma]$ is the zero element. ∎

While $\mathrm{Ext}(\mathfrak{A})$ is not always a group, this is the case for nice C*-algebras. The proof for $\mathcal{O}_n$ will emerge from our analysis. We will compute both $\mathrm{Ext}(\mathcal{O}_n)$ and $\mathrm{Ext}_w(\mathcal{O}_n)$. This latter group will turn out to distinguish the Cuntz algebras from one another.

Lemma V.6.4 *Let v in $\mathcal{Q}(\mathcal{H}_1, \mathcal{H}_2)$ be a partial isometry in the Calkin algebra; and suppose that P in $\mathcal{B}(\mathcal{H}_2)$ and Q in $\mathcal{B}(\mathcal{H}_1)$ be projections such that $\pi(P) = vv^*$ and $\pi(Q) = v^*v$. Then, there is a partial isometry V in $\mathcal{B}(\mathcal{H}_1, \mathcal{H}_2)$ of the form $V = PVQ$ such that $\pi(V) = v$. Moreover, the integer*

$$\dim(Q - V^*V) - \dim(P - VV^*)$$

is defined independent of the choice of V.

Proof. Lift v arbitrarily to an operator T, and let V be the partial isometry in the polar decomposition of $PTQ = VA$ where $A = |PTQ|$. Then $\pi(V)\pi(A)$ is the polar decomposition of $\pi(PTQ) = v$. Hence $\pi(V) = v$.

Considering V as an operator in $\mathcal{B}(P\mathcal{H}, Q\mathcal{H})$, one sees that since the choice of V is unique up to a compact operator, its Fredholm index is well defined independent the the choice. This index is

$$\text{null}(V) - \text{null}(V^*) = \text{null}(V^*V) - \text{null}(VV^*)$$
$$= \dim(Q - V^*V) - \dim(P - VV^*). \blacksquare$$

If τ is a unital monomorphism of $\mathcal{O}_n$ into $\mathcal{Q}(\mathcal{H})$, define a partial isometry v_τ in $\mathcal{Q}(\mathcal{H}^{(n)}, \mathcal{H})$ by

$$v_\tau = \begin{bmatrix} \tau(S_1) & \tau(S_2) & \cdots & \tau(S_n) \end{bmatrix}.$$

Define an integer valued function on extensions by $f(\tau) = \text{ind } v_\tau$. By the lemma above, this function is given by

$$f(\tau) = \dim(I^{(n)} - V_\tau^* V_\tau) - \dim(I - V_\tau V_\tau^*),$$

where V_τ is any partially isometric lifting of v_τ.

If τ and τ' are equivalent extensions, then there is a unitary operator U such that $\tau' = \text{Ad } \pi(U) \tau$. Thus $v_{\tau'} = \pi U v_\tau \pi U^{(n)*}$. Hence

$$f(\tau') = \text{ind } v_{\tau'} = \text{ind } \pi U + \text{ind } v_\tau + \text{ind } \pi U^{(n)*} = f(\tau).$$

Therefore f induces a well defined map $\tilde{f}$ from $\text{Ext}(\mathcal{O}_n)$ into $\mathbb{Z}$.

Theorem V.6.5 *The function $\tilde{f}$ is an isomorphism of $\text{Ext}(\mathcal{O}_n)$ onto $\mathbb{Z}$. In particular, $\text{Ext}(\mathcal{O}_n)$ is a group.*

Proof. First we show that $\tilde{f}$ is additive. For if σ and τ are extensions, then

$$v_{\sigma \oplus \tau} = \begin{bmatrix} \sigma(S_1) & 0 & \sigma(S_2) & 0 & \cdots & \sigma(S_n) & 0 \\ 0 & \tau(S_1) & 0 & \tau(S_2) & \cdots & 0 & \tau(S_n) \end{bmatrix} \simeq v_\sigma \oplus v_\tau.$$

Hence

$$\tilde{f}[\sigma \oplus \tau] = \text{ind } v_\sigma \oplus v_\tau = \text{ind } v_\sigma + \text{ind } v_\tau = \tilde{f}[\sigma] + \tilde{f}[\tau].$$

Therefore $\tilde{f}$ is a semigroup isomorphism.

Suppose that $\tilde{f}[\tau] = 0$. Choose a lifting V_τ of v_τ. As $\text{ind } V_\tau = 0$, there is a finite rank isometry W of $\ker V_\tau$ onto $\ker V_\tau^*$. So $V_\tau + W$ is unitary and $\pi(V_\tau + W) = v_\tau$. We could have taken this as our choice for V_τ. So we may suppose that V_τ is unitary. Write it as a $1 \times n$ matrix in $\mathcal{B}(\mathcal{H}^{(n)}, \mathcal{H})$ as

$$V_\tau = \begin{bmatrix} V_1 & V_2 & \cdots & V_n \end{bmatrix}.$$

Then $I^{(n)} = V_\tau^* V_\tau = [V_i^* V_j]$. This implies that $V_i^* V_i = I$ for $1 \leq i \leq n$, and thus each V_i is an isometry. And $I = V_\tau V_\tau^* = \sum_{i=1}^n V_i V_i^*$. Hence by Corollary V.4.7, the map σ taking S_i to V_i is a faithful representation of $\mathcal{O}_n$ such that $\pi \sigma = \tau$. Therefore τ is trivial.

V.6. Classification of Cuntz Algebras

Conversely, if $\tau = \pi\sigma$ is a trivial extension, define $V_i = \sigma(S_i)$. The operator $V_\tau = \begin{bmatrix} V_1 & V_2 & \ldots & V_n \end{bmatrix}$ is a unitary lifting of v_τ. So $f(\tau) = \operatorname{ind} V_\tau = 0$. Theorem V.6.3 shows that the trivial elements form a zero element for $\operatorname{Ext}(\mathcal{O}_n)$. Thus we have shown that $\ker \widetilde{f} = \{0\}$.

We will show that the range of $\widetilde{f}$ is $\mathbb{Z}$. Clearly, it suffices to construct two extensions $\tau_\pm$ such that $f(\tau_\pm) = \pm 1$. Let $V_1, V_2, \ldots, V_n$ be isometries satisfying ($\ddagger$), and let $S = SV_1V_1^* + (I - V_1V_1^*)$ be an isometry in $\mathcal{B}(\mathcal{H})$ of index -1. Then let

$$\tau_\pm(S_1) = \pi(S)^{\mp 1}\pi(V_1) \quad \text{and} \quad \tau_\pm(S_i) = \pi(V_i) \quad \text{for} \quad 2 \le i \le n.$$

For this choice, we have the partially isometric liftings

$$V_+ = \begin{bmatrix} S^*V_1 & V_2 & \ldots & V_n \end{bmatrix} \quad \text{and} \quad V_- = \begin{bmatrix} SV_1 & V_2 & \ldots & V_n \end{bmatrix}.$$

Then it is readily checked that $f(\tau_\pm) = \operatorname{ind} V_\pm = \pm 1$.

This shows that $\widetilde{f}$ is an isomorphism and that $\operatorname{Ext}(\mathcal{O}_n)$ is a group. For if $[\tau]$ belongs to $\operatorname{Ext}(\mathcal{O}_n)$ and $\widetilde{f}[\tau] = n$, let $[\tau']$ be chosen so that $\widetilde{f}[\tau'] = -n$. Then

$$\widetilde{f}([\tau] + [\tau']) = n - n = 0.$$

Hence $[\tau \oplus \tau']$ is the trivial element. Whence $[\tau']$ is an inverse for $[\tau]$. So $\widetilde{f}$ is an isomorphism of $\operatorname{Ext}(\mathcal{O}_n)$ onto $\mathbb{Z}$. ∎

So far, this does not differentiate the Cuntz algebras. However, with a little more work, we obtain the result that we want.

Theorem V.6.6 *The weak extension group $\operatorname{Ext}_w(\mathcal{O}_n)$ of $\mathcal{O}_n$ for $n \ge 2$ is isomorphic to $\mathbb{Z}_{n-1}$.*

Proof. $\operatorname{Ext}_w(\mathcal{O}_n)$ is the quotient of $\operatorname{Ext}(\mathcal{O}_n)$ by the subgroup of elements weakly equivalent to the zero element. Suppose that $\tau = \pi\sigma$ is a trivial element with $V_i = \sigma(S_i)$. Let U be an isometry of index -1. Then let $\tau' = \operatorname{Ad} \pi(U)\,\tau$. Then $v_{\tau'}$ has the lifting

$$V_{\tau'} = \begin{bmatrix} UV_1U^* & UV_2U^* & \ldots & UV_nU^* \end{bmatrix} = UV_\tau U^{(n)*}.$$

Thus

$$f(\tau') = \operatorname{ind} U + \operatorname{ind} V_\tau + \operatorname{ind} U^{(n)*} = -1 + 0 + n = n - 1.$$

If x is a unitary in $\mathcal{Q}(\mathcal{H})$ with $\operatorname{ind}(x) = -k$, then there is a unitary operator W in $\mathcal{B}(\mathcal{H})$ such that $\pi(W)\pi(U)^k = x$. Thus $(\operatorname{Ad} x)\tau = (\operatorname{Ad} \pi W)(\operatorname{Ad} \pi U)^k\tau$. So

$$f((\operatorname{Ad} x)\tau) = f((\operatorname{Ad} \pi U)^k \tau) = k(n-1).$$

Thus the extensions weakly equivalent to the trivial extensions are identified with the subgroup $(n-1)\mathbb{Z}$. Therefore $\operatorname{Ext}_w(\mathcal{O}_n)$ is isomorphic to $\mathbb{Z}_{n-1}$. ∎

Corollary V.6.7 *If $n \ne m$, then $\mathcal{O}_n$ and $\mathcal{O}_m$ are not isomorphic.*

V.7 Real Rank Zero

A unital C*-algebra $\mathfrak{A}$ is said to have **real rank zero** if the invertible Hermitian elements are dense in $\mathfrak{A}_{sa}$. When $\mathfrak{A}$ is not unital, say it is real rank zero if the unitization $\mathfrak{A}^\sim$ is real rank zero. This is a property that parallels zero-dimensionality in the commutative case. The following elementary proposition justifies this view.

Proposition V.7.1 *If X is totally disconnected, then $C(X)$ is real rank zero. If X contains an arc, then $C(X)$ is not real rank zero.*

Proof. When X is totally disconnected, $C(X)$ is spanned by its projections. The subspace of finite real linear combinations of characteristic functions is therefore dense in the subspace of real valued functions. Clearly, a function taking a finite set of values is easily perturbed slightly to a non-zero function. So $C(X)$ is real rank zero.

On the other hand, if X contains an arc J, consider any real valued function f in $C(X)$ which takes both values ± 1 on J. Any real valued function within distance 1 of f takes both positive and negative values on J. Hence it has a zero by the intermediate value theorem. So the ball of radius 1 about f contains no invertible elements. ■

It is clear that every von Neumann algebra is real rank zero. Indeed, approximate A within ε by $A + \varepsilon E_A(-\varepsilon/2, \varepsilon/2)$. In particular, finite dimensional C*-algebras have real rank zero. It is also very easy to see that inductive limits of real rank zero C*-algebras are still real rank zero. In particular:

Proposition V.7.2 *AF algebras have real rank zero.*

In fact, many other algebras also have real rank zero including Cuntz algebras, Bunce-Deddens algebras and irrational rotation algebras. The following theorem provides several useful formulations of this property.

Theorem V.7.3 *For a C*-algebra $\mathfrak{A}$, the following are equivalent:*

[RR0] *$\mathfrak{A}$ has real rank zero.*
[FS]   *The elements of $\mathfrak{A}_{sa}$ with finite spectrum are dense in $\mathfrak{A}_{sa}$.*
[HP]   *Every hereditary subalgebra of $\mathfrak{A}$ has an approximate unit of projections (not necessarily increasing).*

Proof. Suppose that [RR0] holds. Fix an element A in $\mathfrak{A}_{sa}$ of norm 1 and an $\varepsilon > 0$. Choose an increasing subset $t_1 = -1, \ldots, t_n = 1$ of non-zero points in $[-1, 1]$ which forms an $\varepsilon/2$ net. Let $\varepsilon_1 = \varepsilon/4$. By the real rank zero property, there is an element A_1 in $\mathfrak{A}_{sa}$ such that $A - t_1 I$ is invertible and $\|A - A_1\| < \varepsilon_1$. Then choose $\varepsilon_2 < \varepsilon/8$ sufficiently small so that $[t_1 - \varepsilon_2, t_1 + \varepsilon_2]$ does not intersect $\sigma(A_1)$. Using [RR0] again, choose A_2 in $\mathfrak{A}_{sa}$ such that $A_2 - t_2 I$ is invertible and $\|A_2 - A_1\| < \varepsilon_2$. Then neither t_1 nor t_2 lies in $\sigma(A_2)$. Repeated use of this argument produces an element A_n in $\mathfrak{A}_{sa}$ such that $t_1, \ldots, t_n$ are in the resolvent

V.7. Real Rank Zero

of A_n and
$$\|A - A_n\| < \sum_{i=1}^{n} \varepsilon_i < \sum_{i=1}^{n} 2^{-i-1}\varepsilon < \varepsilon/2.$$

Therefore the Riemann sum
$$B = -E_{A_n}(-1 - \varepsilon/2, -1] + \sum_{i=2}^{n} t_i E_{A_n}(t_{i-1}, t_i] + E_{A_n}(1, 1 + \varepsilon/2).$$

belongs to $\mathfrak{A}$. Evidently B has finite spectrum and
$$\|B - A\| \leq \|B - A_n\| + \|A_n - A\| < \varepsilon/2 + \varepsilon/2 = \varepsilon.$$

Now assume [FS] and suppose that $\mathfrak{B}$ is a hereditary subalgebra of $\mathfrak{A}$. Given $B_1, \ldots, B_n$ in $\mathfrak{B}$ and $0 < \varepsilon < 1$, it suffices to find a projection P in $\mathfrak{B}$ such that $\|B_i(I - P)\| < \varepsilon$ for $1 \leq i \leq n$. Let $B = \sum_{i=1}^{n} B_i^* B_i$. Then
$$\|B_i(I - P)\|^2 = \|(I - P)B_i^* B_i(I - P)\| \leq \|B(I - P)\|.$$

So we work with B, which we may assume to have norm 1.

Pick a positive number δ with $\delta < (\varepsilon - \varepsilon^2)/6$. Then choose an integer n so large that $\delta^{2/n} > 1 - \delta$. By property [FS], there is a positive element C in $\mathfrak{A}$ with finite spectrum such that $\|B^{1/n} - C\| < \delta/n$ and $\|C\| \leq 1$. Thus $A = C^n$ satisfies
$$\|A - B\| = \|\sum_{k=0}^{n-1} C^{n-1-k}(C - B^{1/n})B^{k/n}\| \leq n\|B^{1/n} - C\| < \delta.$$

The projection $Q = E_A[\delta, 1]$ belongs to $\mathfrak{A}$. From the functional calculus, we get the estimates
$$\|A(I - Q)\| < \delta \quad \text{and} \quad \|A^{1/n}QA^{1/n} - Q\| \leq 1 - \delta^{2/n} < \delta.$$

Since $\mathfrak{B}$ is hereditary, the element $X = B^{1/n}QB^{1/n}$ belongs to $\mathfrak{B}$. An easy estimate shows that
$$\|X - Q\| \leq 2\|B^{1/n} - A^{1/n}\| + \|A^{1/n}QA^{1/n} - Q\| < 3\delta.$$

Therefore (compare with Lemma III.3.1)
$$\|X - X^2\| = \|(I - Q)(X - Q) - (X - Q)X\| \leq 6\delta < \varepsilon - \varepsilon^2.$$

Hence $\sigma(X)$ is contained in $[0, \varepsilon] \cup [1 - \varepsilon, 1]$. So the projection $P = E_X[1 - \varepsilon, 1]$ lies in $\mathfrak{B}$ and $\|P - X\| \leq \varepsilon$. Finally we compute
$$\|B(I - P)\| \leq \|P - Q\| + \|A - B\| + \|A(I - Q)\| \leq \varepsilon + 5\delta < 2\varepsilon.$$

To complete the circuit, assume [HP] and let A be a self-adjoint element of norm 1. Write $A = A_+ - A_-$ be the Hahn decomposition of Corollary I.4.2.

Set $\mathfrak{B} = \overline{A_+ \mathfrak{A} A_+}$. By property [HP], there is a projection P in $\mathfrak{B}$ such that $\|A_+ P^\perp\| < \varepsilon$. Since $A_- A_+ = 0$, it follows that $A_- P = 0$. Let

$$\begin{aligned} B &= PAP + 2\varepsilon P + P^\perp A P^\perp - 2\varepsilon P^\perp \\ &= A - (PAP^\perp + P^\perp AP) + 2\varepsilon(P - P^\perp). \end{aligned}$$

Then

$$\|A - B\| \leq \|PAP^\perp + P^\perp AP\| + 2\varepsilon \|P - P^\perp\| \leq \varepsilon + 2\varepsilon = 3\varepsilon.$$

Notice that $PBP = PA_+ P + 2\varepsilon P \geq 2\varepsilon P$ and

$$\begin{aligned} P^\perp B P^\perp &= P^\perp A_- P^\perp + P^\perp A_+ P^\perp - 2\varepsilon P^\perp \\ &\leq 0 + \varepsilon P^\perp - 2\varepsilon P^\perp = -\varepsilon P^\perp. \end{aligned}$$

As B commutes with P, it follows that B is invertible. So $\mathfrak{A}$ is real rank zero. ∎

Now we show that there is a non-trivial class of real rank zero algebras. In fact the definition of purely infinite algebras suggests something of the [HP] condition.

Theorem V.7.4 *If $\mathfrak{A}$ is a purely infinite simple C*-algebra, then it has real rank zero.*

Proof. Suppose that A belongs to $\mathfrak{A}_{sa}$ and $\varepsilon > 0$. Define

$$f_\varepsilon(t) = \begin{cases} 0 & \text{if } |t| \leq \varepsilon \\ t - \varepsilon & \text{if } t \geq \varepsilon \\ t + \varepsilon & \text{if } t \leq -\varepsilon \end{cases} \quad \text{and} \quad g_\varepsilon(t) = \max\{\varepsilon - |t|, 0\}.$$

Since $\mathfrak{A}$ is purely infinite, there is an infinite projection P in the hereditary subalgebra $\mathfrak{B} = \overline{g_\varepsilon(A) \mathfrak{A} g_\varepsilon(A)}$. By Lemma V.5.4, the projection $I - P$ is equivalent to a subprojection of P. Thus there is a partial isometry S in $\mathfrak{A}$ such that

$$S^*S = I - P \quad \text{and} \quad SS^* = Q \leq P.$$

Notice that $f_\varepsilon(A) = (I - P) f_\varepsilon(A) (I - P)$. Define

$$B = f_\varepsilon(A) + \varepsilon(S + S^*) + \varepsilon(P - Q) \simeq \begin{bmatrix} f_\varepsilon(A) & \varepsilon & 0 \\ \varepsilon & 0 & 0 \\ 0 & 0 & \varepsilon \end{bmatrix}.$$

The matrix comes from the decomposition $(I - P)\mathcal{H} \oplus Q\mathcal{H} \oplus (P - Q)\mathcal{H}$ with the matrix unit $E_{21} = S$. From this matrix form, it is evident that B is invertible. Finally,

$$\|B - A\| \leq \|f_\varepsilon(A) - A\| + \varepsilon \|S + S^* + (P - Q)\| \leq 2\varepsilon. \quad \blacksquare$$

Corollary V.7.5 *The Cuntz algebras $\mathcal{O}_n$ for $n \geq 2$ are real rank zero.*

V.7. Real Rank Zero

We consider the class AT of a **limit circle algebras** which are direct limits of algebras of the form $\sum \oplus_{i=1}^{k} \mathcal{M}_{p_i}(C(\mathbb{T}))$. There is an obvious parallel with AF algebras. However, as the Bunce-Deddens algebras have this form, they can exhibit different properties from AF algebras. Since $\mathbb{T}$ is connected, $C(\mathbb{T})$ is very far from being real rank zero. Nevertheless, it is possible for AT algebras to have real rank zero. A subset $\mathcal{P}$ of a C*-algebra **separates** the traces of $\mathfrak{A}$ if two traces τ_i satisfying $\tau_1(P) = \tau_2(P)$ for all P in $\mathcal{P}$ are equal.

If $F = F^*$ is in $\mathcal{M}_p(C(\mathbb{T}))$, let $\lambda_j(F,t)$ denote the jth smallest eigenvalue of $F(t)$. The **variation of the eigenvalues** of F is

$$\sup_{s,t \in \mathbb{T}} \max_{1 \leq j \leq p} |\lambda_j(F,s) - \lambda_j(F,t)|.$$

The **variation of the trace** is

$$\sup_{s,t \in \mathbb{T}} p^{-1} |\operatorname{Tr}(F(s)) - \operatorname{Tr}(F(t))| = \sup_{s,t \in \mathbb{T}} p^{-1} \left| \sum_{j=1}^{p} \lambda_j(F,s) - \lambda_j(F,t) \right|.$$

Lemma V.7.6 *The set of self-adjoint elements of $\mathcal{M}_k(C(\mathbb{T}))$ which have k distinct eigenvalues at each $t \in \mathbb{T}$ is dense in $\mathcal{M}_k(C(\mathbb{T}))_{sa}$.*

Proof. Every continuous function from $\mathbb{T}$ into $\mathcal{M}_k$ may be approximated by a continuous piecewise linear function such that the "corner" points have distinct eigenvalues. Therefore it suffices to approximate a linear function $f(x) = A + xB$ on $[-1,1]$ such that $f(\pm 1)$ has distinct eigenvalues. In this case, the characteristic polynomial $p(x, \lambda) = \det(f(x) - \lambda I)$ is algebraic. It follows that the k eigenvalues and corresponding eigenprojections are analytic, even at the finitely many points where the eigenvalues cross. It is easy to make a small perturbation, say at $f(0)$ so that for each x, at most one pair of eigenvalues coincide. Thus by splitting the interval into smaller intervals if necessary, we may suppose that there is a single point x_0 in $(-1, 1)$ at which a single pair of eigenvalues $\lambda_j(x_0)$ and $\lambda_{j+1}(x_0)$ coincide.

Indeed, we may suppose that $x_0 = 0$ and $\lambda_j(0) = 0$. As the eigenprojections are analytic, after a unitary equivalence, it may be supposed that the corresponding eigenvectors $e_j(t)$ and $e_{j+1}(t)$ are constant. Then ignoring all the other eigenvalues, f is (up to a scalar) just the diagonal function with roots $x_j(t)$ and $x_{j+1}(t)$ which are distinct except for $t = 0$. Let $\varepsilon(t)$ be a real valued function on $[-1,1]$ such that $\|\varepsilon(t)\| < \varepsilon$, $\varepsilon(-1) = \varepsilon(1) = 0$ and $\varepsilon(0) > 0$. Then write

$$f(x) = \begin{bmatrix} x_j(t) & 0 \\ 0 & x_{j+1}(t) \end{bmatrix} \quad \text{and} \quad g(x) = \begin{bmatrix} x_j(t) & \varepsilon(t) \\ \varepsilon(t) & x_{j+1}(t) \end{bmatrix}.$$

A simple calculation shows that $g(x)$ has no repeated roots for any t, and it takes the same values as f at the endpoints. ∎

Now we obtain a useful condition implying real rank zero for this class of algebras. Let us write $\mathfrak{A} = \varinjlim \mathfrak{A}_n$ where $\mathfrak{A}_n = \sum \oplus_{i=1}^{k_n} \mathcal{M}_{p_{n,i}}(C(\mathbb{T}))$, and let

$\varphi_{n,m}$ denote the homomorphism of $\mathfrak{A}_n$ into $\mathfrak{A}_m$ for $n < m$. For a self-adjoint element $F = F_1 \oplus \cdots \oplus F_{k_n}$ in $\mathfrak{A}_n$, the variation of the eigenvalues of F, or of the trace of F, is the maximum of this quantity for each summand.

Theorem V.7.7 *If $\mathfrak{A}$ is a simple AT algebra, then $\mathfrak{A}$ is real rank zero if and only if the projections of $\mathfrak{A}$ separate the traces of $\mathfrak{A}$.*

Proof. If $\mathfrak{A}$ has real rank zero, then every Hermitian element is in the closed real span of the projections. Thus the span of the projections is dense in $\mathfrak{A}$. Hence the projections separate traces.

So suppose that the projections separate traces. We first show that if $F = F^*$ in $\mathfrak{A}_n$ and $\varepsilon > 0$, then there an integer $m > n$ so that the variation of the trace of $\varphi_{n,m}(F)$ is less than ε. Suppose, to the contrary, that there is an F and an $\varepsilon > 0$ so that for every $m > n$, there is a summand $\mathcal{M}_{p_{m,i}}(C(\mathbb{T}))$ of $\mathfrak{A}_m$ and points s_m and t_m in $\mathbb{T}$ such that

$$p_{m,i}^{-1} \left| \text{Tr}(\varphi_{n,m}(F)_i(s_m)) - \text{Tr}(\varphi_{n,m}(F)_i(t_m)) \right| \geq \varepsilon.$$

For each $m > n$, let σ_m be any state of $\mathfrak{A}$ extending the trace on $\mathfrak{A}_m$ given by $\sigma_m(G) = p_{m,i}^{-1} \text{Tr}(G_i(s_m))$; and let τ_m be a state extending the trace on $\mathfrak{A}_m$ given by $\tau_m(G) = p_{m,i}^{-1} \text{Tr}(G_i(t_m))$. Choose a subnet Λ of $\mathbb{N}$ such that $\sigma = \lim_{m \in \Lambda} \sigma_m$ and $\tau = \lim_{m \in \Lambda} \tau_m$ exist as weak-$*$ limits. These are easily seen to be traces on $\mathfrak{A}$ such that $|\sigma(F) - \tau(F)| \geq \varepsilon$. In particular, $\sigma \neq \tau$.

Now suppose that P is a projection in $\mathfrak{A}$. By Lemma III.3.1, P is the limit of a sequence of projections P_k in $\mathfrak{A}_k$. For any $m > k$, P_k is a projection in $\mathfrak{A}_m$, and thus has constant rank on each component. As the trace of a projection in $\mathcal{M}_p$ depends only on rank, it follows that $\sigma_m(P_k) = \tau_m(P_k)$. Therefore σ and τ agree on each P_k, and hence on every projection in $\mathfrak{A}$. This contradicts the hypothesis that projections separate traces.

We claim that if F and G are positive contractions in $\mathfrak{A}_n$ such that $FG = F$, then there is an integer $m > n$ so that for every summand $\mathcal{M}_{p_{m,i}}(C(\mathbb{T}))$ of $\mathfrak{A}_m$ and $s, t \in \mathbb{T}$,

$$\text{rank}(\varphi_{n,m}(F)_i(s)) \leq \text{rank}(\varphi_{n,m}(G)_i(t)).$$

Since $FG = F$ implies that this inequality holds for $s = t$, the result holds if the right hand side is constant; in particular, this is the case when G is a projection. So we may suppose that $\sigma(G)$ intersects $(0,1)$. Let k be a continuous function on $[0,1]$ such that $0 \leq k(x) \leq x$, $k(0) = k(1) = 0$ and $K = k(G) \neq 0$. Then since $F \leq E_G\{1\}$, we have $KF = FK = 0$. Define $H = G - K$.

Since $\mathfrak{A}$ is simple, there are elements A_j and B_j in $\mathfrak{A}$ such that

$$\sum_{j=1}^{k} A_j K B_j = I.$$

V.7. Real Rank Zero

We can arrange that the A_j's and B_j's lie in some $\mathfrak{A}_{m_0}$ for $m_0 \geq n$ sufficiently large. Indeed, choose A'_j and B'_j in some $\mathfrak{A}_{m_0}$ so that

$$\|A'_j - A_j\| < (4k\|K\|\|B_j\|)^{-1} \quad \text{and} \quad \|B'_j - B_j\| < (4k\|K\|\|A'_j\|)^{-1}$$

for $1 \leq j \leq k$. Then a routine estimate yields $\|I - \sum_{j=1}^k A'_j K B'_j\| < 1/2$. So $X = \sum_{j=1}^k A'_j K B'_j$ is invertible in $\mathfrak{A}_{m_0}$. Therefore $\sum_{j=1}^k (X^{-1} A'_j) K B'_j = I$.

If τ is any trace on $\mathfrak{A}_{m_0}$, it follows that

$$\tau(K) \geq \varepsilon := \left(\sum_{j=1}^k \|A_j\|\|B_j\|\right)^{-1}.$$

Take $m \geq m_0$ so large that the variation of the trace of $\varphi_{n,m}(H)$ is less than ε. For any s in $\mathbb{T}$, the identity $FH = F$ shows that $\varphi_{n,m}(F)_i(s) \leq E_{\varphi_{n,m}(H)_i(s)}\{1\}$ and hence for any t in $\mathbb{T}$

$$\operatorname{rank}(\varphi_{n,m}(F)_i(s)) \leq \operatorname{Tr}(\varphi_{n,m}(H)_i(s)) \leq \operatorname{Tr}(\varphi_{n,m}(H)_i(t)) + \varepsilon$$
$$\leq \operatorname{Tr}(\varphi_{n,m}(H)_i(t)) + \operatorname{Tr}(\varphi_{n,m}(K)_i(t))$$
$$= \operatorname{Tr}(\varphi_{n,m}(G)_i(t)) \leq \operatorname{rank}(\varphi_{n,m}(G)_i(t)).$$

We use this result to show that if $F = F^*$ is in $\mathfrak{A}_n$ and $\varepsilon > 0$, then there is an $m > n$ so that the variation of the eigenvalues of $\varphi_{n,m}(F)$ is less than ε. Fix an integer $r > 3\varepsilon^{-1}$. For $0 \leq j \leq r$, define

$$f_j(x) = \begin{cases} 1 & x \leq j/r \\ j+1-rx & j/r \leq x \leq (j+1)/r \\ 0 & (j+1)/r \leq x \end{cases}.$$

Set $F_j = f_j(F)$. From the functional calculus, we see that $F_j F_k = F_j$ for $j < k$. Hence we have an integer $m > n$ so that for every summand $\mathcal{M}_{p_{m,i}}(C(\mathbb{T}))$ of $\mathfrak{A}_m$ and $s, t \in \mathbb{T}$,

$$\operatorname{rank}(\varphi_{n,m}(F_j)_i(s)) \leq \operatorname{rank}(\varphi_{n,m}(F_{j+1})_i(t)) \quad \text{for} \quad 0 \leq j \leq r-1.$$

Fix i and a point s_0 in $\mathbb{T}$; and set $d_j = \operatorname{rank}(\varphi_{n,m}(F_j)_i(s_0))$. Then this is an increasing sequence such that $d_r = p_{m,i}$. Then

$$d_j \leq \max_{s \in \mathbb{T}} \operatorname{rank}(\varphi_{n,m}(F_j)_i(s)) \leq \min_{t \in \mathbb{T}} \operatorname{rank}(\varphi_{n,m}(F_{j+1})_i(t)) \leq d_{j+1}.$$

Now $\operatorname{rank}(\varphi_{n,m}(F_j)_i(t))$ is the number of eigenvalues of $\varphi_{n,m}(F)_i(t))$ in the interval $[0, (j+1)/r)$. This figure lies in the interval $[d_{j-1}, d_{j+1}]$. Conversely, if $d_{j-1} \leq k \leq d_j$, then $\lambda_k(\varphi_{n,m}(F)_i(t))$ lies in $[(j-1)/r, (j+1)/r]$. So the variation of the eigenvalues is at most $3/r < \varepsilon$.

Now given $A = A^*$ in $\mathfrak{A}$ and $\varepsilon > 0$, approximate A to within ε by an $F = F^*$ in $\mathfrak{A}_n$. By replacing n with a larger integer mi, we may suppose that the variation of the eigenvalues of F is at most ε. By Lemma V.7.6, approximate F within ε by an element G of $\mathfrak{A}_m$ which has distinct eigenvalues for every $1 \leq i \leq k_m$ and

t in $\mathbb{T}$. Clearly, the eigenvalues of G have variation at most 3ε. Let $P_{m,i,j}$ for $1 \leq i \leq k_m$ and $1 \leq j \leq p_{m,i}$ be the eigenprojection of G corresponding to the eigenvalue $\lambda_j(G_i)$ (which is well defined because the eigenvalues never intersect each other). Then define

$$H = \sum_{i=1}^{k_m} \oplus \sum_{j=1}^{p_{m,i}} \lambda_j(G_i)(s_0) P_{m,i,j}.$$

Clearly H has finite spectrum and $\|G - H\| < 3\varepsilon$. Hence $\|A - H\| < 5\varepsilon$. Therefore $\mathfrak{A}$ has real rank zero. ∎

Corollary V.7.8 *The Bunce-Deddens algebras have real rank zero.*

Proof. The Bunce-Deddens algebras are simple by Theorem V.3.3 and have a unique trace by Theorem V.3.6. Therefore the projections separate the traces; whence the algebra has real rank zero. ∎

Exercises

V.1 Verify that for h in H^2, $\lim_{r \to 1^-} \|\hat{h}(re^{i\theta})\|_2 = \|h\|_2$.

V.2 Show that H^∞ is a weak-$*$ closed subalgebra of L^∞.

V.3 (*Beurling's Theorem*) Suppose that $\mathcal{M}$ is a proper invariant subspace for T_z. Show that there is an **inner** function ω (i.e. $\omega \in H^\infty$ and $|\omega| = 1$ a.e.) such that $\mathcal{M} = \omega H^2$.
HINT: Pick a unit vector ω in $\mathcal{M} \ominus T_z \mathcal{M}$. Consider $(\omega, \omega z^k)$ for $k \geq 1$.

V.4 Show that if h in H^2 is non-zero, then $|h| > 0$ a.e. on $\mathbb{T}$.
HINT: Let $E = \{t \in \mathbb{T} : h(t) = 0\}$. Apply Beurling's Theorem to the subspace $\mathcal{M} = \{g \in H^2 : g|_E = 0\}$.

V.5 (a) Show that if T_g is invertible, then g is invertible in L^∞.
(b) Hence show that $\sigma(M_g) \subset \sigma(T_g)$ for all g in L^∞.

V.6 (a) Show that if g is in L^∞ and $\operatorname{Re} g \geq 1$, then T_g is invertible.
HINT: Find $c > 0$ such that $\|c - g\|_\infty < c$.
(b) Hence show that $\sigma(T_g) \subset \operatorname{conv}(\sigma(M_g))$.
(c) If $g = \bar{g}$, show that $\sigma(T_g) = [\operatorname{ess\,inf} g, \operatorname{ess\,sup} g]$.

V.7 (a) Show that if f, g belong to L^∞ and at every point $t \in \mathbb{T}$ one of f or g is continuous, then $T_f T_g - T_{fg}$ is compact.
HINT: Use a partition of unity p_i and consider $\sum_{i=1}^n T_f T_{p_i} T_g$.
(b) Let PC denote the C*-algebra generated by the piecewise continuous functions. Show that $\mathcal{T}(\mathrm{PC})/\mathfrak{K}$ is abelian.
(c) Show that the maximal ideal space of $\mathcal{T}(\mathrm{PC})/\mathfrak{K}$ is a cylinder $\mathbb{T} \times [0,1]$ with a non-standard topology.

Exercises

V.8 (a) Show that an automorphism α of $\mathcal{T}(C(\mathbb{T}))$ induces an automorphism $\tilde{\alpha}(f) = f \circ h$ of $C(\mathbb{T})$, where h is an orientation preserving homeomorphism of $\mathbb{T}$.
HINT: Consider the Fredholm index.

(b) If h is an orientation preserving homeomorphism of $\mathbb{T}$, find a compact operator K such that $T_h + K$ is unitarily equivalent to T_z. Hence show that there is an automorphism of $\mathcal{T}(C(\mathbb{T}))$ which induces h as in (a).

V.9 Show that if a unital C*-algebra $\mathfrak{A}$ has no proper closed ideals, then it has no proper algebraic ideals either. Then show that if A is positive and non-zero in a simple C*-algebra, then there are elements X_i in $\mathfrak{A}$ such that $\sum_{i=1}^{n} X_i^* A X_i = I$.
HINT: Show that $XAY + Y^*AX^* \leq XAX^* + Y^*AY$.

V.10 Show that Bunce–Deddens algebras are stably finite.
HINT: Exercise IV.9.

V.11 Show that $\mathcal{T}(C(\mathbb{T}))$ is infinite but not properly infinite.
HINT: Consider the symbol of projections in $\mathcal{T}(C(\mathbb{T}))$.

V.12 Prove that the set of invertible elements of $\mathcal{T}(C(\mathbb{T}))$ is connected.

V.13 Consider the weighted shift $Se_n = a_n e_n$ where $a_n = 2 - \gcd(n, 2^n)^{-1}$ for $n \geq 1$.

(a) Show that $C^*(S)$ contains the unilateral shift, and hence all compact operators.
HINT: The weights are bounded below.

(b) Show that $C^*(S)$ contains all 2^n-periodic weighted shifts.
HINT: Use the functional calculus on S^*S.

(c) Show that S is a limit of 2^n-periodic weighted shifts. Hence deduce that $C^*(S)$ is a Bunce–Deddens algebra.

(d) Show that each Bunce–Deddens algebra is singly generated.

V.14 Define maps from $\mathcal{O}_n$ onto $\mathfrak{F}^n$ by
$$\Phi_i(X) = \begin{cases} \Phi_0(XS_1^{*i}) & \text{for } i \geq 0 \\ \Phi_0(S_1^{-i}X) & \text{for } i < 0 \end{cases}.$$

(a) Show that if Y in $\mathcal{O}_n$ is in the algebraic span of words, then
$$Y = \sum_{i \geq 0} \Phi_i(Y) S_1^i + \sum_{i < 0} S^{*|i|} \Phi_i(Y).$$

(b) Show that X in $\mathcal{O}_n$ satisfies $\Phi_i(X) = 0$ for all $i \in \mathbb{Z}$ if and only if $X = 0$.
HINT: Represent $\mathcal{O}_n$ on a Hilbert space. Consider the Fourier series of the function $f(\lambda) = (\rho_\lambda(X)x, y)$ for arbitrary vectors x and y.

V.15 The C*-algebra $\mathcal{O}_\infty$ is the universal C*-algebra generated by countably many isometries S_k such that $\sum_{k=1}^n S_k S_k^* < I$ for all $n \geq 1$. Show that $\mathcal{O}_\infty$ is simple.

V.16 (a) Show that $\{S_1, S_2 S_1, S_2^2\}$ generate a copy of $\mathcal{O}_3$ as a subalgebra of $\mathcal{O}_2$.
(b) Generalize this to show that $\mathcal{O}_{k(n-1)+1}$ can be unitally imbedded in $\mathcal{O}_n$ for all $k \geq 1$.

V.17 If k divides n, show that $\mathcal{M}_k(\mathcal{O}_n)$ is isomorphic to $\mathcal{O}_n$.
HINT: For $0 \leq j < n/k$ and $1 \leq i \leq k$, consider the operator T_{kj+i} which is a $k \times k$ matrix with $\begin{bmatrix} S_{kj+1} & S_{kj+2} & \cdots & S_{kj+k} \end{bmatrix}$ in the i-th row and all other entries equal to 0.

V.18 Show that $\mathcal{M}_3(\mathcal{O}_2)$ is isomorphic to $\mathcal{O}_2$.
HINT: Show that
$$T_1 = \begin{bmatrix} 0 & 0 & 0 \\ I & 0 & 0 \\ 0 & S_1 & S_2 \end{bmatrix} \quad \text{and} \quad T_2 = \begin{bmatrix} S_1 & S_2 S_1 & S_2^2 \\ 0 & 0 & 0 \\ 0 & 0 & 0 \end{bmatrix}$$
generate $\mathcal{M}_3(\mathcal{O}_2)$.

V.19 Generalize the previous question to show that $\mathcal{M}_n(\mathcal{O}_2)$ is isomorphic to $\mathcal{O}_2$ for all $n \geq 1$.

V.20 Let P be a projection in a C*-algebra $\mathfrak{A}$. Show that $\mathfrak{A}$ has real rank zero if and only if both $P\mathfrak{A}P$ and $P^\perp \mathfrak{A} P^\perp$ are real rank zero.
HINT: An element $X = X^*$ in $\mathfrak{A}$ has a matrix form $\begin{bmatrix} A & B \\ B^* & C \end{bmatrix}$ with respect to $P \oplus P^\perp$. If A is invertible, factor this as
$$\begin{bmatrix} A^{1/2} & 0 \\ B^* A^{-1/2} & I \end{bmatrix} \begin{bmatrix} I & 0 \\ 0 & D \end{bmatrix} \begin{bmatrix} A^{1/2} & A^{-1/2} B \\ 0 & I \end{bmatrix}.$$

V.21 If $\mathfrak{A}$ has real rank zero, show that $\mathcal{M}_n(\mathfrak{A})$ also has real rank zero.
HINT: Use the previous exercise.

V.22 Show that a separable C*-algebra of real rank zero has an *increasing* sequence of projections forming an approximate unit.
HINT: Use a strictly positive element A. Given $P_1 < \cdots < P_n$, choose a projection $Q < P_n^\perp$ such that $\|(P_n^\perp - Q)A\| < 1/n$.

Notes and Remarks.

The study of Toeplitz operators is one of the richest areas of study in operator theory. An excellent introduction to the theory is Douglas [1972]. Theorem V.2.2 is due to Coburn [1967]. Bunce–Deddens algebras were introduced in Bunce–Deddens [1975]. Cuntz algebras were introduced in Cuntz [1977]. The basic results about these two classes come from those original papers. Theorem V.5.5 is due to Cuntz [1981]. The classification of Cuntz algebras by Ext

Exercises

is due to Paschke and Salinas [1979], and simultaneously to Pimsner and Popa [1978]. Cuntz [1981] classifies them as well, but using K-theory. The notion of a real rank zero C*-algebra is taken from Brown and Pedersen [1991], who prove Theorem V.7.3. Zhang [1988] showed that purely infinite C*-algebras are real rank zero. That Bunce–Deddens algebras are real rank zero is due to Blackadar and Kumjian [1985]. The proof given here for limit circle algebras is taken from Blackadar, Bratteli, Elliott and Kumjian [1992]. A stronger result due to Blackadar, Dadarlat and Rørdam [1991] shows that every simple C*-algebra which is the direct limit of finite sums of matrix algebras over continuous functions on spaces of bounded dimension is real rank zero if and only if projections separate traces.

CHAPTER VI

Irrational Rotation Algebras

In this chapter, we will study a class of C*-algebras that has received a lot of special attention in recent years. The canonical model acts on the circle $\mathbb{T}$ which we will think of as $\mathbb{R}/\mathbb{Z}$ via the map taking t to $z(t) = e^{2\pi i t}$. Fix an irrational number θ. Let $\mathcal{H} = L^2(\mathbb{R}/\mathbb{Z})$ and consider two unitary operators on $\mathcal{H}$, the operator $U = M_{z(t)}$ of multiplication by the unimodular function $z(t)$ and V, the operator of rotation by θ. That is

$$Uf(t) = z(t)f(t) \quad \text{and} \quad Vf(t) = f(t - \theta).$$

A simple calculation yields

$$VUf(t) = (Uf)(t - \theta) = z(t - \theta)f(t - \theta)$$
$$= e^{-2\pi i \theta}z(t)(Vf)(t) = e^{-2\pi i \theta}UVf(t)$$

Hence

$$UV = e^{2\pi i \theta}VU \qquad (\dagger)$$

We wish to study the **universal** C*-algebra satisfying ($\dagger$). A C*-algebra $\mathcal{A}_\theta$ is universal for the relation ($\dagger$) provided that it is generated by two unitaries $\widetilde{U}$ and $\widetilde{V}$ satisfying ($\dagger$) and whenever $\mathfrak{A} = C^*(U, V)$ is another C*-algebra satisfying ($\dagger$), there is a homomorphism of $\mathcal{A}_\theta$ onto $\mathfrak{A}$ which carries $\widetilde{U}$ to U and $\widetilde{V}$ to V. Since we know that there are unitaries satisfying the relation, we may consider the collection of all irreducible pairs of unitaries (U_α, V_α) in $\mathcal{B}(\mathcal{H})$ satisfying ($\dagger$). Then form the operators

$$\widetilde{U} = \sum \oplus U_\alpha \quad \text{and} \quad \widetilde{V} = \sum \oplus V_\alpha.$$

Let $\mathcal{A}_\theta = C^*(\widetilde{U}, \widetilde{V})$.

In order to see that $\mathcal{A}_\theta$ is the desired universal algebra, let $\mathfrak{A} = C^*(U, V)$ be another C*-algebra satisfying ($\dagger$). To verify that there is a well defined homomorphism $\varphi : \mathcal{A}_\theta \to \mathfrak{A}$ such that $\varphi(\widetilde{U}) = U$ and $\varphi(\widetilde{V}) = V$, it suffices to show that

$$\|p(U, V, U^*, V^*)\| \leq \|p(\widetilde{U}, \widetilde{V}, \widetilde{U}^*, \widetilde{V}^*)\|$$

for every non-commutative polynomial in four variables. Fix a polynomial p; and let $A = p(U, V, U^*, V^*)$. By the GNS construction (Theorem I.9.12), there is an

VI. Irrational Rotation Algebras

irreducible representation π of $\mathfrak{A}$ such that $\|\pi(A)\| = \|A\|$. Consider the pair $U' = \pi(U)$ and $V' = \pi(V)$. Then (U', V') is an irreducible pair satisfying (†). Hence by construction, we see that

$$\|p(\widetilde{U}, \widetilde{V}, \widetilde{U}^*, \widetilde{V}^*)\| \geq \|p(U', V', U'^*, V'^*)\| = \|p(U, V, U^*, V^*)\|.$$

Therefore φ is well defined and contractive on the $*$-algebra generated by $\widetilde{U}$ and $\widetilde{V}$ into $\mathfrak{A}$. So it extends by continuity to a homomorphism of $\mathcal{A}_\theta$ onto $\mathfrak{A}$.

The C*-algebra $\mathcal{A}_\theta$ for irrational values of θ is called an **irrational rotation algebra**. The argument above produces a universal C*-algebra for any family of relations.

We apply this universal property to obtain certain special automorphisms of $\mathcal{A}_\theta$. For any constants λ, μ on the unit circle ($|\lambda| = |\mu| = 1$), the unitary pair $(\lambda\widetilde{U}, \mu\widetilde{V})$ satisfies (†). Thus there is an endomorphism $\rho_{\lambda,\mu}$ of $\mathcal{A}_\theta$ such that

$$\rho_{\lambda,\mu}(\widetilde{U}) = \lambda\widetilde{U} \quad \text{and} \quad \rho_{\lambda,\mu}(\widetilde{V}) = \mu\widetilde{V}.$$

Let $\sigma = \rho_{\overline{\lambda},\overline{\mu}}\rho_{\lambda,\mu}$. Since $\sigma(U) = U$ and $\sigma(V) = V$, we have $\sigma = \text{id}$. Thus $\rho_{\lambda,\mu}$ is an automorphism.

For each fixed A in $\mathcal{A}_\theta$, the map from $\mathbb{T}^2$ to $\mathcal{A}_\theta$ given by $f(\lambda, \mu) = \rho_{\lambda,\mu}(A)$ is norm continuous. To verify this, notice that it is true for all non-commuting polynomials in $\widetilde{U}, \widetilde{V}, \widetilde{U}^*$ and $\widetilde{V}^*$. These are dense and automorphisms are contractive; so the rest follows from a simple approximation argument.

Define two maps of $\mathcal{A}_\theta$ into itself by the formulae

$$\Phi_1(A) = \int_0^1 \rho_{1,e^{2\pi i t}}(A)\, dt \quad \text{and} \quad \Phi_2(A) = \int_0^1 \rho_{e^{2\pi i t},1}(A)\, dt.$$

These integrals make sense as Riemann sums because the integrand is a norm continuous function. Some of the nice properties of these maps are captured in the following theorem.

An **expectation** of a C*-algebra onto a subalgebra is a positive, unital idempotent map. Expectations occur frequently in the study of operator algebras, and have many nice general properties that will not be developed here. The point of this next theorem is to show that Φ_1 and Φ_2 are expectations. Recall that a map Φ is **contractive** if $\|\Phi\| \leq 1$, **idempotent** if $\Phi^2 = \Phi$, and a positive map is **faithful** if $A \geq 0$ and $\Phi(A) = 0$ implies that $A = 0$.

Theorem VI.1.1 Φ_1 *is positive contractive idempotent and faithful, and maps $\mathcal{A}_\theta$ onto* $C^*(\widetilde{U})$. *Moreover,*

$$\Phi_1(f(\widetilde{U})Ag(\widetilde{U})) = f(\widetilde{U})\Phi_1(A)g(\widetilde{U})$$

for all f, g in $C(\mathbb{T})$. *For any finite linear combination of* $\{\widetilde{U}^k\widetilde{V}^\ell : k, \ell \in \mathbb{Z}\}$,

$$\Phi_1\left(\sum_{k,\ell} a_{k\ell}\widetilde{U}^k\widetilde{V}^\ell\right) = \sum_k a_{k0}\widetilde{U}^k.$$

In addition, for every A in $\mathcal{A}_\theta$,

$$\Phi_1(A) = \lim_{n\to\infty} \frac{1}{2n+1} \sum_{j=-n}^n \widetilde{U}^j A \widetilde{U}^{-j}.$$

Proof. Since $\Phi_1(A)$ is a convex combination of $\{\rho_{1,e^{2\pi it}}(A) : 0 \le t \le 1\}$, and since $\|\rho_{1,e^{2\pi it}}(A)\| = \|A\|$, it is clear that $\|\Phi_1\| \le 1$. As $\Phi_1(I) = I$, this is equality.

Since $\rho_{1,e^{2\pi it}}(\widetilde{U}) = \widetilde{U}$ for all $t \in \mathbb{R}$, $\rho_{1,e^{2\pi it}}$ is the identity map on all of $C^*(\widetilde{U})$, which is canonically isomorphic to $C(\mathbb{T})$ by the functional calculus Corollary I.3.3. Hence we obtain

$$\Phi_1(f(\widetilde{U})Ag(\widetilde{U})) = \int_0^1 \rho_{1,e^{2\pi it}}(f(\widetilde{U}))\rho_{1,e^{2\pi it}}(A)\rho_{1,e^{2\pi it}}(g(\widetilde{U}))\,dt$$

$$= f(\widetilde{U}) \int_0^1 \rho_{1,e^{2\pi it}}(A)\,dt\, g(\widetilde{U}) = f(\widetilde{U})\Phi_1(A)g(\widetilde{U})$$

On the other hand,

$$\Phi_1(\widetilde{U}^k\widetilde{V}^\ell) = \widetilde{U}^k \int_0^1 \rho_{1,e^{2\pi it}}(\widetilde{V}^\ell)\,dt = \widetilde{U}^k \int_0^1 e^{2\pi i\ell t}\,dt\,\widetilde{V}^\ell = \begin{cases} 0 & \ell \ne 0 \\ \widetilde{U}^k & \ell = 0 \end{cases}$$

It follows from examining the (dense) set of polynomials in $\widetilde{U}^{\pm 1}$ and $\widetilde{V}^{\pm 1}$ that the range of Φ_1 is exactly $C^*(\widetilde{U})$. As Φ_1 is the identity on $C^*(\widetilde{U})$, it follows that $\Phi_1^2 = \Phi_1$.

If A is positive and non-zero, then $\rho_{1,e^{2\pi it}}(A)$ is positive and non-zero for all t. Thus the integral $\Phi_1(A)$ is positive and non-zero. Hence Φ_1 is positive and faithful.

Again considering a monomial $\widetilde{U}^k\widetilde{V}^\ell$,

$$\lim_{n\to\infty} \frac{1}{2n+1} \sum_{j=-n}^n \widetilde{U}^j(\widetilde{U}^k\widetilde{V}^\ell)\widetilde{U}^{-j} = \lim_{n\to\infty} \frac{1}{2n+1} \sum_{j=-n}^n e^{2\pi ij\ell\theta}\widetilde{U}^k\widetilde{V}^\ell$$

$$= \lim_{n\to\infty} \frac{1}{2n+1} \left(\frac{\sin(2n+1)\pi\ell\theta}{\sin\pi\ell\theta}\right)\widetilde{U}^k\widetilde{V}^\ell = \delta_{\ell 0}\widetilde{U}^k = \Phi_1(\widetilde{U}^k\widetilde{V}^\ell)$$

By linearity and continuity, this formula is valid for all A in $\mathcal{A}_\theta$. ∎

The corresponding results for Φ_2 also hold. Combining them, we obtain:

Corollary VI.1.2 *The map $\tau = \Phi_1\Phi_2 = \Phi_2\Phi_1$ is a faithful unital scalar valued trace on $\mathcal{A}_\theta$.*

Proof. First apply $\Phi_1\Phi_2$ and $\Phi_2\Phi_1$ to a monomial $\widetilde{U}^k\widetilde{V}^\ell$. We have

$$\Phi_1\Phi_2(\widetilde{U}^k\widetilde{V}^\ell) = \delta_{k0}\Phi_1(\widetilde{V}^\ell) = \begin{cases} I & \text{if } k = \ell = 0 \\ 0 & \text{otherwise} \end{cases}$$

VI. Irrational Rotation Algebras

The same formula holds for $\Phi_2\Phi_1$, and hence they are equal (to τ). Moreover, the range of τ is contained in the scalars.

Since both Φ_1 and Φ_2 are positive, faithful and contractive, τ is also positive, faithful and contractive. And since $\tau(I) = I$, $\|\tau\| = 1$.

To verify that τ is a trace, we again compare on monomials

$$\tau((\widetilde{U}^k\widetilde{V}^\ell)(\widetilde{U}^m\widetilde{V}^n)) = e^{-2\pi i \ell m \theta}\tau(\widetilde{U}^{k+m}\widetilde{V}^{\ell+n})$$

$$= \begin{cases} e^{-2\pi i \ell m \theta} & \text{if } k+m = \ell+n = 0 \\ 0 & \text{otherwise} \end{cases}$$

$$\tau((\widetilde{U}^m\widetilde{V}^n)(\widetilde{U}^k\widetilde{V}^\ell)) = e^{-2\pi i k n \theta}\tau(\widetilde{U}^{k+m}\widetilde{V}^{\ell+n})$$

$$= \begin{cases} e^{-2\pi i k n \theta} & \text{if } k+m = \ell+n = 0 \\ 0 & \text{otherwise} \end{cases}$$

When $k + m = \ell + n = 0$, we also have $kn = \ell m$ and thus τ takes the same value on these two products. By linearity, we obtain $\tau(AB) = \tau(BA)$ for all words in $\widetilde{U}^{\pm 1}$ and $\widetilde{V}^{\pm 1}$. By continuity, this extends to all of $\mathcal{A}_\theta$. So τ is a trace. ∎

We have enough structure to show that τ is in fact the only trace on $\mathcal{A}_\theta$.

Proposition VI.1.3 *τ is the unique trace on $\mathcal{A}_\theta$.*

Proof. Suppose that σ is another trace on $\mathcal{A}_\theta$. Then for any A in $\mathcal{A}_\theta$, we have $\sigma(A) = \sigma(\widetilde{U}^j A \widetilde{U}^{-j})$. So by Theorem VI.1.1,

$$\sigma(A) = \lim_{n\to\infty} \sigma\left(\frac{1}{2n+1}\sum_{j=-n}^{n} \widetilde{U}^j A \widetilde{U}^{-j}\right) = \sigma(\Phi_1(A)).$$

Similarly,

$$\sigma(A) = \sigma(\Phi_2(A)) = \sigma(\Phi_1\Phi_2(A)) = \sigma(\tau(A)) = \tau(A)$$

because $\sigma(I) = 1$ and $\tau(A)$ is always a scalar. ∎

Now we are prepared to prove the main result of this section, which is the uniqueness of the C*-algebra generated by unitaries satisfying (†).

Theorem VI.1.4 *$\mathcal{A}_\theta$ is simple. Thus if U and V are any unitary elements satisfying* (†), *then* $C^*(U,V)$ *is canonically isomorphic to $\mathcal{A}_\theta$.*

Proof. Suppose that $\mathcal{J}$ is a non-zero ideal of $\mathcal{A}_\theta$. Then there is a positive, non-zero element X in $\mathcal{J}$. Since $\widetilde{U}^j X \widetilde{U}^{-j}$ belongs to $\mathcal{J}$, the limit formula for Φ_i shows that they map ideals into themselves. Hence $\tau(X)$ belongs to $\mathcal{J}$. But since τ is a faithful trace, $\tau(X)$ is a non-zero multiple of the identity. Therefore $\mathcal{J} = \mathcal{A}_\theta$.

If U and V are any unitary elements satisfying (†), then there is a canonical homomorphism of $\mathcal{A}_\theta$ onto $C^*(U,V)$ taking $\widetilde{U}$ to U and $\widetilde{V}$ to V. Since $\mathcal{A}_\theta$ is simple, this homomorphism must be an isomorphism. ∎

From now on we will drop the tilde, and use the symbols U and V for the generators of $\mathcal{A}_\theta$. Because of the simplicity of $\mathcal{A}_\theta$, this no longer causes any ambiguity.

VI.2 Projections in $\mathcal{A}_\theta$

Our goal is to decide when two irrational rotation algebras are isomorphic. Of course, only the value of θ mod 1 matters. So we restrict our attention to θ in $(0, 1)$. Also since the pair (V, U) satisfies (†) for $-\theta$ when U, V satisfy it for θ, we see that $\mathcal{A}_\theta$ and $\mathcal{A}_{1-\theta}$ are isomorphic. To proceed further, we need some invariant to tell the algebras apart. Our experience with AF algebras suggests trying to compute the K_0 groups.

However, it is not immediately evident that there are any non-trivial projections at all in $\mathcal{A}_\theta$. The two generating subalgebras $C^*(U)$ and $C^*(V)$ are isomorphic to $C(\mathbb{T})$, the space of continuous functions on the circle. Since $\mathbb{T}$ is connected, neither contains any proper projections. The irrational rotation algebras have often been referred to as non-commutative versions of $C(\mathbb{T}^2)$, because this algebra is universal C*-algebra for two commuting unitaries. As the torus $\mathbb{T}^2$ is connected, $C(\mathbb{T}^2)$ does not contain any projections either. So perhaps we should not expect to find any in $\mathcal{A}_\theta$.

Nevertheless, we have the following result due to **Rieffel**.

Theorem VI.2.1 *For every α in $(\mathbb{Z} + \mathbb{Z}\theta) \cap [0, 1]$, there is a projection P in $\mathcal{A}_\theta$ such that $\tau(P) = \alpha$.*

Proof. We may assume that $0 < \theta < 1/2$. We use the representation of $\mathcal{A}_\theta$ from the first section with $U = M_z$ and V equal to the rotation operator by θ. We will write M_f for the multiplication operator on $L^2(\mathbb{T})$ by the continuous function f in $C(\mathbb{T})$. Then $C^*(U) = \{M_f : f \in C(\mathbb{T})\}$. Notice that

$$VM_f h(t) = f(t - \theta)h(t - \theta) = M_{f_\theta} V h(t)$$

where f_θ denotes the translated function Vf. As before, we identify $\mathbb{T}$ with $\mathbb{R}/\mathbb{Z}$.

A dense set of elements of $\mathcal{A}_\theta$ can be represented by a finite sum of the form $A = \sum_i M_{f_i} V^i$. The functions f_i in $C(\mathbb{T})$ can be thought of as Fourier coefficients of A. Indeed, for any A in $\mathcal{A}_\theta$, we may define f_i by the formula $M_{f_i} = \Phi_1(AV^{-i})$. From this formula, we see that the coefficients in a finite sum are uniquely determined. So we may compare two such terms by comparing coefficients. Also note that on $C^*(M_z)$, the value of the trace $\tau(M_f)$ (which agrees with $\Phi_2(M_f)$) reads off the zero Fourier coefficient of f. So

$$\tau(M_f) = \int_0^1 f(t)\, dt.$$

VI.2. Projections in $\mathcal{A}_\theta$

Look for projections of the special form $P = M_g V + M_f + M_h V^*$. Since $P = P^*$, this forces

$$M_g V + M_f + M_h V^* = V^* M_{\bar{g}} + M_{\bar{f}} + V M_{\bar{h}} = M_{\bar{g}_{-\theta}} V^* + M_{\bar{f}} + M_{\bar{h}_\theta} V.$$

By comparing coefficients, we see that $f = \bar{f}$ is a real valued function; and that $h(t) = \overline{g(t+\theta)}$ or equivalently $h(t-\theta) = \overline{g(t)}$. Since $P = P^2$, we also get

$$M_g V + M_f + M_h V^*$$
$$= M_g V M_g V + M_g V M_f + M_g V M_h V^* + M_{fg} V + M_{f^2} +$$
$$+ M_{fh} V^* + M_h V^* M_g V + M_h V^* M_f + M_h V^* M_h V^*$$
$$= M_{gg_\theta} V^2 + M_{g(f+f_\theta)} V + M_{gh_\theta + f^2 + hg_{-\theta}} + M_{h(f+f_{-\theta})} V^* + M_{hh_{-\theta}} V^{*2}$$

By comparing coefficients and replacing h's with g's using the relation between them, we arrive at the necessary and sufficient conditions:

$$g(t)g(t-\theta) = 0 \tag{1}$$
$$g(t)(1 - f(t) - f(t-\theta)) = 0 \tag{2}$$
$$f(t) - f(t)^2 = |g(t)|^2 + |g(t-\theta)|^2 \tag{3}$$

(The other two coefficients yield the same identities.)

These equations can be explicitly solved. (See Figure VI.1.) Pick any positive ε such that $\theta + \varepsilon < 1/2$. Define f to be the piece-wise linear function

$$f(t) = \begin{cases} \varepsilon^{-1} t & \text{for } 0 \leq t \leq \varepsilon \\ 1 & \text{for } \varepsilon \leq t \leq \theta \\ \varepsilon^{-1}(\theta + \varepsilon - t) & \text{for } \theta \leq t \leq \theta + \varepsilon \\ 0 & \text{for } \theta + \varepsilon \leq t \leq 1 \end{cases}$$

and define

$$g(t) = \begin{cases} \sqrt{f(t) - f(t)^2} & \text{for } \theta \leq t \leq \theta + \varepsilon \\ 0 & \text{otherwise} \end{cases}.$$

Clearly (1) holds. Since $f(t) + f(t-\theta) = 1$ on $[\varepsilon, 2\theta]$ which includes the support of g, (2) holds. Finally, $f(t) - f(t)^2$ is non-zero on $(0, \varepsilon)$ and $(\theta, \theta + \varepsilon)$ where by design, we have the identity (3). So this determines a projection P in $\mathcal{A}_\theta$. We compute the trace by

$$\tau(P) = \tau(M_f) = \int_0^1 f(t)\, dt = \theta.$$

We also get the projection $I - P$ with trace $\tau(I - P) = 1 - \theta$.

Now notice that $UV^k = e^{2\pi i k\theta} V^k U$ for k in $\mathbb{Z}$. Hence $\mathcal{A}_\theta$ contains a copy of $\mathcal{A}_{k\theta}$ for every k. Replacing V by V^k and θ by the fractional part $\{k\theta\}$ of $k\theta$ yields a projection P in this subalgebra with trace $\{k\theta\}$. Hence we obtain every value in $(\mathbb{Z} + \mathbb{Z}\theta) \cap [0, 1]$ as the trace of a projection in $\mathcal{A}_\theta$. ∎

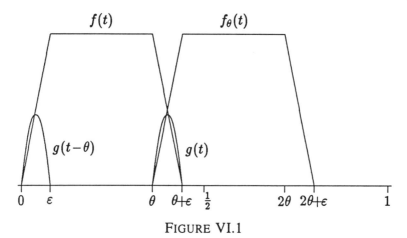

FIGURE VI.1

VI.3 An AF Algebra

At this stage, we have identified a large class of projections in $\mathcal{A}_\theta$ which is directly related to θ. However, we need to know that these are the only possibilities, and that these numbers are not merely a reflection of the construction. The key step, due to **Pimsner** and **Voiculescu**, is to imbed $\mathcal{A}_\theta$ into an AF algebra with the same (apparent) K_0 group

$$(G, G^+, \Gamma(G)) = (\mathbb{Z} + \mathbb{Z}\theta, (\mathbb{Z} + \mathbb{Z}\theta) \cap \mathbb{R}_+, (\mathbb{Z} + \mathbb{Z}\theta) \cap [0,1]).$$

We need some information about continued fractions. If θ is an irrational real number, it has a continued fraction expansion $[a_0, a_1, a_2, \ldots]$ where $a_0 \in \mathbb{Z}$ and $a_i \in \mathbb{N}$ are the unique choices so that

$$\theta = \lim_{n \to \infty} [a_0, a_1, \ldots, a_n] = \lim_{n \to \infty} a_0 + \cfrac{1}{a_1 + \cfrac{1}{a_2 + \cfrac{1}{a_3 + \cfrac{1}{\ddots + \cfrac{1}{a_n}}}}}$$

The rational approximations $p_n/q_n = [a_0, a_1, \ldots, a_n]$ are determined by the recursion formulae

$$p_0 = a_0 \qquad q_0 = 1$$
$$p_1 = a_0 a_1 + 1 \qquad q_1 = a_1$$
$$p_n = a_n p_{n-1} + p_{n-2} \qquad q_n = a_n q_{n-1} + q_{n-2} \quad \text{for} \quad n \geq 2$$

Thus

$$\begin{bmatrix} p_n & q_n \\ p_{n-1} & q_{n-1} \end{bmatrix} = \begin{bmatrix} a_n & 1 \\ 1 & 0 \end{bmatrix} \begin{bmatrix} p_{n-1} & q_{n-1} \\ p_{n-2} & q_{n-2} \end{bmatrix}$$

VI.3. An AF Algebra

In particular, we obtain the identity

$$p_n q_{n-1} - p_{n-1} q_n = \det \begin{bmatrix} p_n & q_n \\ p_{n-1} & q_{n-1} \end{bmatrix} = (-1)^{n-1}.$$

From this identity, it follows that p_n/q_n is an alternating series satisfying

$$\frac{p_{2n-2}}{q_{2n-2}} < \frac{p_{2n}}{q_{2n}} < \frac{p_{2n+1}}{q_{2n+1}} < \frac{p_{2n-1}}{q_{2n-1}}.$$

We also can easily deduce that the p's and q's grow geometrically fast, and thus $\sum_{n\geq 1} q_n^{-1} < \infty$. See Hardy and Wright, *Theory of Numbers* for more details.

The AF algebra we construct will be the increasing union $\mathfrak{A}_\theta = \overline{\bigcup_{n\geq 1} \mathfrak{A}_n}$ of subalgebras $\mathfrak{A}_n = \mathcal{M}_{q_n} \oplus \mathcal{M}_{q_{n-1}}$ and the partial multiplicities of the imbedding $\alpha_{n-1,n}$ are given by $A_n := \begin{bmatrix} a_n & 1 \\ 1 & 0 \end{bmatrix}$.

The computation of $K_0(\mathfrak{A}_\theta)$ follows in the same manner as the Fibonacci algebra of Example IV.3.6. Thus

$$K_0(\mathfrak{A}_\theta) = \varinjlim((\mathbb{Z}^2, \mathbb{Z}_+^2), A_n) = (\mathbb{Z}^2, P_\theta)$$

where P_θ is the positive cone of the limit. Let

$$T_n := A_0^{-1} A_1^{-1} \cdots A_n^{-1} = \begin{bmatrix} p_n & q_n \\ p_{n-1} & q_{n-1} \end{bmatrix}^{-1} = (-1)^{n-1} \begin{bmatrix} q_{n-1} & -q_n \\ -p_{n-1} & p_n \end{bmatrix}.$$

Thus we get a commutative diagram

$$K_0(\mathfrak{A}_n) = \mathbb{Z}^2 \xrightarrow{A_{n+1}} K_0(\mathfrak{A}_{n+1}) = \mathbb{Z}^2$$
$$\searrow T_n \qquad \downarrow T_{n+1}$$
$$K_0(\mathfrak{A}) = \mathbb{Z}^2$$

So $P_\theta = \bigcup_{n\geq 1} T_n \mathbb{Z}_+^2$. Now $T_n \mathbb{Z}_+^2$ is the cone generated by the vectors

$$\begin{pmatrix} (-1)^{n-1} q_{n-1} \\ (-1)^n p_{n-1} \end{pmatrix} \quad \text{and} \quad \begin{pmatrix} (-1)^n q_n \\ (-1)^{n-1} p_n \end{pmatrix}.$$

Notice that $\begin{pmatrix} -q_{2n-1} \\ p_{2n-1} \end{pmatrix}$ lives in the second quadrant, and has slope $\frac{-p_{2n-1}}{q_{2n-1}}$ which increases to $-\theta$; while $\begin{pmatrix} q_{2n} \\ -p_{2n} \end{pmatrix}$ lives in the fourth quadrant, and has slope $\frac{-p_{2n}}{q_{2n}}$ which decreases to $-\theta$. Therefore

$$P_\theta = \{\begin{pmatrix} x \\ y \end{pmatrix} \in \mathbb{Z}^2 : \theta x + y \geq 0\}.$$

The order unit of $\mathfrak{A}_n$ is $\begin{pmatrix} q_n \\ q_{n-1} \end{pmatrix}$. And so $T_n \begin{pmatrix} q_n \\ q_{n-1} \end{pmatrix} = \begin{pmatrix} 0 \\ 1 \end{pmatrix}$ is the order unit of $\mathfrak{A}_\theta$. $K_0(\mathfrak{A}_\theta) = (\mathbb{Z}^2, P_\theta)$ is a total order with a unique state $\sigma_*(\begin{pmatrix} x \\ y \end{pmatrix}) = \theta x + y$. The map σ_* is an order isomorphism of $K_0(\mathfrak{A}_\theta)$ onto G. By Theorem IV.5.3, there is a unique trace on $\mathfrak{A}_\theta$, which we denote by σ.

VI.4 Berg's Technique

Our plan is to imbed the irrational rotation algebra $\mathcal{A}_\theta$ into the AF algebra $\mathfrak{A}_\theta$ by approximating the unitaries U and V by their finite dimensional analogues. So if we take the standard orthonormal basis for L^2 to be $e_k(t) = e^{2\pi i k t}$ for $k \in \mathbb{Z}$, the representation of section VI.1 is seen to have the matrix forms

$$U e_k = e_{k+1} \quad \text{and} \quad V e_k = e^{2\pi i k \theta} e_k.$$

The rational number p_n/q_n approximating θ suggests defining the two unitaries on $\mathbb{C}^{q_n} = L^2(\mathbb{Z}/\mathbb{Z}_{q_n})$ with basis $e_k^{(n)}$ for $k \in \mathbb{Z}/\mathbb{Z}_{q_n}$ by

$$U_n e_k^{(n)} = e_{k+1}^{(n)} \quad \text{and} \quad V_n e_k^{(n)} = e^{2\pi i k p_n/q_n} e_k^{(n)}.$$

It is routine to verify the identity $U_n V_n = e^{2\pi i p_n/q_n} V_n U_n$.

We will consider the pairs $(U_n \oplus U_{n-1}, V_n \oplus V_{n-1})$ in $\mathfrak{A}_n = \mathcal{M}_{q_n} \oplus \mathcal{M}_{q_{n-1}}$. The difficulty is to arrange the imbeddings of $\mathfrak{A}_n$ into $\mathfrak{A}_{n+1}$ so that these pairs are Cauchy. Once this is done, the limits will be the desired unitaries.

In order to arrange the imbeddings, we need to know how to approximate shift operators by direct sums of shifts of smaller order. A method for doing this is called **Berg's technique**.

Theorem VI.4.1 *Let e_j and f_j, for $0 \leq j \leq n$, form an orthonormal set. Suppose that T in $\mathcal{B}(\mathcal{H})$ satisfies $T e_j = e_{j+1}$ and $T f_j = f_{j+1}$ for $0 \leq j \leq n-1$. Then there is an operator S in $\mathcal{B}(\mathcal{H})$ such that $Sx = Tx$ for every x orthogonal to $\mathrm{span}\{e_j, f_j : 0 \leq j \leq n-1\}$ and $S \, \mathrm{span}\{e_j, f_j\} = \mathrm{span}\{e_{j+1}, f_{j+1}\}$ for $0 \leq j \leq n-1$ such that $S^n e_0 = f_n$, $S^n f_0 = e_n$ and*

$$\|S - T\| \leq 2 \sin \tfrac{\pi}{2n} < \tfrac{\pi}{n}.$$

Proof. We introduce a *twist* through an angle $\pi/2n$ using the matrix

$$\Theta = \begin{bmatrix} 1/\sqrt{2} & 1/\sqrt{2} \\ 1/\sqrt{2} & -1/\sqrt{2} \end{bmatrix} \begin{bmatrix} 1 & 0 \\ 0 & e^{\pi i/n} \end{bmatrix} \begin{bmatrix} 1/\sqrt{2} & 1/\sqrt{2} \\ 1/\sqrt{2} & -1/\sqrt{2} \end{bmatrix}$$

$$= e^{\pi i/2n} \begin{bmatrix} \cos(\tfrac{\pi}{2n}) & i \sin(\tfrac{\pi}{2n}) \\ i \sin(\tfrac{\pi}{2n}) & \cos(\tfrac{\pi}{2n}) \end{bmatrix}.$$

Then

$$\Theta^n = \begin{bmatrix} 1/\sqrt{2} & 1/\sqrt{2} \\ 1/\sqrt{2} & -1/\sqrt{2} \end{bmatrix} \begin{bmatrix} 1 & 0 \\ 0 & -1 \end{bmatrix} \begin{bmatrix} 1/\sqrt{2} & 1/\sqrt{2} \\ 1/\sqrt{2} & -1/\sqrt{2} \end{bmatrix} = \begin{bmatrix} 0 & 1 \\ 1 & 0 \end{bmatrix}.$$

Think of $T|\mathrm{span}\{e_j, f_j\} \to \mathrm{span}\{e_{j+1}, f_{j+1}\}$ as the identity matrix with respect to these bases. Let $S|\mathrm{span}\{e_j, f_j\} \to \mathrm{span}\{e_{j+1}, f_{j+1}\}$ be given by the matrix Θ. Then $S^n e_0 = f_n$ and $S^n f_0 = e_n$ as desired. Moreover,

$$\|S - T\| = \|\Theta - I\| = |e^{\pi i/n} - 1| = 2 \sin \tfrac{\pi}{2n} < \tfrac{\pi}{n}. \qquad \blacksquare$$

VI.4. Berg's Technique

Remark VI.4.2 One of the significant points about this approximation method is that the domain and range of the perturbation is limited to

$$\text{span}\{e_j, f_j : 0 \leq j \leq n-1\} \quad \text{and} \quad \text{span}\{e_j, f_j : 1 \leq j \leq n\}$$

respectively. So several applications of this method on pairwise orthogonal pieces will not increase the norm of the perturbation.

Think of T as a right shift acting on two parallel sets of vectors e_j and f_j for $0 \leq j \leq n$. The twist introduced in S produces a cross-over so that S is also a shift on another basis that moves e_0 gradually across to f_n and f_0 gradually across to e_n. Graphically, we may represent this by the picture in Figure VI.2.

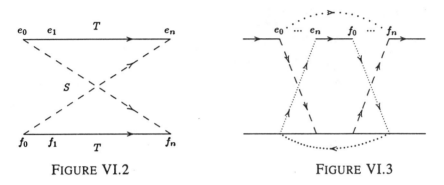

FIGURE VI.2   FIGURE VI.3

However, in our application, the two segments of T will be part of a longer shift. So we may picture this as in Figure VI.3. The interchange between the vectors $\{e_0, \ldots, e_n\}$ and $\{f_0, \ldots, f_n\}$ is represented by the two sets of crossing lines. The dashed lines represent the interchange from e_0 to f_n. The result is a *leap* in n steps from e_0 to f_n denoted by the dotted curve above the diagram. At the same time, the two dotted lines crossing the dashed lines represent the corresponding interchange from f_0 to e_n. This results in pulling out a direct summand which is a unitary shift which follows T from e_n to f_0 and returns along the dotted lines in n steps.

In this way, we may repeatedly cut out unitary summands. Provided that the intertwinings occur on orthogonal pieces, the resultant perturbation is just the maximum of all the changes. So we will be able to split a long shift into a direct sum of smaller shifts with a reasonably small perturbation. The example below demonstrates this in the case of interest to us.

Example VI.4.3 Consider the shift U_{n+1} acting on $L^2(\mathbb{Z}/q_{n+1}\mathbb{Z})$ as above. From the relation $q_{n+1} = a_{n+1}q_n + q_{n-1}$, we see that we should approximate U_{n+1} by the direct sum of a_{n+1} copies of U_n plus one copy of U_{n-1}. Let $b := \lfloor q_n/2 \rfloor$ and $b' = q_n - b = \lceil q_n/2 \rceil$; and set $s = \lceil q_{n-1}/2 \rceil$. Perform the interchange between the vectors $\{e_{-kb+j} : 0 \leq j \leq s\}$ and $\{e_{kb'+j} : 0 \leq j \leq s\}$ for $1 \leq k \leq a_{n+1}$. The following figure illustrates this:

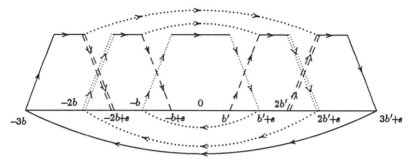

FIGURE VI.4

Notice that the intertwining between

$$\{e_{-b+j} : 0 \le j \le s\} \quad \text{and} \quad \{e_{b'+j} : 0 \le j \le s\}$$

produces a summand $U_{n,1}$ unitarily equivalent to U_n. Indeed, the perturbed operator S takes $g_j^{(1)} := e_{-b+j}$ to $g_{j+1}^{(1)} := e_{-b+j+1}$ for $s \le j \le a_n$ and takes a vector $g_j^{(1)}$ in span$\{e_{-b+j}, e_{b'+j}\}$ to $g_{j+1}^{(1)}$ for $0 < j < s$ starting with $g_0^{(1)} = g_{a_n}^{(1)} = e_{b'}$ and eventually arriving at $g_s^{(1)} = e_{-b+s}$.

Similarly, the interchange between

$$\{e_{-kb+j} : 0 \le j \le s\} \quad \text{and} \quad \{e_{kb'+j} : 0 \le j \le s\}$$

results in a summand which is a shift on kq_n vectors. But out of this, the shift between $\{e_{-(k-1)b+j} : 0 \le j \le s\}$ and $\{e_{(k-1)b'+j} : 0 \le j \le s\}$ took the middle $(k-1)q_n$ basis vectors out into smaller shifts. What remains is a summand $U_{n,k}$ unitarily equivalent to U_n.

Note that the summand $U_{n,k}$ shifts $g_j^{(k)} := e_{-kb+j}$ to $g_{j+1}^{(k)} := e_{-kb+j+1}$ for $s \le j \le b$. Then because of the $(k-1)$st intertwining, it shifts a vector

$$g_j^{(k)} \in \text{span}\{e_{-(k-1)b+(j-b)}, e_{(k-1)b'+(j-b)}\}$$

to $g_{j+1}^{(k)}$ for $b < j < b+s$ until arriving at $g_{b+s}^{(k)} = e_{(k-1)b'+s}$. Then it shifts $g_j^{(k)} := e_{(k-1)b'+(j-b)}$ to $g_{j+1}^{(k)}$ for $b+s \le j \le b+b' = q_n$. Finally, the kth interchange produces a shifting from $g_j^{(k)}$ in span$\{e_{-kb+(j-b)}, e_{kb'+(j-b)}\}$ to $g_{j+1}^{(k)}$ for $0 < j < s$ such that $g_0^{(k)} = e_{kb'}$ and $g_s^{(k)} = e_{-kb+s}$.

Finally, since the interchange between

$$\{e_{-a_{n+1}b+j} : 0 \le j \le s\} \quad \text{and} \quad \{e_{a_{n+1}b'+j} : 0 \le j \le s\}$$

pulls out a shift on $a_{n+1}q_n$ vectors, it must leave behind a summand $U_{n-1,0}$ unitarily equivalent to U_{n-1}. The basis on which $U_{n-1,0}$ acts will be denoted as $g_j^{(0)}$ for $1 \le j \le q_{n-1}$, where $g_j^{(0)} = e_{a_{n+1}b'+j}$ for $s \le j \le q_{n-1}$ and $g_j^{(0)}$ belongs to span$\{e_{-a_{n+1}b+j}, e_{a_{n+1}b'+j}\}$ for $1 \le j < s$.

The total norm perturbation of these interchanges is the maximum of each individual perturbation, which are all bounded by $\pi/s \leq 2\pi/q_{n-1}$.

Let us identify the unitary operator that maps $U_n^{(a_{n+1})} \oplus U_{n-1}$ to
$$\sum_{k=1}^{a_{n+1}} \oplus U_{n,k} \oplus U_{n-1,0}.$$

We don't wish to prejudge where each shift begins, so we allow a rotation of each term. Let $e_j^{(k)}$ for $j \in \mathbb{Z}/\mathbb{Z}_{q_n}$ denote the basis on which the kth summand equivalent to U_n acts, and let $e_j^{(0)}$ for $j \in \mathbb{Z}/\mathbb{Z}_{q_{n-1}}$ be the basis for U_{n-1}. Then define a unitary W_n from $\mathbb{C}^{q_n(a_{n+1})} \oplus \mathbb{C}_{q_{n-1}}$ to $\mathbb{C}^{q_{n+1}}$ by

$$W_n e_j^{(k)} = g_{j+c_k \pmod{q_n}}^{(k)} \quad \text{for} \quad 0 \leq j < q_n$$
$$W_n e_j^{(0)} = g_{j+c_0 \pmod{q_{n-1}}}^{(0)} \quad \text{for} \quad 0 \leq j < q_{n-1}.$$

The values of the integers c_k will be determined later to suit our needs. With any choice of these c_k's, we obtain
$$\|W_n(U_n^{(a_{n+1})} \oplus U_{n-1})W_n^* - U_{n+1}\| < \tfrac{2\pi}{q_{n-1}}.$$

VI.5 Imbedding $\mathcal{A}_\theta$ into $\mathfrak{A}_\theta$

We follow the plan outlined at the beginning of the last section. Indeed, our example shows how to imbed $\mathfrak{A}_n = \mathcal{M}_{q_n} \oplus \mathcal{M}_{q_{n-1}}$ into $\mathcal{M}_{q_{n+1}}$ in such a way that $U_n \oplus U_{n-1}$ is mapped to an operator unitarily equivalent to $U_n^{(a_{n+1})} \oplus U_{n-1}$ close to the shift U_{n+1}. The constants c_k will now be chosen so that $V_n \oplus V_{n-1}$ is simultaneously mapped to an operator unitarily equivalent to $V_n^{(a_{n+1})} \oplus V_{n-1}$ close to V_{n+1}. The reason that this is possible is that V_{n+1} is a diagonal operator with eigenvalues that are q_{n+1} periodic, and thus almost q_n periodic and q_{n-1} periodic due to the nature of the rational approximations. This allows us to match up the eigenvalues of the V_n's with those of V_{n+1} within a reasonable accuracy. How close we get is a fairly straightforward calculation.

For convenience, let us write $\lambda_n = e^{2\pi i p_n/q_n}$. Now $V_{n+1} e_j = \lambda_{n+1}^j e_j$ and hence $V_{n+1} e_{-kb+j} = \lambda_{n+1}^{-kb+j} e_{-kb+j}$. Now λ_n^m is close to λ_{n+1}^m. Since we wish to identify V_k with a summand $V_{n,k}$ acting on the basis $\{g_j^{(k)} : 0 \leq j < q_n\}$, this suggests that we send $e_j^{(k)}$ to $g_{j+kb}^{(k)}$. That is, choose $c_k = kb$ for the definition of W_n in Example VI.4.3 for $1 \leq k \leq a_n$. For the imbedding of the copy of V_{n-1}, choose an integer c_0 so that
$$\left|\lambda_{n-1}^{c_0} - \lambda_{n+1}^{a_{n+1}b'}\right| < \tfrac{\pi}{q_{n-1}}.$$

The key computation is the estimate:

Lemma VI.5.1 *With the choices for c_k made above, we have*
$$\|W_n(V_n^{(a_{n+1})} \oplus V_{n-1})W_n^* - V_{n+1}\| < \tfrac{4\pi}{q_n} + \tfrac{\pi}{q_{n-1}}.$$

Proof. Let $V'_{n+1} := W_n(V_n^{(a_{n+1})} \oplus V_{n-1})W_n^*$. Note that both V_{n+1} and V'_{n+1} have eigenvectors $e_{-kb+j} = g_j^{(k)}$ for $s \leq j \leq b$ and $e_{(k-1)b'+j} = g_{j+b}^{(k)}$ for $s \leq j \leq b'$ and $1 \leq k \leq a_{n+1}$; and eigenvectors $e_{a_{n+1}b'+j} = g_j^{(0)}$ for $s \leq j \leq q_{n-1}$. In addition, they have common two dimensional reducing subspaces
$$\mathcal{E}_{k,j} := \operatorname{span}\{e_{-kb+j}, e_{kb'+j}\} = \operatorname{span}\{g_j^{(k)}, g_{j+b}^{(k+1)}\}$$

for $0 < j < s$ and $1 \leq k < a_{n+1}$ and
$$\mathcal{E}_{0,j} := \operatorname{span}\{e_{-a_{n+1}b+j}, e_{a_{n+1}b'+j}\} = \operatorname{span}\{g_j^{(a_{n+1})}, g_j^{(0)}\}$$

for $0 < j < s$. Thus it suffices to estimate the difference of norms on all of these subspaces and take the maximum (as their difference is also a direct sum).

On $e_{-kb+j} = g_j^{(k)}$ for $s \leq j \leq b$ we have (using $c_k = -kb$)
$$\|(V'_{n+1} - V_{n+1})e_{-kb+j}\| = |\lambda_n^{j-kb} - \lambda_{n+1}^{j-kb}|$$
$$= 2\left|\sin\left(\pi(j-kb)\left(\tfrac{p_n}{q_n} - \tfrac{p_{n+1}}{q_{n+1}}\right)\right)\right|$$
$$\leq 2\sin\left(\tfrac{\pi a_{n+1} b}{q_n q_{n+1}}\right) < \tfrac{\pi}{q_n}$$

Similarly,
$$\|(V'_{n+1} - V_{n+1})e_{kb'+j}\| < \tfrac{\pi}{q_n}.$$

On $\mathcal{E}_{k,j}$, we have
$$V'_{n+1} g_j^{(k)} = \lambda_n^{j-kb} g_j^{(k)}$$

and
$$V'_{n+1} g_{j+b}^{(k+1)} = \lambda_n^{j+b-(k+1)b} g_{j+b}^{(k+1)} = \lambda_n^{j-kb} g_{j+b}^{(k+1)}.$$

Therefore $V'_{n+1}|\mathcal{E}_{k,j} = \lambda_n^{j-kb} I_2$ is scalar. Hence as above, we obtain the estimates
$$\|(V'_{n+1} - V_{n+1})e_{-kb+j}\| = |\lambda_n^{j-kb} - \lambda_{n+1}^{j-kb}| < \tfrac{\pi}{q_n}$$

and
$$\|(V'_{n+1} - V_{n+1})e_{kb'+j}\| = |\lambda_n^{kb'+j} - \lambda_{n+1}^{kb'+j}| < \tfrac{\pi}{q_n}.$$

VI.5. Imbedding $\mathcal{A}_\theta$ into $\mathfrak{A}_\theta$

The estimates relating to the shorter shift $U_{n-1,0}$ are not quite as good, but still suffice. On $e_{a_{n+1}b'+j} = g_j^{(0)}$ for $s \leq j \leq q_{n-1}$, we have

$$\|(V'_{n+1} - V_{n+1})e_{a_{n+1}b'+j}\| = |\lambda_{n-1}^{j+c_0} - \lambda_{n+1}^{a_{n+1}b'+j}|$$
$$\leq |\lambda_{n-1}^{c_0} - \lambda_{n+1}^{a_{n+1}b'}| + |\lambda_{n-1}^{j} - \lambda_{n+1}^{j}|$$
$$\leq \tfrac{\pi}{q_{n-1}} + 2\sin \pi j \left|\tfrac{p_{n-1}}{q_{n-1}} - \tfrac{p_{n+1}}{q_{n+1}}\right|$$
$$\leq \tfrac{\pi}{q_{n-1}} + 2\sin\left(\tfrac{\pi a_{n+1}}{q_{n+1}}\right) < \tfrac{\pi}{q_{n-1}} + \tfrac{2\pi}{q_n}.$$

Finally, on $\mathcal{E}_{0,j}$, we have

$$\|V_{n+1}|\mathcal{E}_{0,j} - \lambda_{n+1}^{j-a_{n+1}b} I_2\| = |\lambda_{n+1}^{a_{n+1}b'+j} - \lambda_{n+1}^{j-a_{n+1}b}| = |\lambda_{n+1}^{a_{n+1}q_n} - 1| < \tfrac{2\pi}{q_n}$$

and

$$\|V'_{n+1}|\mathcal{E}_{0,j} - \lambda_{n+1}^{j-a_{n+1}b} I_2\| = \max\{|\lambda_n^{j-a_{n+1}b} - \lambda_{n+1}^{j-a_{n+1}b}|, |\lambda_{n-1}^{j+c_0} - \lambda_{n+1}^{j-a_{n+1}b}|\}$$
$$\leq \max\{\tfrac{2\pi}{q_n}, \tfrac{\pi}{q_{n-1}} + \tfrac{2\pi}{q_n}\} = \tfrac{\pi}{q_{n-1}} + \tfrac{2\pi}{q_n}$$

Thus we obtain the total estimate

$$\|V'_{n+1} - V_{n+1}\| < \tfrac{\pi}{q_{n-1}} + \tfrac{4\pi}{q_n}.$$

This establishes the lemma. ∎

Theorem VI.5.2 *There is a $*$-monomorphism ρ of the irrational rotation algebra $\mathcal{A}_\theta$ into the AF algebra $\mathfrak{A}_\theta$ corresponding to the continued fraction expansion of θ such that ρ_* is a homomorphism of $K_0(\mathcal{A}_\theta)$ onto $K_0(\mathfrak{A}_\theta)$. Moreover if τ and σ denote the unique traces on $\mathcal{A}_\theta$ and $\mathfrak{A}_\theta$ respectively, we have $\tau_* = \sigma_*\rho_*$ is an order homomorphism of $K_0(\mathcal{A}_\theta)$ onto $\mathbb{Z} + \mathbb{Z}\theta$.*

Proof. Define $*$-monomorphisms of $\mathfrak{A}_n$ into $\mathfrak{A}_{n+1}$ by

$$\varphi_{n,n+1}(A \oplus B) := W_n(A^{(a_{n+1})} \oplus B)W_n^* \oplus A.$$

Clearly, this is an imbedding with partial multiplicities $A_n := \begin{bmatrix} a_n & 1 \\ 1 & 0 \end{bmatrix}$. So the direct limit is $\mathfrak{A}_\theta$.

From the estimates in Example VI.4.3 and Lemma VI.5.1, we see that

$$\|\varphi_n(U_n \oplus U_{n-1}) - \varphi_{n+1}(U_{n+1} \oplus U_n)\| < \tfrac{2\pi}{q_{n-1}}$$

and

$$\|\varphi_n(V_n \oplus V_{n-1}) - \varphi_{n+1}(V_{n+1} \oplus V_n)\| < \tfrac{4\pi}{q_n} + \tfrac{4\pi}{q_{n-1}}.$$

Since $\sum_{n\geq 1} q_n^{-1} < \infty$, these two sequences are Cauchy. Let

$$U := \lim_{n \to \infty} \varphi_n(U_n \oplus U_{n-1}) \quad \text{and} \quad V := \lim_{n \to \infty} \varphi_n(V_n \oplus V_{n-1}).$$

As $U_n V_n = \lambda_n V_n U_n$, it follows that in the limit
$$UV = \lim_{n \to \infty} \lambda_n V_n U_n = e^{2\pi i \theta} VU.$$
Hence there is a $*$-homomorphism ρ which takes the generators of $\mathcal{A}_\theta$ to U and V respectively. Since $\mathcal{A}_\theta$ is simple, this is a monomorphism.

If σ is the unique trace on $\mathfrak{A}_\theta$, then clearly $\sigma\rho$ is a trace on $\mathcal{A}_\theta$, and thus $\tau = \sigma\rho$. Hence $\tau_* = \sigma_*\rho_*$ is an order homomorphism of $K_0(\mathcal{A}_\theta)$ into $\mathbb{Z} + \mathbb{Z}\theta$. By Theorem VI.2.1, this map is surjective. ∎

In Example VIII.5.2, we will show that τ_* is an isomorphism. However, we already have sufficient information to distinguish between different rotation algebras.

Corollary VI.5.3 *Two irrational rotation algebras A_θ and A_η are isomorphic if and only if $\eta \equiv \pm\theta \mod \mathbb{Z}$.*

Proof. Two isomorphic algebras will have the same K_0 group and the same unique trace. Hence we obtain
$$\mathbb{Z} + \mathbb{Z}\theta = \mathbb{Z} + \mathbb{Z}\eta,$$
from which it follows that $\eta \equiv \pm\theta \mod \mathbb{Z}$. ∎

Exercises

VI.1 Consider the rational rotation algebra $\mathcal{A}_{m/n}$, the universal C*-algebra for the relation $VU = e^{2\pi i m/n} UV$, where $\gcd(m, n) = 1$.

(a) Show that U^n and V^n lie in the centre.

(b) Hence show that each irreducible representation acts on an n-dimensional space.

HINT: Find the relationship between V and the spectral projections of U.

(c) Describe all irreducible representations of $\mathcal{A}_{m/n}$.

VI.2 Show that any two irreducible representations of $\mathcal{A}_\theta$ are approximately unitarily equivalent.

HINT: Apply Voiculescu's Theorem.

VI.3 Show that there are irreducible representations of $\mathcal{A}_\theta$ which are not unitarily equivalent.

HINT: Let U be the bilateral shift, and let V be a diagonal operator.

VI.4 Let Φ be an expectation of $\mathcal{B}(\mathcal{H})$ onto a von Neumann subalgebra $\mathfrak{A}$.

(a) Imitate the proof of Lemma I.9.9 to show that Φ is self-adjoint and order preserving.

(b) Show that $\Phi(PT) = P\Phi(T)$ for all T in $\mathcal{B}(\mathcal{H})$ and projections P in $\mathfrak{A}$. Hence show that $\Phi(ATB) = A\Phi(T)B$ for all T in $\mathcal{B}(\mathcal{H})$ and A, B in $\mathfrak{A}$.

(c) Show that $\Phi(T)^*\Phi(T) \leq \Phi(T^*T)$.

Exercises

VI.5 Show that $C(\mathbb{T}^2)$ is the universal C*-algebra for a pair of commuting unitaries.

VI.6 Let $\mathfrak{A}$ be the universal C*-algebra generated by two projections. Show that every irreducible representation of $\mathfrak{A}$ is at most two dimensional. Hence show that $\mathfrak{A}$ is isomorphic to the subalgebra of $C([0, 1], \mathcal{M}_2)$ consisting of those functions F such that $F(0)$ and $F(1)$ are diagonal.

Notes and Remarks.

Irrational rotation algebras were first systematically studied by Rieffel [1981], who constructed the projections of Theorem VI.2.1. Then Pimsner and Voiculescu [1980a] imbedded $\mathcal{A}_\theta$ in the AF algebra of the continued fraction of θ in order to classify them. The proof here is taken from Davidson [1984]. Berg's technique is valid in much greater generality for operators which are *block tridiagonal* (see Berg–Davidson [1991]). As we shall see in the chapter on crossed products, Pimsner and Voiculescu [1980b] show that the map τ_* is an isomorphism on $K_0(\mathcal{A}_\theta)$. Rieffel [1983b] also showed that the irrational rotation algebras satisfy cancellation. Putnam [1990b] has shown that the invertible elements are dense in $\mathcal{A}_\theta$. Recently, Elliott and Evans [1993] have proven that $\mathcal{A}_\theta$ is a limit circle algebra, and hence has real rank zero.

CHAPTER VII

Group C*-algebras

Many groups which arise in harmonic analysis yield important examples of C*-algebras. The C*-algebra of a group encodes all the information about unitary representations of the group. Groups also arise naturally as subgroups of automorphism groups of C*-algebras. This in turn leads to a general C*-algebra construction which will be discussed in the next chapter.

We will be concerned with locally compact (Hausdorff) groups. A **unitary representation** of a group G is a homomorphism of G into the unitary group of $\mathcal{B}(\mathcal{H})$ which is continuous in the strong operator topology, meaning that the map from s in G to $\pi(s)x$ is continuous for every vector x in $\mathcal{H}$. A representation is irreducible if the range does not commute with any proper projections. It is easy to see that this is equivalent to saying that $C^*(\pi(G))$ is irreducible.

A locally compact group G supports a regular Borel measure μ_G which is invariant under left translation; meaning that $\mu_G(sE) = \mu_G(E)$ for every s in G and Borel subset E of G. It is unique up to a scalar multiple, and is known as **left Haar measure**. This measure is finite when G is compact. In this case, we normalize it so that $\mu_G(G) = 1$. If G is infinite and discrete, then we normalize so that $\mu_G(\{e\})=1$, where e is the identity element of G. We will write ds instead of $d\mu_G(s)$ when there is no ambiguity.

In general, left Haar measure need not be right translation invariant. There is a continuous homomorphism Δ of G into $\mathbb{R}_+$ known as the **modular function** such that $\mu_G(Es) = \Delta(s)\mu_G(E)$. A group G is called **unimodular** if Δ is trivial. This occurs when G is abelian or discrete or compact, for example. In particular, we have the formula

$$d\mu_G(t^{-1}) = \Delta(t)^{-1} d\mu_G(t).$$

The space $L^1(G)$ of absolutely integrable functions with respect to Haar measure becomes a $*$-algebra with the operations of convolution and inversion:

$$f * g(t) = \int f(s)g(s^{-1}t)\,ds$$
$$f^*(t) = \Delta(t)^{-1}\overline{f(t^{-1})}$$

The $L^1(G)$ norm is not a C*-algebra norm however.

VII. Group C*-algebras

The algebra $L^1(G)$ is unital only if G is discrete; and then we write $\ell^1(G)$. In this case, δ_e, the characteristic function of the identity element, is the unit. Moreover, the group algebra $\mathbb{C}G$ consisting of all finite sums $\sum_{s \in G} \alpha_s \delta_s$ forms a dense subalgebra of $\ell^1(G)$. For many purposes, it suffices to work with the group algebra rather than $\ell^1(G)$.

In general, $L^1(G)$ has a norm one approximate identity. One may be obtained as follows—for each open neighbourhood U of e, choose a positive function f_U in $L^1(G)$ supported in U such that $f_U^* = f_U$ and $\|f_U\|_1 = 1$. This net is ordered by containment of sets. When G is metrizable, we can obtain a sequential approximate unit. The details are left as an exercise.

When π is a unitary representation of G, it induces a representation of $L^1(G)$ by integration:

$$\widetilde{\pi}(f) = \int f(t)\pi(t)\,dt.$$

One readily verifies that

$$\|\widetilde{\pi}(f)\| \le \int |f(t)|\,dt = \|f\|_1.$$

This is a homomorphism because, by Fubini's Theorem and left invariance,

$$\widetilde{\pi}(f*g) = \int\int f(s)g(s^{-1}t)\,ds\,\pi(t)\,dt$$
$$= \int f(s)\pi(s)\int g(s^{-1}t)\pi(s^{-1}t)\,dt\,ds$$
$$= \int f(s)\pi(s)\int g(u)\pi(u)\,du\,ds = \widetilde{\pi}(f)\widetilde{\pi}(g).$$

To see that it preserves adjoints, calculate (with substitution $u = t^{-1}$)

$$(\widetilde{\pi}(f)^*x, y) = (x, \widetilde{\pi}(f)y) = \int \overline{f(t)}(x, \pi(t)y)\,dt$$
$$= \int \overline{f(t)}(\pi(t^{-1})x, y)\,dt = \int \overline{f(u^{-1})}(\pi(u)x, y)\,\Delta(u)^{-1}du$$
$$= (\int f^*(u)\pi(u)x\,du, y) = (\widetilde{\pi}(f^*)x, y).$$

Conversely, if $\widetilde{\pi}$ is a representation of $L^1(G)$ which is **non-degenerate**, meaning that $\overline{\widetilde{\pi}(L^1(G))\mathcal{H}} = \mathcal{H}$, then it determines a unique unitary representation of G. To see this, let f_λ, $\lambda \in \Lambda$, be a norm one approximate identity for $L^1(G)$. Then

$$\lim_{\lambda \in \Lambda} \widetilde{\pi}(f_\lambda)\widetilde{\pi}(g)x = \widetilde{\pi}(g)x$$

for every g in $L^1(G)$ and x in $\mathcal{H}$. Hence the contractions $\widetilde{\pi}(f_\lambda)$ converge strongly to the identity operator. Define

$$\pi(s)\widetilde{\pi}(g)x = \widetilde{\pi}(g_s)x \quad \text{where} \quad g_s(t) = g(s^{-1}t).$$

It is easy to verify that $\pi(s) = \text{SOT-lim}_{\lambda \in \Lambda} \widetilde{\pi}((f_\lambda)_s)$. Moreover, this implies that π is a contractive homomorphism of G into $\mathcal{B}(\mathcal{H})$. Since $\|\pi(s)\| \leq 1$ and $\|\pi(s)^{-1}\| \leq 1$, it follows that $\pi(s)$ is unitary. These formulae are clearly necessary if $\widetilde{\pi}$ is to be induced by π; so this establishes uniqueness. Standard notation now is to suppress the tilde, and refer to the extension of π to $L^1(G)$ as π as well.

Every locally compact group G has a distinguished representation called the **left regular representation** on $L^2(G)$. This is defined by

$$\lambda(s)g(t) = g_s(t) = g(s^{-1}t).$$

This map is unitary because left translation is isometric due to the translation invariance of Haar measure. The **reduced group C*-algebra** of G is defined to be $C_r^*(G) := \overline{\lambda(L^1(G))}$.

The **group C*-algebra** of G is the closure of the universal representation of $L^1(G)$. That is, take π_u to be a direct sum of all irreducible representations (up to unitary equivalence) of G. Then $C^*(G)$ is the norm closure of $\pi_u(L^1(G))$. Equivalently, define a C*-norm on $L^1(G)$ by

$$\|f\| = \sup\{\|\pi(f)\| : \pi \text{ is a *-representation of } L^1(G)\}.$$

This collection of representations is non-empty because any irreducible representation of $C_r^*(G)$ provides an irreducible representation of G; and such representations exist by the GNS construction. Since $\|f\| \leq \|f\|_1$, this supremum is well-defined. The group C*-algebra $C^*(G)$ is the completion of $L^1(G)$ in this norm. By construction, there is a one-to-one correspondence between the irreducible representations of $C^*(G)$ and $L^1(G)$, and hence with the irreducible representations of G. In particular, there is a representation λ of $C^*(G)$ onto $C_r^*(G)$ extending the left regular representation.

When G is an abelian group, the dual group $\widehat{G} = \text{Hom}(G, \mathbb{T})$ is the group of **characters** on G. The following easy result computes $C^*(G)$ in this case.

Proposition VII.1.1 *If G is an abelian group, then*

$$C^*(G) = C_r^*(G) = C_0(\widehat{G}).$$

Proof. For abelian groups, $L^1(G)$ is an abelian Banach algebra. So by the Gelfand theory, the irreducible representations of $L^1(G)$ are the one-dimensional representations corresponding to multiplicative linear functionals. These functionals correspond by the argument above to the one-dimensional unitary representations of G, namely the elements of $\widehat{G}$. Thus the Gelfand map sends f in $L^1(G)$ to its Fourier transform $\hat{f}$ in $C_0(\widehat{G})$ given by

$$\hat{f}(\chi) = \mathcal{F}f(\chi) = \int_G f(t)\chi(t)\,dt \quad \text{for all } \chi \in \widehat{G}.$$

The range is self-adjoint and separates points; and thus is dense in $C_0(\widehat{G})$ by the Stone–Weierstrass Theorem.

VII.2. Amenability

The Fourier transform $\mathcal{F}$ extends to a unitary operator U from $L^2(G)$ onto $L^2(\hat{G})$. The left regular representation is now understood by conjugating it by U:

$$U\lambda(f)U^*\hat{g} = U\lambda(f)g = \mathcal{F}(f*g) = \hat{f}\hat{g} = M_{\hat{f}}\hat{g}$$

for all f in $L^1(G)$ and g in $L^2(G) \cap L^1(G)$. Thus each $\hat{f}$ in $C_0(G)$ is sent to the multiplication operator $M_{\hat{f}}$. This map is isometric; whence λ is an isometric isomorphism. Therefore $C^*(G) = C_r^*(G)$. ∎

VII.2 Amenability

A group G is called **amenable** if there is a left translation invariant mean for G. A **mean** is just a state m on $L^\infty(G)$; left invariance indicates that $m(g_s) = m(g)$ for all g in $L^\infty(G)$ and s in G. Compact groups are amenable—an invariant mean is given by integration against Haar measure.

Abelian groups are also amenable. The existence of an invariant mean follows from an application of the following well known fixed point theorem due to **Markov** and **Kakutani**.

Theorem VII.2.1 *Let $\mathcal{T}$ be a commuting family of continuous linear maps of a topological vector space $\mathcal{X}$ into itself. Suppose that K is a compact convex subset of $\mathcal{X}$ such that $TK \subset K$ for every T in $\mathcal{T}$. Then there is a point x in K such that $Tx = x$ for all T in $\mathcal{T}$.*

Proof. Expand $\mathcal{T}$ to include all finite products of its elements. For $n \geq 1$ and T in $\mathcal{T}$, let $T^{(n)} = \frac{1}{n}(I + T + T^2 + \cdots + T^{n-1})$; and set $K(n,T) = T^{(n)}K$, which is a compact convex subset of K. Since $\mathcal{T}$ is commutative, for any subset $\{T_i : 1 \leq i \leq k\}$ of $\mathcal{T}$,

$$\bigcap_{i=1}^k K(n_i, T_i) \supset \left(\prod_{i=1}^k T_i^{(n_i)}\right) K = K\left(1, \prod_{i=1}^k T_i^{(n_i)}\right).$$

Thus the family of all $K(n,T)$ for all $n \geq 1$ and all T in $\mathcal{T}$ satisfies the finite intersection property. So by compactness, there is a point x lying in the intersection.

Let $\mathcal{O}$ be an open neighbourhood of the origin in $\mathcal{X}$, and let T belong to $\mathcal{T}$. Since K is compact, so is $K - K$, and thus there in an integer n so that $K - K$ is contained in $n\mathcal{O}$. Now x belongs to $K(n,T)$, and therefore there is a point y in K such that $T^{(n)}y = x$. Hence

$$Tx - x = \tfrac{1}{n}(I + T + T^2 + \cdots + T^{n-1})(Ty - y)$$
$$= \tfrac{1}{n}(T^n y - y) \in \tfrac{1}{n}(K - K) \subset \mathcal{O}.$$

This holds for every open set $\mathcal{O}$ containing 0, and hence $Tx = x$. ∎

Corollary VII.2.2 *Every abelian locally compact group G is amenable.*

Proof. Consider the state space $\mathcal{S}$ of $L^\infty(G)$ endowed with the weak-∗ topology, which is a compact convex set. For each s in G, consider the left translation operator $T_s(f) = f_s$ on $L^\infty(G)$. The dual operator T_s^* acts on the dual of $L^\infty(G)$ and is weak-∗ continuous. Moreover, it is easy to see that if φ is a state, then $T_s^*\varphi$ is also a state. Indeed, it is positive because
$$T_s^*\varphi(f) = \varphi(f_s) \geq 0 \quad \text{for all} \quad f \in L^\infty_+(G).$$
Hence
$$\|T_s^*\varphi\| = \|T_s^*\varphi(1)\| = \|\varphi(1)\| = 1.$$
It follows that each operator T_s^* maps $\mathcal{S}$ into itself. As G is abelian, the family $\{T_s^* : s \in G\}$ is abelian. Therefore by the Markov–Kakutani Theorem VII.2.1, there is a common fixed point m in $\mathcal{S}$. Hence
$$m(f_s) = (T_s^*m)(f) = m(f) \quad \text{for all} \quad f \in L^\infty(G).$$
So m is an invariant mean for G. ∎

For discrete groups, many standard group constructions preserve amenability.

Proposition VII.2.3 *For discrete groups, amenability is preserved under taking subgroups, quotients, direct limits and extensions.*

Proof. Suppose that H is a subgroup of a discrete amenable group G with invariant mean m_G. Fix a set of elements $\{s_\lambda : \lambda \in \Lambda\}$ with one element from each left coset of H in G. There is a positive, unital, isometric imbedding η of $L^\infty(H)$ into $L^\infty(G)$ by
$$(\eta f)(ts_\lambda) = f(t) \quad \text{for all} \quad t \in H, \lambda \in \Lambda.$$
Notice that this is indeed positive, unital and isometric. Moreover, a simple calculation shows that $\eta(f_t) = (\eta f)_t$ for all t in H. Thus $m_H := m_G \eta$ is an invariant mean on H.

If H is a normal subgroup, let G/H be the quotient group. Define a positive, unital, isometric imbedding q of $L^\infty(G/H)$ into $L^\infty(G)$ by $qf(s) = f(sH)$. Then another computation shows that $m_{G/H} := m_G q$ is an invariant mean on G/H. Indeed, if tH is in G/H and f belongs to $L^\infty(G/H)$, then
$$q(f_{tH})(s) = f((tH)^{-1}sH) = f(t^{-1}sH) = (qf)_t(s).$$
Hence
$$m_{G/H}(f) = m_G q(f_{tH}) = m_G((qf)_t) = m_G(qf) = m_{G/H}(f).$$

Suppose that $G = \varinjlim G_n$, where each G_n is amenable. By replacing each G_n by its image in G, we may assume that the imbeddings of G_n into G are injective. This image is a quotient of G_n, and thus is amenable. Let $m_n = m_{G_n} r_n$ where r_n is the restriction map of $L^\infty(G)$ onto $L^\infty(G_n)$, and m_{G_n} is an invariant mean on G_n. Let m be any weak-∗ cluster point of this net of states. Every s in G lies in G_n for n sufficiently large. Thus is follows that m is translation invariant.

VII.2. Amenability

Suppose that H is an amenable normal subgroup of G with amenable quotient G/H. Let m_H and $m_{G/H}$ be translation invariant means on these groups. Define a map Φ of $L^\infty(G)$ into $L^\infty(G/H)$ by

$$\Phi f(sH) = m_H(f_{s^{-1}}|_H),$$

which is well defined because of the translation invariance of m_H. Then set $m_G(f) = m_{G/H}(\Phi f)$. This is positive because Φ is positive; and is norm one because $\Phi 1 = 1$, whence $m_G(1) = m_{G/H}(1) = 1$. Finally, it is translation invariant because

$$(\Phi f_t)(sH) = m_H(f_{s^{-1}t}|_H) = (\Phi f)_{tH}(sH).$$

Whence

$$m_G(f_t) = m_{G/H}(\Phi f_t) = m_{G/H}(\Phi f) = m_G(f). \blacksquare$$

Example VII.2.4 The free group is a prototypical example of a non-amenable group. To see this, consider the set $\mathcal{U}_0$ (and $\mathcal{U}_1$) of elements of $\mathbb{F}_2$ which, in reduced form, begin with an even (odd) power of u followed by e or a word beginning with a power of v. Similarly, define $\mathcal{V}_0$, $\mathcal{V}_1$ and $\mathcal{V}_2$ beginning with a power of v congruent to 0, 1 or 2 modulo 3. Clearly, $\mathcal{U}_1 = u\mathcal{U}_0$ and $\mathcal{V}_j = v^j \mathcal{V}_0$ for $j = 0, 1, 2$. Also, $\mathbb{F}_2$ is the disjoint union of $\mathcal{U}_0$ and $\mathcal{U}_1$, and also of $\mathcal{V}_0$, $\mathcal{V}_1$ and $\mathcal{V}_2$. Hence a translation invariant mean must satisfy

$$1 = m(\chi_{\mathbb{F}_2}) = 2m(\chi_{\mathcal{U}_1}) = 3m(\chi_{\mathcal{V}_0}).$$

However, $\mathcal{U}_1$ is a proper subset of $\mathcal{V}_0$. Thus

$$\tfrac{1}{2} = m(\chi_{\mathcal{U}_1}) \leq m(\chi_{\mathcal{V}_0}) = \tfrac{1}{3}$$

which is absurd. Therefore $\mathbb{F}_2$ is not amenable.

We will see examples of C*-algebras for both amenable and non-amenable groups in this chapter. A deep result that lies at the intersection of harmonic analysis and C*-algebras is the following, which we will not prove here. We will not need to quote this theorem for the analysis of our examples because we will be able to establish the representation theory explicitly in these cases.

Theorem VII.2.5 *If G is a locally compact group, then the left regular representation λ of $C^*(G)$ onto $C^*_r(G)$ is an isomorphism if and only if G is amenable.*

We will provide a proof of half of this theorem for discrete groups. This should give the reader a feeling for the kind of arguments involved relating positive definite functions on the group with states on the group algebra.

A function φ on a discrete group G is called **positive definite** if

$$\sum_{i=1}^n \sum_{j=1}^n \alpha_i \overline{\alpha_j} \varphi(s_j^{-1} s_i) \geq 0 \quad \text{for all} \quad n \geq 1,\ \alpha_i \in \mathbb{C},\ s_i \in G,\ 1 \leq i \leq n.$$

Let $\mathcal{P}(G)$ denote the set of all positive definite functions on G such that $\varphi(e) = 1$. Any state Φ on $C^*(G)$ determines a positive definite function in $\mathcal{P}(G)$ by setting $\varphi(s) = \Phi(\delta_s)$ for all s in G. Indeed, $\varphi(e) = 1$; and if $\alpha_i \in \mathbb{C}$ and $s_i \in G$ for $1 \leq i \leq n$, then setting $f = \sum_{i=1}^{n} \alpha_i \delta_{s_i}$, we obtain

$$\sum_{i=1}^{n} \sum_{j=1}^{n} \alpha_i \overline{\alpha_j} \varphi(s_j^{-1} s_i) = \Phi(f^* * f) \geq 0.$$

Conversely, if φ is a positive definite function on G, define a functional on $\mathbb{C}G$ by

$$\Phi(\sum_{i=1}^{n} \alpha_i \delta_{s_i}) = \sum_{i=1}^{n} \alpha_i \varphi(s_i).$$

This is a unital positive linear functional since if $f = \sum_{i=1}^{n} \alpha_i \delta_{s_i}$ is in $\mathbb{C}G$, then

$$\Phi(f^* * f) = \sum_{i=1}^{n} \sum_{j=1}^{n} \alpha_i \overline{\alpha_j} \varphi(s_j^{-1} s_i) \geq 0.$$

The proof of Lemma I.9.5 shows that Φ is continuous with respect to the $C^*(G)$ norm. Thus Φ extends by continuity to a state on $C^*(G)$. This establishes a bijective pairing between $\mathcal{P}(G)$ and the state space of $C^*(G)$. Consequently, for any element $f = \sum_{i=1}^{n} \alpha_i \delta_{s_i}$ in $\mathbb{C}G$, the norm in $C^*(G)$ is determined by

$$\|f\|_{C^*(G)} = \sup_{\Phi \in \mathcal{S}(C^*(G))} \Phi(f^* * f)^{1/2} = \sup_{\varphi \in \mathcal{P}(G)} \Big(\sum_{i=1}^{n} \sum_{j=1}^{n} \alpha_i \overline{\alpha_j} \varphi(s_j^{-1} s_i)\Big)^{1/2}.$$

Lemma VII.2.6 *If φ_1 and φ_2 belong to $\mathcal{P}(G)$, then the product $\varphi_1 \varphi_2$ belongs to $\mathcal{P}(G)$.*

Proof. Let $s_1, \ldots, s_n$ in G be given. Let A_k denote the $n \times n$ matrix with coefficients $a_{ij}^{(k)} = \varphi_k(s_j^{-1} s_i)$ for $1 \leq i, j \leq n$ and $k = 1, 2$. This is a positive matrix since for α in $\mathbb{C}^n$, let $f = \sum_{i=1}^{n} \alpha_i \delta_{s_i}$ belong to $\mathbb{C}G$; and note that

$$(A_i \alpha, \alpha) = \varphi(f^* * f) \geq 0.$$

Let $A = A_1 \circ A_2$ denote the matrix obtained from pointwise (Schur) product of the matrix entries $a_{ij} = a_{ij}^{(1)} a_{ij}^{(2)}$. This is a positive matrix (see Exercise VII.3). Thus

$$\sum_{i=1}^{n} \sum_{j=1}^{n} \alpha_i \overline{\alpha_j} \varphi_1(s_j^{-1} s_i) \varphi_2(s_j^{-1} s_i) = (A \alpha, \alpha) \geq 0.$$

Since $\varphi_1 \varphi_2(e) = 1$, the product lies in $\mathcal{P}(G)$. ∎

A crucial fact is that positive definite functions with finite support are determined by states on the left regular representation. This will reduce the problem to approximating positive definite functions by functions of finite support.

Lemma VII.2.7 *Suppose that φ in $\mathcal{P}(G)$ has finite support. Then there is a unit vector x in $\ell^2(G)$ such that $\varphi(s) = (\lambda(s)x, x)$ for s in G.*

Proof. Define a linear map on $\mathbb{C}G$ by $Tf = f * \varphi$. As φ has finite support, it is evident that T is bounded by $\sum_{s \in G} |\varphi(s)|$ when $\mathbb{C}G$ is endowed with the $\ell^2(G)$ norm. Thus T may be extended to a bounded operator on $\ell^2(G)$ by continuity. Furthermore, if $f = \sum_{i=1}^{n} \alpha_i \delta_{s_i}$ belongs to $\mathbb{C}G$, then

$$(Tf, f) = \sum_{s \in G} \sum_{t \in G} f(t)\varphi(t^{-1}s)\overline{f(s)} = \sum_{i=1}^{n} \sum_{j=1}^{n} \alpha_j \overline{\alpha_i} \varphi(s_j^{-1} s_i) \geq 0.$$

Hence T is positive. Next notice that T commutes with $\lambda(G)$ because

$$\lambda(s)Tf = \delta_s * (f * \varphi) = (\delta_s * f) * \varphi = T\lambda(s)f.$$

Set $x = T^{1/2}\delta_e$. Then

$$(\lambda(s)x, x) = (\lambda(s)T^{1/2}\delta_e, T^{1/2}\delta_e) = (\delta_s, T\delta_e) = (\delta_s, \varphi) = \varphi(s).$$

Finally, $\|x\|^2 = \varphi(e) = 1$. ∎

We are now prepared to use amenability to approximate positive definite functions by positive definite functions of finite support. In this way, we will establish the following theorem.

Theorem VII.2.8 *If G is a discrete amenable group, then $C_r^*(G) = C^*(G)$.*

Proof. Let m be an invariant mean on $\ell^\infty(G)$. Since $\ell^1(G)$ is the predual of $\ell^\infty(G)$, Goldstine's Theorem states that the unit ball of $\ell^1(G)$ is weak-$*$ dense in the unit ball of $\ell^\infty(G)^*$. Since $\mathbb{C}G$ is dense in $\ell^1(G)$, there is a net f_γ, $\gamma \in \Gamma$, in the $\ell^1(G)$ unit ball of $\mathbb{C}G$ converging weak-$*$ to m. First we show that we may assume that each f_γ is positive and satisfies $(f_\gamma, \mathbf{1}) = 1$, where $\mathbf{1}$ is the constant function $\mathbf{1}(s) = 1$. (Such functions will be called finite means.) Indeed,

$$|(f_\gamma, \mathbf{1})| = |\sum_{s \in G} f_\gamma(s)| \leq \|f_\gamma\|_1 \leq 1.$$

But $\lim_\gamma (f_\gamma, \mathbf{1}) = m(\mathbf{1}) = 1$. So it is apparent that each f_γ may be replaced with the function $g_\gamma(s) = |f_\gamma(s)|/\|f_\gamma\|_1$ which will differ from f_γ by small norm for γ sufficiently large.

Since m is left translation invariant, for every s in G,

$$\text{w*-}\lim_\gamma \delta_s * g_\gamma - g_\gamma = \delta_s * m - m = 0$$

Thus $\delta_s * g_\gamma - g_\gamma$ converges to 0 weakly in $\ell^1(G)$. By the Hahn–Banach Theorem, for any finite set $s_1, \ldots, s_n$ in G, the norm closed convex hull of

$$\{(\delta_{s_i} * g_\gamma - g_\gamma)_{1 \leq i \leq n} : \gamma \in \Gamma\}$$

in $\ell^1(G)^n$ contains 0. Therefore we obtain a sequence g_n of finite means in the convex hull of the g_γ's such that

$$\lim_{n\to\infty} \|\delta_s * g_n - g_n\|_1 = 0 \quad \text{for each} \quad s \in G.$$

Then $h_n = g_n^{1/2}$ are positive functions in $\ell^2(G)$ with $\|h_n\|_2 = 1$. Note the simple inequality $|a - b|^2 \leq |a - b||a + b| = |a^2 - b^2|$ for all $a, b \geq 0$. Therefore, since h_n is positive,

$$\lim_{n\to\infty} \|\lambda(s)h_n - h_n\|_2^2 = \lim_{n\to\infty} \sum_{t \in G} |h_n(st) - h_n(t)|^2$$

$$\leq \lim_{n\to\infty} \sum_{t \in G} |g_n(st) - g_n(t)| = \lim_{n\to\infty} \|\delta_s * g_n - g_n\|_1 = 0$$

for every s in G. Since $(\lambda(s)h_n, h_n)$ is real, we obtain

$$\lim_{n\to\infty} (\lambda(s)h_n, h_n) = \lim_{n\to\infty} (\lambda(s)h_n, h_n) + \tfrac{1}{2}\|\lambda(s)h_n - h_n\|_2^2$$

$$= \frac{1}{2} \lim_{n\to\infty} \left(\|\lambda(s)h_n\|_2^2 + \|h_n\|_2^2 \right) = 1.$$

In conclusion, $\varphi_n(s) = (\lambda(s)h_n, h_n)$ are positive definite functions in $\mathcal{P}(G)$ for $n \geq 1$. These functions have finite support because each h_n has finite support; and they converge pointwise to the constant function **1**.

Now let φ belong to $\mathcal{P}(G)$. Then by Lemma VII.2.6, the functions $\varphi\varphi_n$ belong to $\mathcal{P}(G)$ and have finite support. Furthermore, they converge pointwise to φ. By Lemma VII.2.7, there are unit vectors x_n in $\ell^2(G)$ such that

$$\varphi(s)\varphi_n(s) = (\lambda(s)x_n, x_n) \quad \text{for all} \quad s \in G.$$

Thus if $f = \sum_{i=1}^n \alpha_i \delta_{s_i}$ belongs to $\mathbb{C}G$, then

$$\varphi(f^* * f) = \sum_{s \in G} \varphi(s)(f^* * f)(s) = \lim_{n\to\infty} \sum_{s \in G} \varphi(s)\varphi_n(s)(f^* * f)(s)$$

$$= \lim_{n\to\infty} (\lambda(f^* * f)x_n, x_n) \leq \|\lambda(f^* * f)\| = \|\lambda(f)\|^2.$$

Taking the supremum over all positive definite functions yields

$$\|f\|_{C^*(G)} = \|\lambda(f)\| \quad \text{for all} \quad f \in \mathbb{C}G.$$

Hence the canonical map of $C^*(G)$ onto $C_r^*(G)$ is an isomorphism. ∎

VII.3 Primitive Ideals

When studying the representation theory of a C*-algebra, two objects are normally used to act as a spectrum. The basic elements are the irreducible representations. The kernel of an irreducible representation is called a **primitive ideal**. The set $\text{Prim}(\mathfrak{A})$ of all primitive ideals of a C*-algebra $\mathfrak{A}$ is called the **primitive ideal space**. This space has a natural topology which we will describe soon. Moreover,

VII.3. Primitive Ideals

this space turns out to be a classifying space for the irreducible representations up to approximate unitary equivalence. The other space that is often used to encode the representation theory of $\mathfrak{A}$ is the **spectrum** $\widehat{\mathfrak{A}}$ consisting of all unitary equivalence classes of irreducible representations. This space contains more precise information, but is generally intractable when it does not coincide with $\text{Prim}(\mathfrak{A})$.

There is a natural map from $\widehat{\mathfrak{A}}$ onto $\text{Prim}(\mathfrak{A})$ given by sending each representation to its kernel. However, the only natural topology put on $\widehat{\mathfrak{A}}$ is obtained by pulling back the topology on $\text{Prim}(\mathfrak{A})$. In fact, there is no good way to topologize it to separate two approximately unitarily equivalent representations which are not unitarily equivalent because they each contain the other in the closure of their unitary orbits. Moreover, a deep result of Glimm shows that $\widehat{\mathfrak{A}}$ is unmanageable (in the sense that it does not have a good Borel structure) except for the very special class known as GCR algebras, when $\widehat{\mathfrak{A}}$ coincides with $\text{Prim}(\mathfrak{A})$. The only simple GCR algebras are $\mathcal{M}_n$ for $n \geq 1$ and $\mathfrak{K}$. So we will content ourselves in this chapter with looking at the primitive ideal space; and will note to what extent this classifies irreducible representations.

The topology on $\text{Prim}(\mathfrak{A})$ is known as the **hull–kernel topology**. The nomenclature comes from the definitions of the **kernel** of a collection $\mathcal{J}$ of ideals as the ideal $\ker(\mathcal{J}) := \bigcap_{\mathfrak{J} \in \mathcal{J}} \mathfrak{J}$; and the **hull** of an ideal $\mathfrak{J}$ is the set

$$\text{hull}(\mathfrak{J}) := \{\mathfrak{P} \in \text{Prim}(\mathfrak{A}) : \mathfrak{P} \supset \mathfrak{J}\}.$$

The closure of a set $\mathcal{J}$ of primitive ideals is $\overline{\mathcal{J}} = \text{hull}(\ker(\mathcal{J}))$.

Lemma VII.3.1 *Primitive ideals are prime.*

Proof. An ideal $\mathfrak{J}$ is called **prime** provided that whenever the product $\mathfrak{J}_1\mathfrak{J}_2$ of two ideals is contained in $\mathfrak{J}$, then either $\mathfrak{J}_1 \subseteq \mathfrak{J}$ or $\mathfrak{J}_2 \subseteq \mathfrak{J}$. Let π be an irreducible representation with kernel $\mathfrak{J}$. Then by Lemma I.9.15, $\pi(\mathfrak{J}_i)''$ is either 0 or $\mathcal{B}(\mathcal{H})$. In the first case, $\mathfrak{J}_i \subset \mathfrak{J}$. While if the second case holds for both $\mathfrak{J}_1$ and $\mathfrak{J}_2$, then

$$\overline{\pi(\mathfrak{J}_1\mathfrak{J}_2)\mathcal{H}} = \pi(\mathfrak{J}_1)''\pi(\mathfrak{J}_2)''\mathcal{H} = \mathcal{H};$$

whence $\mathfrak{J}_1\mathfrak{J}_2$ is not contained in $\mathfrak{J}$. ∎

Proposition VII.3.2 *The hull–kernel topology makes* $\text{Prim}(\mathfrak{A})$ *into a* T_0 *space.*

Proof. First, to establish that this is a topology, we verify the Kuratowski axioms for a closure operation. The closure of the empty set is evidently empty. It is also evident that $\mathcal{J} \subset \overline{\mathcal{J}}$; and that $\ker(\overline{\mathcal{J}}) = \ker(\mathcal{J})$. Hence $\overline{\overline{\mathcal{J}}} = \overline{\mathcal{J}}$. The fourth condition concerns $\overline{\mathcal{J}_1 \cup \mathcal{J}_2}$. Let $\mathfrak{J}_i = \ker \mathcal{J}_i$; and note that $\ker(\mathcal{J}_1 \cup \mathcal{J}_2) = \mathfrak{J}_1 \cap \mathfrak{J}_2$. Thus $\mathfrak{P}$ belongs to $\overline{\mathcal{J}_1 \cup \mathcal{J}_2}$ if and only if $\mathfrak{P}$ contains $\mathfrak{J}_1 \cap \mathfrak{J}_2$, which in turn contains $\mathfrak{J}_1\mathfrak{J}_2$. Then by Lemma VII.3.1, $\mathfrak{P}$ contains either $\mathfrak{J}_1$ or $\mathfrak{J}_2$, and therefore belongs to $\overline{\mathcal{J}_1}$ or $\overline{\mathcal{J}_2}$ respectively. It follows that $\overline{\mathcal{J}_1 \cup \mathcal{J}_2} = \overline{\mathcal{J}_1} \cup \overline{\mathcal{J}_2}$.

A T_0 space satisfies the weak separation axiom that given two points $\mathfrak{P}_1$ and $\mathfrak{P}_2$ in $\text{Prim}(\mathfrak{A})$, there is an open neighbourhood of one that is disjoint from the other. Equivalently, it must be shown that if each primitive ideal is in the closure of the other, then they are equal. But $\mathfrak{P}_1$ belongs to the closure of $\{\mathfrak{P}_2\}$ only if $\mathfrak{P}_2$ is contained in $\mathfrak{P}_1$; and conversely. ∎

Example VII.3.3 Quite nice C*-algebras show that one cannot expect the topology of the primitive ideal space to be better than T_0. Consider the Toeplitz algebra $\mathcal{T}(C(\mathbb{T}))$ from section V.1. This algebra contains the ideal of compact operators. If π is an irreducible representation such that $\pi(\mathfrak{K}) \neq 0$, then by Lemma I.9.15, $\pi|\mathfrak{K}$ is irreducible. Hence by Lemma I.10.2, $\pi|\mathfrak{K}$ is unitarily equivalent to the identity representation. Thus by Lemma I.9.14, π is unitarily equivalent to the identity representation.

The other irreducible representations of $\mathcal{T}(C(\mathbb{T}))$ annihilate $\mathfrak{K}$; and thus they factor through the quotient $\mathcal{T}(C(\mathbb{T}))/\mathfrak{K} \simeq C(\mathbb{T})$. As $C(\mathbb{T})$ is abelian, the irreducible representations are the point evaluations for λ in $\mathbb{T}$. They correspond to the maximal ideals

$$\mathfrak{J}_\lambda = \{T_f + K : f \in C(\mathbb{T}), f(\lambda) = 0, K \in \mathfrak{K}\}.$$

So

$$\text{Prim}(\mathcal{T}(C(\mathbb{T}))) = \{\mathfrak{K}\} \cup \{\mathfrak{J}_\lambda : \lambda \in \mathbb{T}\} \simeq \{\mathfrak{k}\} \cup \mathbb{T}.$$

The closure of $\mathfrak{k}$ is the whole space because each $\mathfrak{J}_\lambda$ contains $\mathfrak{K}$. The circle $\mathbb{T}$ is closed, and the relative topology is the usual one. Indeed, it is evident that

$$\bigcap_{\lambda \in S} \mathfrak{J}_\lambda = \{T_f + K : f \in C(\mathbb{T}), f|_{\overline{S}} = 0, K \in \mathfrak{K}\}$$

for any subset $S \subset \mathbb{T}$; whence we deduce that

$$\text{hull ker}\{\mathfrak{J}_\lambda : \lambda \in S\} = \{\mathfrak{J}_\lambda : \lambda \in \overline{S}\}.$$

This space is not T_1 because the point $\mathfrak{k}$ cannot be separated from any point on the circle. This leads to the peculiar fact that the constant sequence $\mathfrak{k}$ converges to every point.

Nevertheless, this is a useful space. In this case, we see that each primitive ideal corresponds to a unique irreducible representation up to unitary equivalence. So $\widehat{\mathcal{T}(C(\mathbb{T}))}$ equals $\text{Prim}(\mathcal{T}(C(\mathbb{T})))$.

Next we establish the connection with approximate unitary equivalence.

Proposition VII.3.4 *Two irreducible representations of a C*-algebra $\mathfrak{A}$ with the same kernel are approximately unitarily equivalent.*

Proof. Let σ_1 and σ_2 be two irreducible representations with a common kernel $\mathfrak{J}$. Then there is a representation $\tau = \sigma_2 \sigma_1^{-1}$ of $\sigma_1(\mathfrak{A})$ onto $\sigma_2(\mathfrak{A})$. Suppose first that $\sigma_1(\mathfrak{A})$ contains non-zero compact operators. Then by Corollary I.10.4, $\sigma_1(\mathfrak{A})$ contains $\mathfrak{K}$. Then by the argument used in the example above, it follows that τ is unitarily equivalent to the identity representation; and therefore σ_1 and

σ_2 are unitarily equivalent. Similarly, we obtain the same conclusion when $\sigma_2(\mathfrak{A})$ intersects $\mathfrak{K}$ non-trivially. Otherwise, we have

$$\ker \sigma_1 = \ker \pi\sigma_1 = \ker \sigma_2 = \ker \pi\sigma_2,$$

where π is the quotient by $\mathfrak{K}$. Thus by Corollary II.5.6, $\sigma_1 \sim_\mathfrak{K} \sigma_2$. ∎

As it is more usual to describe a topology in terms of its open sets, we do that now.

Proposition VII.3.5 *For A in $\mathfrak{A}$, the map taking σ in $\widehat{\mathfrak{A}}$ to $\|\sigma(A)\|$ is lower semicontinuous. If $\mathcal{A}$ is a dense subset of $\mathfrak{A}$, then the sets*

$$\mathcal{O}_A = \{\sigma \in \widehat{\mathfrak{A}} : \|\sigma(A)\| > 1\} \quad \text{for} \quad A \in \mathcal{A}$$

form a base for the hull-kernel topology. A net σ_α has ρ in $\widehat{\mathfrak{A}}$ as a limit point if and only if

$$\liminf_\alpha \|\sigma_\alpha(A)\| \geq \|\rho(A)\| \quad \text{for all} \quad A \in \mathfrak{A}.$$

Proof. Let us show that $\mathcal{J} = \{\sigma \in \widehat{\mathfrak{A}} : \|\sigma(A)\| \leq r\}$ is closed. Notice that $\mathcal{J}$ consists of all σ such that $\sigma(A^*A) \leq r^2$, which is the same as those sigma such that $\sigma(f(A^*A)) = 0$, where $f(x) = (r^2 \vee x) - r^2$. Let $\mathfrak{J}$ be the ideal generated by $f(A^*A)$. Then $\mathcal{J} = \text{hull}(\mathfrak{J})$, and therefore it is closed. Consequently the sets $\mathcal{O}_A$ are open.

Suppose that $\mathcal{J}$ is a closed set in $\widehat{\mathfrak{A}}$ disjoint from σ. Then σ does not vanish on $\mathfrak{J} = \ker \mathcal{J}$. Hence there is an element B in $\mathfrak{J}$ such that $\|\sigma(B)\| > 2$. Clearly the set $\mathcal{O}_B$ is an open set containing σ which is disjoint from $\mathcal{J}$. Let A in $\mathcal{A}$ be chosen so that $\|A - B\| < 1$. Then $\|\sigma(A)\| > 1$ and $\|\rho(B)\| < 1$ for all ρ in $\mathcal{J}$. Hence $\mathcal{O}_A$ is also a neighbourhood of σ disjoint from $\mathcal{J}$.

Thus ρ in $\widehat{\mathfrak{A}}$ is a limit point of σ_α if and only if for every A in $\mathfrak{A}$ and every $r < \|\rho(A)\|$, there is an α_0 so that $r < \|\sigma_\alpha(A)\|$ for all $\alpha > \alpha_0$. Clearly, this is equivalent to the statement: $\liminf_\alpha \|\sigma_\alpha(A)\| \geq \|\rho(A)\|$ for all A in $\mathfrak{A}$. ∎

VII.4 A Crystallographic Group

The first group we will examine is the symmetry group of a tiling pattern of the plane. Such groups are called **crystallographic groups**. There are seventeen such groups, and they are all extensions of an abelian group of translations isomorphic to $\mathbb{Z}^2$ by a finite group. Thus they are amenable by Proposition VII.2.3. The group C*-algebra admits a very explicit description while displaying some of the features that can appear in the C*-algebras of nonabelian groups.

The particular example we will focus on is the symmetry group $\mathfrak{pg}$ of the planar crystal or wallpaper pattern shown in Figure VII.1. It is assumed to continue indefinitely in all directions. The origin and the unit square are to fix the coordinates, and are not part of the pattern. It is evident that $\mathfrak{pg}$ contains the group of translations $A \simeq \mathbb{Z}^2$ given by $\tau_{m,n}(x,y) = (x+m, y+n)$ for (m,n) in $\mathbb{Z}^2$. This pattern

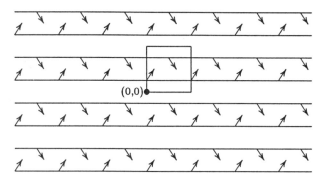

FIGURE VII.1

is also preserved by a glide reflection $\sigma(x,y) = (x+\frac{1}{2},-y)$. A moment's thought shows that these elements determine the full symmetry group of this pattern. Every element has the form $\tau_{m,n}$ or $\sigma\tau_{m,n}$ for (m,n) in $\mathbb{Z}^2$. A simple computation shows that $\tau_{m,n}\sigma = \sigma\tau_{m,-n}$ and $\sigma^2 = \tau_{1,0}$. In particular, A is a subgroup of $\mathfrak{pg}$ of index 2. Notice that $\mathfrak{pg}$ contains no elements of finite order; so this is not a semidirect product.

The first order of business is to obtain a useful description of the reduced C*-algebra of $\mathfrak{pg}$. We know from Theorem VII.2.5 that this equals the full C*-algebra. However, we will be able to establish this without recourse to that result. Much of the detail is the same for any abelian by finite group. So let G be a discrete group with an abelian normal subgroup A such that $D = G/A$ is a finite group of order n. Let the quotient map be denoted by π. Fix a cross-section $\gamma : D \to G$ so that $\pi\gamma = \text{id}$ and $\gamma(e_D) = e$, the identity element of G. Sometimes γ can be chosen to be an isomorphism of D onto a finite subgroup of G; in which case, G is isomorphic to a semidirect product of D with A. In general, γ cannot be taken to be a homomorphism. But once γ is fixed, each element s in G can be uniquely represented in the form $s = \gamma(d)a$, for d in D and a in A.

The quotient group D acts on A by conjugation:
$$\theta_d(a) := \gamma(d)a\gamma(d)^{-1} \quad \text{for} \quad d \in D,\ a \in A.$$
This action is independent of the choice of γ because any other element in the same coset has the form $\gamma(d)b$ for some b in A; and this is readily seen to yield the same action. The map θ is a homomorphism of D into $\text{Aut}(A)$.

There is a cocycle map $\alpha : D \times D \to A$ associated with the cross-section γ. For c,d in D, the element $\gamma(c)\gamma(d)$ belongs to the coset of cd. Thus, there is a unique element $\alpha(c,d)$ in A such that
$$\gamma(c)\gamma(d) = \gamma(cd)\alpha(c,d).$$
Calculate for b,c,d in D,
$$\gamma(b)\bigl(\gamma(c)\gamma(d)\bigr) = \gamma(b)\gamma(cd)\alpha(c,d) = \gamma(bcd)\alpha(b,cd)\alpha(c,d)$$

VII.4. A Crystallographic Group

and

$$(\gamma(b)\gamma(c))\gamma(d) = \gamma(bc)\alpha(b,c)\gamma(d)$$
$$= \gamma(bc)\gamma(d)\theta_{d^{-1}}(\alpha(b,c)) = \gamma(bcd)\alpha(bc,d)\theta_{d^{-1}}(\alpha(b,c))$$

Therefore we obtain the 2-cocycle identity

$$\alpha(b,cd)\alpha(c,d) = \alpha(bc,d)\theta_{d^{-1}}(\alpha(b,c)).$$

The group product has a useful expression for two elements $s_i = \gamma(d_i)a_i$ in G,

$$s_1 s_2 = \gamma(d_1)a_1 \gamma(d_2)a_2 = \gamma(d_1 d_2)\alpha(d_1,d_2)\theta_{d_2^{-1}}(a_1)a_2.$$

Let U be the unitary taking $\ell^2(A)$ onto $L^2(\widehat{A})$ which extends the Fourier transform. Since A is a discrete abelian group, $\widehat{A}$ is a compact abelian topological group. Each function h in $\ell^2(G)$ is determined by its restrictions to the n cosets of A. So define h_d in $\ell^2(A)$ by $h_d(a) = h(\gamma(d)a)$ for a in A and d in D. Then define a unitary map Ψ of $\ell^2(G)$ into $L^2(\widehat{A})^D := L^2(\widehat{A})^{(n)}$, the direct sum of n copies of $L^2(\widehat{A})$ indexed by elements of D, by

$$\Psi(h) = (Uh_d)_{d \in D} = (\widehat{h_d})_{d \in D}.$$

The main strategy for describing $C_r^*(G)$ is to compute $\Psi \lambda(f) \Psi^*$ for f in $\ell^1(G)$. Let $\mathcal{M}_D(C(\widehat{A}))$ denote the C*-algebra of $n \times n$ matrices indexed by D with coefficients in $C(\widehat{A})$ acting on $L^2(\widehat{A})^D$. The typical element is a matrix $F = [F_{c,d}]$ indexed by $D \times D$.

Theorem VII.4.1 *Let G be an extension of an abelian group A by a finite group D. Then $C_r^*(G)$ is isomorphic to the subalgebra of $\mathcal{M}_D(C(\widehat{A}))$ consisting of those elements $F = [F_{c,d}]$ such that*

$$F_{c,d}(\chi) = \chi(\alpha(cd^{-1},d))F_{cd^{-1},e}(\theta_d^*\chi) \quad \text{for} \quad \chi \in \widehat{A},$$

where α is the cocycle on $D \times D$ and $\theta_d^*\chi(a) := \chi(\theta_d^{-1}a)$.

Proof. Fix c in D, b in A, h in $\ell^2(G)$ and f in $\ell^1(G)$. Then

$$(\lambda(f)h)_c(b) = (f * h)(\gamma(c)b) = \sum_{s \in G} f(s)h(s^{-1}\gamma(c)b).$$

Make the substitution $s = \gamma(c)a\gamma(d)^{-1}$ for a in A and d in D. Notice that

$$\gamma(cd^{-1}) = \gamma(c)\alpha(cd^{-1},d)\gamma(d)^{-1} = \gamma(c)\gamma(d)^{-1}\theta_d(\alpha(cd^{-1},d)).$$

Then we obtain

$$(\lambda(f)h)_c(b) = \sum_{d \in D} \sum_{a \in A} f(\gamma(c)a\gamma(d)^{-1})h(\gamma(d)a^{-1}b)$$
$$= \sum_{d \in D} \sum_{a \in A} f(\gamma(c)\gamma(d)^{-1}\theta_d(a))h_d(a^{-1}b)$$

$$= \sum_{d \in D} \sum_{a \in A} f\bigl(\gamma(cd^{-1})\theta_d(\alpha(cd^{-1}, d)^{-1}a)\bigr) h_d(a^{-1}b)$$

$$= \sum_{d \in D} (g_{c,d} * h_d)(b)$$

where

$$g_{c,d}(a) = f\bigl(\gamma(cd^{-1})\theta_d(\alpha(cd^{-1}, d)^{-1}a)\bigr) = f_{cd^{-1}}\bigl(\theta_d(\alpha(cd^{-1}, d)^{-1}a)\bigr)$$

Taking the Fourier transform, it follows that

$$\Psi\lambda(f)\Psi^* = [\widehat{g_{c,d}}] \in \mathcal{M}_D(C(\widehat{A})).$$

Finally, compute

$$F_{c,d}(\chi) := \widehat{g_{c,d}}(\chi) = \sum_{a \in A} g_{c,d}(a)\chi(a)$$

$$= \sum_{a \in A} g_{c,d}(\alpha(cd^{-1}, d)a)\chi(\alpha(cd^{-1}, d)a)$$

$$= \sum_{a \in A} f_{cd^{-1}}(\theta_d(a))\chi(\alpha(cd^{-1}, d))\chi(a)$$

$$= \chi(\alpha(cd^{-1}, d)) \sum_{a' \in A} f_{cd^{-1}}(a')\chi(\theta_d^{-1}(a'))$$

$$= \chi(\alpha(cd^{-1}, d))\widehat{f_{cd^{-1}}}(\theta_d^* \chi) = \chi(\alpha(cd^{-1}, d)) F_{cd^{-1}, e}(\theta_d^* \chi) \quad\blacksquare$$

This result will be applied to the group $\mathfrak{pg}$. For each z in $\mathbb{T}$, let $\mathfrak{R}_z$ denote the C*-subalgebra of $\mathcal{M}_2$ consisting of all matrices of the form $\begin{bmatrix} a & zb \\ b & a \end{bmatrix}$ for arbitrary scalars a, b in $\mathbb{C}$.

Corollary VII.4.2 $C_r^*(\mathfrak{pg})$ *is isomorphic (via λ') to*

$$\mathfrak{A} = \{F \in \mathcal{M}_2(\mathbb{T} \times [0,1]) : F(z,t) \in \mathfrak{R}_z \;\; \text{for all} \;\; z \in \mathbb{T},\, t \in \{0,1\}\}$$

The generators of the group are sent to

$$\lambda'(\sigma)(z,t) = \begin{bmatrix} 0 & z \\ 1 & 0 \end{bmatrix} \qquad \lambda'(\tau_{1,0})(z,t) = \begin{bmatrix} z & 0 \\ 0 & z \end{bmatrix}$$

and

$$\lambda'(\tau_{0,1})(z,t) = \begin{bmatrix} e^{\pi i t} & 0 \\ 0 & e^{-\pi i t} \end{bmatrix}.$$

Proof. As we noted at the beginning of this section, $\mathfrak{pg}$ is generated by an abelian subgroup $A \simeq \mathbb{Z}^2$ and an element σ which has non-trivial image in the quotient $\mathfrak{pg}/A \simeq \mathbb{Z}_2$. Define a section γ by $\gamma(1) = e$ and $\gamma(-1) = \sigma$. We compute the cocycle

$$\alpha(1,1) = \alpha(1,-1) = \alpha(-1,1) = e \quad \text{and} \quad \alpha(-1,-1) = \sigma^2 = \tau_{1,0}.$$

VII.4. A Crystallographic Group

The dual group of $\mathbb{Z}^2$ is $\mathbb{T}^2$, and $\theta_{-1}^*(z, w) = (z, \overline{w})$. Therefore Theorem VII.4.1 shows that $C_r^*(\mathfrak{p}\mathfrak{g})$ is isomorphic to the subalgebra of $\mathcal{M}_2(C(\mathbb{T}^2))$ consisting of elements $F = [F_{i,j}]_{i,j \in \mathbb{Z}_2}$ such that

$$F_{1,-1}(z,w) = (z,w)(\alpha(-1,-1))\, F_{-1,1}(\theta_{-1}^*(z,w)) = zF_{-1,1}(z,\overline{w})$$

and

$$F_{-1,-1}(z,w) = (z,w)(\alpha(1,-1))\, F_{1,1}(\theta_{-1}^*(z,w)) = F_{1,1}(z,\overline{w}).$$

That is,

$$F(z,w) = \begin{bmatrix} f(z,w) & zg(z,\overline{w}) \\ g(z,w) & f(z,\overline{w}) \end{bmatrix} \quad \text{for} \quad f,g \in C(\mathbb{T}^2).$$

Note that $F(z, \pm 1)$ belongs to $\mathfrak{R}_z$ for all z in $\mathbb{T}$; and setting $U_z = \begin{bmatrix} 0 & z \\ 1 & 0 \end{bmatrix}$,

$$U_z^* F(z,w) U_z = \begin{bmatrix} 0 & 1 \\ \overline{z} & 0 \end{bmatrix} \begin{bmatrix} f(z,w) & zg(z,\overline{w}) \\ g(z,w) & f(z,\overline{w}) \end{bmatrix} \begin{bmatrix} 0 & z \\ 1 & 0 \end{bmatrix}$$

$$= \begin{bmatrix} f(z,\overline{w}) & zg(z,w) \\ g(z,\overline{w}) & f(z,w) \end{bmatrix} = F(z,\overline{w})$$

Thus F is determined by its values on the cylinder $\{(z,w) \in \mathbb{T}^2 : \text{Im}\, w \geq 0\}$. This cylinder is homeomorphic to $\mathbb{T} \times [0,1]$ by the identification of (z,t) with $(z, e^{\pi i t})$ for $0 \leq t \leq 1$, which completes the identification with $\mathfrak{A}$.

If $f = \delta_s$ is the characteristic function of s in G, the formulae for $\lambda(f)$ from the proof of the previous theorem may be applied. For $s = \sigma = \gamma(-1)e$ and $\chi = (z,w)$, one computes that $\widehat{g_{1,1}} = \widehat{g_{-1,-1}} = 0$,

$$\widehat{g_{-1,1}}(\chi) = \chi(\alpha(-1,1)e) = \chi(e) = 1$$

and

$$\widehat{g_{1,-1}}(\chi) = \chi(\alpha(-1,-1)\theta_{-1}(e)) = \chi(\tau_{1,0}) = z.$$

Similarly, for $\tau_{0,1} = \gamma(1)\tau_{0,1}$, one obtains $\widehat{g_{1,-1}} = \widehat{g_{-1,1}} = 0$,

$$\widehat{g_{1,1}}(\chi) = \chi(\alpha(1,1)\theta_1(\tau_{0,1})) = \chi(\tau_{0,1}) = w$$

and

$$\widehat{g_{-1,-1}}(\chi) = \chi(\alpha(1,-1)\theta_{-1}(\tau_{0,1})) = \chi(\tau_{0,-1}) = \overline{w}.$$

Therefore

$$\lambda(\sigma)(z,w) = \begin{bmatrix} 0 & z \\ 1 & 0 \end{bmatrix} \qquad \lambda(\tau_{0,1})(z,w) = \begin{bmatrix} w & 0 \\ 0 & \overline{w} \end{bmatrix}$$

and

$$\lambda(\tau_{1,0})(z,w) = \lambda(\sigma^2)(z,w) = \begin{bmatrix} z & 0 \\ 0 & z \end{bmatrix}.$$

The identification with $\mathfrak{A}$ merely replaces w by $e^{\pi i t}$. ∎

The structure of the algebras on the two boundary circles is interesting, and it will be reflected in the primitive ideal structure. The algebra $\mathfrak{R}_z$ is spanned by a pair of orthogonal projections

$$P_{\pm\alpha} = \begin{bmatrix} 1/2 & \pm\alpha/2 \\ \pm\overline{\alpha}/2 & 1/2 \end{bmatrix}$$

where $\pm\alpha$ are the two square roots of z. The map taking α in $\mathbb{T}$ to P_α in $\mathfrak{R}_{\alpha^2}$ is a continuous map into the boundary circle of $\mathfrak{A}$ that wraps twice around. Moreover, there is a unique continuous path of projections in the boundary circle starting at P_1 in $\mathfrak{R}_1$. This path arrives at P_{-1} when it first returns to $\mathfrak{R}_1$ again.

Now it is possible to establish that the full C*algebra of $\mathfrak{pg}$ is the same as the reduced C*-algebra, and to compute the primitive ideal space and spectrum.

Theorem VII.4.3 *If π is an irreducible representation of $\mathfrak{pg}$, then either*

(i) *π is a two-dimensional representation determined up to unitary equivalence by points $z \in \mathbb{T}$ and $t \in (0,1)$ given by*

$$\pi(\sigma) = \begin{bmatrix} 0 & z \\ 1 & 0 \end{bmatrix} \quad \text{and} \quad \pi(\tau_{0,1}) = \begin{bmatrix} e^{\pi it} & 0 \\ 0 & e^{-\pi it} \end{bmatrix}.$$

or

(ii) *π is a one-dimensional representation uniquely determined by a character of $\mathfrak{pg}$ given by $\omega(\sigma) = \alpha$ for some $\alpha \in \mathbb{T}$ and $\omega(\tau_{0,1}) = \pm 1$.*

In particular, two irreducible representations with the same kernel are unitarily equivalent.

Proof. Let $\pi(\sigma) = U$ and $\pi(\tau_{0,1}) = V$. Since $\sigma^2 = \tau_{1,0}$ lies in the centre of $\mathfrak{pg}$, it follows from Lemma I.9.1 that U^2 is scalar, say $U^2 = \alpha^2 I$ for some α in $\mathbb{T}$. Thus the spectrum of U is contained in $\{\pm\alpha\}$. Consider the spectral projection $E = E_U(\alpha)$, and note that $U = \alpha E - \alpha E^\perp$. Write $V = \begin{bmatrix} A & B \\ C & D \end{bmatrix}$ with respect to the decomposition of $\mathcal{H} = E\mathcal{H} \oplus E^\perp\mathcal{H}$. Since $\sigma^{-1}\tau_{0,1}\sigma = \tau_{0,-1}$, one has $V^* = U^*VU$; whence

$$\begin{bmatrix} A^* & C^* \\ B^* & D^* \end{bmatrix} = \begin{bmatrix} \overline{\alpha}I & 0 \\ 0 & -\overline{\alpha}I \end{bmatrix} \begin{bmatrix} A & B \\ C & D \end{bmatrix} \begin{bmatrix} \alpha I & 0 \\ 0 & -\alpha I \end{bmatrix} = \begin{bmatrix} A & -B \\ -C & D \end{bmatrix}.$$

Therefore $A = A^*$, $D = D^*$, and $C = -B^*$. Moreover,

$$A^2 + BB^* = I = B^*B + D^2.$$

Write $B = W|B|$ in its polar decomposition. Then it is easy to see that the ranges of the projections $E - WW^*$, $E^\perp - W^*W$ and $WW^* + W^*W$ are invariant for both U and V, and thus for $\pi(\mathfrak{pg})$. As they sum to the identity and π is irreducible, one equals the identity.

In the first two cases, U is scalar and V is a self-adjoint unitary. As π is irreducible, $V = \pm 1$ and $U = \alpha$ (in the first case) or $U = -\alpha$ (in the second case) are one-dimensional scalars determining a character ω.

VII.4. A Crystallographic Group

In the remaining case, W is a unitary operator from $E^\perp \mathcal{H}$ onto $E\mathcal{H}$. So with this identification on the two subspaces, we may suppose that $B > 0$ (since it one-to-one with dense range). Then this reduces to

$$U = \begin{bmatrix} \alpha I & 0 \\ 0 & -\alpha I \end{bmatrix} \quad \text{and} \quad V = \begin{bmatrix} A & B \\ -B & D \end{bmatrix}$$

Since $A^2 = D^2 = I - B^2$, it is clear that A and D belong to $\{B\}'$; and since $AB = BD$, it follows that $A = D$. If F is a projection in $\{B\}''$, it is evident that the range of $F \oplus F$ is invariant for $\pi(\mathfrak{pg})$. Hence B is a positive scalar. So if F is now any projection in $\{A\}'$, then again $F \oplus F$ is invariant for $\pi(\mathfrak{pg})$. So A is scalar and one-dimensional. As $B > 0$ and $A^2 + B^2 = I$, there is a unique $0 < t \leq 1$ so that $A = \cos(\pi t)$ and $B = \sin(\pi t)$. With respect to the basis $\binom{1}{i}$, $\binom{\alpha}{-\alpha i}$, we obtain the desired matrix forms

$$U \simeq \begin{bmatrix} 0 & \alpha^2 \\ 1 & 0 \end{bmatrix} \quad \text{and} \quad V \simeq \begin{bmatrix} e^{\pi i t} & 0 \\ 0 & e^{-\pi i t} \end{bmatrix}.$$

When $t = 1$, this pair is reducible, and splits into two one-dimensional representations already noted above. ∎

Corollary VII.4.4 *The canonical map λ of $\mathrm{C}^*(\mathfrak{pg})$ onto $\mathrm{C}_r^*(\mathfrak{pg})$ is an isomorphism; and the canonical map of the spectrum $\widehat{\mathrm{C}^*(\mathfrak{pg})}$ onto $\mathrm{Prim}(\mathrm{C}^*(\mathfrak{pg}))$ is a homeomorphism. The primitive ideal space consists of an cylinder $\mathbb{T} \times (0,1)$ with the usual topology together with two circles that are glued two to one onto the boundary circles.*

Proof. It is evident from the explicit descriptions of the irreducible representations in Theorem VII.4.3 and of $\mathrm{C}_r^*(\mathfrak{pg})$ in Corollary VII.4.2 that the two-dimensional irreducible representations are obtained by evaluations of $\mathfrak{A}$ at (z,t) for z in $\mathbb{T}$ and $0 < t < 1$. The one-dimensional representations corresponding to $\omega(\sigma) = \pm\alpha$ and $\omega(\tau_{0,1}) = \pm 1$ correspond to the characters of $\mathfrak{A}(\alpha^2, \pm 1) = \Re_{\alpha^2}$ on the two boundary circles. The norm in $\mathrm{C}^*(\mathfrak{pg})$ of f in $\ell^1(\mathfrak{pg})$ is the supremum of $\|\pi(f)\|$ as π runs over all irreducible representations of $\mathfrak{pg}$. Hence it follows that λ is isometric; whence $\mathrm{C}^*(\mathfrak{pg}) = \mathrm{C}_r^*(\mathfrak{pg})$.

Note that each point $(\alpha^2, \pm 1)$ of the closed cylinder corresponds to two points in $\mathrm{Prim}(\mathrm{C}^*(\mathfrak{pg}))$. Proposition VII.3.5 and consideration of functions in $\pi(\tau_{1,0})$ and $\pi(\tau_{0,1})$ show that the topology on the open cylinder is the standard one, and that the projection onto the closed cylinder is continuous. Consideration of $\pi(\sigma)$ shows that the relative topology on each boundary circle is the standard one, but that the neighbourhood of a point $(\alpha, 0)$ contains a neighbourhood of the point $(\alpha^2, 0)$ intersected with the open cylinder. A similar statement is true for the other boundary circle. ∎

Another notable property of $\mathfrak{pg}$ is that it is torsion free. There is an important conjecture concerning the C*-algebras of torsion free discrete groups.

Conjecture VII.4.5 (Kadison–Kaplansky) *If G is a torsion free discrete group, then the reduced C^*-algebra $C_r^*(G)$ has no projections other than 0 and 1.*

This conjecture remains open; although, it has been verified for large classes of discrete groups. Using our knowledge of the primitive ideal structure in $C_r^*(\mathfrak{pg})$, we can give a simple proof of the conjecture in this case.

Theorem VII.4.6 *There are no proper projections in $C^*(\mathfrak{pg})$.*

Proof. Let us work in the concrete realization $\mathfrak{A}$ of $C^*(\mathfrak{pg})$. Let P in $\mathfrak{A}$ be a projection. Since $\mathbb{T} \times [0,1]$ is connected and $P(z,t)$ is continuous, rank $P(z,t)$ is a constant of value 0, 1 or 2 on the cylinder. A constant rank of 2 implies that P is the identity in $\mathfrak{A}$ and rank 0 implies that $P = 0$. So assume that rank $P(z,t) = 1$ for all (z,t) in $\mathbb{T} \times [0,1]$. Since $\mathfrak{R}_z$ is two dimensional, $P(z,0)$ is one of the two projections $P_{\pm\alpha}$. But $P(z,0)$ is continuous as a function of z. If $P(1,0) = P_1$ say, then following the projection once around the boundary circle leads to the conclusion that $P(1,0) = P_{-1}$ instead. This is a contradiction. Therefore, P is either zero or the identity.

VII.5 The Discrete Heisenberg Group

Our next example is a little more complicated. The Heisenberg group is still amenable. It will follow from direct computation that the left regular representation is faithful on the full C*-algebra. However, the spectrum is not homeomorphic to the primitive ideal space.

The **discrete Heisenberg group** is the multiplicative group $\mathbb{H}_3$ of matrices of the form

$$\begin{bmatrix} 1 & a & c \\ 0 & 1 & b \\ 0 & 0 & 1 \end{bmatrix} \quad a,b,c \in \mathbb{Z}.$$

This group is generated by

$$u = \begin{bmatrix} 1 & 1 & 0 \\ 0 & 1 & 0 \\ 0 & 0 & 1 \end{bmatrix} \quad \text{and} \quad v = \begin{bmatrix} 1 & 0 & 0 \\ 0 & 1 & 1 \\ 0 & 0 & 1 \end{bmatrix}.$$

Notice that

$$w := uvu^{-1}v^{-1} = \begin{bmatrix} 1 & 0 & 1 \\ 0 & 1 & 0 \\ 0 & 0 & 1 \end{bmatrix}.$$

A simple calculation shows that w commutes with u and v, and generates the centre $Z(\mathbb{H}_3)$ which is isomorphic to $\mathbb{Z}$. It follows that $\mathbb{H}_3$ is an extension of $\mathbb{Z}$ by $\mathbb{Z}^2$. Hence it is amenable by Proposition VII.2.3.

To understand $C^*(\mathbb{H}_3)$, it suffices to know enough about the irreducible representations.

VII.5. The Discrete Heisenberg Group

Theorem VII.5.1 *Let π be an irreducible representation of $\mathbb{H}_3$. Then there is a real number $\theta \in (-1, 1]$ such that $\pi(w) = e^{2\pi i\theta} I$.*

(i) *When θ is irrational, $C^*(\pi(u), \pi(v))$ is canonically isomorphic to the irrational rotation algebra $\mathcal{A}_\theta$. The representation is determined up to approximate unitary equivalence.*

(ii) *When $\theta = k/n$ and $\gcd(k, n) = 1$, then $C^*(\pi(u), \pi(v))$ is isomorphic to $\mathcal{M}_n$. Let $\gamma = e^{2\pi i\theta}$. There are complex numbers α and β in $\mathbb{T}$ such that $\pi(u)$ and $\pi(v)$ are (simultaneously) unitarily equivalent to the pair*

$$\alpha \begin{bmatrix} 1 & 0 & \cdots & 0 & 0 \\ 0 & \gamma & \cdots & 0 & 0 \\ \vdots & \vdots & \ddots & \vdots & \vdots \\ 0 & 0 & \cdots & \gamma^{n-2} & 0 \\ 0 & 0 & \cdots & 0 & \gamma^{n-1} \end{bmatrix} \quad \text{and} \quad \beta \begin{bmatrix} 0 & 0 & \cdots & 0 & 1 \\ 1 & 0 & \cdots & 0 & 0 \\ 0 & 1 & \cdots & 0 & 0 \\ \vdots & \vdots & \ddots & \vdots & \vdots \\ 0 & 0 & \cdots & 1 & 0 \end{bmatrix}.$$

The pair (α^n, β^n) is uniquely determined in $\mathbb{T}^2$ by π; and this pair determines π up to unitary equivalence.

Proof. Since $\pi(w)$ lies in the centre of an irreducible C*-algebra, it is a scalar multiple of the identity by Lemma I.9.1. Thus there is a unique number θ in $(-1, 1]$ such that that $\pi(w) = e^{2\pi i\theta} I =: \gamma I$. Let $U = \pi(u)$ and $V = \pi(v)$. They satisfy

$$UVU^*V^* = \pi(w) = \gamma I. \tag{1}$$

Let $\mathfrak{I}_\theta$ denote the ideal of $C^*(\mathbb{H}_3)$ generated by the image of $w - \gamma e$. Then $C^*(\mathbb{H}_3)/\mathfrak{I}_\theta$ is generated by two unitaries satisfying (1). Hence by Theorem VI.1.4, when θ is irrational, there is a canonical isomorphism of $\mathcal{A}_\theta$ onto $C^*(\mathbb{H}_3)/\mathfrak{I}_\theta$ taking the generators onto U and V. As $\mathcal{A}_\theta$ is simple, every irreducible representation π such that $\pi(w) = \gamma I$ has kernel $\mathfrak{I}_\theta$. So by Proposition VII.3.4, they are all approximately unitarily equivalent.

Now suppose that $\theta = k/n$ is rational. Then

$$U^n V = \gamma^n V U^n = V U^n.$$

Thus U^n lies in the centre of $\mathfrak{A}$, and so is scalar. Similarly, V^n is scalar; and the pair (U^n, V^n) is clearly an invariant of the representation. Choose scalars α and β in $\mathbb{T}$ so that $U^n = \alpha^n I$ and $V^n = \beta^n I$. Since γ is a primitive n-th root of unity, the possible eigenvalues of U are $\alpha\gamma^j$ for $0 \leq j < n$. Let $E_j = E_U(\alpha\gamma^j)$ be the corresponding spectral projections. We may write

$$U = \sum_{j=0}^{n-1} \alpha\gamma^j E_j.$$

It follows (using the finite Fourier transform) that

$$E_i = \frac{1}{n} \sum_{k=0}^{n-1} (\alpha\gamma^i)^{-k} U^k.$$

Therefore, for $0 \leq i < n$,

$$VE_i = \frac{1}{n}\sum_{k=0}^{n-1}(\alpha\gamma^i)^{-k}VU^k = \frac{1}{n}\sum_{k=0}^{n-1}(\alpha\gamma^i)^{-k}\gamma^{-k}U^kV$$

$$= \frac{1}{n}\sum_{k=0}^{n-1}(\alpha\gamma^{i+1})^{-k}U^kV = E_{i+1}V.$$

Thus we may define partial isometries $E_{ij} = (\overline{\beta}V)^{i-j}E_j$ for $0 \leq i,j < n$. Since $(\overline{\beta}V)^n = I$, it is routine to verify that these form a set of matrix units for a copy of $\mathcal{M}_n$. Moreover,

$$V = \sum_{j=0}^{n-1}VE_j = \beta\sum_{j=0}^{n-1}E_{i+1,i}$$

where we interpret $E_{n,n-1}$ as $E_{0,n}$. Hence $C^*(\pi(\mathbb{H}_3))$ is isomorphic to $\mathcal{M}_n$. By Theorem I.10.6, the only irreducible representation of $\mathcal{M}_n$ is the identity representation. Thus E_i are all one dimensional. The matrix forms follow from the formulae derived above for U and V in terms of the matrix units.

It is also apparent that conjugation of this representation by U leaves U fixed and sends V to γV. Similarly, conjugation by V fixes V and sends U to $\overline{\gamma}U$. Thus the constants α and β are only determined up to a multiple by an n-th root of unity; which is to say that the pair (α^n, β^n) is the complete unitary invariant. So π has been determined up to unitary equivalence by $\theta = k/n$ and the pair (α^n, β^n).

Conversely, given $\theta = k/n$, α and β, it is evident that these formulae yield an irreducible representation of $\mathbb{H}_3$. So we have a complete description of $C^*(\pi(\mathbb{H}_3))$ for every irreducible representation π. (Note that we have not classified the irreducible representations of $\mathcal{A}_\theta$ up to unitary equivalence.) ∎

Remark VII.5.2 This theorem allows us to describe the primitive ideal space. There is a canonical projection p of $\text{Prim}(C^*(\mathbb{H}_3))$ onto $\mathbb{T}$ given by evaluation at w. The inverse image of irrational points is a singleton; but the inverse image of each rational point is a copy of $\mathbb{T}^2$. The basic neighbourhoods of any irrational point is the inverse image under p of a neighbourhood on the circle. While for a point on a torus over a rational point, the neighbourhoods consist of a neighbourhood on the torus together with the inverse image under p of a punctured neighbourhood of the rational point on the circle. The details will be left as an exercise.

The spectrum of $C^*(\mathbb{H}_3)$ projects onto $\text{Prim}(C^*(\mathbb{H}_3))$. The fibres are singletons over each point on the torus at each rational point of the circle. However, over each irrational point of the circle, there is a very large space corresponding to the unitary equivalence classes of representations of $\mathcal{A}_\theta$. (See Exercise VI.3.)

Next we will show that $C_r^*(\mathbb{H}_3)$ has the same irreducible representations. This will imply that $C_r^*(\mathbb{H}_3)$ is isomorphic to $C^*(\mathbb{H}_3)$.

Theorem VII.5.3 *The set of irreducible representations of $C^*_r(\mathbb{H}_3)$ are the same as the set of irreducible representations of $C^*(\mathbb{H}_3)$.*

Proof. First we identify $W = \lambda(w)$. For convenience of notation, let us write $\delta_{a,b,c}$ for the basis vector of $\ell^2(\mathbb{H}_3)$ corresponding to $\begin{bmatrix} 1 & a & c \\ 0 & 1 & b \\ 0 & 0 & 1 \end{bmatrix}$. Notice that

$$W\delta_{a,b,c} = \delta_{a,b,c+1} \quad \text{for all} \quad a,b,c \in \mathbb{Z}.$$

Hence W is a bilateral shift of infinite multiplicity. In particular, its spectrum is the whole unit circle. For each γ in $\mathbb{T}$, let $\mathfrak{I}_\gamma$ be the ideal of $C^*_r(\mathbb{H}_3)$ generated by $W - \gamma I$. Since W lies in the centre, this is a proper ideal. Consider the quotient map q_γ of $C^*_r(\mathbb{H}_3)$ onto $\mathfrak{A}_\gamma := C^*_r(\mathbb{H}_3)/\mathfrak{I}_\gamma$. Then $q_\gamma(W) = \gamma I$. So any irreducible representation ρ of $\mathfrak{A}_\gamma$ determines an irreducible representation $\pi = \rho q_\gamma \lambda$ of $\mathbb{H}_3$ such that $\pi(w) = \gamma I$. The proof of the previous theorem now applies, and we deduce that $C^*_r(\mathbb{H}_3)$ has the same irreducible representations as $C^*(\mathbb{H}_3)$. ∎

Corollary VII.5.4 *The canonical homomorphism of $C^*(\mathbb{H}_3)$ onto $C^*_r(\mathbb{H}_3)$ is an isomorphism.*

Proof. From the GNS construction Theorem I.9.12, it follows that the norm of A in a C*-algebra $\mathfrak{A}$ is determined by $\sup \|\pi(A)\|$ as the sup runs over all irreducible representations of $\mathfrak{A}$. Since the quotient algebra $C^*_r(\mathbb{H}_3)$ has the same irreducible representations as $C^*(\mathbb{H}_3)$, it follows that the quotient map is isometric. ∎

VII.6 The Free Group

A very important example of a non-amenable group is the free group $\mathbb{F}_2$ on two generators u and v. In the next few sections, we will examine this C*-algebra in detail. This group is discrete, and therefore $\ell^1(\mathbb{F}_2)$ is spanned by the characteristic functions of elements of $\mathbb{F}_2$. Hence $C^*(\mathbb{F}_2)$ is generated by a pair of *universal unitaries* $(\pi_u(u), \pi_u(v))$. Any representation π takes u and v to unitary operators $U = \pi(u)$ and $V = \pi(v)$. Since there are no relations on u and v, every unitary pair U, V determines a representation of $\mathbb{F}_2$. The irreducible pairs determine irreducible representations. Let $\{(U_\alpha, V_\alpha) : \alpha \in A\}$ be a collection of irreducible unitary pairs, one from each unitary equivalence class. (Irreducible representations of $\mathbb{F}_2$ act on separable Hilbert spaces. See Exercise VII.6.) Then the group C*-algebra is isomorphic to $C^*(\mathbf{U}, \mathbf{V})$ where

$$\mathbf{U} = \sum \oplus U_\alpha \quad \text{and} \quad \mathbf{V} = \sum \oplus V_\alpha.$$

This is the C*-algebra which is universal with respect to the property that if U, V is any unitary pair, then there is a *-homomorphism π of $C^*(\mathbf{U}, \mathbf{V})$ onto $C^*(U, V)$ such that $\pi(\mathbf{U}) = U$ and $\pi(\mathbf{V}) = V$.

We will establish the existence of some separating families of irreducible representations to obtain some interesting faithful representations of $C^*(\mathbb{F}_2)$.

Proposition VII.6.1 $C^*(\mathbb{F}_2)$ *has a family* π_n, $n \geq 1$, *of finite dimensional representations such that* $\sum \oplus_{n \geq 1} \pi_n$ *is faithful.*

Proof. First notice that there is a pair (U, V) of unitaries on a separable Hilbert space $\mathcal{H}$ such that the canonical map σ of $C^*(\mathbb{F}_2)$ onto $C^*(U, V)$ is an isomorphism. Since $\mathbb{F}_2$ is countable, $\ell^1(\mathbb{F}_2)$ has a countable dense subset. Hence the image under π_u yields a countable dense subset of $C^*(\mathbb{F}_2)$. So this is a separable C*-algebra; and therefore it has a faithful separable representation which we will also denote by σ. Let $U = \sigma(u)$ and $V = \sigma(v)$.

Let P_n be an increasing sequence of finite rank projections in $\mathcal{B}(\mathcal{H})$ tending strongly to the identity. Let $A_n = P_n U|_{P_n \mathcal{H}}$ and $B_n = P_n V|_{P_n \mathcal{H}}$. Let I_n denote the identity in $\mathcal{B}(P_n \mathcal{H})$, and define operators on $\mathcal{H}_n = P_n \mathcal{H} \oplus P_n \mathcal{H}$ by

$$U_n = \begin{bmatrix} A_n & (I_n - A_n A_n^*)^{1/2} \\ (I_n - A_n^* A_n)^{1/2} & -A_n^* \end{bmatrix}$$

and

$$V_n = \begin{bmatrix} B_n & (I_n - B_n B_n^*)^{1/2} \\ (I_n - B_n^* B_n)^{1/2} & -B_n^* \end{bmatrix}.$$

These are unitary operators acting on a finite dimensional Hilbert space. Let π_n be the representation determined by $\pi_n(\mathbf{U}) = U_n$ and $\pi_n(\mathbf{V}) = V_n$.

Set $\pi = \sum \oplus_{n \geq 1} \pi_n$. If we think of U_n and V_n as acting on a subspace of $\mathcal{H} \oplus \mathcal{H}$, then it is evident from the definition of A_n and B_n that

$$\text{SOT-lim}_{n \to \infty} U_n = \begin{bmatrix} U & 0 \\ 0 & -U^* \end{bmatrix} \quad \text{and} \quad \text{SOT-lim}_{n \to \infty} V_n = \begin{bmatrix} V & 0 \\ 0 & -V^* \end{bmatrix}.$$

Therefore, if $f(s) = \sum \alpha_s \delta_s$ is a *finite* sum (in the group algebra $\mathbb{C}\mathbb{F}_2$ as a subset of $\ell^1(\mathbb{F}_2)$), it follows that

$$\text{SOT-lim}_{n \to \infty} \pi_n(f) = \begin{bmatrix} \sigma(f) & 0 \\ 0 & * \end{bmatrix}$$

Therefore

$$\|\pi(f)\| = \sup \|\pi_n(f)\| \geq \|\sigma(f)\| = \|\pi_u(f)\|.$$

Since $\mathbb{C}\mathbb{F}_2$ is dense in $C^*(\mathbb{F}_2)$, it follows that π is isometric. ∎

A C*-algebra $\mathfrak{A}$ is called **quasidiagonal** if it has a faithful representation π for which there exists an increasing sequence P_n of finite rank projections tending strongly to the identity such that

$$\lim_{n \to \infty} \|P_n \pi(A) - \pi(A) P_n\| = 0 \quad \text{for all} \quad A \in \mathfrak{A}.$$

So we immediately obtain:

Corollary VII.6.2 *The group C*-algebra $C^*(\mathbb{F}_2)$ is quasidiagonal.*

We also obtain some information about traces.

VII.6. The Free Group

Corollary VII.6.3 *The group C*-algebra* $C^*(\mathbb{F}_2)$ *has a faithful trace.*

Proof. Let π be the representation constructed in Theorem VII.6.1. Let τ_n be the normalized trace on the matrix algebra $\mathcal{B}(P_n\mathcal{H})$. Then
$$\tau(A) := \sum_{n \geq 1} 2^{-n} \tau_n(\pi_n(A))$$
is a trace on $C^*(\mathbb{F}_2)$. If $A \geq 0$, then $\pi_n(A)$ is non-zero for n sufficiently large; whence $\tau_n(\pi_n(A)) > 0$. Therefore $\tau(A) > 0$, and thus τ is faithful. ∎

Next we twist this representation a bit to get a faithful irreducible representation. A simple preliminary lemma is needed.

Lemma VII.6.4 *Suppose that* $D = \mathrm{diag}(d_n)$ *is a diagonal operator with respect to a basis* $\{e_n\}$ *with distinct eigenvalues* d_n *for* $n \geq 1$. *If* A *is any operator such that* $(Ae_1, e_n) \neq 0$ *for all* $n \geq 2$, *then* $C^*(D, A)$ *is irreducible.*

Proof. The commutant of the diagonal operator D is the masa of all diagonal operators (see Exercise VII.7). Clearly, A does not commute with any proper diagonal projection. Thus $C^*(D, A)$ is irreducible. ∎

Theorem VII.6.5 *The group C*-algebra* $C^*(\mathbb{F}_2)$ *has a faithful irreducible representation.*

Proof. Let U and V be a pair of unitaries of joint infinite multiplicity acting on a separable Hilbert space such that $C^*(U, V) \simeq C^*(\mathbb{F}_2)$. Let σ be the canonical isomorphism of $C^*(\mathbb{F}_2)$ onto $C^*(U, V)$. (One can use $U^{(\infty)}$ and $V^{(\infty)}$ where U and V were those used in the previous theorem.) By the Weyl–von Neumann–Berg Theorem (Corollary II.4.2), there is a compact perturbation $U_1 = U + K$ of U which is diagonal with respect to a basis $\{e_n\}_{n \geq 1}$. With a further compact diagonal perturbation, it can be arranged that $U_1 = \mathrm{diag}(u_n)$ where u_n are *distinct* points on the unit circle. Let V_1 be a unitary operator which is a finite rank perturbation of V such that $(V_1 e_1, e_n) \neq 0$ for all $n \geq 2$. For example, one can change V on the two dimensional subspace span$\{e_1, V^*x\}$, where $x = \sum_{k \geq 1} 2^{-k/2} e_k$, so that $V_1 e_1 = x$.

By the previous lemma, $C^*(U_1, V_1)$ is irreducible. If σ_1 is the canonical homomorphism of $C^*(\mathbb{F}_2)$ onto $C^*(U_1, V_1)$ and π is the quotient map of $\mathcal{B}(\mathcal{H})$ onto $\mathcal{B}(\mathcal{H})/\mathfrak{K}$, then for each A in $C^*(\mathbb{F}_2)$,
$$\|\sigma_1(A)\| \geq \|\pi\sigma_1(A)\| = \|\pi\sigma(A)\| = \|\sigma(A)\| = \|A\|.$$
Therefore σ_1 is an isomorphism. ∎

We finish this section with an easy argument showing that $C^*(\mathbb{F}_2)$ is projectionless. This is also the case, as we shall see, for the reduced C*-algebra. But the proof in that case is much more difficult.

Theorem VII.6.6 *The group C*-algebra* $C^*(\mathbb{F}_2)$ *contains no proper projections.*

Proof. We will imbed $C^*(\mathbb{F}_2)$ into the C*-algebra

$$\mathfrak{A} = \{\Phi \in C([0,1], \mathcal{B}(\mathcal{H})) : \Phi(0) \in \mathbb{C}I\}.$$

Let U and V be a universal pair of unitaries on a separable space, and let σ be the corresponding isomorphism of $C^*(\mathbb{F}_2)$. By the spectral theorem, there are bounded Hermitian operators A and B such that $U = e^{iA}$ and $V = e^{iB}$. So one may define a pair of unitary elements in $\mathfrak{A}$ by $\widetilde{U}(t) = e^{itA}$ and $\widetilde{V} = e^{itB}$. Therefore there is a *-homomorphism φ of $C^*(\mathbb{F}_2)$ into $\mathfrak{A}$ such that $\varphi(\mathbf{U}) = \widetilde{U}$ and $\varphi(\mathbf{V}) = \widetilde{V}$. If δ_1 is the homomorphism on $\mathfrak{A}$ of evaluation at 1, we see that $\delta_1 \varphi = \sigma$ is an isomorphism. Hence φ is also an isomorphism.

The C*-algebra $\mathfrak{A}$ has no proper projections. To see this, note that Φ is a projection in $\mathfrak{A}$ if and only if $\Phi(t)$ is a projection for all $t \in [0,1]$. However, Φ is evidently homotopic to the constant projection $\Phi_0(t) = \Phi(0)$ which is scalar. The only projection homotopic to 0 (or I) is 0 (or I). Hence there are no proper projections in $\mathfrak{A}$; and *a fortiori*, there are no projections in $C^*(\mathbb{F}_2)$. ∎

Corollary VII.6.7 *If π is a faithful representation of $C^*(\mathbb{F}_2)$, then there are no non-zero compact operators in the range of π.*

Proof. A C*-algebra containing a non-zero compact operator contains a non-zero finite rank projection. ∎

VII.7 The reduced C*-algebra of the free group

We write $\ell^2(\mathbb{F}_2)$ instead of $L^2(\mathbb{F}_2)$ because the Haar measure just assigns mass one to each point. Let δ_t denote the characteristic function of t in $\mathbb{F}_2$ as an element of $\ell^2(\mathbb{F}_2)$. Clearly, these vectors form an orthonormal basis for $\ell^2(\mathbb{F}_2)$. Notice that

$$\lambda(s)\delta_t(r) = \delta_t(s^{-1}r) = \delta_{st}(r) \quad \text{for} \quad r, s, t \in \mathbb{F}_2.$$

Thus $\lambda(s)\delta_t = \delta_{st}$ implements a permutation of these basis vectors. It is evident that this left Haar measure is also right translation invariant. So $\mathbb{F}_2$ is unimodular. Consider the right regular representation on $\ell^2(\mathbb{F}_2)$ given by

$$(\rho(s)g)(t) = g(ts) \quad \text{whence} \quad \rho(s)\delta_t = \delta_{ts^{-1}}.$$

These two representations commute since

$$\rho(s)\lambda(t)\delta_r = \rho(s)\delta_{tr} = \delta_{trs^{-1}} = \lambda(t)\delta_{rs^{-1}} = \lambda(t)\rho(s)\delta_r.$$

Proposition VII.7.1 *The map $\tau(A) = (A\delta_e, \delta_e)$ is a faithful trace on $C_r^*(\mathbb{F}_2)$. Thus the map $\Phi(A) = \tau(A)I$ is a faithful expectation of $C_r^*(\mathbb{F}_2)$ onto the scalars.*

VII.7. The reduced C*-algebra of the free group

Proof. Consider $f = \sum_{s \in \mathbb{F}_2} \alpha_s \delta_s$ and $g = \sum_{s \in \mathbb{F}_2} \beta_s \delta_s$ in $\mathbb{C}\mathbb{F}_2$ (i.e. finite sums). Then $\tau(\lambda(f)) = \alpha_e$ and $\tau(\lambda(g)) = \beta_e$. Calculate

$$\tau(\lambda(f*g)) = \tau(\sum_{s \in \mathbb{F}_2} \sum_{t \in \mathbb{F}_2} \alpha_s \beta_t (\lambda(st)\delta_e, \delta_e)$$

$$= \sum_{s \in \mathbb{F}_2} \alpha_s \beta_{s^{-1}} = \sum_{s \in \mathbb{F}_2} \beta_s \alpha_{s^{-1}} = \tau(\lambda(g*f)).$$

So by continuity, τ is a trace on $C_r^*(\mathbb{F}_2)$.

Suppose that A in $C_r^*(\mathbb{F}_2)$ is positive and $\tau(A) = 0$. Then

$$(A\delta_s, \delta_s) = (A\lambda(s)\delta_e, \lambda(s)\delta_e) = (\lambda(s^{-1})A\lambda(s)\delta_e, \delta_e)$$
$$= \tau(\lambda(s^{-1})A\lambda(s)) = \tau(A) = 0.$$

Hence

$$|(A\delta_s, \delta_t)| \le (A\delta_s, \delta_s)^{1/2}(A\delta_t, \delta_t)^{1/2} = 0$$

for all s, t in $\mathbb{F}_2$. Therefore $A = 0$; and so τ is faithful.

It is then evident that Φ is a faithful expectation onto the scalars. ∎

We wish to show that $C_r^*(\mathbb{F}_2)$ is simple with a unique trace. As for the irrational rotation algebras, we first construct an explicit formula for the expectation Φ.

Lemma VII.7.2 Let $\mathcal{H} = \mathcal{M} \oplus \mathcal{M}^\perp$. Suppose that B has the form $\begin{bmatrix} 0 & * \\ * & * \end{bmatrix}$ and U_i are unitary operators such that $U_i U_j^*$ have the form $\begin{bmatrix} * & * \\ * & 0 \end{bmatrix}$ when $i \ne j$. Then

$$\|\frac{1}{n}\sum_{i=1}^n U_i B U_i^*\| \le \frac{2}{\sqrt{n}}\|B\|.$$

Proof. First suppose that B and C have the forms

$$B = \begin{bmatrix} 0 & 0 \\ * & * \end{bmatrix} \quad \text{and} \quad C = \begin{bmatrix} * & * \\ 0 & 0 \end{bmatrix}.$$

Then

$$\|B + C\|^2 = \|(B^* + C^*)(B+C)\| = \|B^*B + C^*C\| \le \|B\|^2 + \|C\|^2.$$

Notice that when $i \ne j$, the operator $(U_i U_j^*)^* B (U_i U_j^*)$ has the form

$$\begin{bmatrix} * & * \\ * & 0 \end{bmatrix} \begin{bmatrix} 0 & 0 \\ * & * \end{bmatrix} \begin{bmatrix} * & * \\ * & 0 \end{bmatrix} = \begin{bmatrix} * & * \\ 0 & 0 \end{bmatrix}.$$

Hence

$$\|\sum_{i=1}^n U_i B U_i^*\|^2 = \|U_1^*(B + \sum_{i=2}^n (U_1^* U_i) B (U_i U_1^*)) U_1\|^2$$

$$\le \|B\|^2 + \|\sum_{i=2}^n (U_1^* U_i) B (U_i U_1^*)\|^2 = \|B\|^2 + \|\sum_{i=2}^n U_i B U_i\|^2.$$

By induction, it follows that $\left\|\sum_{i=1}^n U_i B U_i^*\right\|^2 \leq n\|B\|^2$. Therefore,

$$\left\|\tfrac{1}{n}\sum_{i=1}^n U_i B U_i^*\right\| \leq \tfrac{1}{\sqrt{n}}\|B\|.$$

The same result holds when B has the form $\begin{bmatrix} 0 & * \\ 0 & * \end{bmatrix}$. For the general case, one may split B as a sum of two terms in the forms $\begin{bmatrix} 0 & 0 \\ * & * \end{bmatrix}+\begin{bmatrix} 0 & * \\ 0 & 0 \end{bmatrix}$. This yields the desired estimate. ∎

Corollary VII.7.3 *For s in $\mathbb{F}_2$,*

$$\lim_{n\to\infty} \tfrac{1}{n}\sum_{i=1}^n \lambda(u^i)\lambda(s)\lambda(u^{-i}) = \begin{cases} \lambda(s) & \text{if } s = u^k \text{ for some } k \in \mathbb{Z} \\ 0 & \text{otherwise} \end{cases}.$$

Proof. An element of $\mathbb{F}_2$ is a product of terms in u, v, u^{-1} and v^{-1}. It is in reduced form if no adjacent terms cancel. If s is not a power of u, then it has some non-zero power of v in its reduced form. Thus there are integers k_0 and ℓ_0 (possibly 0) such that $s = u^{k_0} s_0 u^{\ell_0}$, where s_0 has the form $v^{\pm 1} t v^{\pm 1}$ or $v^{\pm 1}$.

Let

$$\mathcal{M} = \operatorname{span}\{\delta_s : s = 1 \text{ or } s = u^{\pm 1} t \text{ in reduced form}\}.$$

Then

$$\mathcal{M}^\perp = \operatorname{span}\{\delta_s : s = v^{\pm 1} t \text{ in reduced form}\}.$$

So $\lambda(s_0)\mathcal{M}$ is contained in $\mathcal{M}^\perp$; while $\lambda(u^k)\mathcal{M}^\perp$ is contained in $\mathcal{M}$ for $k \neq 0$. Hence with respect to this decomposition, $\lambda(s_0) = \begin{bmatrix} 0 & * \\ * & * \end{bmatrix}$ and $\lambda(u^k)\lambda(u^{\ell *})$ has the form $\begin{bmatrix} * & * \\ * & 0 \end{bmatrix}$ when $k \neq \ell$. By the previous lemma, for $s = u^{k_0} s_0 u^{\ell_0}$,

$$\lim_{n\to\infty} \tfrac{1}{n}\sum_{i=1}^n \lambda(u^i)\lambda(s)\lambda(u^{-i})$$

$$= \lambda(u^{k_0})\left(\lim_{n\to\infty} \tfrac{1}{n}\sum_{i=1}^n \lambda(u^i)\lambda(s_0)\lambda(u^{-i})\right)\lambda(u^{\ell_0}) = 0.$$

On the other hand, if $s = u^k$, then evidently

$$\tfrac{1}{n}\sum_{i=1}^n \lambda(u^i)\lambda(s)\lambda(u^{-i}) = \lambda(s) \quad \text{for all} \quad n \geq 1. \qquad \blacksquare$$

Now we can construct a useful method for computing the expectation Φ.

Theorem VII.7.4 *For all A in $C_r^*(\mathbb{F}_2)$,*

$$\lim_{m\to\infty}\lim_{n\to\infty} \tfrac{1}{mn}\sum_{i=1}^m\sum_{j=1}^n \lambda(u^i v^j) A \lambda(v^{-j} u^{-i}) = \tau(A)I.$$

VII.7. The reduced C*-algebra of the free group

Proof. Consider $f = \sum_{s \in \mathbb{F}_2} \alpha_s \delta_s$ in $\mathbb{C}\mathbb{F}_2$. By the previous corollary,

$$\lim_{n \to \infty} \frac{1}{n} \sum_{j=1}^{n} \lambda(v^j) \lambda(f) \lambda(v^{-j}) = \sum \alpha_{v^k} \lambda(v^k) =: f_0.$$

Hence the desired limit equals

$$\lim_{m \to \infty} \frac{1}{m} \sum_{i=1}^{m} \lambda(u^i) \lambda(f_0) \lambda(u^{-i}) = \alpha_e I = \tau(A) I = \Phi(A).$$

By continuity, this identity extends to every element of $C_r^*(\mathbb{F}_2)$. A routine norm estimate shows that this limit exists as m and n tend to infinity independently of each other. ∎

Corollary VII.7.5 *The reduced group C*-algebra $C_r^*(\mathbb{F}_2)$ is simple.*

Proof. If $A \neq 0$, then $A^*A \neq 0$ is positive. Hence $\tau(A^*A) > 0$. By the previous theorem, the ideal generated by A contains

$$\lim_{m \to \infty} \lim_{n \to \infty} \frac{1}{mn} \sum_{i=1}^{m} \sum_{j=1}^{n} \lambda(u^i v^j) A^* A \lambda(v^{-j} u^{-i}) = \tau(A^*A) I.$$

Hence this ideal is all of $C^*(\mathbb{F}_2)$. ∎

Corollary VII.7.6 $C_r^*(\mathbb{F}_2)$ *has a unique trace.*

Proof. If τ' is a trace on $C_r^*(\mathbb{F}_2)$, then by linearity and the trace property,

$$\tau'(A) = \tau'\Big(\frac{1}{mn} \sum_{i=1}^{m} \sum_{j=1}^{n} \lambda(u^i v^j) A \lambda(v^{-j} u^{-i})\Big).$$

But the right hand side converges to $\tau'(\tau(A)I) = \tau(A)$. ∎

The non-amenability of the free group is evidenced by the following simple consequence.

Corollary VII.7.7 *The reduced group C*-algebra $C_r^*(\mathbb{F}_2)$ is not isomorphic to the full group C*-algebra $C^*(\mathbb{F}_2)$.*

Proof. $C_r^*(\mathbb{F}_2)$ is simple, while $C^*(\mathbb{F}_2)$ has many quotients including finite dimensional ones. ∎

Another property distinguishing $C^*(\mathbb{F}_2)$ from $C_r^*(\mathbb{F}_2)$ is quasidiagonality. This also depends on non-amenability.

Proposition VII.7.8 *A countable discrete group G such that $C_r^*(G)$ is quasidiagonal is amenable.*

Proof. Suppose that there are finite rank projections P_n in $\mathcal{B}(\ell^2(G))$ increasing to the identity such that
$$\lim_{n\to\infty} \|P_n\lambda(s) - \lambda(s)P_n\| = 0 \quad \text{for all} \quad s \in G.$$
Let $\ell^\infty(G)$ act on $\ell^2(G)$ by multiplication: $M_f x(s) = f(s)x(s)$ for f in $\ell^\infty(G)$, x in $\ell^2(G)$ and s in G. Notice that for t in G,
$$\lambda(t)M_f\lambda(t)^* x(s) = M_f \lambda(t)^* x(t^{-1}s)$$
$$= f(t^{-1}s)\big(\lambda(t)^* x\big)(t^{-1}s) = f(t^{-1}s)x(s).$$
Hence $\lambda(t)M_f\lambda(t)^* = M_{f_t}$.

Let $d_n = \operatorname{rank}(P_n)$; and let Tr be the usual trace on the finite rank operators. Choose any free ultrafilter $\mathcal{U}$ on $\mathbb{N}$, and define a state on $\ell^\infty(G)$ by
$$m(f) = \lim_{n\in\mathcal{U}} d_n^{-1}\operatorname{Tr}(P_n M_f P_n).$$
Then
$$|m(f_t) - m(f)| = \lim_{n\in\mathcal{U}} d_n^{-1}\big|\operatorname{Tr}(P_n\lambda(t)M_f\lambda(t)^* P_n) - \operatorname{Tr}(P_n M_f P_n)\big|$$
$$= \lim_{n\in\mathcal{U}} d_n^{-1}\big|\operatorname{Tr}(P_n\lambda(t)M_f\lambda(t)^* P_n) - \operatorname{Tr}(\lambda(t)P_n M_f P_n\lambda(t)^*)\big|$$
$$= \lim_{n\in\mathcal{U}} d_n^{-1}\big|\operatorname{Tr}\big([P_n\lambda(t) - \lambda(t)P_n]M_f\lambda(t)^* P_n\big)$$
$$+ \operatorname{Tr}\big(\lambda(t)P_n M_f[P_n\lambda(t) - \lambda(t)P_n]\big)\big|$$
$$\leq \lim_{n\in\mathcal{U}} 2\|P_n\lambda(t) - \lambda(t)P_n\|\,\|f\|_\infty = 0.$$
Thus we have constructed a left invariant mean on $\ell^\infty(G)$; and therefore G is amenable. ∎

Corollary VII.7.9 $C_r^*(\mathbb{F}_2)$ *is not quasidiagonal.*

VII.8 $C_r^*(\mathbb{F}_2)$ is projectionless

We have already seen an example of a simple unital projectionless C*-algebra in section IV.8. However, $C_r^*(\mathbb{F}_2)$ provides a "naturally occurring" example of this phenomenon. Nevertheless, the first known examples were constructed "artificially". This proof uses a dense subalgebra which is an analogue of the C^∞ functions in $C(X)$. Recall that every element of $\mathbb{F}_2$ can be uniquely written as a reduced word in $u^{\pm 1}$ and $v^{\pm 1}$.

To define this subalgebra, we first need another representation of $C_r^*(\mathbb{F}_2)$. Let $\mathcal{H}'$ be a Hilbert space spanned by orthonormal basis vectors $\xi_{\{x,y\}}$ where $\{x,y\}$ runs over all the *unordered* pairs $\mathcal{P}$ of elements of $\mathbb{F}_2$ such that $x^{-1}y$ has length 1 in reduced form (i.e. $x^{-1}y \in \{u^{\pm 1}, v^{\pm 1}\}$). Define a representation σ of $\mathbb{F}_2$ on $\mathcal{H}'$ by
$$\sigma(s)\xi_{\{x,y\}} = \xi_{\{sx,sy\}}.$$

VII.8. $C_r^*(\mathbb{F}_2)$ is projectionless

This is easily seen to be a unitary representation.

Lemma VII.8.1 *The representation σ is unitarily equivalent to $\lambda \oplus \lambda$. Hence σ extends to a $*$-representation of $C_r^*(\mathbb{F}_2)$.*

Proof. Let $\mathcal{H}_1' = \text{span}\{\xi_{\{x,y\}} : \{x,y\} \in \mathcal{P}_u\}$, where $\mathcal{P}_u$ consists of those pairs $\{x,y\}$ such that $x^{-1}y = u^{\pm 1}$. Then $\mathcal{H}_2' := \mathcal{H}_1'^{\perp} = \text{span}\{\xi_{\{x,y\}} : \{x,y\} \in \mathcal{P}_v\}$, where $\mathcal{P}_v = \mathcal{P} \setminus \mathcal{P}_u$ consists of those pairs $\{x,y\}$ such that $x^{-1}y = v^{\pm 1}$. It is clear that $\mathcal{H}_1'$ and $\mathcal{H}_2'$ are invariant for σ. Let $\sigma_i = \sigma|_{\mathcal{H}_i'}$ for $i = 1, 2$, so that $\sigma \simeq \sigma_1 \oplus \sigma_2$.

For each pair $\{x,y\}$ in $\mathcal{P}_u$, one of the words has the *reduced* form $su^{\pm 1}$ such that the other equals s in reduced form. Let W_1 be the unitary map in $\mathcal{B}(\ell^2(\mathbb{F}_2), \mathcal{H}_1')$ given by $W_1 \delta_t = \xi_{\{t, tu\}}$. It is routine to verify that this is indeed unitary. Moreover,

$$W_1 \lambda(s) \delta_t = W_1 \delta_{st} = \xi_{\{st, stu\}} = \sigma_1(s)\xi_{\{t,tu\}} = \sigma_1(s) W_1 \delta_t.$$

Thus $\sigma_1 = \text{Ad}\, W_1 \lambda$. Similarly, there is a unitary operator W_2 of $\ell^2(\mathbb{F}_2)$ onto $\mathcal{H}_2'$ such that $\sigma_2 = \text{Ad}\, W_2 \lambda$. Hence we may extend σ to the $*$-representation $\text{Ad}\, W_1 \oplus \text{Ad}\, W_2$ of $C_r^*(\mathbb{F}_2)$ into $\mathcal{B}(\mathcal{H}')$. ∎

The sneaky part of our argument is to use another map which almost intertwines λ and σ. For each element s in $\mathbb{F}_2 \setminus \{e\}$, define $\gamma(s)$ to be the element of $\mathbb{F}_2$ with the last term deleted from the reduced word representing s. Define a map S in $\mathcal{B}(\ell^2(\mathbb{F}_2), \mathcal{H}')$ by

$$S\delta_e = 0 \quad \text{and} \quad S\delta_s = \xi_{\{s, \gamma(s)\}} \quad \text{when } s \neq e.$$

It is easy to verify that S is a co-isometry of index 1.

Lemma VII.8.2 *With the definitions above, $\lambda(s) - S^* \sigma(s) S$ is finite rank for every s in $\mathbb{F}_2$.*

Proof. Note that $\gamma(st) = s\gamma(t)$ unless s completely cancels t, which occurs only if $s = a_n a_{n-1} \cdots a_1$ and $t = a_1^{-1} \cdots a_i^{-1}$ for $0 \leq i \leq n$; where a_i belong to $\{u^{\pm 1}, v^{\pm 1}\}$. So for $s = a_n a_{n-1} \cdots a_1$, compute

$$S^* \sigma(s) S \delta_t = \begin{cases} 0 & \text{if } t = e \\ S^* \xi_{\{st, s\gamma(t)\}} = \delta_{st} & \text{if } \gamma(st) = s\gamma(t) \\ S^* \xi_{\{a_n a_{n-1} \cdots a_{i+1}, a_n a_{n-1} \cdots a_i\}} = \delta_{a_n a_{n-1} \cdots a_i} & \text{if } t = a_1^{-1} \cdots a_i^{-1}, \\ & 1 \leq i \leq n. \end{cases}$$

Therefore

$$(\lambda(s) - S^* \sigma(s) S) \delta_t = \begin{cases} \delta_s & \text{if } t = e \\ 0 & \text{if } \gamma(st) = s\gamma(t) \\ \delta_{a_n a_{n-1} \cdots a_{i+1}} - \delta_{a_n a_{n-1} \cdots a_i} & \text{if } t = a_1^{-1} \cdots a_i^{-1}, \\ & 1 \leq i \leq n. \end{cases}$$

Consequently, $\lambda(s) - S^* \sigma(s) S$ is finite rank. ∎

Thus $A - S^*\sigma(A)S$ is finite rank for every A in $\lambda(\mathbb{C}\mathbb{F}_2)$. Define $\mathcal{A}_\infty$ to be the set of elements A in $C_r^*(\mathbb{F}_2)$ such that $A - S^*\sigma(A)S$ is in the trace class. Let Tr denote the trace on the space $\mathcal{C}_1$ of trace class operators. The following lemma outlines the main properties of $\mathcal{A}_\infty$.

Lemma VII.8.3 $\mathcal{A}_\infty$ *is a dense unital $*$-subalgebra of $C_r^*(\mathbb{F}_2)$ which is inverse closed. The function*

$$f_A(\alpha) = (\alpha I - A)^{-1} - S^*\sigma((\alpha I - A)^{-1})S$$

is continuous from the resolvent of A into $\mathcal{C}_1$. For A in $\mathcal{A}_\infty$,

$$\mathrm{Tr}(A - S^*\sigma(A)S) = \tau(A).$$

Proof. Clearly $\mathcal{A}_\infty$ is a unital self-adjoint subspace of $C_r^*(\mathbb{F}_2)$. It is an algebra because if A, B belong to $\mathcal{A}_\infty$, then

$$AB - S^*\sigma(AB)S = (A - S^*\sigma(A)S)B + S^*\sigma(A)S(B - S^*\sigma(B)S)$$

is trace class. It is dense because $\lambda(\mathbb{C}\mathbb{F}_2)$ is already dense. $\mathcal{A}_\infty$ is inverse closed because of the identity

$$A^{-1} - S^*\sigma(A^{-1})S = A^{-1}(S^*\sigma(A)S - A)S^*\sigma(A)^{-1}S + A^{-1}(I - S^*S)$$

since $I - S^*S = \delta_e\delta_e^*$ is rank one.

Replacing A by $\alpha I - A$ in this identity yields

$$(\alpha I - A)^{-1} - S^*\sigma(\alpha I - A)^{-1}S$$
$$= (\alpha I - A)^{-1}\left((S^*\sigma(\alpha I - A)S - (\alpha I - A))S^*\sigma(\alpha I - A)^{-1}S + (I - S^*S)\right)$$
$$= (\alpha I - A)^{-1}(A - S^*\sigma(A)S - \alpha\delta_e\delta_e^*)S^*\sigma(\alpha I - A)^{-1}S + (\alpha I - A)^{-1}\delta_e\delta_e^*$$

which is continuous in the trace norm.

From the formula computed in the previous lemma, one may deduce that

$$((\lambda(s) - S^*\sigma(s)S)\delta_t, \delta_t) = \begin{cases} 0 & t \neq e \\ \delta_{se} = \tau(\lambda(s)) & t = e \end{cases}.$$

This uses the observation that a_i^{-1} does not equal a_i or a_{i+1} because s is written in reduced form. By linearity and *norm* continuity,

$$((A - S^*\sigma(A)S)\delta_t, \delta_t) = \begin{cases} 0 & t \neq e \\ \tau(A) & t = e \end{cases}.$$

for all A in $C_r^*(\mathbb{F}_2)$. Thus if A is in $\mathcal{A}_\infty$, it makes sense to compute the trace in two ways

$$\mathrm{Tr}(A - S^*\sigma(A)S) = \sum_{t \in G}((A - S^*\sigma(A)S)\delta_t, \delta_t)$$
$$= ((A - S^*\sigma(A)S)\delta_e, \delta_e) = \tau(A). \blacksquare$$

Lemma VII.8.4 *If P is a projection in $C_r^*(\mathbb{F}_2)$ and $0 < \varepsilon < 1$, then there is a projection Q in $\mathcal{A}_\infty$ such that $\|P - Q\| < \varepsilon$.*

Proof. Pick $A = A^*$ in $\mathcal{A}_\infty$ such that $\|P - A\| < \varepsilon/2$. Then the spectrum of A is contained in $(-\varepsilon/2, \varepsilon/2) \cup (1 - \varepsilon/2, 1 + \varepsilon/2)$. Let $Q = E_A(1 - \varepsilon/2, 1 + \varepsilon/2)$. Then Q is a projection such that $\|P - Q\| \leq \|P - A\| + \|A - Q\| < \varepsilon$. From the Riesz functional calculus,

$$Q = \frac{1}{2\pi} \int_0^{2\pi} \left((1 + e^{it}/2)I - A\right)^{-1} dt.$$

Hence $Q - S^*\sigma(Q)S$ is given by the expression

$$\frac{1}{2\pi} \int_0^{2\pi} \left((1 + e^{it}/2)I - A\right)^{-1} - S^*\left((1 + e^{it}/2)I - A\right)^{-1} S \, dt,$$

which is the Riemann integral of a continuous $\mathcal{C}_1$ valued function, and therefore is also trace class. So Q belongs to $\mathcal{A}_\infty$. ∎

We have now reduced the problem to analyzing the projections in $\mathcal{A}_\infty$. We need a simple lemma from operator theory.

Lemma VII.8.5 *If P and Q are projections such that $P - Q$ is trace class, then $\mathrm{Tr}(P - Q)$ is an integer.*

Proof. Notice that

$$P(P - Q)^2 = P - PQP = (P - Q)^2 P.$$

Similarly, Q commutes with $(P - Q)^2$ which is a positive trace class operator. By the spectral theorem for compact operators, there are distinct positive eigenvalues λ_i and finite rank projections E_i such that $(P - Q)^2 = \sum_i \lambda_i E_i$. The kernel of $(P - Q)^2$ is $E_0 = (\sum_i E_i)^\perp$. Thus $PE_0 = QE_0 = R_0$ and

$$P = \sum_i PE_i + R_0 \quad \text{and} \quad Q = \sum_i QE_i + R_0.$$

Therefore there is a convergent sum

$$\mathrm{Tr}(P - Q) = \sum_i \mathrm{Tr}\big((P - Q)E_i\big).$$

But $(P - Q)E_i$ is the difference of two finite rank projections. Hence its trace is rank PE_i − rank QE_i which is an integer. As the sum converges, this difference equals 0 except finitely often; and $\mathrm{Tr}(P - Q)$ is an integer. ∎

Combining all these results, we obtain the main result of this section.

Theorem VII.8.6 $C_r^*(\mathbb{F}_2)$ *contains no proper projections.*

Proof. Suppose that there is a proper projection in $C_r^*(\mathbb{F}_2)$. By Lemma VII.8.4, there is a proper projection P in $\mathcal{A}_\infty$. Hence $Q = S^*\sigma(P)S$ is also a projection; and $P - Q$ is trace class. Therefore by Lemma VII.8.3, $\tau(P) = \text{Tr}(P - Q)$. By Lemma VII.8.5, it follows that $\tau(P)$ is an integer, necessarily 0 or 1. But by Proposition VII.7.1, τ is faithful. Thus if $\tau(P) = 0$, then $P = 0$; while if $\tau(P) = 1$, then $\tau(I - P) = 0$ and so $P = I$. Hence there are no proper projections in $C_r^*(\mathbb{F}_2)$. ∎

Exercises

VII.1 Verify that $L^1(G)$ has a norm one approximate identity. When G is metrizable, show that there is a sequential approximate identity.

VII.2 Use the spectral theorem to prove Stone's Theorem that every unitary representation of $\mathbb{R}$ is associated to a spectral measure E on $\mathbb{R}$ such that
$$\pi(t) = \int_\mathbb{R} e^{ist} E(ds).$$

VII.3 Show that the Schur product of two positive matrices is positive.
HINT: Write each positive matrix as a sum of positive rank one matrices.

VII.4 Identify the topology on the primitive ideal space of the discrete Heisenberg group.

VII.5 The continuous Heisenberg group H is the set of all matrices of the form $(x,y,z) = \begin{bmatrix} 1 & x & z \\ 0 & 1 & y \\ 0 & 0 & 1 \end{bmatrix}$ for $x,y,z \in \mathbb{R}$ with the usual topology on $\mathbb{R}^3$.
(a) Show that unitary operators on $L^2(\mathbb{R})$ determined for each real number s by $U_s(x,y,z)f(t) = e^{is(xt+z)}f(t-y)$ yields an irreducible representation of H.
HINT: The operators of multiplication by e^{ixt} generate a masa.
(b) Show that $C_r^*(H) = C^*(H)$.

VII.6 Show that any countable family $\{T_n\}_{n \geq 1}$ of operators on a Hilbert space has a non-zero separable reducing subspace.
HINT: Fix a vector $x \neq 0$ and consider $\overline{C^*(\{T_n\})x}$.

VII.7 Show that if D is a diagonal operator with distinct eigenvalues, then $\{D\}'$ is the diagonal algebra.

VII.8 Show that $\lambda(\mathbb{F}_2)' = \rho(\mathbb{F}_2)''$.
HINT: Show that T in $\{\lambda(\mathbb{F}_2), \rho(\mathbb{F}_2)\}'$ must be scalar.

VII.9 Verify that the operators W_1 and S defined in Section VII.8 are a unitary and a co-isometry of index 1 respectively.

VII.10 Let $\mathcal{H}_0$ be a subspace of $\mathcal{H}$ of infinite dimension and co-dimension. Let U be a unitary such that $U\mathcal{H}_0 = \mathcal{H}_0^\perp$ and $U^2 = I$; and let V be a unitary operator such that $\mathcal{H}_0^\perp = V\mathcal{H}_0 \oplus V^2\mathcal{H}_0$ and $V^3 = I$. Consider the C*-algebra $C^*(U, V)$.

(a) By mimicking the proof for $C_r^*(\mathbb{F}_2)$, show that $C^*(U, V)$ is simple, and has a unique trace.

(b) Let E be the projection onto $\mathcal{H}_0$. Show that $C^*(U, V, E)$ is isomorphic to $\mathcal{M}_2(\mathcal{O}_2)$, which is isomorphic to $\mathcal{O}_2$ by Exercise V.17.

HINT: Think of U as $\begin{bmatrix} 0 & I \\ I & 0 \end{bmatrix}$ and V as $\begin{bmatrix} 0 & S_2^* \\ S_1 & S_2 S_1^* \end{bmatrix}$ w.r.t. $\mathcal{H}_0 \oplus \mathcal{H}_0^\perp$.

(c) Show that $C^*(U, V)$ is not quasidiagonal.

HINT: Use the identities $I = E + UEU = E + VEV^* + V^*EV$. If P is a finite rank projection almost commuting with U and V, show that the trace of PEP should be close to both $\text{rank}(P)/2$ and $\text{rank}(P)/3$.

VII.11 Let $G = \mathbb{Z}_2 * \mathbb{Z}_3$ denote the free product of $\mathbb{Z}_2$ and $\mathbb{Z}_3$. This is the group generated by two elements u and v subject to the relations $u^2 = e = v^3$. Let K be the Cantor set consisting of all reduced infinite words in u and v. That is, $x = u^k v^{\pm 1} u v^{\pm 1} u v^{\pm 1} u \cdots$ for $k = 0, 1$. Then G acts on K by left multiplication. Pick any element x in K and define S to be the set $\{t \in G : tx \text{ begins with a } u\}$. Show that $G = S \,\dot\cup\, uS = S \,\dot\cup\, vS \,\dot\cup\, v^{-1}S$. Hence show that $C_r^*(G)$ is isomorphic to the C*-algebra of the previous exercise.

Notes and Remarks.

The classical source for general information of the C*-algebraic approach to group representations is Dixmier [1964], which also contains a good treatment of the primitive ideal space and spectrum of a C*-algebra. A proof that amenability of G is equivalent to $C^*(G) = C_r^*(G)$ may be found in Pedersen [1979] or Paterson[1988]. The proof for discrete groups given here was provided by Alan Paterson (private communication). The observation that the primitive ideal space classifies representations up to approximate unitary equivalence is due to Hadwin [1980]. The treatment of the crystal groups was shown to me by Keith Taylor (private communication). The Kadison–Kaplansky conjecture has been verified for groups of polynomial growth by Ji [1992]. The special representations of the free group were found by Choi [1980]. The simplicity of $C_r^*(\mathbb{F}_2)$ and its unique trace are due to Powers [1975], but the proof given here is taken from Choi [1979] where he actually discusses the group $\mathbb{Z}_2 * \mathbb{Z}_3$ (see the exercises). Cohen [1979] showed that $C^*(\mathbb{F}_2)$ is projectionless. Pimsner and Voiculescu [1982] showed that $C_r^*(\mathbb{F}_2)$ is projectionless by K-theoretic methods. The proof given here is due to Connes [1986], which was exposited by Cohen–Figa-Talamanca [1988] and Effros [1989]. That quasidiagonality implies amenability is due to Rosenberg [1987].

CHAPTER VIII

Discrete Crossed Products

Crossed product C*-algebras were introduced as a tool for making a systematic study of groups acting on C*-algebras as automorphisms. They provide a larger algebra which encodes the original C*-algebra and the group action. The consequence has also been to produce new classes of interesting algebras, and new ways of looking at old ones.

A **C*-dynamical system** $(\mathfrak{A}, G, \alpha)$ consists of a C*-algebra $\mathfrak{A}$ together with a homomorphism α of a locally compact group G into $\mathrm{Aut}(\mathfrak{A})$. We will denote by α_s the automorphism $\alpha(s)$ for s in G. Given a C*-dynamical system, a **covariant representation** is a pair (π, U) where π is a $*$-representation of $\mathfrak{A}$ on a Hilbert space $\mathcal{H}$ and $s \to U_s$ is a unitary representation of G on the same space such that

$$U_s \pi(A) U_s^* = \pi(\alpha_s(A)) \quad \text{for all} \quad A \in \mathfrak{A},\ s \in G.$$

In these notes, we will restrict our attention to crossed products by discrete groups. Because the Haar measure is then just counting measure, the notions simplify in this setting. We will outline the general situation later. So let G be a countable discrete group, and let $(\mathfrak{A}, G, \alpha)$ be a dynamical system. The space of continuous compactly supported $\mathfrak{A}$-valued functions on G is just the algebra $\mathfrak{A}G$ of all finite sums $f = \sum_{t \in G} A_t t$ with coefficients in $\mathfrak{A}$. Multiplication is determined by the formal rule $tAt^{-1} = \alpha_t(A)$. Whence if $g = \sum_{u \in G} B_u u$ is another finite sum, then

$$fg = \sum_{t \in G} \sum_{u \in G} A_t t B_u u \quad = \sum_{t \in G} \sum_{u \in G} A_t (t B_u t^{-1}) tu$$
$$= \sum_{t \in G} \sum_{u \in G} A_t \alpha_t(B_u) tu = \sum_{s \in G} \Big(\sum_{t \in G} A_t \alpha_t(B_{t^{-1}s}) \Big) s$$

which is just a twisted convolution product. The adjoint is determined by the rule $s^* = s^{-1}$, so that

$$(As)^* = s^* A^* = s^{-1} A^* s s^{-1} = \alpha_s^{-1}(A^*) s^{-1}.$$

Hence

$$f^* = \sum_{t \in G} \alpha_t(A_{t^{-1}}^*) t.$$

VIII. Discrete Crossed Products

Notice that a covariant representation (π, U) of $(\mathfrak{A}, G, \alpha)$ yields a $*$-representation of $\mathfrak{A}G$ by
$$\sigma(f) = \sum_{t \in G} \pi(A_t) U_t.$$
Indeed,
$$\sigma(f)^* = \sum_{t \in G} U_t^* \pi(A_t)^* = \sum_{t \in G} U_{t^{-1}} \pi(A_t^*) U_t U_{t^{-1}}$$
$$= \sum_{s \in G} \pi\big(\alpha_s(A_{s^{-1}}^*)\big) U_s = \sigma(f^*)$$
and
$$\sigma(f)\sigma(g) = \sum_{t \in G}\sum_{u \in G} \pi(A_t) U_t \pi(B_u) U_u$$
$$= \sum_{t \in G}\sum_{u \in G} \pi(A_t)\big(U_t \pi(B_u) U_t^*\big) U_t U_u$$
$$= \sum_{t \in G}\sum_{u \in G} \pi(A_t) \pi\big(\alpha_t(B_u)\big) U_{tu}$$
$$= \sum_{s \in G} \Big(\sum_{t \in G} \pi\big(A_t \alpha_t(B_{t^{-1}s})\big)\Big) U_s = \sigma(fg).$$

Conversely, when $\mathfrak{A}$ is unital, a $*$-representation of $\mathfrak{A}G$ yields a covariant representation of $(\mathfrak{A}, G, \alpha)$ simply by the restrictions
$$\pi(A) = \sigma(Ae) \quad \text{and} \quad U_s = \sigma(s).$$
For indeed,
$$U_s \pi(A) U_s^* = \sigma(s)\sigma(Ae)\sigma(s)^* = \sigma(sAs^{-1}) = \sigma(\alpha_s(A)e) = \pi(\alpha_s(A)).$$
When $\mathfrak{A}$ does not have a unit, let E_n be an approximate unit, and let
$$U_s = \lim_{n \to \infty} \sigma(E_n s).$$
The proof that this converges is left as an exercise.

The **crossed product** $\mathfrak{A} \times_\alpha G$ is the **enveloping C*-algebra** of $\mathfrak{A}G$. That is, one defines a C*-algebra norm by
$$\|f\| = \sup_\sigma \|\sigma(f)\|$$
as σ runs over all $*$-representations of $\mathfrak{A}G$. The supremum is always bounded by
$$\|f\|_1 = \sum_{t \in G} \|A_t\|.$$
This supremum is taken over a non-empty family of representations because certain representations can be explicitly constructed. Indeed, let π be any $*$-representation of $\mathfrak{A}$ on $\mathcal{H}$. Then we form the tensor product of this representation with the left

regular representation of G. To this end, form the Hilbert space $\ell^2(G, \mathcal{H})$ of all square summable functions x of G into $\mathcal{H}$ with the norm

$$\|x\|_2^2 = \sum_{t \in G} \|x(t)\|^2.$$

Define a covariant representation $(\tilde{\pi}, \Lambda)$ of $(\mathfrak{A}, G, \alpha)$ by

$$(\tilde{\pi}(A)x)(s) = \pi(\alpha_s^{-1}(A))(x(s))$$
$$(\Lambda_t x)(s) = x(t^{-1}s)$$

for all A in $\mathfrak{A}$, x in $\ell^2(G, \mathcal{H})$ and s, t in G. It is easily verified that $\tilde{\pi}$ is a $*$-representation of $\mathfrak{A}$ and Λ is a unitary representation of G. To verify the covariance condition, compute

$$(\Lambda_t \tilde{\pi}(A) \Lambda_t^* x)(s) = (\tilde{\pi}(A) \Lambda_t^* x)(t^{-1}s) = \pi(\alpha_{t^{-1}s}^{-1}(A))(\Lambda_t^* x(t^{-1}s))$$
$$= \pi(\alpha_s^{-1} \alpha_t(A))(x(s)) = (\tilde{\pi}(\alpha_t(A))x)(s)$$

In particular, since there are faithful representations of $\mathfrak{A}$ by the GNS construction, we obtain an isometric imbedding of $\mathfrak{A}$ into $\mathfrak{A} \times_\alpha G$ by sending A to Ae. And in the unital case, $\mathfrak{A} \times_\alpha G$ contains a unitary subgroup isomorphic to G. This crossed product has the universal property: if (π, U) is any covariant representation of $(\mathfrak{A}, G, \alpha)$, then there is representation of $\mathfrak{A} \times_\alpha G$ into C$^*(\pi(\mathfrak{A}), U(G))$ obtained by setting

$$\sigma(f) = \sum_{t \in G} \pi(A_t) U_t \quad \text{for} \quad f = \sum_{t \in G} A_t t \in \mathfrak{A} G,$$

and extending by continuity. In the unital case, this map is surjective.

For general discrete groups, the collection of representations constructed above using the left regular representation of G are not sufficient to determine the norm in $\mathfrak{A} \times_\alpha G$. This restricted class of representations determines the **reduced crossed product** $\mathfrak{A} \times_{\alpha r} G$. For example, if G is any discrete group, $\mathbb{C} \times G$ is just the group C*-algebra. (The action of G on $\mathbb{C}$ is the trivial one.) The reduced crossed product is just the reduced group C*-algebra. It is a theorem that when G is amenable, the reduced crossed product equals the full crossed product. The cases we will examine are crossed products by the integers, and usually we will be able to show that the crossed product is simple. Since there is a canonical homomorphism of the crossed product onto the reduced crossed product, equality will then follow.

Example VIII.1.1 The irrational rotation algebra is a good example of a crossed product. Recall that this is the universal C*-algebra $\mathcal{A}_\theta$ generated by two unitaries satisfying the relation $UV = e^{2\pi i \theta} VU$. In particular, $VUV^* = e^{-2\pi i \theta} U$. This implies that the spectrum of U is invariant under rotation R_θ through the irrational angle $2\pi\theta$; whence the spectrum is the whole unit circle $\mathbb{T}$. For any polynomial

VIII. Discrete Crossed Products

$p(z) = \sum_{k=-N}^{N} a_k z^k$, one has

$$Vp(U)V^* = \sum_{k=-N}^{N} a_k(VUV^*)^k = \sum_{k=-N}^{N} e^{-2\pi ik\theta} a_k U^k = \tau_\theta(p)(U)$$

where $\tau_\theta(f)(z) = f(e^{-2\pi i\theta} z) = f \circ R_\theta^{-1}(z)$ is the automorphism of $C(\mathbb{T})$ induced by the rotation homeomorphism R_θ.

We claim that $\mathcal{A}_\theta$ is the crossed product $C(\mathbb{T}) \times_{\tau_\theta} \mathbb{Z}$. Indeed, the crossed product is generated by the image Z of the coordinate function z and a unitary W implementing the automorphism τ_θ, whence

$$WZW^* = \tau_\theta(z)(Z) = e^{-2\pi i\theta} Z.$$

Therefore by the universal property of $\mathcal{A}_\theta$, there is a homomorphism of $\mathcal{A}_\theta$ onto $C(\mathbb{T}) \times_{\tau_\theta} \mathbb{Z}$ taking U onto Z and V onto W. Conversely, we saw above that $\mathcal{A}_\theta$ provides a covariant representation of $(C(\mathbb{T}), \mathbb{Z}, \tau_\theta)$. Therefore by the universal property of the crossed product, there is a homomorphism of $C(\mathbb{T}) \times_{\tau_\theta} \mathbb{Z}$ onto $\mathcal{A}_\theta$ taking Z to U and W to V. Clearly these maps are inverses.

Later in this chapter, we will use this connection to extract some additional information about $\mathcal{A}_\theta$.

Example VIII.1.2 Recall from Proposition V.4.2 that the Cuntz algebra $\mathcal{O}_n$ contains a UHF subalgebra $\mathfrak{F}^n$ of type n^∞. Consider the direct limit C*-algebra $\mathfrak{A}$ of the algebras $\mathfrak{A}_{-k} := \mathcal{M}_{n^k}(\mathfrak{F}_n)$ given by the multiplicity one imbeddings $\alpha_{-k}(A) = E_{11} \otimes A$ of $\mathfrak{A}_{-k}$ into $\mathfrak{A}_{-k-1}$. Then it is evident that the limit algebra is just $\mathfrak{K} \otimes \mathfrak{F}^n$, where $\mathfrak{K}$ is the algebra of compact operators. We will show how this can be extended to a doubly infinite sequence. Since $\mathfrak{F}^n$ is the direct limit of $\mathcal{M}_{n^k}$ with multiplicity n imbeddings, define $\mathfrak{A}_k$, $k \geq 1$, to be the compression $\mathfrak{A}_k$ of $\mathfrak{F}^n$ to the range of the rank one projection $E_{11}^{(k)}$ in $\mathcal{M}_{n^k}$. Evidently, these algebras are all isomorphic to $\mathfrak{F}^n$. Thus the direct limit sequence for $\mathfrak{A}$ extends to the doubly infinite sequence

$$\cdots \to \mathfrak{A}_2 \to \mathfrak{A}_1 \to \mathfrak{A}_0 \to \mathfrak{A}_{-1} \to \mathfrak{A}_{-2} \to \cdots$$

where all the imbeddings of $\mathfrak{A}_k$ into $\mathfrak{A}_{k-1}$ are given by the multiplicity one maps $\alpha_k(A) = E_{11} \otimes A$. Let $\alpha_{k,\infty}$ denote the maps from $\mathfrak{A}_k$ into $\mathfrak{A}$ determining the limit.

Define an automorphism σ of $\mathfrak{A}$ by "shifting to the left" in this sequence. In other words, since every $\mathfrak{A}_k$ is isomorphic to $\mathfrak{F}^n$ and all the maps α_k are equivalent, there is a sequence of isomorphisms σ_k from $\mathfrak{A}_k$ onto $\mathfrak{A}_{k+1}$ for k in $\mathbb{Z}$ such that $\alpha_{k+1}\sigma_k = \sigma_{k-1}\alpha_k$. Therefore there is a (unique) automorphism σ of $\mathfrak{A}$ such that

$\sigma\alpha_{k,\infty} = \alpha_{k+1,\infty}\sigma_k$ for all $k \in \mathbb{Z}$. This yields the commutative diagram

$$\begin{array}{ccccccccc}
\cdots & \longrightarrow & \mathfrak{A}_{k+1} & \xrightarrow{\alpha_{k+1}} & \mathfrak{A}_k & \xrightarrow{\alpha_k} & \mathfrak{A}_{k-1} & \xrightarrow{\alpha_{k-1}} & \cdots \longrightarrow \mathfrak{A} \\
& & \uparrow \sigma_k & & \uparrow \sigma_{k-1} & & \uparrow \sigma_{k-2} & & \uparrow \sigma \\
\cdots & \longrightarrow & \mathfrak{A}_k & \xrightarrow{\alpha_k} & \mathfrak{A}_{k-1} & \xrightarrow{\alpha_{k-1}} & \mathfrak{A}_{k-2} & \xrightarrow{\alpha_{k-2}} & \cdots \longrightarrow \mathfrak{A}
\end{array}$$

Consider the crossed product $\mathfrak{A} \times_\sigma \mathbb{Z}$. This algebra contains a unitary element U such that $UAU^* = \sigma(A)$ for all A in $\mathfrak{A}$. Let P_k denote the unit of the algebra $\mathfrak{A}_k$. Then $UP_kU^* = P_{k+1}$. Consider the algebra $\mathfrak{B} = P_0(\mathfrak{A} \times_\sigma \mathbb{Z})P_0$. This contains $\mathfrak{A}_0 = P_0\mathfrak{A}P_0$ and $V = UP_0$ (since $UP_0 = P_1U = P_0UP_0$). Notice that V is a proper isometry in $\mathfrak{B}$.

We claim that $\mathfrak{B}$ is generated by $\mathfrak{A}_0$ and V. First note that for $k \geq 0$, $P_0U^kP_0 = U^kP_0 = V^k$ and thus $P_0U^{-k}P_0 = P_0U^{-k} = V^{*k}$. A dense set of elements of $\mathfrak{A}$ are given by the finite sums of the form

$$X = \sum_{k=-N}^{N} A_kU^k = \sum_{k=0}^{N} A_kU^k + \sum_{k=1}^{N} U^{-k}A'_{-k}$$

where $A'_{-k} = U^kA_{-k}U^{*k}$. Hence the elements of the form

$$P_0XP_0 = \sum_{k=0}^{N} P_0A_kU^kP_0 + \sum_{k=1}^{N} P_0U^{-k}A'_{-k}P_0$$
$$= \sum_{k=0}^{N} P_0A_kP_0U^kP_0 + \sum_{k=1}^{N} P_0U^{-k}P_0A'_{-k}P_0$$
$$= \sum_{k=0}^{N} P_0A_kP_0V^k + \sum_{k=1}^{N} V^{*k}P_0A'_{-k}P_0.$$

provide a dense subset of $\mathfrak{B}$. So $\mathfrak{A}_0$ and V generate $\mathfrak{B}$.

Think of $\mathfrak{A}_0$ as $\mathcal{M}_n(\mathfrak{A}_1)$. Let E_{ij} be set of matrix units for $\mathcal{M}_n(\mathbb{C}P_1)$ be chosen so that $E_{11} = P_1$. Let $S_i = E_{i1}V$ for $1 \leq i \leq n$. Then S_i are isometries in $\mathfrak{B}$ such that $S_iS_i^* = E_{ii}$. Therefore $C^*(S_1,\ldots,S_n)$ is isomorphic to $\mathcal{O}_n$. We claim that $S_1,\ldots,S_n$ generate all of $\mathfrak{B}$. Since $V = S_1$, it suffices to show that $\mathfrak{A}_0$ is also generated by $S_1,\ldots,S_n$. Indeed, $\mathfrak{A}_0$ may be thought of as $n^k \times n^k$ matrices over $\mathfrak{A}_k$. Those matrices with scalar entries from $\mathfrak{A}_k$ form a copy $\mathfrak{M}_k$ of $\mathcal{M}_{n^k}$ in $\mathfrak{A}_0$. The union of these subalgebras is dense in the full UHF algebra $\mathfrak{A}_0$. Recall from Section V.4 that the elements of the form $S_\mu S_\nu^*$, for words μ, ν in $\mathcal{W}_k^n$ where $\mathcal{W}_k^n$ are the words of length k in the symbols $1,\ldots,n$, generate a copy $\mathfrak{W}_k$ of $\mathcal{M}_{n^k}$. To see that $\mathfrak{W}_k = \mathfrak{M}_k$, notice that the matrix units for $P_{k-1}\mathfrak{M}_kP_{k-1}$ are just

$$\sigma^{k-1}(E_{ij}) = U^{k-1}(S_iS_j^*)U^{*k-1} = (S_1^{k-1}S_i)(S_1^{k-1}S_j)^* \quad \text{for} \quad 1 \leq i,j \leq k.$$

VIII. Discrete Crossed Products

The rest of the matrix units are obtained in a similar fashion. Hence $\mathfrak{A}_0$ belongs to $C^*(S_1, \ldots, S_n)$; whence $\mathfrak{B}$ is isomorphic to $\mathcal{O}_n$.

Likewise $\mathfrak{B}_k := P_k \mathfrak{A} P_k$ is isomorphic to $\mathcal{O}_n$ for all $k \in \mathbb{Z}$. The imbedding of $\mathfrak{B}_k$ into $\mathfrak{B}_{k-1}$ is just the multiplicity one imbedding $\beta_k(B) = E_{11} \otimes B$. The full crossed product $\mathfrak{A}$ is the direct limit of the $\mathfrak{B}_k$; so the crossed product $(\mathfrak{K} \otimes \mathfrak{F}^n) \times_\sigma \mathbb{Z}$ is seen to be isomorphic to $\mathfrak{K} \otimes \mathcal{O}_n$. ∎

For the interested reader, we now outline how the crossed product is constructed in the general locally compact case. Form the algebra $L^1(G, \mathfrak{A})$ as follows. First consider all continuous, compactly supported functions f from G into $\mathfrak{A}$ and complete with respect to the norm

$$\|f\|_1 = \int_G \|f(s)\| \, d\mu_G(s).$$

Multiplication is defined by a convolution twisted by α:

$$f * g(s) = \int_G f(t)\alpha_t(g(t^{-1}s)) \, d\mu_G(t).$$

An adjoint operation is obtained by the formula

$$f^*(s) = \Delta(s)^{-1} \alpha_s(f(s^{-1})^*).,$$

where Δ is the modular function relating left and right Haar measure. It will be left to the interested reader to verify that this is a $*$-algebra.

The crossed product of $\mathfrak{A}$ by G is the enveloping C*-algebra $\mathfrak{A} \times_\alpha G$ of $L^1(G, \mathfrak{A})$. That is, define a C*-norm on $L^1(G, \mathfrak{A})$ by

$$\|f\| = \sup_\pi \|\pi(f)\|$$

where this supremum is taken over all the non-degenerate $*$-representations of $L^1(G, \mathfrak{A})$. The crossed product is the completion of $L^1(G, \mathfrak{A})$ in this norm.

As above, covariant representations always exist. Let π be any $*$-representation of $\mathfrak{A}$ on a Hilbert space $\mathcal{H}$. Form the Hilbert space $L^2(G, \mathcal{H})$ of all square integrable $\mathcal{H}$-valued functions on G. Then define a covariant representation $(\tilde{\pi}, \Lambda)$ of $(\mathfrak{A}, G, \mathcal{H})$ on $L^2(G, \mathcal{H})$ as follows:

$$(\tilde{\pi}(A)f)(s) = \pi(\alpha_s^{-1}(A))(f(s))$$
$$(\Lambda_t f)(s) = f(t^{-1}s)$$

for all A in $\mathfrak{A}$, f in $L^2(G, \mathcal{H})$ and s, t in G. It is easily verified that yields a covariant representation. The completion of $L^1(G, \mathfrak{A})$ with respect to this restricted class of representations is called the reduced crossed product.

Any $*$-representation of $\mathfrak{A} \times_\alpha G$ yields (by restriction) a representation of $L^1(G, \mathfrak{A})$; and by definition, every representation of $L^1(G, \mathfrak{A})$ is a representation of $\mathfrak{A} \times_\alpha G$. By an argument essentially identical to the group C*-algebra case, there is a bijective correspondence between the covariant representations of

($\mathfrak{A}, G, \alpha$) and $*$-representations of $L^1(G, \mathfrak{A})$. Thus the representation theory of $\mathfrak{A} \times_\alpha G$ encodes the covariant representation theory of ($\mathfrak{A}, G, \alpha$).

VIII.2 Crossed Products by Z

Consider a single automorphism α in $\text{Aut}(\mathfrak{A})$. This gives rise to an action of the integers by $\alpha_n := \alpha^n$. For convenience, assume that $\mathfrak{A}$ is unital so that there is a unitary u in the crossed product $\mathfrak{A} \times_\alpha \mathbb{Z}$ such that $uAu^* = \alpha(A)$ for every A in $\mathfrak{A}$. We shall consider $\mathfrak{A}$ as a subalgebra of $\mathfrak{A} \times_\alpha \mathbb{Z}$.

There is always a faithful expectation from $\mathfrak{A} \times_\alpha \mathbb{Z}$ onto $\mathfrak{A}$ analogous to computing the zero-th Fourier coefficient of a function on the circle. This will prove to be a very useful tool.

Theorem VIII.2.1 *Let* ($\mathfrak{A}, \mathbb{Z}, \alpha$) *be a C*-dynamical system. Then there is a canonical faithful expectation Φ of $\mathfrak{A} \times_\alpha \mathbb{Z}$ onto $\mathfrak{A}$.*

Proof. For every scalar λ of modulus 1, λu also determines a unitary representation of $\mathbb{Z}$ (by $n \to \lambda^n u^n$) such that

$$(\lambda u)^n A (\lambda u)^{-n} = u^n A u^{-n} = \alpha^n(A).$$

Thus $(\text{id}, \lambda u)$ is another covariant representation of $(\mathfrak{A}, \mathbb{Z}, \alpha)$. Clearly, $\mathfrak{A}$ and λu generate $\mathfrak{A} \times_\alpha \mathbb{Z}$. From the universal property of the crossed product, there is a homomorphism ρ_λ of $\mathfrak{A} \times_\alpha \mathbb{Z}$ onto itself such $\rho_\lambda(A) = A$ for all A in $\mathfrak{A}$ and $\rho_\lambda(u) = \lambda u$. Evidently ρ_λ is an automorphism.

It is easy to check that for X in $\mathfrak{A} \times_\alpha \mathbb{Z}$, the function $f(t) = \rho_{e^{2\pi i t}}(X)$ is norm continuous. Indeed, one may verify it on the dense subalgebra $\mathfrak{A}\mathbb{Z}$, and extend it to the closure by a simple approximation argument. Then define a map Φ on $\mathfrak{A} \times_\alpha \mathbb{Z}$ by

$$\Phi(X) = \int_0^1 \rho_{e^{2\pi i t}}(X)\, dt.$$

Notice that since each ρ_λ is a faithful (completely) positive isometric map, it follows that Φ is a faithful (completely) positive contraction. Next, consider the effect on AXB for A, B in $\mathfrak{A}$.

$$\Phi(AXB) = \int_0^1 \rho_{e^{2\pi i t}}(AXB)\, dt = A \int_0^1 \rho_{e^{2\pi i t}}(X)\, dt\, B = A\Phi(X) B.$$

So Φ is an $\mathfrak{A}$–bimodule map. In particular, $\Phi(A) = A$ for all A in $\mathfrak{A}$. Also,

$$\Phi(u^k) = \int_0^1 \rho_{e^{2\pi i t}}(u^k)\, dt = \int_0^1 e^{2\pi i k t} u^k\, dt = 0 \quad \text{for} \quad k \neq 0.$$

Hence on an element of $\mathfrak{A}\mathbb{Z}$ such as a finite sum $\sum_n A_n u^n$, it follows that

$$\Phi(\sum_n A_n u^n) = \sum_n A_n \Phi(u^n) = A_0.$$

This lies in $\mathfrak{A}$. By the density of $\mathfrak{A}\mathbb{Z}$ and the continuity of Φ, it follows that the range of Φ lies in $\mathfrak{A}$. Since Φ is the identity map on $\mathfrak{A}$, it follows that Φ is an expectation. ∎

Theorem VIII.2.1 permits the definition of a Fourier series for an element of $\mathfrak{A} \times_\alpha \mathbb{Z}$ by

$$\Phi_n(X) := \Phi(Xu^{-n}).$$

For every element $X = \sum_n A_n u^n$ in $\mathfrak{A}\mathbb{Z}$, it follows that $\Phi_n(X) = A_n$; and hence $X = \sum_n \Phi_n(X) u^n$. In fact, we obtain the following analogue of **Fejér**'s Theorem. Define the **Cesàro sums**

$$\Sigma_n(X) = \sum_{j=-n}^n (1 - \tfrac{|j|}{n+1}) \Phi_j(X) u^j.$$

Theorem VIII.2.2 *For every X in $\mathfrak{A} \times_\alpha \mathbb{Z}$, the Cesàro sums $\Sigma_n(X)$ converge in norm to X.*

Proof. The proof is the same as the scalar case. Indeed,

$$\Sigma_n(X) = \sum_{j=-n}^n (1 - \tfrac{|j|}{n+1}) \Phi(Xu^{-j}) u^j$$

$$= \int_0^1 \sum_{j=-n}^n (1 - \tfrac{|j|}{n+1}) e^{-2\pi i j t} \rho_{e^{2\pi it}}(X)\, dt = \int_0^1 \rho_{e^{2\pi it}}(X) \sigma_n(t)\, dt$$

where $\sigma_n(t) = \sum_{j=-n}^n (1 - \tfrac{|j|}{n+1}) e^{-2\pi i j t}$ is the Fejér kernel function. It is well known that $\sigma_n(t) dt$ are probability measures on $\mathbb{T} = \mathbb{R}/\mathbb{Z}$ converging weak-∗ to the point mass at 0. In particular, each Σ_n is a contractive (completely) positive map.

The proof can be completed in the standard way, but since we already know that $\mathfrak{A}\mathbb{Z}$ is dense, it follows more simply. A routine calculation shows that for Y in $\mathfrak{A}\mathbb{Z}$, $Y = \lim_{n\to\infty} \Sigma_n(Y)$. Thus for any X in $\mathfrak{A} \times_\alpha \mathbb{Z}$ and $\varepsilon > 0$, choose Y in $\mathfrak{A}\mathbb{Z}$ with $\|X - Y\| < \varepsilon$ and an integer N so that $\|Y - \Sigma_n(Y)\| < \varepsilon$ for $n \geq N$. Then for $n \geq N$,

$$\|X - \Sigma_n(X)\| \leq \|X - Y\| + \|Y - \Sigma_n(Y)\| + \|\Sigma_n(Y - X)\| < 3\varepsilon. \quad \blacksquare$$

VIII.3 Minimal Dynamical Systems

A **classical dynamical system** consists of a compact Hausdorff space X together with a homeomorphism σ of X onto itself. This determines a C*-dynamical system $(C(X), \mathbb{Z}, \sigma)$ where

$$\sigma_n(f) := f \circ \sigma^{-n}.$$

In this section, we will develop some properties of the crossed product $C(X) \times_\sigma \mathbb{Z}$.

We start with a basic result from ergodic theory. A finite Borel measure μ on X is **translation invariant** if $\mu(\sigma^{-1}(E)) = \mu(E)$ for every Borel subset of X. The existence of such measures follows from an application of the Markov–Kakutani fixed point theorem.

Theorem VIII.3.1 *Let (X, σ) be a classical dynamical system. Then there is a Borel probability measure on X which is translation invariant for σ.*

Proof. Let C_σ denote the automorphism $C_\sigma f = f \circ \sigma^{-1}$ of $C(X)$. Clearly this is a norm one linear operator. Thus the adjoint C_σ^* is a norm one linear map on the dual space $M(X)$ of all finite regular Borel measures on X. Moreover, since C_σ is positive, $\mu \geq 0$ implies that $C_\sigma^* \mu \geq 0$. Finally, if μ is a positive measure, then

$$\|C_\sigma^* \mu\| = C_\sigma^* \mu(X) = \mu(\sigma^{-1}(X)) = \mu(X) = \|\mu\|.$$

Combining these observations shows that C_σ^* maps the weak-∗ compact convex set

$$\mathcal{P} = \{\mu \in M(X) : \mu(X) = \|\mu\| = 1\}$$

of all probability measures into itself. Thus by the Markov–Kakutani fixed point Theorem VII.2.1, C_σ^* has a fixed point μ_0. In other words,

$$\mu_0(\sigma^{-1}(E)) = (C_\sigma^* \mu_0)(E) = \mu_0(E).$$

So μ_0 is a translation invariant measure. ∎

Certain of these invariant measures have an additional minimality property. The system (X, σ, μ) is **ergodic** if μ is translation invariant, and whenever E is a translation invariant measurable set, then $\mu(E) = 0$ or 1.

Proposition VIII.3.2 *Every dynamical system (X, σ) has an ergodic measure.*

Proof. By Theorem VIII.3.1, there are invariant probability measures for (X, σ). It is easy to check that the set $\mathcal{T}$ of all invariant probability measures is convex and weak-∗ closed. So by the Krein–Milman Theorem, $\mathcal{T}$ is the closed convex hull of its extreme points. We will show that every extreme point of this set is ergodic. This will establish the proposition.

If μ in $\mathcal{T}$ is not ergodic, there is a translation invariant set E such that

$$0 < \mu(E) < 1.$$

Evidently E^c is also essentially translation invariant. Let

$$\mu_1(A) = \mu(E)^{-1}(A \cap E) \quad \text{and} \quad \mu_2(A) = \mu(E^c)^{-1}(A \cap E^c).$$

Then both μ_i belong to $\mathcal{T}$; and they are distinct as they have disjoint supports. Since

$$\mu = \mu(E)\mu_1 + (1 - \mu(E))\mu_2,$$

it follows that μ is not extreme. ∎

VIII.3. Minimal Dynamical Systems

These measures allow us to construct a useful covariant representation of the dynamical system (X, σ). Let μ be a translation invariant measure, and form the Hilbert space $L^2(\mu)$. Represent $C(X)$ on $L^2(\mu)$ as multiplication operators M_f^μ, and define an operator U_μ by $U_\mu h = h \circ \sigma^{-1}$. Then the translation invariance of μ implies that U_μ is isometric: for g, h in the dense subspace $C(X)$,

$$(U_\mu g, U_\mu h) = \int C_\sigma(g\overline{h})\, d\mu = \int g\overline{h}\, dC_\sigma^*\mu = \int g\overline{h}\, d\mu = (g, h).$$

Since U_μ is clearly invertible, it is a unitary operator. Finally,

$$U_\mu M_f^\mu U_\mu^* h(\xi) = (M_f^\mu U_\mu^* h)(\sigma^{-1}\xi)$$
$$= f(\sigma^{-1}\xi)(U_\mu^* h)(\sigma^{-1}\xi) = M_{f\circ\sigma^{-1}}^\mu h(\xi)$$

for f in $C(X)$, h in $L^2(\mu)$ and ξ in X. Thus (M^μ, U_μ) is a covariant representation of $(C(X), \sigma)$. Let $\mathcal{A}_\sigma^\mu$ denote the algebra generated by this covariant representation.

A dynamical system (X, σ) is **minimal** if X has no proper closed σ-invariant subsets. This notion corresponds to simplicity of the crossed product. In this case, we will obtain a lot of additional structure. We begin with the easy direction.

Proposition VIII.3.3 *If F is a proper closed invariant subset of (X, σ), then the ideal $\mathfrak{C}_F = \{f \in C(X) : f|_F = 0\}$ generates a proper ideal $\mathfrak{J}_F$ of $C(X) \times_\sigma \mathbb{Z}$.*

Proof. Let $\mathfrak{J}_F^0$ be the set of elements in $C(X)\mathbb{Z}$ consisting of all finite sums of the form $\sum_n J_n u^n$ for J_n in $\mathfrak{C}_F$. This is an ideal in $C(X)\mathbb{Z}$ because of the invariance of F. Hence its closure $\mathfrak{J}_F$ is the ideal of $C(X) \times_\sigma \mathbb{Z}$ generated by $\mathfrak{C}_F$. Let Φ be the expectation of $C(X) \times_\sigma \mathbb{Z}$ onto $C(X)$ of Theorem VIII.2.1. Since $\Phi(\mathfrak{J}_F^0) = \mathfrak{C}_F$, continuity shows that $\Phi(\mathfrak{J}_F) = \mathfrak{C}_F$ as well. Thus $\mathfrak{J}_F$ is a proper ideal. ∎

We need another basic result from ergodic theory known as **Rohlin's Lemma**.

Lemma VIII.3.4 *Let (X, σ, μ) be an ergodic dynamical system such that the support of μ is not finite. Then for every $\varepsilon > 0$ and positive integer n, there is a Borel subset F of X such that $\sigma^i(F)$ are disjoint for $0 \leq i \leq n - 1$ and*

$$\mu\Big(\bigcup_{i=0}^{n-1} \sigma^i(F)\Big) > 1 - \varepsilon.$$

Proof. Clearly the set

$$X_n = \{x \in X : \sigma^n(x) = x \neq \sigma^j(x) \text{ for } 1 \leq j < n\}$$

is translation invariant for each $n \geq 1$. By ergodicity, each has measure 0 or 1. If $\mu(X_n) = 1$ and cannot be divided into invariant subsets of proper measure, then it is easy to see that X_n must consist of precisely n point masses of μ contrary to hypothesis. Thus we may assume that off a set of measure zero (which we discard), $\sigma^n(x) \neq x$ for all $n \geq 1$.

Choose an integer $k > \varepsilon^{-1}$. We will construct a measurable subset E of X of positive measure such that $\sigma^j(E)$ are pairwise disjoint for $0 \leq j < kn$. This is done by induction. Suppose that E is a set of positive measure such that $\sigma^j(E)$ are pairwise disjoint for $0 \leq j < m$. Ergodicity and the previous paragraph imply that there is a non-null subset E' of E such that E' and $\sigma^m(E')$ are disjoint. Thus $\sigma^j(E')$ are disjoint for $0 \leq j \leq m$. Eventually we obtain $m = kn$.

Now apply Zorn's Lemma to obtain a maximal set (up to a null set) with this property. If E_λ is an increasing chain of measurable subsets with this property, it is easy to extract a sequence E_{λ_n} such that $\sup_{n \geq 1} \mu(E_{\lambda_n}) = \sup_\lambda \mu(E_\lambda)$. Then $E = \cup_{n \geq 1} E_{\lambda_n}$ contains all the E_λ modulo null sets. Therefore Zorn's Lemma provides a maximal set E such that $\sigma^j(E)$ are pairwise disjoint for $0 \leq j < kn$.

Let $G = \sigma^{kn-1}(E)$. For each x in G, let $N(x)$ denote the smallest positive integer such that $\sigma^{N(x)}x$ belongs to E and set it equal to ∞ if it never returns to E. Notice that $\sigma^i(x)$ does not belong to $\cup_{j=0}^{kn-1} \sigma^j(E)$ for $i < N(x)$. For if $\sigma^i(x)$ belongs to $\sigma^j(E)$ and $i > j$, then $\sigma^{i-j}(x)$ lies in E; and if $i \leq j$, then x lies in $\sigma^{j-i}(E)$, neither of which is possible. Define

$$G_j = \{x \in G : N(x) = j\} \text{ for } 1 \leq j \leq kn$$

and

$$G' = \{x \in G : N(x) > kn\}.$$

The set G' is a null set; for otherwise $E' = E \cup \sigma(G')$ determines a larger set with $\sigma^j(E')$ pairwise disjoint for $0 \leq j < kn$. The disjoint union

$$\bigcup_{j=0}^{kn-1} \sigma^j(E) \cup \bigcup_{j=1}^{kn} \bigcup_{i=1}^{j-1} \sigma^i(G_j)$$

is invariant for σ, and thus has full measure by ergodicity.

The idea is to form a set F from every nth translate of E and each G_j. Define

$$F = \bigcup_{s=0}^{k-1} \sigma^{sn}(E) \cup \bigcup_{j=n+1}^{kn} \bigcup_{sn<j} \sigma^{(s-1)n+1} G_j.$$

It follows by construction that $\sigma^i(F)$ are disjoint for $0 \leq i < n$. Moreover,

$$X \setminus \bigcup_{i=0}^{n-1} \sigma^i(F) = \bigcup_{s=1}^{k} \bigcup_{1 \leq i \leq j \leq n} \sigma^{(s-1)n+i}(G_{(s-1)n+j}).$$

This has measure

$$\sum_{s=1}^{k} \sum_{1 \leq i \leq j \leq n} \mu(G_{(s-1)n+j}) \leq \sum_{s=1}^{k} \sum_{1 \leq j \leq n} n\mu(G_{(s-1)n+j}) = n\mu(G) = n\mu(E).$$

The sets $\sigma^j(E)$ for $0 \leq j < kn$ are disjoint and all have the same measure. Thus $n\mu(E) < k^{-1} < \varepsilon$ as desired. ∎

VIII.3. Minimal Dynamical Systems

Now we are in a position to show that the algebra $\mathcal{A}_\sigma^\mu$ constructed above is in fact isomorphic to the crossed product when σ is ergodic (and X is infinite). The key is showing that the expectation Φ onto C(X) factors through $\mathcal{A}_\sigma^\mu$.

Lemma VIII.3.5 *Let (X, σ, μ) be an ergodic dynamical system such that the support of μ is not finite. On the algebra $\mathcal{A}_\sigma^\mu$, the map*

$$\Psi\Big(\sum_{j=-n}^{n} M_{f_j}^\mu (U_\mu)^j\Big) = M_{f_0}^\mu$$

is contractive, and thus extends to an expectation of $\mathcal{A}_\sigma^\mu$ onto C(X).

Proof. Fix an element $A = \sum_{j=-n}^{n} M_{f_j}^\mu U_\mu^j$ and an $\varepsilon > 0$. Let

$$E = \{x \in X : |f_0| > \|f_0\|_\infty - \varepsilon\}.$$

Apply Rohlin's Lemma VIII.3.4 (for $2n+1$) to obtain pairwise disjoint non-null sets F_j for $-n \leq j \leq n$ such that $\sigma(F_j) = F_{j+1}$ for $-n \leq j < n$. By the ergodicity of σ, there is an integer m such that $\sigma^m(F_0) \cap E$ has positive measure. By replacing each F_j by $\sigma^m(F_j)$, we may assume that $m = 0$.

Let g be any unit vector in $L^2(\mu)$ supported on $E \cap F_0$. Then $M_{f_j}^\mu U_\mu^j g$ is supported on F_j for all $|j| \leq n$. In particular, they are pairwise orthogonal. Thus

$$\|A\|^2 \geq \|Ag\|^2 = \sum_{j=-n}^{n} \|f_j U_\mu^j g\|^2$$

$$\geq \|f_0 g\|^2 > (\|f_0\|_\infty - \varepsilon)^2 = (\|M_{f_0}^\mu\| - \varepsilon)^2.$$

Therefore $\|\Psi\| = 1$. It is now evident that the map Ψ extends by continuity to an expectation of $\mathcal{A}_\sigma^\mu$ onto C(X). ∎

Corollary VIII.3.6 *When (X, σ, μ) is ergodic and X is infinite, the algebra $\mathcal{A}_\sigma^\mu$ is isomorphic to the crossed product $C(X) \times_\sigma \mathbb{Z}$.*

Proof. There is a canonical homomorphism φ of $C(X) \times_\sigma \mathbb{Z}$ onto $\mathcal{A}_\sigma^\mu$ which carries each element $\sum_{j=-n}^{n} f_j u^j$ in $C(X)\mathbb{Z}$ to $\sum_{j=-n}^{n} M_{f_j}^\mu U_\mu^j$. It is evident that $\Psi\varphi$ agrees with Φ on this dense subalgebra, whence $\Phi = \Psi\varphi$. However, Φ is faithful and therefore φ is faithful and so must be an isomorphism. ∎

Now we will use Rohlin's Lemma again to establish uniqueness of the expectation and simplicity when σ is minimal.

Lemma VIII.3.7 *If (X, σ) is a minimal dynamical system on an infinite compact Hausdorff space X, then for each element A in $C(X) \times_\sigma \mathbb{Z}$ and $\varepsilon > 0$, there are unimodular functions $\theta_1, \ldots, \theta_m$ in $C(X)$ such that*

$$\Big\|\Phi(A) - \frac{1}{m}\sum_{s=1}^{m} \theta_s A \overline{\theta}_s\Big\| < \varepsilon.$$

Proof. Let μ be an ergodic measure for (X, σ). First we show that if $\varepsilon > 0$ and $k \neq 0$, there is an open subset F of X with $\mu(F) > 1 - \varepsilon$ and a unimodular function ψ in $C(X)$ such that $\text{Re } \psi\overline{(\psi \circ \sigma^{-k})} = 0$ on F. Clearly we may suppose that $k > 0$. To this end, let $N > 2k\varepsilon^{-1}$ and apply Rohlin's Lemma VIII.3.4 to obtain pairwise disjoint sets F_j for $0 \leq j \leq N$ such that $\sigma(F_j) = F_{j+1}$ for $0 \leq j \leq N$ and $\mu\big(\cup_{j=0}^N F_j\big) > 1 - \varepsilon/2$. Since μ is regular, we may replace each F_j by a closed subset with the same properties. Then using the continuity of σ and the disjointness of the F_j, we may find an open neighbourhood of the F_j's with these same properties and disjoint closures. Let these open sets be renamed F_j for convenience.

Set $F = \cup_{j=0}^{N-k} F_j$. Since $\mu(F_j) < N^{-1} < \varepsilon/2k$, it follows that

$$\mu(F) > 1 - \varepsilon/2 - k\varepsilon/2k = 1 - \varepsilon.$$

Let $\lambda = e^{\pi i/2k}$. Define ψ to equal λ^j on each F_j for $0 \leq j \leq N$, and extend it to a unimodular function in $C(X)$. Then $\psi\overline{(\psi \circ \sigma^{-k})} = -i$ on each F_j for $0 \leq j \leq N - k$ which is all of F. Notice that

$$\psi u^k \psi^{-1} + \psi^{-1} u^k \psi = 2\big(\text{Re } \psi\overline{(\psi \circ \sigma^k)}\big) u^k$$

vanishes on F.

Consider an element $A = \sum_{j=-n}^n f_j u^j$ in $C(X)\mathbb{Z}$ such that $\Phi(A) = f_0 = 0$. For each $1 \leq |k| \leq n$, let ψ_k be constructed as above so that $\text{Re } \psi_k\overline{(\psi_k \circ \sigma^{-k})} = 0$ on an open set F_k such that $\mu(F_k) > 1 - (2n)^{-1}$. This implies that the open set $F = \cap_{1 \leq |k| \leq n} F_k$ has positive measure, and thus is non-empty. Let

$$\theta_s = \prod_{1 \leq |k| \leq n} \psi_k^{s_k} \quad \text{for} \quad s \in \prod_{1 \leq |k| \leq n} \{1, -1\}.$$

For each $|k| \leq n$, consider s and s' which agree except on the kth coordinate, which differ; so that there is a function φ such that $\theta_s = \varphi\psi_k$ and $\theta_{s'} = \varphi\psi_k^{-1}$. Then

$$\theta_s f_k u^k \overline{\theta_s} + \theta_{s'} f_k u^k \overline{\theta_{s'}} = 2\varphi f_k \left(\text{Re } \psi_k\overline{(\psi_k \circ \sigma^{-k})}\right) u^k \overline{\varphi}$$

which vanishes on F_k. Thus summing over all values of s yields the same conclusion. Applying this to each k in turn shows that $2^{-2n} \sum_s \theta_s A \overline{\theta_s}$ vanishes on F.

Now the sets $\sigma^j(F)$ form an open cover of X by the minimality of σ, and hence there is an integer M so that the first M of these sets cover X. Let $\theta_s^{(j)} := \theta_s \circ \sigma^{-j}$ for $1 \leq j \leq M$. Consider the family of functions

$$\theta_{s_1,\ldots,s_M} = \prod_{j=1}^M \theta_{s_j}^{(j)} \quad \text{for} \quad s_j \in \prod_{1 \leq |k| \leq n} \{-1, 1\}, \, 1 \leq j \leq M.$$

VIII.3. Minimal Dynamical Systems

Then the calculation of the previous paragraph now shows that

$$2^{-2nM} \sum_{s=(s_1,\ldots,s_M)} \theta_s A \bar{\theta}_s$$

vanishes on $\cup_{j=1}^M \sigma^j F = X$.

Finally, $C(X)\mathbb{Z}$ is dense in $C(X) \times_\sigma \mathbb{Z}$ and both Φ and the averaging operator are contractive and unital. Therefore, a simple approximation argument completes the proof. ∎

Corollary VIII.3.8 *When (X, σ) is minimal and X is infinite, the crossed product $C(X) \times_\sigma \mathbb{Z}$ has a faithful trace which is obtained from the expectation Φ onto $C(X)$ followed by integration with respect to any ergodic measure μ. If (X, σ) has a unique invariant measure, then the trace is unique.*

Proof. It is easy to verify that $\tau(A) := \int_X \Phi(A)\, d\mu$ is a trace on $C(X) \times_\sigma \mathbb{Z}$. Indeed, if $A = \sum_m f_m u^m$ and $B = \sum_n g_n u^n$ are elements of $C(X)\mathbb{Z}$, then

$$\tau(AB) = \sum_n \int_X f_n g_{-n}\, d\mu = \tau(BA).$$

By continuity, the trace condition extends to $C(X) \times_\sigma \mathbb{Z}$.

We show that any trace τ' must factor through Φ. By the previous lemma, for any $\varepsilon > 0$ and A in $C(X) \times_\sigma \mathbb{Z}$, we obtain unimodular functions θ_s in $C(X)$ such that

$$\varepsilon > \left|\tau'(\Phi(A)) - \tau'\left(m^{-1}\sum_{s=1}^m \theta_s A \bar{\theta}_s\right)\right|$$
$$= \left|\tau'(\Phi(A)) - m^{-1}\sum_{s=1}^m \tau'(A)\right| = \left|\tau'(\Phi(A)) - \tau'(A)\right|.$$

Hence τ' factors as $\tau' = \varphi'\Phi$, where φ' is the restriction of τ' to $C(X)$. By the Riesz Representation Theorem, there is a unique probability measure μ' on X such that $\tau'(f) = \int_X f\, d\mu'$. Moreover this measure is translation invariant because $\tau'(f) = \tau'(UfU^*) = \tau'(\sigma(f))$.

Conversely, any invariant probability measure yields a trace. Thus the uniqueness of this measure implies the uniqueness of the trace. ∎

Theorem VIII.3.9 *Let (X, σ) be a dynamical system on an infinite compact Hausdorff space X. Then the crossed product $C(X) \times_\sigma \mathbb{Z}$ is simple if and only if σ is minimal.*

Proof. Suppose that σ is minimal and $\mathfrak{J}$ is a non-zero ideal of $C(X) \times_\sigma \mathbb{Z}$. Let A be a non-zero positive element of $\mathfrak{J}$. By Lemma VIII.3.7, it follows that $\Phi(A) = f_0$

belongs to $\mathfrak{J}$. Since Φ is faithful, f_0 is non-zero. Hence the ideal contains

$$f_n := \sum_{j=0}^{n} u^j(f_0) u^{j*} = \sum_{j=0}^{n} \sigma^j(f_0).$$

Now $f_0 > 0$ on a non-empty open set $\mathcal{O}$. As in the proof of Lemma VIII.3.7, there is an integer n sufficiently large that the sets $\sigma^j(\mathcal{O})$ for $0 \leq j \leq n$ cover X. Therefore f_n is strictly positive, and thus invertible. So the ideal $\mathfrak{J}$ is all of $C(X) \times_\sigma \mathbb{Z}$.

The converse is contained in Proposition VIII.3.3. ∎

We remark that if X is finite, then a minimal action of $\mathbb{Z}$ on X is a cyclic permutation. In this case, the crossed product $C(X) \times_\sigma \mathbb{Z}$ is isomorphic to $\mathcal{M}_n(C(\mathbb{T}))$, where $n = |X|$. This is not simple.

VIII.4 Odometers

In this section, we will consider a well known homeomorphism on the Cantor set studied in ergodic theory. Perhaps surprisingly, the crossed product algebra turns out to be one we have already studied.

Let $\{n_i\}$ be a sequence of integers $n_i \geq 2$. Let $X_i = \{0, 1, \ldots, n_i - 1\}$, and let μ_i be the measure on X_i assigning mass n_i^{-1} to each point. Form the Cantor set $X = \prod_{i \geq 1} X_i$; and let μ be the product measure of the μ_i. Think of each element $\mathbf{a} = \{a_i\}$ as a formal sum $\sum_{i \geq 1} a_i N_i$ where $N_1 = 1$ and $N_{i+1} = n_i N_i$. A group operation may be defined on $\bar{X}$ by formal addition with carries. That is, the sum $\{a_i\} + \{b_i\} = \{c_i\}$ where c_i are the unique integers in X_i such that

$$\sum_{i=1}^{n}(a_i + b_i) N_i \equiv \sum_{i=1}^{n} c_i N_i \mod N_{i+1} \quad \text{for all} \quad i \geq 1.$$

In particular, $\mathbf{b} = -\mathbf{a}$ is obtained as follows. Let i_0 be the least integer such that $a_{i_0} \neq 0$ (or ∞ when all $a_i = 0$). Then $b_i = 0$ for $i < i_0$, $b_{i_0} = n_{i_0} - a_{i_0}$ and $b_i = n_i - 1 - a_i$ for all $i > i_0$. It is readily verified that the group operations are continuous. Also $\mathbb{Z}$ imbeds as a dense subgroup by sending each n in $\mathbb{N}$ to the corresponding finite sum of the same total. Then the negative integers are sent to sequences such that $a_i = n_i - 1$ for all sufficiently large i.

The **odometer action** σ on X is obtained by addition of 1. The name comes from the parallel with the way a car odometer works. When the ith digit rolls over from $n_i - 1$ to 0, the $i+1$st digit in incremented by 1. Since addition is continuous, this is a homeomorphism. Consider the cylinder sets

$$J(x_1, \ldots, x_k) = \{\mathbf{a} : a_i = x_i \text{ for } 1 \leq i \leq k\}.$$

VIII.4. Odometers

Then it is easy to see that $\sigma(J(x_1, \ldots, x_k)) = J(y_1, \ldots, y_k)$ where

$$1 + \sum_{i=1}^{k} x_i N_i \equiv \sum_{i=1}^{k} y_i N_i \mod N_{k+1}.$$

Therefore any invariant measure for σ must agree with μ on every cylinder set. As these cylinders generate the Borel sets as a σ-algebra, it follows that μ is the unique invariant measure for σ. Hence it is also the Haar measure for the group X.

Form the crossed product $\mathfrak{A} = C(X) \times_\sigma \mathbb{Z}$. We wish to determine the structure of this algebra. Let u represent the universal unitary element implementing σ. The action is minimal because the semi-orbit of any point $\mathbf{a}$ is $\mathbf{a} + \mathbb{N}$ which is dense in X. Therefore by Theorem VIII.3.9 and Corollary VIII.3.8, this is a simple C*-algebra with a unique trace.

Theorem VIII.4.1 *Let $X = \prod_{i \geq 1} X_i$ where each X_i has $n_i \geq 2$ points, and let σ be the odometer action. Then the crossed product $C(X) \times_\sigma \mathbb{Z}$ is isomorphic to the Bunce–Deddens algebra corresponding to the supernatural number $\mathbf{n} = \prod_{i \geq 1} n_i$.*

Proof. The proof is accomplished by identifying a nested sequence of subalgebras isomorphic to $M_{N_k}(C(\mathbb{T}))$ in the crossed product $\mathfrak{A}$, and verifying that the imbeddings are the right ones. Map $C(\mathbb{T})$ into $\mathfrak{A}$ by the functional calculus of the unitary u inducing σ. That is, $\rho_1(f) = f(u)$. Then for each k, let $\mathfrak{A}_k$ be the subalgebra generated by the unitary u and the characteristic functions $\chi_{(x_1, \ldots, x_k)}$ of the cylinder sets $J(x_1, \ldots, x_k)$. Since $\sigma(J(x_1, \ldots, x_k)) = J(y_1, \ldots, y_k)$ with notation as above, it follows that

$$u \chi_{(x_1, \ldots, x_k)} = \chi_{(y_1, \ldots, y_k)} u.$$

Thus $E_{j0}^{(k)} := u^j \chi_{(0,\ldots,0)}$ are partial isometries with initial projection $\chi_{(0,\ldots,0)}$ and range projection $\chi_{(x_1, \ldots, x_k)}$ where $j = \sum_{i=1}^k x_i N_i$. They generate a set of matrix units for $\mathcal{M}_{N_k}$ given by $E_{ij}^{(k)} = E_{i0}^{(k)} E_{j0}^{(k)*}$. With respect to this decomposition, the span of the cylinder sets of size n is the set of diagonal scalar matrices. And the unitary u is given by the matrix U_k with copies of the identity in the $(j+1, j)$ entries for $1 \leq j < N_k$, in the $(0, N_k - 1)$ entry there is a coefficient V_k which we must analyze; all other entries are 0.

To understand V_k, note that it represents the matrix entry given by

$$V_k E_{0, N_k - 1}^{(k)} = \chi_{(0,\ldots,0)} u \chi_{(n_1 - 1, \ldots, n_k - 1)}.$$

Thus

$$V_k = E_{00}^{(k)} u E_{N_k - 1, 0}^{(k)} = \chi_{(0,\ldots,0)} u^{N_k} \chi_{(0,\ldots,0)}.$$

This is a unitary element in the algebra $\chi_{(0,\ldots,0)} \mathfrak{A} \chi_{(0,\ldots,0)}$. Hence there is a homomorphism ρ_k which carries $\mathcal{M}_{n_k}(C(\mathbb{T}))$ onto $\mathfrak{A}_k = C^*(\{E_{ij}^{(k)}\}, V_k)$ by

$$\rho_k([f_{ij}]) = [f_{ij}(V_k)].$$

Let α_k denote the injection of $\mathfrak{A}_k$ into $\mathfrak{A}_{k+1}$. Likewise recall the injections β_k of $\mathfrak{B}(N_k) = \mathcal{M}_{N_k}(C(\mathbb{T}))$ into $\mathfrak{B}_{N_{k+1}}$ in the construction of the Bunce–Deddens algebra $\mathfrak{B}(\mathbf{n})$. We need to verify that $\alpha_k \rho_k = \rho_{k+1} \beta_k$. Once this is established, it follows that there is a well-defined homomorphism ρ of $\mathfrak{B}(\mathbf{n})$ onto $\mathfrak{A}$ given by $\rho(A) = \rho_k(A)$ for every A in $\mathfrak{B}(N_k)$. Moreover, ρ must be an isomorphism because $\mathfrak{B}(\mathbf{n})$ is simple.

Let us determine the structure of the map α_k. Notice that each cylinder set $J(x_1, \ldots, x_k)$ splits as the disjoint union of $J(x_1, \ldots, x_n, j)$ for $0 \leq j < n_{k+1}$. Thus we have the decomposition

$$E_{ii}^{(k)} = \sum_{j=0}^{n_{k+1}-1} E_{i+jN_k, i+jN_k}^{(k+1)}.$$

The unitary u is represented as U_k and U_{k+1} respectively in the two subalgebras. If we think of U_k as an $N_k \times N_k$ matrix whose entries are $n_{k+1} \times n_{k+1}$ matrices obtained from this decomposition in the natural order, then the entries on the first subdiagonal are equal to $I_{n_{k+1}}$, most entries are 0, and the $(0, N_k - 1)$ entry V_k maps each $E_{N_k-1+jN_k, N_k-1+jN_k}^{(k+1)}$ onto $E_{(j+1)N_k, (j+1)N_k}^{(k+1)}$ for $0 \leq j \leq n_{k+1} - 2$ by the canonical identification (I) and it takes $E_{N_{k+1}-1, N_{k+1}-1}^{(k+1)}$ to $E_{00}^{(k+1)}$ by V_{k+1}. Thus

$$V_k = \begin{bmatrix} 0 & 0 & \cdots & 0 & V_{k+1} \\ I & 0 & \cdots & 0 & 0 \\ 0 & I & \cdots & 0 & 0 \\ \vdots & \vdots & \ddots & \vdots & \vdots \\ 0 & 0 & \cdots & I & 0 \end{bmatrix}.$$

Now recall that the imbedding $C(\mathbb{T})$ into $\mathcal{M}_{N_k}(C(\mathbb{T}))$ in the Bunce–Deddens construction takes z to the matrix A_k with 1's on the first subdiagonal, z in the $(0, N_k - 1)$ entry, and 0's elsewhere. The map ρ_k takes z to V_k. Hence it is evident that $\rho_k(A_k) = U_k$. Hence

$$\alpha_k \rho_k(A_k) = V_{k+1} = \rho_{k+1}(A_{k+1}) = \rho_{k+1}\beta_k(A_k).$$

The restriction of ρ_k to the $N_k \times N_k$ matrices with scalar entries is the identity map. Moreover, the imbeddings ρ_k and β_k agree on these scalar matrices. The scalar matrices and A_k generate $\mathcal{M}_{N_k}(C(\mathbb{T}))$. Thus the two mappings agree. ∎

VIII.5 K-theory for Crossed Products

There is an important tool for computing the K-theory of crossed products. Unfortunately, the proof of this fact requires the development of much more general K-theory than is contained in these notes. As its use in these note is limited, we have decided to state them without proof.

VIII.5. K-theory for Crossed Products

We will need to define the **K$_1$ group** of a C*-algebra $\mathfrak{A}$. Let $GL_n(\mathfrak{A})$ denote the group of invertible elements of $\mathcal{M}_n(\mathfrak{A})$; and let $GL_n^0(\mathfrak{A})$ denote the connected component of the identity, which is a clopen normal subgroup of $GL_n(\mathfrak{A})$. There is a natural imbedding of $GL_n(\mathfrak{A})$ into $GL_{n+1}(\mathfrak{A})$ given by sending A to $\begin{bmatrix} A & 0 \\ 0 & I \end{bmatrix}$. It is clear that $GL_n^0(\mathfrak{A})$ is carried into $GL_{n+1}^0(\mathfrak{A})$. Thus this induces a homomorphism of the quotient $GL_n(\mathfrak{A})/GL_n^0(\mathfrak{A})$ into $GL_{n+1}(\mathfrak{A})/GL_{n+1}^0(\mathfrak{A})$. The direct limit of this sequence of groups is denoted by $K_1(\mathfrak{A})$. It is always an abelian group even though the quotient groups need not be abelian. (See the exercises.)

When $\mathfrak{A}$ is not unital, we define $GL_n(\mathfrak{A})$ to be the group of those invertible $n \times n$ matrices with coefficients in $\mathfrak{A}^\sim$ such that the off-diagonal entries lie in $\mathfrak{A}$ and the diagonal entries belong to $I + \mathfrak{A}$. (That is, matrices congruent to I_n modulo $\mathcal{M}_n(\mathfrak{A})$.)

In a C*-algebra, every invertible element is connected in the invertibles to the unitary operator in its polar decomposition. Thus it is also easy to see that one could use the unitary groups $\mathcal{U}(\mathcal{M}_n(\mathfrak{A}))$ instead of $GL_n(\mathfrak{A})$ in the definition of K_1. Notice that the direct limit of $\mathcal{M}_n(\mathfrak{A})$ under multiplicity one imbeddings is $\mathfrak{K} \otimes \mathfrak{A}$. An element of $GL_1((\mathfrak{K} \otimes \mathfrak{A})^\sim)$ is connected to a unitary in the image of $GL_n(\mathfrak{A})$ for n sufficiently large. It isn't difficult to show that

$$K_1(\mathfrak{A}) = K_1(\mathfrak{K} \otimes \mathfrak{A}) = GL_1((\mathfrak{K} \otimes \mathfrak{A})^\sim)/GL_1^0((\mathfrak{K} \otimes \mathfrak{A})^\sim).$$

In any Banach algebra, it is elementary to check that when $\|A - I\| < 1$, there is an element L in the algebra such that $A = e^L$. So $GL_n^0(\mathfrak{A})$ is the set of all finite products of exponentials e^L for L in $\mathcal{M}_n(\mathfrak{A})$. In the commutative case, the product of exponentials is the exponential of the sum. So $GL_1^0(\mathrm{C}(X)) = e^{\mathrm{C}(X)}$.

Any invertible element with spectrum in the right half plane is connected to the identity by a straight line. More generally, any invertible element with 0 in the unbounded component of the resolvent is connected to the identity by using the functional calculus to construct a path to an element with spectrum in the right half plane. In particular, any invertible element with finite spectrum is in the connected component of the identity.

It follows that $K_1(\mathfrak{A}) = 0$ for every finite dimensional C*-algebra. Since $GL_n^0(\mathfrak{A})$ is open and closed in $GL_n(\mathfrak{A})$, it follows that when $\mathfrak{A} = \varinjlim \mathfrak{A}_n$, then $K_1(\mathfrak{A}) = \varinjlim K_1(\mathfrak{A}_n)$. Hence $K_1(\mathfrak{A}) = 0$ for every AF-algebra.

If X is a totally disconnected space such as the Cantor set, every element of $GL_n(\mathrm{C}(X))$ can be approximated by a function taking only finitely many values in $\mathcal{M}_n$. Such functions have finite spectrum and consequently lie in $GL_n^0(\mathrm{C}(X))$. Therefore $K_1(\mathrm{C}(X)) = 0$. On the other hand, $K_1(\mathrm{C}(\mathbb{T})) = \mathbb{Z}$. The map is obtained by sending each element of $GL_n(\mathrm{C}(\mathbb{T}))$ to the winding number of its determinant function.

As in the case of K_0, a homomorphism α from $\mathfrak{A}$ into $\mathfrak{B}$ induces a homomorphism α_* of $K_1(\mathfrak{A})$ into $K_1(\mathfrak{B})$. This makes K_1 a covariant functor from the category of C*-algebras into the category of abelian groups. This is barely sufficient for us to state the **Pimsner–Voiculescu short exact sequence**. In fact, we do not give sufficient information here to define the vertical maps. However, in our applications, this extra detail will not be needed.

Theorem VIII.5.1 *Suppose that α is an automorphism of a C*-algebra $\mathfrak{A}$. Then there is a cyclic six term exact sequence*

$$\begin{array}{ccccc}
K_0(\mathfrak{A}) & \xrightarrow{\mathrm{id}_*-\alpha_*} & K_0(\mathfrak{A}) & \xrightarrow{\iota_*} & K_0(\mathfrak{A} \times_\alpha \mathbb{Z}) \\
\uparrow & & & & \downarrow \\
K_1(\mathfrak{A} \times_\alpha \mathbb{Z}) & \xleftarrow{\iota_*} & K_1(\mathfrak{A}) & \xleftarrow{\mathrm{id}_*-\alpha_*} & K_1(\mathfrak{A})
\end{array}$$

Example VIII.5.2 Consider the irrational rotation algebra $\mathcal{A}_\theta$, which is shown to be a crossed product in Example VIII.1.1. First compute the K-theory for $C(\mathbb{T})$. Since $\mathbb{T}$ is connected, every projection valued function in $C(\mathbb{T}, \mathcal{M}_n)$ has constant rank. It is easy to find a homotopy of projections that connects any projection P in $C(\mathbb{T}, \mathcal{M}_n)$ to one with a fixed unit vector in the range of $P(t)$ for every t in $\mathbb{T}$. Thus by induction, one shows that the rank is the only homotopy invariant. Hence $K_0(C(\mathbb{T})) = \mathbb{Z}$. As noted above, $K_1(\mathbb{T}) = \mathbb{Z}$ as well.

Now consider the automorphism θ on $C(\mathbb{T})$ induced by rotation through the angle $2\pi\theta$. This does not effect the rank of a projection, nor the winding number of the determinant of an invertible function. Therefore $\theta_* = \mathrm{id}_*$. Plugging this into the P-V sequence, we obtain

$$0 \to K_0(C(\mathbb{T})) = \mathbb{Z} \to K_0(\mathcal{A}_\theta) \to K_1(C(\mathbb{T})) = \mathbb{Z} \to 0$$

and

$$0 \to K_1(C(\mathbb{T})) = \mathbb{Z} \to K_1(\mathcal{A}_\theta) \to K_0(C(\mathbb{T})) = \mathbb{Z} \to 0.$$

Therefore $K_0(\mathcal{A}_\theta) \simeq K_1(\mathcal{A}_\theta) \simeq \mathbb{Z}^2$. It follows that the order homomorphism τ_* of $K_0(\mathcal{A}_\theta)$ onto $\mathbb{Z} + \mathbb{Z}\theta$ of Theorem VI.5.2 induced by the unique trace τ is an isomorphism.

Example VIII.5.3 Consider the crossed product representation of $\mathfrak{K} \otimes \mathcal{O}_n$ from Example VIII.1.2. This is the crossed product $(\mathfrak{K} \otimes \mathfrak{F}^n) \times_\sigma \mathbb{Z}$ where $\mathfrak{F}^n$ is the UHF algebra of type n^∞ and σ is a left shift. We will use the P-V exact sequence to compute the K-theory of the Cuntz algebras. As noted above, $K_*(\mathfrak{K} \otimes \mathfrak{A}) = K_*(\mathfrak{A})$. Thus $K_0(\mathfrak{K} \otimes \mathcal{O}_n) = K_0(\mathcal{O}_n)$ and $K_1(\mathfrak{K} \otimes \mathcal{O}_n) = K_1(\mathcal{O}_n)$. Because $\mathfrak{F}^n$ is an AF algebra, $K_1(\mathfrak{K} \otimes \mathfrak{F}^n) = K_1(\mathfrak{F}^n) = 0$. As in Example IV.3.4, we may easily compute that $K_0(\mathfrak{K} \otimes \mathfrak{F}^n) = K_0(\mathfrak{F}^n) = \mathbb{Z}\left[\frac{1}{n}\right]$. By Exercise IV.7, the map τ_* induced by the unique trace is an isomorphism. The automorphism σ satisfies $\sigma(P_k) = P_{k+1}$. Since $\tau(P_k) = n^{-k}$, we deduce that σ_* is multiplication by n^{-1}.

VIII.6. AF subalgebras of Crossed Products

Putting these facts together, we obtain the exact sequence

$$\begin{array}{ccccccccc}
K_1(\mathfrak{F}^n) & \to & K_1(\mathfrak{K}\otimes\mathcal{O}_n) & \to & K_0(\mathfrak{F}^n) & \stackrel{\mathrm{id}_*-\sigma_*}{\to} & K_0(\mathfrak{F}^n) & \to & K_0(\mathfrak{K}\otimes\mathcal{O}_n) & \to & K_1(\mathfrak{F}^n) \\
\| & & \| & & \| & & \| & & \| & & \| \\
0 & \to & K_1(\mathcal{O}_n) & \to & \mathbb{Z}\left[\tfrac{1}{n}\right] & \stackrel{1-n^{-1}}{\to} & \mathbb{Z}\left[\tfrac{1}{n}\right] & \to & K_0(\mathcal{O}_n) & \to & 0
\end{array}$$

Consider the group homomorphism on $\mathbb{Z}\left[\tfrac{1}{n}\right]$ of multiplication by $1 - n^{-1}$. This consists of the automorphism of multiplication by n^{-1} followed by multiplication α by $n-1$. Since $\mathbb{Z}\left[\tfrac{1}{n}\right]$ is an integral domain, multiplication is injective. We claim that the range of α consists of all elements of the form $n^{-k}a$ such that $n-1$ divides a. Indeed, if $a = (n-1)b$, then $\alpha(n^{-k}b) = n^{-k}a$. Conversely, suppose that

$$n^{-k}a = \alpha(n^{-\ell}b) = n^{-\ell}(n-1)b.$$

Then since n and $n-1$ are relatively prime, unique factorization in $\mathbb{Z}$ shows that $n-1$ divides a. We conclude that

$$K_0(\mathcal{O}_n) \simeq K_1(\mathfrak{F}^n)/\operatorname{Im}(\mathrm{id}_* - \sigma_*) = \mathbb{Z}_{n-1}$$

and

$$K_1(\mathcal{O}_n) = \ker(\mathrm{id}_* - \sigma_*) = 0.$$

This provides another method of distinguishing Cuntz algebras from one another.

VIII.6 AF subalgebras of Crossed Products

In this section, we will show that certain natural subalgebras of crossed products for minimal dynamical systems on Cantor sets are AF. Then we will extend this analysis to obtain several useful structural results about these algebras including an imbedding into one of these AF algebras analogous to the Pimsner–Voiculescu imbedding of $\mathcal{A}_\theta$ into an AF algebra.

For this section, let X be a Cantor set and let σ be a minimal homeomorphism. We will call (X, σ) a **minimal Cantor system**. A **partition** of X is a finite collection $\mathcal{P}$ of pairwise disjoint clopen subsets of X which cover X. Let $\mathrm{C}(\mathcal{P})$ denote the span of the characteristic functions of the members of $\mathcal{P}$. Let u denote the unitary element of $\mathrm{C}(X) \times_\sigma \mathbb{Z}$ which implements σ on $\mathrm{C}(X)$.

Lemma VIII.6.1 *Let Y be a non-empty clopen subset of X and let $\mathcal{P}$ be a partition of X into clopen sets. Then the C*-subalgebra of $\mathrm{C}(X) \times_\sigma \mathbb{Z}$ generated by $\mathrm{C}(\mathcal{P})$ and $u\chi_{X\setminus Y}$ is finite dimensional.*

Proof. Define the *first return* function for Y by

$$n_Y(y) := \min\{n \geq 1 : \sigma^n(y) \in Y\} \quad \text{for } y \in Y.$$

Since σ is minimal, $\{\sigma^n(y) : n \geq 1\}$ is dense in X. Thus since Y is open, n_Y is finite for every y in Y and is upper semi-continuous. Indeed, if $n_Y(y) = n$, then $\sigma^n(y')$ belongs to Y for all y' in a sufficiently small neighbourhood of y; and thus $n_Y(y') \leq n$ on this neighbourhood. Similarly, since Y is closed, n_Y is lower semi-continuous. Hence n_Y is continuous on Y. Then because Y is compact, n_Y takes only finitely many values, say $\{n_1, \ldots, n_K\}$.

For $1 \leq k \leq K$ and $1 \leq j \leq n_k$, define $Y(k, j) = \sigma^j(n_Y^{-1}(n_k))$. Then the sets $Y(k, j)$ are pairwise disjoint clopen sets (why?). Notice that the sets $Y(k, n_k)$ must form a cover of Y because each y in Y is the first return of the first point $\sigma^{-s}(y)$, $s \geq 1$, to lie in Y. It follows that the union of all the $Y(k, j)$ is invariant for σ because

$$Y = \cup_{k=1}^{K} Y(k, n_k), \qquad \sigma(Y) = \cup_{k=1}^{K} Y(k, 1)$$

and

$$\sigma(Y(k, j)) = Y(k, j+1) \quad \text{for} \quad 1 \leq j < n_k.$$

As σ is minimal, the sets $Y(k, j)$ form a partition $\mathcal{Y}$ of X. The collection

$$\mathcal{Y}_k = \{Y(k, j) : 1 \leq j \leq n_k\}$$

is a *tower* of height n_k for each $1 \leq k \leq K$.

It is a routine matter to further subdivide these towers vertically in order to arrange that $\{Y(k, j)\}$ refines $\mathcal{P}$. Indeed, if P is in $\mathcal{P}$ and $A = P \cap Y(k, j_0)$ is a proper subset of $Y(k, j_0)$, construct the sets $A(k, j) = \sigma^{j-j_0}(A)$ for $1 \leq j \leq n_k$. This forms a tower of height n_k, as do the sets $Y(k, j) \setminus A(k, j)$; splitting the tower into two of the same height. Continuing this process for each such proper intersection yields a finer partition $\mathcal{Y}'$ of X than $\mathcal{Y} \vee \mathcal{P}$ with the additional feature that it consists of towers $Y'(k, j)$ for $1 \leq j \leq n'_k$ and $1 \leq k \leq K'$ so that

$$\sigma(Y(k, j)) = Y(k, j+1) \quad \text{for} \quad 1 \leq j < n'_k,$$

$$\cup_{k=1}^{K'} Y'(k, 1) = \sigma(Y) \qquad \text{and} \qquad \cup_{k=1}^{K'} Y'(k, n'_k) = Y.$$

We will work with this partition, but will suppress the primes for notational simplicity.

Now we define a finite dimensional subalgebra $\mathfrak{A}(Y, \mathcal{Y})$ of $C(X) \times_\sigma \mathbb{Z}$ isomorphic to $\sum_{k=1}^{K} \oplus \mathcal{M}_{n_k}$ determined by the matrix units

$$E_{ij}^{(k)} = u^{i-j} \chi_{Y(k,j)} = \chi_{Y(k,i)} u^{i-j}$$

for $1 \leq k \leq K$ and $1 \leq i, j \leq n_k$. These do form a system of matrix units because of the identities

$$u \chi_{Y(k,j)} u^* = \chi_{\sigma(Y(k,j))} = \chi_{Y(k,j+1)} \quad \text{for} \quad 1 \leq j < n_k.$$

The diagonal elements $E_{jj}^{(k)}$ span $C(\mathcal{Y})$ which contains $C(\mathcal{P})$. The off diagonal elements $E_{ij}^{(k)}$ have the form $(u\chi_{X \setminus Y})^{i-j} \chi_{Y(k,j)}$ when $1 \leq j < i \leq n_k$. They

VIII.6. AF subalgebras of Crossed Products

evidently lie in the algebra generated by $u\chi_{X\setminus Y}$ and $C(\mathcal{Y})$. Conversely,

$$u\chi_{X\setminus Y} = \sum_{k=1}^{K}\sum_{j=1}^{n_k-1} E_{j+1,j}^{(k)}.$$

Therefore $\mathfrak{A}(Y,\mathcal{Y})$ is the C*-algebra generated by $C(\mathcal{Y})$ and $u\chi_{X\setminus Y}$. Since the C*-algebra generated by $C(\mathcal{P})$ and $u\chi_{X\setminus Y}$ is contained in this one, it is also finite dimensional. ∎

Theorem VIII.6.2 *Let (X,σ) be a minimal Cantor system; and let Y be a closed subset of X. Then the C*-subalgebra $\mathfrak{A}_Y$ of $C(X)\times_\sigma \mathbb{Z}$ generated by $C(X)$ and $uC_0(X\setminus Y)$ is AF. Moreover the injection j of $C(X)$ into $\mathfrak{A}_Y$ induces a surjective unital order homomorphism j_* of K_0 groups which carries $K_0(C(X))_1^+$ onto $K_0(\mathfrak{A}_Y)_1^+$.*

Proof. Start with a decreasing sequence Y_n of clopen sets with intersection Y; and with an increasing sequence of partitions $\mathcal{P}_n$ whose union generates the topology of X. Thus $\cup_{n\geq 1}C(\mathcal{P}_n)$ is dense in $C(X)$. We then construct an increasing sequence of partitions $\mathcal{Y}_n$ into towers built from Y_n as in the preceding lemma so that $\mathcal{Y}_n$ refines both $\mathcal{P}_n$ and $\mathcal{Y}_{n-1}$. Then form the sequence of finite dimensional algebras $\mathfrak{A}_n = \mathfrak{A}(Y_n,\mathcal{Y}_n)$. Notice that this is an increasing chain of algebras because $C(\mathcal{Y}_n)$ is contained in $C(\mathcal{Y}_{n+1})$ and

$$u\chi_{X\setminus Y_n} = (u\chi_{X\setminus Y_{n+1}})\chi_{X\setminus Y_n} \in \mathfrak{A}_{n+1}.$$

The closed union of the $\mathfrak{A}_n$ contains $C(\mathcal{P}_n)$ for all $n\geq 1$, and thus it contains $C(X)$. Also any function f in $C_0(X\setminus Y)$ (thought of as an element of $C(X)$ vanishing on Y) can be approximated by a function g in some $C(\mathcal{Y}_n)$ which vanishes on Y_n. But then ug belongs to $\mathfrak{A}_n$ and approximates uf. Conversely, it is evident that each $\mathfrak{A}_n$ lies in $\mathfrak{A}_Y$. Therefore $\mathfrak{A}_Y$ is AF.

The injection j_n of $C(\mathcal{Y}_n)$ into $\mathfrak{A}_n$ maps onto the diagonal algebra $\mathfrak{D}_n$ spanned by the diagonal matrix units. Every projection in $\mathfrak{A}_n$ is unitarily equivalent to a diagonal projection. Therefore j_{n*} maps

$$K_0^+(C(\mathcal{Y}_n))_1 = \{[p]_{C(\mathcal{Y}_n)} : p = p^2 = p^* \in C(\mathcal{Y}_n)\}$$

onto the corresponding set

$$K_0^+(\mathfrak{A}_n)_1 = \{[p]_{\mathfrak{A}_n} : p = p^2 = p^* \in \mathfrak{A}_n\}.$$

By Theorem IV.1.6, these sets are the order intervals $[0,1]$ in the two K_0 groups. Hence j_{n*} is an order isomorphism. Let α_n be the injection of $\mathfrak{A}_n$ into $\mathfrak{A}_Y$; and let

β_n be the injection of $C(\mathcal{Y}_n)$ into $C(X)$. Then

$$K_0^+(\mathfrak{A}_Y)_1 = \bigcup_{n\geq 1} \alpha_{n*}K_0^+(\mathfrak{A}_n)_1 = \bigcup_{n\geq 1} \alpha_{n*}j_{n*}K_0^+(C(\mathcal{Y}_n))_1$$

$$= \bigcup_{n\geq 1} j_*\beta_{n*}K_0^+(C(\mathcal{Y}_n))_1 = j_*K_0^+(C(X))_1.$$

Thus j_* is a surjection which takes the unit order interval of $K_0(C(X))$ onto the unit order interval of $K_0(\mathfrak{A}_Y)$. ∎

Example VIII.6.3 Recall the crossed product representation of the Bunce–Deddens algebra from Theorem VIII.4.1. Take Y to be the single point $0 = (0,0,\ldots)$. In the construction above, take $Y_k = J(0,\ldots,0)$ where the sequence has k zeros; and let the partition be

$$\mathcal{Y}_k = \{J(x_1,\ldots,x_k) : 0 \leq x_i < n_i,\ 1 \leq i \leq k\}.$$

From the proof and notation of that theorem,

$$u\chi_{X\setminus Y_k} = \sum_{j=0}^{N_k-2} E_{j,j+1}^{(k)}.$$

It is evident that the subalgebra $\mathfrak{A}(Y_n,\mathcal{Y}_n)$ is isomorphic to $\mathcal{M}_{N_k}$. The imbeddings are all unital, and therefore the algebra $\mathfrak{A}_0$ is isomorphic to the UHF algebra of type $\mathbf{n} = \prod_{i>1} n_i$.

It will be shown in the next section that the injection j of $\mathfrak{A}_0$ into $\mathfrak{A}$ induces an isomorphism j_* of K_0 groups.

VIII.7 Crossed Product subalgebras of AF algebras

We will show that for a minimal Cantor system (X,σ) and an arbitrary point y in X, there is an imbedding of the crossed product $C(X) \times_\sigma \mathbb{Z}$ into $\mathfrak{A}_{\{y\}}$ in a way which induces an isomorphism of K_0 groups. This is analogous to the Pimsner–Voiculescu imbedding of the irrational rotation algebra $\mathcal{A}_\theta$ into the AF algebra of the continued fraction of θ. In fact, our imbedding can be used to obtain the Pimsner–Voiculescu imbedding.

Fix an arbitrary point y in X. As in the previous section, construct an increasing sequence $\mathfrak{A}_n = \mathfrak{A}(Y_n,\mathcal{Y}_n)$ of finite dimensional subalgebras corresponding to a decreasing sequence Y_n of clopen sets with intersection $\{y\}$ and increasing partitions $\mathcal{Y}_n$ of X. By dropping to a subsequence if necessary, we may suppose that the sets $\sigma^j(Y_n)$ are pairwise disjoint for $1 \leq j \leq 2^n$ and so that each is contained in a single element of $\mathcal{Y}_{n-1}$. (This is easily achieved because the points $\sigma^j(y)$ are distinct and each is contained in some element of the partition $\mathcal{Y}_{n-1}$ for $1 \leq j \leq 2^n$. Hence these properties remain valid for a sufficiently small neighbourhood of $\{y\}$.) Following the notation of the previous section, let $E_{ij}^{(k)}$ for $1 \leq k \leq K,\ 1 \leq i,j \leq n_k$ be the matrix units of $\mathfrak{A}_n$.

VIII.7. Crossed Product subalgebras of AF algebras

Define a unitary v_n in $\mathfrak{A}_n$ by

$$v_n = u\chi_{X\setminus Y_n} + \sum_{k=1}^{K} u^{1-n_k}\chi_{Y(k,n_k)} = \sum_{k=1}^{K}\left(E_{1,n_k}^{(k)} + \sum_{j=1}^{n_k-1} E_{j+1,j}^{(k)}\right).$$

Notice that $v_n\chi_{X\setminus Y_n} = u\chi_{X\setminus Y_n}$. Also v_n mimics the role of u in an important way.

Lemma VIII.7.1 *If f in $C(\mathcal{Y}_n)$ is constant on Y_n, then $v_n f v_n^* = \sigma(f)$. In particular, $v_n\chi_{Y_n}v_n^* = \chi_{\sigma(Y_n)}$ and $v_n f v_n^* = \sigma(f)$ for every f in $C(\mathcal{Y}_{n-1})$.*

Proof. If $f\chi_{Y_n} = 0$, then

$$v_n f v_n^* = (v_n\chi_{X\setminus Y_n})f(v_n\chi_{X\setminus Y_n})^* = ufu^* = \sigma(f).$$

Also from the definition of v_n, we obtain

$$v_n\chi_{Y_n}v_n^* = v_n\sum_{k=1}^{K}\chi_{Y(k,n_k)}v_n^* = \sum_{k=1}^{K}\chi_{Y(k,1)} = \chi_{\sigma(Y)}.$$

Thus the result follows. Since Y_n is contained in a single element of the partition $\mathcal{Y}_{n-1}$, every function in $C(\mathcal{Y}_{n-1})$ is constant on Y_n. ∎

Now v_n and v_{n+1} are direct sums of weighted shifts. We are looking for a unitary w_n in $\mathfrak{A}_{n+1}$ which commutes with $C(\mathcal{Y}_{n-1})$ such that $w_n v_{n+1} w_n^*$ is close to v_n. In the case of the irrational rotation algebra, this type of approximation was managed by Berg's technique. The same principle is used here, but it will be done in a more algebraic way. Notice that

$$v_{n+1}\chi_{X\setminus Y_n} = u\chi_{X\setminus Y_n} = v_n\chi_{X\setminus Y_n}$$
$$= \chi_{X\setminus\sigma(Y_n)}u = \chi_{X\setminus\sigma(Y_n)}v_{n+1} = \chi_{X\setminus\sigma(Y_n)}v_n.$$

Therefore

$$\chi_{X\setminus\sigma(Y_n)}v_n v_{n+1}^* = \chi_{X\setminus\sigma(Y_n)} = v_n v_{n+1}^*\chi_{X\setminus\sigma(Y_n)}.$$

Thus we also obtain

$$\chi_{\sigma(Y_n)}v_n v_{n+1}^* = v_n v_{n+1}^*\chi_{\sigma(Y_n)}.$$

Since $\mathfrak{A}_{n+1}$ is finite dimensional, the unitary element $v_n v_{n+1}^*$ has finite spectrum in $\mathbb{T}$. Hence by the continuous functional calculus, we obtain a unitary $z = h(v_n v_{n+1}^*)$ where $h(e^{i\theta}) = e^{i2^{-n}\theta}$ for $-\pi < \theta \leq \pi$. Therefore

$$z^{2^n} = v_n v_{n+1}^* \qquad \qquad \|z-1\| < 2^{-n}\pi$$
$$z\chi_{\sigma(Y_n)} = \chi_{\sigma(Y_n)}z \quad \text{and} \quad z\chi_{X\setminus\sigma(Y_n)} = \chi_{X\setminus\sigma(Y_n)} = \chi_{X\setminus\sigma(Y_n)}z.$$

Recall that $\sigma^j(Y_n)$ are pairwise disjoint for $1 \leq j \leq 2^n$. From the fact that z commutes with $\chi_{\sigma(Y_n)}$, we see that $u^{j-1}z^\ell u^{1-j}$ commutes with $\chi_{\sigma^j(Y_n)}$. Set

$Z = X \setminus \bigcup_{j=1}^{2^n} \sigma^j(Y_n)$. Define a unitary element

$$w_n = \chi_Z + \sum_{j=1}^{2^n} u^{j-1} z^{2^n+1-j} u^{1-j} \chi_{\sigma^j(Y_n)}.$$

The unitary z can be thought of as a unitary twist through 2^{-n}th of the unitary $v_n v_{n+1}^*$. The unitary w_n is the identity on χ_Z and acts by z^j on $\chi_{\sigma^{2^n+1-j}(Y_n)}$. Thus there is a gradual shifting from I to $v_{n+1} v_n^*$ as one shifts by σ along the blocks. This is the analogy to Berg's technique.

The key approximation lemma is the following.

Lemma VIII.7.2 *The unitary w_n belongs to the algebra $\mathfrak{A}_{n+1}$, commutes with $C(\mathcal{Y}_{n-1})$ and satisfies*

$$\|w_n v_{n+1} w_n^* - v_n\| < 2^{-n}\pi.$$

Proof. Since z belongs to $\mathfrak{A}_{n+1}$ and χ_Z and $\chi_{\sigma^j(Y_n)}$ belong to $C(\mathcal{Y}_n)$, the containment of w_n in $\mathfrak{A}_{n+1}$ follows from the fact that

$$\chi_{\sigma^j(Y_n)} u^{j-1} = u^{j-1} \chi_{\sigma(Y_n)} = v_n^{j-1} \chi_{\sigma(Y_n)}$$

also lies in $\mathfrak{A}_n$. So

$$u^{j-1} z^{2^n+1-j} u^{1-j} \chi_{\sigma^j(Y)} = u_n^{j-1} z^{2^n+1-j} \chi_{\sigma(Y)} v_n^{1-j}$$
$$= u_n^{j-1} \chi_{\sigma(Y)} z^{2^n+1-j} v_n^{1-j} = v_n^{j-1} \chi_{\sigma(Y)} z^{2^n+1-j} v_n^{1-j}$$

belongs to $\mathfrak{A}_{n+1}$. Now $w_n \chi_Z = \chi_Z$, and w_n commutes with each $\chi_{\sigma^j(Y_n)}$ for $1 \leq j \leq 2^n$. As any function in $C(\mathcal{Y}_{n-1})$ is constant on each $\chi_{\sigma^j(Y_n)}$, it will commute with w_n.

Notice that for $1 \leq j \leq 2^n$, Lemma VIII.7.1 shows that

$$(v_n w_n - w_n v_{n+1}) \chi_{\sigma^j(Y_n)} = v_n \chi_{\sigma^j(Y_n)} u^{j-1} z^{2^n+1-j} u^{1-j} - w_n \chi_{\sigma^{j+1}(Y_n)} u$$
$$= \chi_{\sigma^{j+1}(Y_n)} \big(u^j z^{2^n+1-j} u^{1-j} - u^j z^{2^n-j} u^{-j} u \big) \chi_{\sigma^j(Y_n)}$$
$$= \chi_{\sigma^{j+1}(Y_n)} u^j z^{2^n-j} (z-1) u^{1-j} \chi_{\sigma^j(Y_n)}$$

Thus

$$\|(v_n w_n - w_n v_{n+1}) \chi_{\sigma^j(Y_n)}\| \leq \|z - 1\| < 2^{-n}\pi.$$

Also

$$(v_n w_n - w_n v_{n+1}) \chi_{Y_n} = v_n \chi_{Y_n} - w_n \chi_{\sigma(Y_n)} v_{n+1}$$
$$= (v_n - z^{2^n} v_{n+1}) \chi_{Y_n} = (v - v_n v_{n+1}^* v_{n+1}) \chi_{Y_n} = 0$$

and since

$$\sigma(Z \setminus Y_n) = \big(X \setminus \bigcup_{j=1}^{2^n} \sigma^j(Y_n)\big) \setminus \sigma(Y_n) = Z \setminus \sigma^{2^n+1}(Y_n),$$

VIII.7. Crossed Product subalgebras of AF algebras

it follows that

$$\begin{aligned}(v_n w_n - w_n v_{n+1})\chi_{Z\setminus Y_n} &= v_n \chi_{Z\setminus Y_n} - w_n \chi_{\sigma(Z\setminus Y_n)} u \chi_{Z\setminus Y_n} \\ &= u\chi_{Z\setminus Y_n} - w_n \chi_{Z\setminus \sigma^{2^n+1}(Y_n)} u \chi_{Z\setminus Y_n} \\ &= (u-u)\chi_{Z\setminus Y_n} = 0.\end{aligned}$$

Because the terms $\chi_{\sigma^{j+1}(Y_n)}(v_n w_n - w_n v_{n+1})\chi_{\sigma^j(Y_n)}$ have pairwise orthogonal domains and ranges, one can compute

$$\|v_n w_n - w_n v_{n+1}\| = \Big\|\sum_{j=1}^{2^n} \chi_{\sigma^{j+1}(Y_n)}(v_n w_n - w_n v_{n+1})\chi_{\sigma^j(Y_n)}\Big\|$$
$$= \max_{1\le j\le 2^n}\|(v_n w_n - w_n v_{n+1})\chi_{\sigma^j(Y_n)}\| < 2^{-n}\pi. \blacksquare$$

Now we are ready to state the main result.

Theorem VIII.7.3 *Let (X,σ) be a minimal dynamical system on a Cantor set X, and let y be a point in X. Then there is a unital imbedding α of $C(X)\times_\sigma \mathbb{Z}$ into $\mathfrak{A}_{\{y\}}$ such that α_* is an ordered group isomorphism of the K_0 groups.*

Proof. For $n\ge 1$, define automorphisms α_n of $\mathfrak{A}_{\{y\}}$ by

$$\alpha_n(A) = \operatorname{Ad}(w_1\ldots w_n)(A) = \alpha_{n-1}(w_n A w_n^*).$$

Each g in $C(\mathcal{Y}_n)$ commutes with w_k for $k>n$. Hence $\alpha_k(g) = \alpha_n(g)$ for all $k>n$. So for an arbitrary f in $C(X)$ and $\varepsilon>0$, there is an integer n sufficiently large and g in $C(\mathcal{Y}_n)$ such that $\|f-g\|<\varepsilon$. Hence for $k,\ell>n$,

$$\|\alpha_k(f) - \alpha_\ell(f)\| \le \|\alpha_k(f-g)\| + \|\alpha_k(g) - \alpha_\ell(g)\| + \|\alpha_\ell(f-g)\| < 2\varepsilon.$$

Thus $\alpha_k(f)$ is Cauchy. So we may define an isometric $*$-homomorphism on $C(X)$ by $\alpha_0(f) = \lim_{n\to\infty}\alpha_n(f)$.

Next notice that by Lemma VIII.7.2

$$\|\alpha_{n-1}(v_n) - \alpha_n(v_{n+1})\| = \|\alpha_{n-1}(v_n - w_n v_{n+1} w_n^*)\| < 2^{-n}\pi.$$

Therefore this is a Cauchy sequence. So define a unitary $v = \lim_{n\to\infty}\alpha_n(v_{n+1})$. Now compute for f in $\bigcup_{k\ge 1} C(\mathcal{Y}_k)$ using Lemma VIII.7.1

$$v\alpha(f)v^* = \lim_{n\to\infty}\alpha_n(v_{n+1})\alpha_n(f)\alpha_n(v_{n+1}^*)$$
$$= \lim_{n\to\infty}\alpha_n(v_{n+1} f v_{n+1}^*) = \alpha(\sigma(f)).$$

This is valid on a dense subset of $C(X)$, and thus extends by continuity to the whole algebra.

Therefore the pair (α_0, v) is a covariant representation of the dynamical system (X,σ). So we may define a representation of the crossed product $C(X)\times_\sigma \mathbb{Z}$ by

$$\alpha(f) = \alpha_0(f) \quad \text{for} \quad f\in C(X) \quad \text{and} \quad \alpha(u) = v.$$

Since $C(X)\times_\sigma \mathbb{Z}$ is simple by Theorem VIII.3.9, α is injective.

To compute the K-theory, let j be the injection of $C(X)$ into $\mathfrak{A}_{\{y\}}$, let i be the injection of $\mathfrak{A}_{\{y\}}$ into $C(X) \times_\sigma \mathbb{Z}$, and let $j_0 = ij$. Consider the commutative diagram

$$\begin{array}{ccc}
 & C(X) & \\
\overset{j}{\swarrow} \ \ \downarrow j_0 & & \searrow \alpha_0 \\
\mathfrak{A}_{\{y\}} \xrightarrow{\ i\ } C(X) & \times_\sigma \mathbb{Z} & \xrightarrow{\ \alpha\ } \mathfrak{A}_{\{y\}}
\end{array}$$

Now α_0 is the pointwise limit of inner automorphisms composed with j. So by Remark IV.5.6, $\alpha_{0*} = j_*$. Therefore

$$j_* = \alpha_{0*} = \alpha_* i_* j_*.$$

By Theorem VIII.6.2, j_* is surjective. Therefore $\alpha_* i_* = \mathrm{id}_{K_0(\mathfrak{A}_{\{y\}})}$. We now need to quote the Pimsner–Voiculescu exact sequence (Theorem VIII.5.1) *only* to obtain the information that j_{0*} is also surjective. Thus i_* is surjective and consequently $\alpha_* = i_*^{-1}$. Both α_* and i_* are unital order homomorphisms, so they are order isomorphisms. ∎

There is another immediate consequence of the construction used in this last proof. Recall from section V.7 that a limit circle algebra (AT algebra) is an inductive limit of algebras of the form $\sum \oplus_{i=1}^{k_n} \mathcal{M}_{p_{n,i}}(C(\mathbb{T}))$. Notice that if some of the summands are replaced with scalar matrices, the limit algebra is still an AT algebra.

Corollary VIII.7.4 *Let (X, σ) be a minimal dynamical system on a Cantor set X. Given a finite partition $\mathcal{P}$ of X and an $\varepsilon > 0$, there is a subalgebra of the crossed product $C(X) \times_\sigma \mathbb{Z}$ of the form $\mathfrak{B} \simeq \mathcal{M}_{p_{n,1}}(C(\mathbb{T})) \oplus \sum \oplus_{i=2}^{k_n} \mathcal{M}_{p_{n,i}}$ containing $C(\mathcal{P})$ and a unitary u' such that $\|u' - u\| < \varepsilon$. Thus any finite set of elements of $C(X) \times_\sigma \mathbb{Z}$ may be simultaneously approximated to any given accuracy by elements of a subalgebra of this form.*

Proof. We use the notation developed throughout this section. We may suppose that $\mathcal{P}$ is contained in $\mathcal{Y}_1$. Let

$$\mathfrak{B}_n = C^*\{C(\mathcal{Y}_{n-1}), w_n^* v_n w_n, v_{n+1}^* u\}.$$

Notice that

$$\mathfrak{C}_n = C^*\{C(\mathcal{Y}_{n-1}), w_n^* v_n w_n\} = w_n^* C^*\{C(\mathcal{Y}_{n-1}), v_n\} w_n$$

is contained in $w_n^* \mathfrak{A}_n w_n$ which is finite dimensional, say

$$\mathfrak{C}_n \simeq \sum_{k=1}^{K} \oplus \mathcal{M}_{p_k}.$$

VIII.7. Crossed Product subalgebras of AF algebras

So this algebra has a tower structure corresponding to a partition in between $\mathcal{Y}_{n-1}$ and $\mathcal{Y}_n$. Moreover, $v_{n+1}^* u \chi_{X \setminus Y_{n+1}} = \chi_{X \setminus Y_{n+1}}$. So there is a unitary element z acting on the support of $\chi_{Y_{n+1}}$ such that

$$v_{n+1}^* u = z \chi_{Y_{n+1}} + \chi_{X \setminus Y_{n+1}}.$$

Since $\chi_{Y_{n+1}}$ is dominated by a minimal projection in $C(\mathcal{Y}_n)$, it is also dominated by a diagonal matrix unit, say $E_{11}^{(1)}$, in $\mathfrak{C}_n$. As z is a unitary supported on this matrix unit, it follows that

$$\mathfrak{B}_n = C^*(\mathfrak{C}_n, v_{n+1}^* u) \simeq \mathcal{M}_{p_1}(C^*(z)) \oplus \sum_{k=2}^{K} \oplus \mathcal{M}_{p_k}.$$

Moreover this algebra contains $u' = (w_n^* v_n w_n)(v_{n+1}^* u)$. Compute

$$\|u' - u\| = \|(w_n^* v_n w_n v_{n+1}^* - 1)u\| = \|v_n - w_n v_{n+1} w_n^*\| < 2^{-n}\pi$$

by Lemma VIII.7.2.

Now $C^*(z)$ is isomorphic to the continuous functions on its spectrum, which is a subset of the circle. Let us show that it is the full circle. The unitary z corresponds to the dynamical system (Y_{n+1}, φ) where φ is the first return map induced on Y_{n+1} by σ. From the tower structure of $\mathfrak{A}_{n+1}$, it is evident that z is a unitary in a covariant representation of (Y_{n+1}, φ). It is easy to check that φ is minimal. For if C is a closed non-empty φ invariant subset of Y_{n+1}, then $C' = \overline{\bigcup_{k \in \mathbb{Z}} \sigma^k(C)}$ is a non-empty closed invariant subset of X. Thus $C' = X$. But then

$$C = C' \cap Y_{n+1} = Y_{n+1}.$$

Therefore the crossed product $C(Y_{n+1}) \times_\varphi \mathbb{Z}$ is simple, and so z is the universal unitary in this algebra. So by Exercise VIII.3, it has spectrum equal to the whole circle.

Any element of the crossed product may be approximated to any desired accuracy by elements of the form $\sum_{i=-N}^{N} f_i u^i$ where f_i belongs to $C(\mathcal{P})$ for some sufficiently fine finite partition. It is clear that if $\mathcal{Y}_n$ contains $\mathcal{P}$ and $\|u - u'\|$ is sufficiently small, the corresponding polynomials in u' provide the desired approximants. ∎

Theorem VIII.7.5 *Let (X, σ) be a minimal system on a Cantor set X. Then the crossed product $\mathfrak{A} = C(X) \times_\sigma \mathbb{Z}$ is a limit circle algebra.*

Proof. The remaining details are routine. It suffices to show that for an appropriate subsequence of the algebras $\mathfrak{B}_n$ constructed above, there are homomorphisms β_k of $\mathfrak{B}_{n_k}$ into $\mathfrak{B}_{n_{k+1}}$ which asymptotically are almost isometric. Suppose that a circle algebra $\mathfrak{B}_n$ and an $\varepsilon > 0$ are given. Let $\delta = \delta(n, \varepsilon)$ be given by Lemma III.3.1, where n is the dimension of the matrix algebra of scalars in $\mathfrak{B}_n$. Choose $m > n$ sufficiently large so that the matrix units $E_{ij}^{(k)}$ and the unitary u' in $\mathfrak{B}_n$ are within δ of elements of $\mathfrak{B}_m$. By Lemma III.3.1, there is a unitary element W_n in $\mathfrak{A}$ such that $\|W_n - I\| < \varepsilon$ so that each matrix unit $E_{ij}^{(k)}$ is contained in $W_n \mathfrak{B}_m W_n^*$. Recall the element z in $\mathfrak{B}_n$ given by the formula $u' = z E_{11}^{(1)} + E_{11}^{(1)\perp}$. There is an element $U' = Z E_{11}^{(1)} + E_{11}^{(1)\perp}$ in $W_n \mathfrak{B}_m W_n^*$ such that $\|u' - U'\| < 2(\delta + \varepsilon) < 3\varepsilon$. Since the spectrum of u' is the whole circle, there is a homomorphism of $C^*(u')$ onto $C^*(U')$ taking u' onto U'. Combining this with the injection on the matrix algebra, one obtains a homomorphism β_n of $\mathfrak{B}_n$ into $W_n \mathfrak{B}_m W_n^*$ which is the identity on the matrix units and satisfies $\|\beta(u') - u'\| < 3\varepsilon$.

Choose an increasing sequence of such algebras and maps such that ε_n converges rapidly to 0. There is a natural sequence j_n of injections of these circle algebras in $\mathfrak{A}$ and automorphisms $\alpha_n = \mathrm{Ad}\, W_n$ of $\mathfrak{A}$ such that $j_{n+1} \beta_n$ and $\alpha_n j_n$ are asymptotically close on the generators. Some routine estimates show that the limit map is an isomorphism of the limit circle algebra onto $\mathfrak{A}$. The details are left as an exercise. ∎

Example VIII.7.6 We wish to show how the Pimsner–Voiculescu imbedding of the irrational rotation algebra into the AF algebra of the continued fraction can be obtained in another way. Consider the Cantor set X obtained from the unit circle $\mathbb{T} = \mathbb{R}/\mathbb{Z}$ by introducing cuts at the points $n\theta$ for all n in $\mathbb{Z}$. More precisely, replace each such point by a left and right limit point. The topology on this space is the order topology induced from the order on $[0, 1]$. This space is a totally disconnected compact metric space; so it is homeomorphic to the Cantor set.

Let σ be the homeomorphism of rotation by θ on X, and consider the crossed product $\mathfrak{B} = C(X) \times_\sigma \mathbb{Z}$. Since $C(\mathbb{T})$ sits inside $C(X)$, there is a natural imbedding j of $\mathcal{A}_\theta$ into $\mathfrak{A}$. Follow this with the imbedding α of $\mathfrak{B}$ into the AF algebra $\mathfrak{B}_{\{y\}}$. To show that this is the right one, it suffices to compute the K-theory. See Exercise VIII.7.

VIII.8 Topological stable rank

For a unital C*-algebra, let $Lg_n(\mathfrak{A})$ and $Rg_n(\mathfrak{A})$ denote the n-tuples of elements of $\mathfrak{A}$ which generate $\mathfrak{A}$ as a left or right ideal respectively. The smallest positive integer n such that $Lg_n(\mathfrak{A})$ is dense in the set $\mathfrak{A}^n$ of n-tuples of $\mathfrak{A}$ is called the **topological stable rank** of $\mathfrak{A}$; write $tsr(\mathfrak{A}) = n$. If no n exists, then $tsr(\mathfrak{A}) = \infty$. Similarly, we could define a right topological stable rank using $Rg_n(\mathfrak{A})$. However, it is evident that $Rg_n(\mathfrak{A}) = Lg_n(\mathfrak{A})^*$; so the two notions coincide.

This notion is another algebraic measure of something akin to dimension. We have already seen the notion of real rank in section V.7. As with real rank, we concentrate on the lowest case—topological stable rank one, and our analysis will be fairly superficial. As a simple example, note that if X is a closed subset of the circle $\mathbb{T}$, then $C(X)$ has topological stable rank one because every function on the circle can be approximated by a non-vanishing one. Another simple example is that AF algebras are topological stable rank one. Indeed, every element can

VIII.8. Topological stable rank

be approximated by an element in a finite dimensional C*-subalgebra. But the invertibles are dense in every $\mathcal{M}_n$.

Proposition VIII.8.1 *A C*-algebra has topological stable rank one if and only if the invertible elements are dense.*

Proof. If the invertible elements are dense, then the set $Lg_1(\mathfrak{A})$ of left invertible elements are also dense. Conversely, suppose that $Lg_1(\mathfrak{A})$ is dense in $\mathfrak{A}$. Let A be a left invertible element with left inverse B. Find a left invertible element C such that $\|B - C\| < \|A\|^{-1}$. Then

$$\|CA - I\| = \|(C - B)A\| \leq \|C - B\|\|A\| < 1.$$

Thus CA is invertible. Therefore C is right invertible with right inverse $A(CA)^{-1}$. Since it is also left invertible, C is invertible. So A is also invertible with inverse $(CA)^{-1}C$. Thus $\mathfrak{A}^{-1} = Lg_1(\mathfrak{A})$ is dense in $\mathfrak{A}$. ∎

Theorem VIII.8.2 *Let $\mathfrak{A}$ be a unital C*-algebra, and let n be a positive integer. The invertible elements are dense in $\mathfrak{A}$ if and only if they are dense in $\mathcal{M}_n(\mathfrak{A})$.*

Proof. Suppose that the invertible elements are dense in $\mathfrak{A}$. We will show by induction that the invertibles are also dense in $\mathcal{M}_n(\mathfrak{A})$. Suppose that the result is valid for n, and consider a matrix T in $\mathcal{M}_{n+1}(\mathfrak{A})$ and a positive real number ε. Write T in the form $T = \begin{bmatrix} a & B \\ C & D \end{bmatrix}$ where $a \in \mathfrak{A}$ and $D \in \mathcal{M}_n(\mathfrak{A})$. By the induction hypothesis, there is an invertible element a_0 in $\mathfrak{A}$ such that $\|a - a_0\| < \varepsilon$ and an element D_0 in $\mathcal{M}_n(\mathfrak{A})$ such that $D_0 - Ca_0^{-1}B$ is invertible and $\|D - D_0\| < \varepsilon$. Hence

$$T_0 := \begin{bmatrix} a_0 & B \\ C & D_0 \end{bmatrix} = \begin{bmatrix} a_0 & 0 \\ C & I_n \end{bmatrix} \begin{bmatrix} 1 & 0 \\ 0 & D_0 - Ca_0^{-1}B \end{bmatrix} \begin{bmatrix} 1 & a_0^{-1}B \\ 0 & I_n \end{bmatrix}.$$

The three matrices on the right are all invertible, and thus T_0 is invertible. Since $\|T - T_0\| < \varepsilon$, it follows that the invertibles are dense in $\mathcal{M}_{n+1}(\mathfrak{A})$.

Conversely, suppose that for a certain n, the invertible elements are dense in $\mathcal{M}_{n+1}(\mathfrak{A})$. Given a in $\mathfrak{A}$ and $0 < \varepsilon < 1$, form the matrix $T = \begin{bmatrix} a & 0 \\ 0 & I_n \end{bmatrix}$ in $\mathcal{M}_{n+1}(\mathfrak{A})$. Let $T_0 = \begin{bmatrix} a_0 & B \\ C & D \end{bmatrix}$ be an invertible element in $\mathcal{M}_{n+1}(\mathfrak{A})$ such that $\|T - T_0\| < \varepsilon/2$. In particular, $\|B\| < \varepsilon/2$, $\|C\| < \varepsilon/2$ and $\|D - I_n\| < \varepsilon/2$. It follows that D is invertible and $\|D^{-1}\| < (1 - \varepsilon/2)^{-1}$. Thus we construct an invertible matrix

$$\begin{bmatrix} 1 & -BD^{-1} \\ 0 & I_n \end{bmatrix} \begin{bmatrix} a_0 & B \\ C & D \end{bmatrix} \begin{bmatrix} 1 & 0 \\ -D^{-1}C & I_n \end{bmatrix} = \begin{bmatrix} a_0 - BD^{-1}C & 0 \\ 0 & D \end{bmatrix}.$$

Hence $a_0 - BD^{-1}C$ is invertible in $\mathfrak{A}$. Finally,

$$\|a - (a_0 - BD^{-1}C)\| \leq \|a - a_0\| + \|B\|\|D^{-1}\|\|C\|$$
$$\leq \varepsilon/2 + (\varepsilon/2)^2(1 - \varepsilon/2)^{-1} = \varepsilon(1 - \varepsilon/2)^{-1}/2 < \varepsilon.$$

So the invertible elements are dense in $\mathfrak{A}$. ∎

The following consequence is immediate.

Corollary VIII.8.3 *A C*-algebra which is a finite direct sum of algebras of the form $\mathcal{M}_n(C(X_n))$ where X_n are homeomorphic to closed subsets of the circle has topological stable rank one.*

This next result is easy, but is significant because we have been able to show that many C*-algebras have this form for non-obvious reasons. The proof consists merely of approximating by elements of a special subalgebra and applying the corollary.

Theorem VIII.8.4 *Every limit circle algebra has topological stable rank one.*

Corollary VIII.8.5 *Let (X, σ) be a minimal system on a Cantor set X. Then the invertible elements are dense in the crossed product $C(X) \times_\sigma \mathbb{Z}$.*

Proof. This follows from Corollary VIII.7.4 and the observations above. ∎

Corollary VIII.8.6 *The invertible elements are dense in every Bunce–Deddens algebra.*

The following connection with K-theory indicates the usefulness of this notion.

Theorem VIII.8.7 *If the invertible elements are dense in $\mathfrak{A}$, then $\mathfrak{A}$ has cancellation.*

Proof. First we show that if P and Q are equivalent projections in $\mathcal{M}_n(\mathfrak{A})$, then they are unitarily equivalent; and hence $P^\perp$ is equivalent to $Q^\perp$. Indeed, suppose that V is a partial isometry such that $P = V^*V$ and $Q = VV^*$. By Theorem VIII.8.2, there is an invertible element A in $\mathcal{M}_n(\mathfrak{A})$ such that $\|V - A\| < 1/6$. Then

$$\|P - A^*A\| \leq \|V^*(V - A)\| + \|(V - A)^*A\| < 1/2.$$

Similarly, $\|Q - AA^*\| < 1/2$. Let $A = U|A|$ be the polar decomposition of A. Then U is a unitary element since A is invertible. Then since $U(A^*A)U^* = AA^*$, we obtain

$$\|UP^\perp U^* - Q^\perp\| = \|UPU^* - Q\| \leq \|U(P - A^*A)U^*\| + \|AA^* - Q\| < 1.$$

By Proposition IV.1.2, $Q^\perp$ is equivalent to $UP^\perp U^*$ which is equivalent to $P^\perp$. Let W be the partial isometry such that $W^*W = P^\perp$ and $WW^* = Q^\perp$. Then $V + W$ is a unitary implementing the equivalence between P and Q.

We must show that if P and Q are stably equivalent projections in $\mathcal{M}_n(\mathfrak{A})$, then they are equivalent. Suppose that P and Q are projections in $\mathcal{M}_n(\mathfrak{A})$ and that R is a projection in $\mathcal{M}_m(\mathfrak{A})$ such that $P \oplus R$ and $Q \oplus R$ are equivalent in $\mathcal{M}_{n+m}(\mathfrak{A})$. Then

$$P \oplus I_m = (P \oplus R) + (0 \oplus R^\perp) \sim (Q \oplus R) + (0 \oplus R^\perp) = Q \oplus I_m.$$

VIII.9 An order 2 automorphism

Hence by the previous paragraph, $P^\perp \oplus 0$ is equivalent to $Q^\perp \oplus 0$. By compressing the implementing partial isometry to $\mathcal{M}_n(\mathfrak{A})$ (which contains its domain and range projections), one obtains an equivalence of $P^\perp$ and $Q^\perp$. A second use of the previous paragraph yields the equivalence of P and Q. ∎

VIII.9 An order 2 automorphism

In this section, we construct a crossed product of the Bunce–Deddens algebra by $\mathbb{Z}_2$ to obtain an AF algebra. This shows that there is an AF algebra and an order 2 automorphism which has a fixed point algebra that is not AF. The first example of such a phenomenon was considered very surprising. The constructions involved are by now familiar to the reader.

The first step is to represent the 2^∞ Bunce–Deddens algebra as a crossed product in a natural way. Let $\mathbf{D}$ denote the group of diadic rationals modulo 1. Think of the circle as $\mathbf{T} = \mathbb{R}/\mathbb{Z}$. Then let τ be the action of $\mathbf{D}$ on $\mathbf{T}$ by translation. We claim that the crossed product of $C(\mathbf{T})$ by this action is the 2^∞ Bunce–Deddens algebra.

Since the action of $\mathbf{D}$ is minimal, the crossed product $\mathfrak{A} = C(\mathbf{T}) \times_\tau \mathbf{D}$ is simple by the analogue of Theorem VIII.3.9. It will be convenient to use the particular representation of $\mathfrak{A}$ on $L^2(\mathbf{T})$ by multiplication and rotation operators. The group $\mathbf{D}$ is the direct limit of the subgroups $\mathbb{Z}_{2^n}$ generated by 2^{-n}. Let $u_n = R_{2^{-n}}$ denote rotation by 2^{-n}.

We claim that $C^*(C(\mathbf{T}), u_n)$ is isomorphic to $\mathcal{M}_{2^n}(C(\mathbf{T}))$. This can be seen as follows. The spectrum of u_n consists of the 2^nth roots of unity λ_n^k for $\lambda_n = e^{2^{1-n}\pi i}$ and $0 \le k < 2^n$. Let the corresponding eigenspaces are

$$\mathcal{H}_{n,k} = \operatorname{span}\{z^{k+j2^n} : j \in \mathbb{Z}\}.$$

Let E_k denote the spectral projection of u_n onto $\mathcal{H}_{n,k}$. Consider the action of multiplication by z on these subspaces. It is evident that

$$M_z z^{k+jn} = z^{k+1+j2^n} \quad \text{for} \quad j \in \mathbb{Z} \text{ and } 0 \le k < 2^n - 1$$

and

$$M_z z^{(2^n-1)+j2^n} = z^{0+(j+1)2^n} \quad \text{for} \quad j \in \mathbb{Z}.$$

This makes it evident that with respect to this decomposition, M_z has the matrix decomposition

$$M_z \simeq \begin{bmatrix} 0 & 0 & \cdots & 0 & U \\ I & 0 & \cdots & 0 & 0 \\ 0 & I & \cdots & 0 & 0 \\ \vdots & \vdots & \ddots & \vdots & \vdots \\ 0 & 0 & \cdots & I & 0 \end{bmatrix}$$

where U is the bilateral shift. Thus it is clear that $E_{ij} = E_i M_z^{i-j} E_j$ forms a set of matrix units for $\mathcal{M}_{2^n}$ which are matrices with scalar entries with respect to this basis. Then $\mathfrak{A}_n = C^*(u_n, M_z)$ is seen to be isomorphic to

$$\mathfrak{A}_n = \mathcal{M}_{2^n}(C^*(U)) \simeq \mathcal{M}_{2^n}(C(\mathbb{T})).$$

Next consider the imbedding of $\mathfrak{A}_n$ into $\mathfrak{A}_{n+1}$. A moment's reflection shows that the operator U becomes the 2×2 matrix $\begin{bmatrix} 0 & U \\ I & 0 \end{bmatrix}$ when $\mathcal{H}_{n,k}$ is split into $\mathcal{H}_{n+1,k} \oplus \mathcal{H}_{n+1,k+2^n}$. This is just the two times around imbedding from section V.3. Thus $\mathfrak{A}$ is the 2^∞ Bunce–Deddens algebra as asserted.

The next step is to consider the order two automorphism of $\mathfrak{A}$ given by

$$\sigma(f)(t) = f(-t) \quad \text{for } f \in C(\mathbb{T}) \quad \text{and} \quad \sigma(R_d) = R_{-d} \quad \text{for all } d \in \mathbf{D}.$$

Covariance is preserved because

$$(\operatorname{Ad} \sigma(R_d))\sigma(f)(t) = R_{-d}\sigma(f)R_d(t) = \sigma(f)(t+d)$$
$$= f(-t-d) = (R_d f R_{-d})(-t) = \sigma((\operatorname{Ad} R_d)f)(t).$$

Consider the algebra $\mathfrak{B} = \mathfrak{A} \times_\sigma \mathbb{Z}_2$; and let v represent the order 2 unitary implementing σ. For $0 \leq k < 2^n$, let $\theta_{n,k} = 2^{-n-1}(2k+1)$. Then $u_n^{2k+1}v$ implements the automorphism associated to the homeomorphism of the reflection which sends $\theta_{n,k} + x$ to $\theta_{n,k} - x$. Indeed,

$$u_n^{2k+1}vfvu_n^{-2k-1}(\theta_{n,k} + x) = (vfv)(x - \theta_{n,k}) = f(\theta_{n,k} - x). \quad (1)$$

Fix a positive integer n and a real number $0 < \delta < 2^{-n-1}$. We will define projections in $\mathfrak{A}_n \times_\sigma \mathbb{Z}_2$ which are analogous to the Rieffel projections in $\mathcal{A}_\theta$. Let $\varphi_{n,k}(x)$ be the linear function defined on the line segment $[\theta_{n,k} - \delta, \theta_{n,k} + \delta]$ such that $\varphi_{n,k}(\theta_{n,k} - \delta) = 0$ and $\varphi_{n,k}(\theta_{n,k} + \delta) = 1$. Define functions $f_{n,k}$ and $g_{n,k}$ in $C(\mathbb{T})$ by

$$f_{n,k}(x) = \begin{cases} \varphi_{n,k-1}(x) & \text{for } \theta_{n,k-1} - \delta \leq x \leq \theta_{n,k-1} + \delta \\ 1 & \text{for } \theta_{n,k-1} + \delta \leq x \leq \theta_{n,k} - \delta \\ 1 - \varphi_{n,k}(x) & \text{for } \theta_{n,k} - \delta \leq x \leq \theta_{n,k} + \delta \\ 0 & \text{otherwise} \end{cases}$$

and

$$g_{n,k}(x) = \begin{cases} (\varphi_{n,k}(x)(1 - \varphi_{n,k}(x)))^{1/2} & \text{for } \theta_{n,k} - \delta \leq x \leq \theta_{n,k} + \delta \\ 0 & \text{otherwise} \end{cases}.$$

Then define

$$p_{n,k} = f_{n,k} + (-1)^k(u_n^{2k+1}vg_{n,k} + u_n^{2k-1}vg_{n,k-1}).$$

The calculation parallels section VI.2 for the irrational rotation algebra. Since $g_{n,k}$ is symmetric about $\theta_{n,k}$, it follows from (1) that

$$u_n^{2k+1}vg_{n,k} = g_{n,k}u_n^{2k+1}v.$$

VIII.9. An order 2 automorphism

Thus
$$p^*_{n,k} = f_{n,k} + (-1)^k(g_{n,k}vu_n^{-2k-1} + g_{n,k-1}vu_n^{-2k+1})$$
$$= f_{n,k} + (-1)^k(g_{n,k}u_n^{2k+1}v + g_{n,k-1}u_n^{2k-1}v)$$
$$= f_{n,k} + (-1)^k(u_n^{2k+1}vg_{n,k} + u_n^{2k-1}vg_{n,k-1}) = p_{n,k}$$

Notice that
$$(u_n^{2k+1}vf_{n,k}(x) + f_{n,k}u_n^{2k+1}v)g_{n,k} = u_n^{2k+1}vg_{n,k}\big(f_{n,k}(x) + f_{n,k}(2\theta_{n,k} - x)\big)$$
$$= u_n^{2k+1}vg_{n,k}.$$

and similarly
$$(u_n^{2k-1}vf_{n,k}(x) + f_{n,k}u_n^{2k-1}v)g_{n,k-1}$$
$$= u_n^{2k-1}vg_{n,k-1}\big(f_{n,k}(x) + f_{n,k}(2\theta_{n,k-1} - x)\big)$$
$$= u_n^{2k-1}vg_{n,k-1}.$$

Then calculate
$$p_{n,k}^2 = f_{n,k}^2 + (u_n^{2k+1}vg_{n,k})^2 + (u_n^{2k-1}vg_{n,k-1})^2 +$$
$$+ u_n^{2k+1}vg_{n,k}g_{n,k-1}u_n^{2k-1}v + u_n^{2k-1}vg_{n,k-1}g_{n,k}u_n^{2k+1}v +$$
$$+ (-1)^k(f_{n,k}u_n^{2k+1}vg_{n,k} + u_n^{2k+1}vg_{n,k}f_{n,k}) +$$
$$+ (-1)^k(f_{n,k}u_n^{2k-1}vg_{n,k-1} + u_n^{2k-1}vg_{n,k-1}f_{n,k})$$
$$= f_{n,k}^2 + g_{n,k}^2 + g_{n,k-1}^2 + (-1)^k u_n^{2k+1}vg_{n,k} + (-1)^k u_n^{2k+1}vg_{n,k} = p_{n,k}.$$

Thus $p_{n,k}$ are projections. A simple calculation shows that
$$\sum_{k=0}^{2^n-1} p_{n,k} = \sum_{k=0}^{2^n-1} f_{n,k} + \sum_{k=0}^{2^n-1} \big((-1)^k + (-1)^{k+1}\big)u^{2k+1}vg_{n,k} = 1$$

So the $p_{n,k}$ form a partition of the identity. Note that

$$u_{n-1}p_{n,k}u_{n-1}$$
$$= u_n^2 f_{n,k}u_n^{-2} + (-1)^k u_n^{2k+3}vg_{n,k}u_n^{-2} + (-1)^k u_n^{2k+1}vg_{n,k-1}u_n^{-2}$$
$$= f_{k+2} + (-1)^{k+2}u_n^{2k+3}vu_n^{-2}g_{n,k+2} + (-1)^{k+2}u_n^{2k+1}vu_n^{-2}g_{n,k+1}$$
$$= f_{k+2} + (-1)^{k+2}u_n^{2k+5}vg_{n,k+2} + (-1)^{k+2}u_n^{2k+3}vg_{n,k+1} = p_{n,k+2}.$$

Also $vp_{n,k}v = p_{n,2^n-k}$ and $vu_{n-1}v = u_{n-1}^*$. So the algebra
$$\mathfrak{B}_n = C^*(u_{n-1}, v, p_{n,k}, 0 \le k < 2^n)$$
is finite dimensional. In fact, it is isomorphic to the direct sum of four copies of $\mathcal{M}_{2^{n-1}}$ (see the Exercises).

To see that the union of the $\mathfrak{B}_n$ is dense in $\mathfrak{B}$, note that the set of polynomials in v and u_n for $n \ge 1$ with coefficients in $C(\mathbb{T})$ is dense in $\mathfrak{B}$. Clearly v and u_n belong to $\mathfrak{B}_{n+1}$. So it suffices to approximate each function f in $C(\mathbb{T})$. This follows from an easy partition of unity type estimate. By uniform continuity, given $\varepsilon > 0$, there is an integer n so that $|f(x) - f(y)| < \varepsilon$ when $|x - y| < 2^{-n}$. The support of the functions $f_{n,k}$, $g_{n,k}$ and $g_{n,k-1}$ are contained in $\{x : |x - 2^{-n}k| < 2^{-n}\}$. Hence we see that $f - f(\frac{k}{2^n})p_{n,k} = \varepsilon_k p_{n,k}$ where ε_k is a function in $C(\mathbb{T})$ supported on the interval $[(k-1)2^{-n}, (k+1)2^{-n}]$ with norm at most ε. Thus $\overline{\varepsilon_\ell}\varepsilon_k = 0$ if $|\ell - k| > 1$. So

$$\left\| f - \sum_{k=0}^{2^n-1} f(\tfrac{k}{2^n})p_{n,k} \right\|^2 = \left\| \sum_{k=0}^{2^n-1} \varepsilon_k p_{n,k} \right\|^2 = \left\| \sum_{k=0}^{2^n-1} \sum_{\ell=0}^{2^n-1} p_{n,\ell}\overline{\varepsilon_\ell}\varepsilon_k p_{n,k} \right\|$$

$$\leq \left\| \sum_{k=0}^{2^n-1} p_{n,k}|\varepsilon_k|^2 p_{n,k} \right\| + 2\left\| \sum_{k=0}^{2^n-1} p_{n,k+1}\overline{\varepsilon_{k+1}}\varepsilon_k p_{n,k} \right\|$$

$$\leq 3\varepsilon^2.$$

Therefore $\mathfrak{B}$ is AF.

It follows that $\mathrm{Ad}\, v$ is an inner automorphism of $\mathfrak{B}$ of order two with $\mathfrak{A}$ as its fixed point algebra. Since this is a Bunce–Deddens algebra, it is not AF by Theorem V.3.4.

Exercises

VIII.1 Suppose that $(\mathfrak{A}, G, \alpha)$ is a discrete dynamical system. Show that if $\mathfrak{A}$ is not unital and E_n is an approximate unit for $\mathfrak{A}$, then any representation σ of $\mathfrak{A}G$ yields a covariant representation of $(\mathfrak{A}, G, \alpha)$ by

$$\pi(A) = \sigma(Ae) \quad \text{and} \quad U_s = \lim_{n\to\infty} \sigma(E_n s).$$

VIII.2 Show that if U, V are in $GL_n(\mathfrak{A})$, then $U \oplus I_n$ and $I_n \oplus U$ are connected in $GL_{2n}(\mathfrak{A})$. Hence show that $UV \oplus I_n$ and $VU \oplus I_n$ are connected. Deduce that $K_1(\mathfrak{A})$ is always abelian.

VIII.3 Show that if $(C(X), u)$ is a covariant representation of a minimal system, then the spectrum of u is the full unit circle.

VIII.4 If X is an n element set and σ is a cyclic permutation, show that $C(X) \times_\sigma \mathbb{Z}$ is isomorphic to $\mathcal{M}_n(C(\mathbb{T}))$.

VIII.5 Use the Pimsner–Voiculescu exact sequence to compute K_1 of the Bunce–Deddens algebras.

VIII.6 Show that for a minimal system (X, σ) on a Cantor set and a closed subset Y of X, there is an exact sequence of groups

$$0 \longrightarrow \mathbb{Z} \xrightarrow{\alpha} C(Y, \mathbb{Z}) \xrightarrow{\beta} K_0(\mathfrak{A}_Y) \xrightarrow{i_*} K_0(C(X) \times_\sigma \mathbb{Z}) \longrightarrow 0$$

where $\alpha(n) = n\chi_Y$, j is the imbedding of $C(X)$ into $\mathfrak{A}_Y$, i is the imbedding of $\mathfrak{A}_Y$ into $C(X) \times_\sigma \mathbb{Z}$ and $\beta(g) = j_*(g - \sigma(g))$.

Exercises

VIII.7 Use the Pimsner–Voiculescu exact sequence to compute K_0 of the algebra $\mathfrak{A}$ of Example VIII.7.6. Hence show that the imbedding of $\mathcal{A}_\theta$ into $\mathfrak{A}$ induces an isomorphism of K_0 groups. Hence deduce that this is an imbedding into the same AF algebra as the Pimsner–Voiculescu imbedding.

VIII.8 Suppose that $\mathfrak{B} = \varinjlim(\mathfrak{B}_n, \beta_n)$, and that there are homomorphisms φ_n of $\mathfrak{B}_n$ into a C*-algebra $\mathfrak{A}$ such that for every element B in $\mathfrak{B}_n$, the sequence $\varphi_n \beta_{k,n}(A)$ is Cauchy. Show that this limit yields a homomorphism of $\mathfrak{B}$ into $\mathfrak{A}$. Hence verify the details of Theorem VIII.7.5.

VIII.9 Show that $\mathfrak{A}$ has topological stable rank one if and only if $\mathfrak{K} \otimes \mathfrak{A}$ has. (Recall that $\mathfrak{K} \otimes \mathfrak{A}$ is the direct limit of $\mathcal{M}_n(\mathfrak{A})$.)

VIII.10 Show that the algebra $\mathfrak{B}_n = C^*(u_{n-1}, v, p_{n,k}, 0 \leq k < 2^n)$ of the section VIII.9 is isomorphic to the direct sum of four copies of $\mathcal{M}_{2^{n-1}}$.
HINT: First show that the sum P_e of the $p_{n,k}$ for even k commutes with u_{n-1} and v; and that the even (odd) projections and $P_e u_{n-1}$ ($P_e^\perp u_{n-1}$) generate a copy of $\mathcal{M}_{2^{n-1}}$. Then note that the action of v on $\mathcal{M}_{2^{n-1}}$ is equivalent to an inner automorphism by a unitary w. Show that the projection $Q = (wv + I)/2$ commutes with both v and $\mathcal{M}_{2^{n-1}}$.

Notes and Remarks.

The best comprehensive text on crossed products of C*-algebras is Pedersen [1979]. The simplicity of C*-algebras of minimal actions by amenable groups is due to Zeller-Meier [1968]. The proof for $\mathbb{Z}$ actions is taken from Power [1978]. The crossed product representation of Bunce–Deddens algebras goes back to their original 1975 paper. Pimsner and Voiculescu [1980b] established the short exact sequence for K-groups of crossed products, and computed these groups for the irrational rotation algebras. Cuntz [1981] computed the K-groups of $\mathcal{O}_n$. The analysis of AF subalgebras of minimal crossed products of Cantor sets is due to Putnam [1989]. Versik [1981] establishes the imbedding of these algebras into AF algebras, but the proof here is taken from Putnam's paper which establishes that it yields an isomorphism of K_0 groups. That this leads to the limit circle algebra structure is due to Putman [1990a]. Pimsner [1983] established a more general criterion for imbedding a crossed product into an AF algebra. His construction does not yield an isomorphism of K_0 groups. Topological stable rank was introduced by Rieffel [1983a] where the theory is developed at length. It was independently introduced in Corach and Laratonda [1984]. Putnam [1989] showed that the minimal crossed products on Cantor sets are topological stable rank one; and used this to prove that the irrational rotation algebras are also stable rank one in Putnam [1990b]. Elliott and Evans [1993] showed that irrational rotation algebras are limit circle algebras. Blackadar [1990] gave the first example of an order two automorphism of an AF algebra with non-AF fixed point algebra. The example given here is due to Kumjian [1988].

CHAPTER IX

Brown–Douglas–Fillmore Theory

An operator T in $\mathcal{B}(\mathcal{H})$ is called **essentially normal** if $T^*T - TT^*$ is compact. Let π be the quotient map of $\mathcal{B}(\mathcal{H})$ onto the Calkin algebra $\mathcal{Q}(\mathcal{H}) := \mathcal{B}(\mathcal{H})/\mathfrak{K}$. Then we can restate this as saying that $\pi(T)$ is a normal element of $\mathcal{Q}(\mathcal{H})$.

The problem is to classify essentially normal operators. As we shall see, this is handled by solving a C*-algebra problem about extensions of the compact operators. The first step is to decide on the right notion of equivalence to use to identify two operators which are essentially the same. A natural choice for operators is unitary equivalence, that is $A \simeq B$ if there is a unitary operator U such that $B = UAU^*$. However, in this case, there is an extra ingredient. If T is an essentially normal operator, then $T + K$ is also essentially normal for any K in $\mathfrak{K}$ (a compact perturbation of T). So we will include this as a trivial change. We say that two operators A and B are **compalent** if there is a unitary operator U such that $B - UAU^*$ is compact.

Another natural choice would be to use unitary equivalence in the Calkin algebra. The notion of compalence of A and B may be reformulated as saying that there is a unitary operator U such that $\pi(B) = \pi(U)\pi(A)\pi(U)^*$. So we might think to use unitary elements of $\mathcal{Q}(\mathcal{H})$ instead, since the elements $\pi(U)$ are precisely the unitary elements in $\mathcal{Q}(\mathcal{H})$ of index 0. Say that A and B are **weakly compalent** if there is a unitary element u in $\mathcal{Q}(\mathcal{H})$ such that $\pi(B) = u\pi(A)u^*$. For essentially normal operators, these two notions coincide. However, as we saw in §V.6, this is not the case in general.

If T is essentially normal, then $C^*(\pi(T))$ is an abelian subalgebra of $\mathcal{Q}(\mathcal{H})$ which is canonically isomorphic to $C(\sigma_e(T))$ by the functional calculus (Corollary I.3.2), where $\sigma_e(T) = \sigma(\pi(T))$ is the **essential spectrum** of T. Let

$$\mathfrak{E}(T) = \pi^{-1}C^*(\pi(T)) = C^*(T) + \mathfrak{K}.$$

Then this algebra contains $\mathfrak{K}$ as an ideal, and has quotient equivalent to $C(\sigma_e(T))$ obtained by identifying $\pi(T)$ with the identity function z. We obtain the exact sequence

$$0 \longrightarrow \mathfrak{K} \xrightarrow{\iota} \mathfrak{E}(T) \xrightarrow{\pi} C(\sigma_e(T)) \longrightarrow 0$$

This is an **extension** of the compact operators by the abelian C*-algebra $C(\sigma_e(T))$.

More generally, when X is a compact metric space, we say that an **extension** of $\mathfrak{K}$ by $C(X)$ is a C*-subalgebra $\mathfrak{E}$ of $\mathcal{B}(\mathcal{H})$ containing the compact operators $\mathfrak{K}$

IX. Brown–Douglas–Fillmore Theory

such that
$$0 \longrightarrow \mathfrak{K} \xrightarrow{\iota} \mathfrak{E} \xrightarrow{\varphi} C(X) \longrightarrow 0$$
is exact. Two extensions $(\mathfrak{E}_1, \varphi_1)$ and $(\mathfrak{E}_2, \varphi_2)$ are said to be **equivalent** if there is a $*$-isomorphism ψ of $\mathfrak{E}_1$ onto $\mathfrak{E}_2$ such that $\varphi_1 = \varphi_2 \psi$. That is, the following diagram commutes:

$$\begin{array}{ccccccccc} 0 & \longrightarrow & \mathfrak{K} & \xrightarrow{\iota} & \mathfrak{E}_1 & \xrightarrow{\varphi_1} & C(X) & \longrightarrow & 0 \\ & & \downarrow \psi|\mathfrak{K} & & \downarrow \psi & & \| & & \\ 0 & \longrightarrow & \mathfrak{K} & \xrightarrow{\iota} & \mathfrak{E}_2 & \xrightarrow{\varphi_2} & C(X) & \longrightarrow & 0 \end{array}$$

By Lemma V.6.1, there is a unitary operator U such that $\psi|\mathfrak{K} = \operatorname{Ad} U$. Thus $\psi = \operatorname{Ad} U$ because of the identity

$$\psi(A)\psi(K) = \psi(AK) = \operatorname{Ad} U(AK)$$
$$= \operatorname{Ad} U(A) \operatorname{Ad} U(K) = \operatorname{Ad} U(A) \psi(K).$$

Since $\mathfrak{K}$ is an essential ideal of $\mathcal{B}(\mathcal{H})$, this implies that $\psi(A) = \operatorname{Ad} U(A)$ for all A in $\mathfrak{E}_1$. So $(\mathfrak{E}_1, \varphi_1)$ and $(\mathfrak{E}_2, \varphi_2)$ are equivalent if and only if there is a unitary operator U such that $\mathfrak{E}_2 = U\mathfrak{E}_1 U^*$ and $\varphi_1 = \varphi_2 \operatorname{Ad} U$.

Returning to the case of a single operator, the metric space $X = \sigma_e(T)$ is contained in $\mathbb{C}$. Thus $C(X)$ has a single generator $z(t) = t$. It follows that two extensions $\mathfrak{E}(T_i)$ are equivalent if and only if T_1 and T_2 are compalent. Indeed, when $T_2 = UT_1U^* + K$ for a unitary operator U and compact operator K, the map $\psi = \operatorname{Ad} U$ is the desired map since

$$\varphi_2\psi(T_1) = \varphi_2(T_2 + K) = z = \varphi_1(T_1).$$

As T_i generate $\mathfrak{E}_i$ modulo $\mathfrak{K}$, this implies that $\varphi_1 = \psi\varphi_2$. Conversely, if $\mathfrak{E}_i$ are equivalent via $\psi = \operatorname{Ad} U$, then it follows that $T_2 - UT_1U^*$ lies in $\ker \varphi_2 = \mathfrak{K}$. So T_1 and T_2 are compalent.

Associated to an extension $(\mathfrak{E}, \varphi)$ of $\mathfrak{K}$ by $C(X)$ is a $*$-monomorphism τ of $C(X)$ into $\mathcal{Q}(\mathcal{H})$ given by $\tau(f) = \pi\varphi^{-1}(f)$, which is the inverse of the identification of the subalgebra $\mathfrak{E}/\mathfrak{K}$ of $\mathcal{Q}(\mathcal{H})$ with $C(X)$ via φ. Conversely, if τ is a $*$-monomorphism of $C(X)$ into $\mathcal{Q}(\mathcal{H})$, let $\mathfrak{E} = \pi^{-1}\tau(C(X))$ and set $\varphi = \tau^{-1}\pi$. Then it is clear that $\mathfrak{E}$ is a C^*-subalgebra of $\mathcal{B}(\mathcal{H})$ containing the compact operators such that $\mathfrak{E}/\mathfrak{K}$ is isomorphic to $C(X)$ via φ. So $\mathfrak{E}$ is an extension of $\mathfrak{K}$ by $C(X)$. With this formulation, τ_1 and τ_2 are **equivalent** if there is a unitary operator U such that $\tau_2 = \operatorname{Ad} \pi(U) \tau_1$. In the context of planar sets, an essentially normal operator T with $\sigma_e(T) = X$ is identified with the monomorphism $\tau(f) = f(\pi(T))$ by the functional calculus for the normal element $\pi(T)$.

We will work with both formulations of an extension as is convenient. The reader should note that the C^*-algebra $C(X)$ may be replaced by any C^*-algebra in the discussion above. Indeed, in chapter V, this was considered for the C*-algebra $\mathcal{O}_n$.

Two extensions τ_1 and τ_2 are called **weakly equivalent** if there is a unitary element u in $\mathcal{Q}(\mathcal{H})$ such that $\tau_2 = \operatorname{Ad} u\, \tau_1$.

We denote by $\operatorname{Ext}(X)$ the collection of all equivalence classes of extensions. The goal of this chapter is to show that $\operatorname{Ext}(X)$ is an abelian group, and that the map Ext is a functor from compact metric spaces into abelian groups that has the properties of a homology theory. This will enable us to compute $\operatorname{Ext}(X)$ for compact subsets of the plane, and hence classify essentially normal operators. These results are known as the **Brown–Douglas–Fillmore (BDF) Theory**.

This is a rather unusual approach to a problem in operator theory that sounds much more pedestrian at first sight. However, twenty years after the original proof, there still is no easy way to obtain a proof (although there is an "operator theoretic" proof). But the real impact of BDF came from the introduction of topological methods into C*-algebras. This connection was one of the most important developments in C*-algebras in this century.

IX.2 An Addition and Zero Element for Ext(X).

The group operation $+$ on $\operatorname{Ext}(X)$ is most easily defined for the monomorphism version. Set
$$[\tau_1] + [\tau_2] = [\tau_1 \oplus \tau_2].$$
It is implicit in this definition that we may identify $\mathcal{M}_2(\mathcal{Q}(\mathcal{H}))$ with $\mathcal{Q}(\mathcal{H})$ because we may identify $\mathcal{H} \oplus \mathcal{H}$ with $\mathcal{H}$ for any separable Hilbert space $\mathcal{H}$, and this identification yields an isomorphism of $\mathcal{M}_2(\mathcal{B}(\mathcal{H}))$ onto $\mathcal{B}(\mathcal{H})$ which takes $\mathcal{M}_2(\mathfrak{K})$ onto $\mathfrak{K}$. It is evident that this is a well defined abelian and associative operation. The details of this routine verification will be left to the reader. In the case of essentially normal operators, the generator of the sum of $[T_1]$ and $[T_2]$ is $[T_1 \oplus T_2]$. See the exercises for the other formulation of sum for extensions $(\mathfrak{E}, \varphi)$.

The first step is to identify a zero element. An extension $(\mathfrak{E}, \varphi)$ is called **trivial** if there is a *-monomorphism σ of $C(X)$ into $\mathfrak{E}$ such that $\varphi \sigma = \operatorname{id}_{C(X)}$. In other words, we obtain the split exact sequence
$$0 \longrightarrow \mathfrak{K} \overset{\iota}{\longrightarrow} \mathfrak{E} \underset{\sigma}{\overset{\varphi}{\rightleftarrows}} C(X) \longrightarrow 0$$
In the planar case, $N = \sigma(z)$ is a normal operator with spectrum and essential spectrum equal to X such that $\mathfrak{E} = C^*(N) + \mathfrak{K}$. Conversely, given such a normal operator N, the map $\sigma(f) = f(N)$ is the desired splitting of the extension defined by N.

The main step in showing that the trivial extensions form a single equivalence class which acts as a zero element for $\operatorname{Ext}(X)$ is the Weyl–von Neumann Theorem (Theorem II.4.6) for representations of $C(X)$. This result also follows from Theorem V.6.3, but we give a more elementary proof here.

IX.2. An Addition and Zero Element for Ext(X).

Theorem IX.2.1 *Let X be a compact metric space. Then $C(X)$ has trivial extensions, and all trivial extensions of $\mathfrak{K}$ by $C(X)$ are equivalent.*

Proof. Existence is easy. Take any countable dense subset $\{\xi_n\}$ of X and define $\sigma(f) = \sum_{n\geq 1} \oplus f(\xi_n)I$. This is a $*$-isomorphism of infinite multiplicity, so the range contains no compact operators. Thus $\pi\sigma$ is also a monomorphism. It determines an extension which, by construction, has the section σ. Thus it is a trivial extension.

Two trivial extensions of $C(X)$ are given by $*$-isomorphisms ρ and σ of $C(X)$ into $\mathcal{B}(\mathcal{H})$ such that $\pi\rho$ and $\pi\sigma$ are still injective. Thus the ranges of ρ and σ contain no compact operators, whence $\operatorname{rank} \rho(f) = \operatorname{rank} \sigma(f) = \infty$ for every non-zero function f in $C(X)$. Hence by Theorem II.4.6, ρ and σ are approximately unitarily equivalent relative to the compact operators. In particular, there is a unitary operator U such that $\operatorname{Ad} U \rho - \sigma$ has range in the compact operators. Hence $\operatorname{Ad} \pi(U) \pi\rho = \pi\sigma$ and so ρ and σ determine equivalent extensions. ∎

To show that the trivial class represents a zero element, we need to show that every extension essentially contains a big diagonal representation as a summand. This is accomplished by pulling out lots of approximate eigenvectors. Again, this follows from Voiculescu's Theorem as in Theorem V.6.3. But we will show that it is more elementary in the commutative case. Recall that if $\{a_1, a_2, \ldots\}$ are elements of a commutative Banach algebra $\mathcal{B}$, then their **joint spectrum** is the set of sequences

$$\{(\varphi(a_1), \varphi(a_2), \ldots) : \varphi \in \mathcal{M}_\mathcal{B}\};$$

where $\mathcal{M}_\mathcal{B}$ is the maximal ideal space of $\mathcal{B}$. In a C*-algebra, it is easy to see that $\lambda = (\lambda_1, \lambda_2, \ldots)$ is in the joint spectrum of $(A_1, A_2, \ldots)$ if and only if

$$\sum_{k\geq 1} 2^{-k} \frac{(A_k - \lambda_k I)^*(A_k - \lambda_k I)}{\|A_k - \lambda_k I\|^2}$$

is not invertible. In particular, if $\mathbf{T} = (T_1, T_2, \ldots)$ is an essentially commuting family of essentially normal operators, then the **joint essential spectrum** of $\mathbf{T}$ is just the joint spectrum of $\pi(\mathbf{T}) = (\pi(T_1), \pi(T_2), \ldots)$. So this consists of those λ such that

$$\sum_{k\geq 1} 2^{-k} \frac{(T_k - \lambda_k I)^*(T_k - \lambda_k I)}{\|T_k - \lambda_k I\|^2}$$

is not Fredholm.

Lemma IX.2.2 *Let $\mathbf{T} = (T_1, T_2, \ldots)$, be an essentially commuting family of essentially normal operators. Suppose that $\lambda_n = (\lambda_{n,1}, \lambda_{n,2}, \ldots)$ is a sequence of points in the joint essential spectrum of the $\mathbf{T}$. Then there is an orthonormal sequence x_n such that*

$$\lim_{n\to\infty} \|T_k x_n - \lambda_{n,k} x_n\| = 0 \quad \text{for all} \quad k \geq 1.$$

Proof. Suppose that we are given a finite dimensional subspace $\mathcal{L}$ of $\mathcal{H}$, $\varepsilon > 0$ and a point $\boldsymbol{\lambda} = (\lambda_1, \lambda_2, \ldots)$ in the joint essential spectrum of $\mathbf{T}$. Then

$$A = \sum_{k \geq 1} 2^{-k} \frac{(T_k - \lambda_k I)^*(T_k - \lambda_k I)}{\|T_k - \lambda_k I\|^2}$$

is not Fredholm. Hence either A has infinite dimensional kernel or it is not bounded below on the complement of its kernel. In either case, there is a unit vector x orthogonal to $\mathcal{L}$ so that $(Ax, x) < \varepsilon^2$. Thus

$$\|T_k x - \lambda_k x\|^2 \leq 2^k \|T_k - \lambda_k I\|^2 (Ax, x)$$
$$\leq 2^k 4 \|T_k\|^2 \varepsilon^2 = (2^{1+k/2} \|T_k\| \varepsilon)^2.$$

Now proceed recursively. At the n-th stage, we suppose that we have constructed an orthonormal set $x_1, \ldots, x_{n-1}$ so that

$$\|T_k x_i - \lambda_{i,k} x_i\| < 2^{-i} \quad \text{for all} \quad 1 \leq k \leq i \quad \text{and} \quad 1 \leq i \leq n-1.$$

Then set $\mathcal{L} = \text{span}\{x_1, \ldots, x_{n-1}\}$ and choose ε sufficiently small so that

$$\max_{1 \leq k \leq n} 2^{1+k/2} \|T_k\| \varepsilon < 2^{-n}.$$

Then applying the previous paragraph, one obtains a unit vector x_n orthogonal to $\mathcal{L}$ so that $\|T_k x_n - \lambda_{n,k} x_n\| < 2^{-n}$ for $1 \leq k \leq n$. This is the desired sequence. ∎

Theorem IX.2.3 *Let X be a compact metric space. The class of trivial extensions for $C(X)$ forms a zero element for $\text{Ext}(X)$.*

Proof. Suppose that τ is a $*$-monomorphism of $C(X)$ into $\mathcal{Q}(\mathcal{H})$ representing an element of $\text{Ext}(X)$. Let $f_1, f_2, \ldots$ be a countable family of norm one positive functions in $C(X)$ which separate points of X. Then the joint spectrum Λ of $(f_1, f_2, \ldots)$ is homeomorphic to X via the map which takes ξ in X to $(f_1(\xi), f_2(\xi), \ldots)$.

Let T_k be positive elements of norm 1 in $\mathcal{B}(\mathcal{H})$ such that $\pi(T_k) = \tau(f_k)$. Then $(T_1, T_2, \ldots)$ is an essentially commuting family of essentially normal operators with joint essential spectrum Λ. Apply the previous lemma to a sequence $\boldsymbol{\lambda}_n$ which is dense in Λ and repeats each isolated point infinitely often to obtain orthogonal unit vectors x_n so that $\|T_k x_n - \lambda_{n,k} x_n\| < 2^{-n}$ for $1 \leq k \leq n$. Let P be the projection onto $\text{span}\{x_1, x_2, \ldots\}$. We will show that $p = \pi(P)$ commutes with every $\tau(f)$ and that $p\tau p$ is a trivial extension into $\mathcal{Q}(P\mathcal{H})$.

IX.2. An Addition and Zero Element for Ext(X).

Indeed, for each k, let D_k denote the diagonal operator on $P\mathcal{H}$ given by $D_k x_n = \lambda_{n,k} x_n$. Then

$$\|T_k P - D_k\|_2^2 = \sum_{n \geq 1} \|(T_k - D_k)x_n\|^2$$

$$\leq (k-1) + \sum_{n \geq k} \|T_k x_k - \lambda_{n,k} x_k\|^2$$

$$\leq k - 1 + \sum_{n \geq k} 4^{-n} < k.$$

Hence $T_k P - D_k$ is a Hilbert-Schmidt operator for every k, and thus is compact. Therefore, $P^\perp T_k P = P^\perp(T_k - D_k)P$ is compact; whence

$$PT_k - T_k P = PT_k P^\perp - P^\perp T_k P = -2i \operatorname{Im}(P^\perp T_k P)$$

is also compact. That means that p commutes with each $\tau(f_k)$. As they generate $C(X)$, p commutes with the range of τ. Moreover, $T_k = (D_k \oplus P^\perp T_k P^\perp) + K_k$ where K_k is compact.

Let ξ_n be the point in X corresponding to the point λ_n in Λ. Evidently, the map $\sigma(f) = \operatorname{diag}(f(\xi_n))$ with respect to the basis x_n of $P\mathcal{H}$ is a trivial extension by $C(X)$ because $\{\xi_n : n \geq 1\}$ is dense in X and isolated points have infinite multiplicity. From the preceding paragraph, we see that

$$\tau(f) = \pi\sigma(f) \oplus (1-p)\tau(f)(1-p).$$

By Theorem IX.2.1, $\pi\sigma$ is unitarily equivalent to $\pi\sigma \oplus \pi\sigma$. We have

$$\tau(f) = \pi\sigma(f) \oplus (1-p)\tau(f)(1-p)$$
$$\simeq \pi\sigma(f) \oplus \pi\sigma(f) \oplus (1-p)\tau(f)(1-p)$$
$$= \pi\sigma(f) \oplus \tau(f).$$

Hence $[\tau] + [\sigma] = [\tau]$. That is, $[\sigma]$ is a zero element for Ext(X). ∎

We can use this result to show that weak equivalence and equivalence coincide in Ext(X). Recall that this is not the case for the Cuntz algebras.

Corollary IX.2.4 *Weakly equivalent extensions by $C(X)$ are equivalent.*

Proof. Let τ be an extension by $C(X)$. It suffices to show that $\operatorname{Ad} u\,\tau$ and τ are equivalent for every unitary element of $\mathcal{Q}(\mathcal{H})$. However, if s is any particular unitary element of index 1, then every unitary u in $\mathcal{Q}(\mathcal{H})$ has the form $\pi(U)s^k$ for some unitary U in $\mathcal{B}(\mathcal{H})$ and integer k in $\mathbb{Z}$. Thus it is enough to find one particular s of index 1 so that $[\operatorname{Ad} s\,\tau] = [\tau]$.

Let σ be the diagonal representation constructed above; and let S be the backward shift with respect to the basis x_n. Then

$$\sigma'(f) := \operatorname{Ad} S\,\sigma(f) = \operatorname{diag}(f(\xi_{n+1}))$$

is another diagonal representation. The density of the sequence ξ_n shows that $\pi\sigma'$ is another trivial extension. By Theorem IX.2.1, $[\pi\sigma'] = [\pi\sigma]$. Hence

$$[\tau] = [\tau \oplus \pi\sigma] = [\tau \oplus \pi\sigma'] = [\operatorname{Ad} \pi(I \oplus S)\, \tau \oplus \pi\sigma] = [\operatorname{Ad} s\, \tau]$$

where s is the unitary of index 1 corresponding to $\pi(I \oplus S)$ under the equivalence between $\tau \oplus \pi\sigma$ and τ. ∎

IX.3 Some Special Cases

Before going on to the general theory, we show how to compute a few important examples.

Theorem IX.3.1 *If $\mathrm{C}(X)$ is generated by a single real valued function, then every extension is trivial and so $\operatorname{Ext}(X) = \{0\}$. In particular, this is the case if X is a subset of $\mathbb{R}$.*

Proof. Let a denote the real generator of $\mathrm{C}(X)$, and let $Y := a(X) = \sigma(a)$. Then $\mathrm{C}(X) = \{f(a) : f \in \mathrm{C}(Y)\}$. For any extension τ of $\mathfrak{K}$ by $\mathrm{C}(X)$, $\tau(a)$ is Hermitian. So there is a Hermitian operator A such that $\pi A = \tau(a)$. Replace A by a compact perturbation if necessary so that $\sigma(A) = \sigma_e(A) = \sigma(\tau(a))$. Define a representation ρ of $\mathrm{C}(X)$ by the functional calculus of A:

$$\rho(f(a)) = f(A) \quad \text{for} \quad f \in \mathrm{C}(Y).$$

It is clear that $\tau(f(a)) = \pi f(A) = \pi\rho(f(a))$. So τ is a trivial extension.

In particular, if X is contained in $\mathbb{R}$, then the coordinate function $z(t) = t$ generates $\mathrm{C}(X)$ by the Stone–Weierstrass Theorem. ∎

Recall that a topological space X is **totally disconnected** if the topology is generated by a family of clopen sets.

Corollary IX.3.2 *If $\mathrm{C}(X)$ is generated by its projections, then $Ext(X) = \{0\}$. This occurs when X is totally disconnected.*

Proof. From the proof of Corollary II.4.5, we saw that if $\mathrm{C}(X)$ is generated by a family $\mathcal{E} = \{E_n\}$ of commuting projections, then $\mathrm{C}(X)$ is generated by the Hermitian operator $A = \sum_{n \geq 1} 3^{-n} E_n$. Projections in $\mathrm{C}(X)$ are characteristic functions of clopen sets. By the Stone–Weierstrass Theorem, they generate $\mathrm{C}(X)$ if and only if they separate points. They separate points exactly when the clopen sets generate the topology. ∎

The case of the unit circle $\mathbb{T}$ is the problem of classifying unitary elements of $\mathcal{Q}(\mathcal{H})$ with spectrum equal to the whole circle up to compalence. Non-trivial extensions may be recognized by non-zero Fredholm index. In particular, the Toeplitz extension determined by $\tau(f) = \pi T_f$ is generated by the unilateral shift $T_z = \tau(z)$ which has index -1. As unitary equivalence and compact perturbations preserve Fredholm index, this extension is not trivial. Similarly, there is an extension

$$\tau'(f) = f(\pi T_z^*) = \pi T_{\bar{f}}$$

IX.4. Positive Maps

where $\tilde{f}(z) = f(\bar{z})$ for f in $C(\mathbb{T})$. This yields the same C*-algebra $\mathcal{T}(C(\mathbb{T}))$ since $C^*(T_z^*) = C^*(T_z)$, but the maps onto $C(\mathbb{T})$ are different. One can see that τ' is a different extension because the Fredholm index $\text{ind}(\tau'(z)) = \text{ind}(T_z^*) = +1$.

Theorem IX.3.3 $\text{Ext}(\mathbb{T}) = \mathbb{Z}$ *is generated by the Toeplitz extension. The Fredholm index* $\gamma([\rho]) = \text{ind}\,\rho(z)$ *is an isomorphism from* $\text{Ext}(\mathbb{T})$ *onto* $\mathbb{Z}$.

Proof. First note that $\gamma([\rho])$ is well defined. For if ρ and ρ' are two representatives of $[\rho]$, then there is a unitary operator U such that $\rho' = \text{Ad}\,U\,\rho$. Hence

$$\text{ind}\,\rho'(z) = \text{ind}\,U\rho(z)U^* = \text{ind}\,\rho(z).$$

This map is a homomorphism because

$$\gamma([\rho_1] + [\rho_2]) = \text{ind}(\rho_1(z) \oplus \rho_2(z))$$
$$= \text{ind}\,\rho_1(z) + \text{ind}\,\rho_2(z) = \gamma([\rho_1]) + \gamma([\rho_2]).$$

Let ρ be any extension by $C(\mathbb{T})$, and let W be chosen so that $\pi W = \rho(z)$, where z is the identity function on $\mathbb{T}$. Let U be the partial isometry in the polar decomposition of W. Then $\pi(W^*W)^{1/2} = \pi I$, so $\pi U = \pi W$. Moreover, $\pi(U^*U) = \pi I = \pi(UU^*)$, so $I - U^*U$ and $I - UU^*$ are finite rank projections and U is Fredholm. Let $k = \text{ind}\,U$.

When $k = 0$, U has a finite rank perturbation U' which is unitary obtained by adding a finite rank partial isometry from $(I - U^*U)\mathcal{H}$ onto $(I - UU^*)\mathcal{H}$. Thus $\sigma(f) = f(U')$ provides a representation of $C(\mathbb{T})$ such that $\pi\sigma = \rho$. Hence ρ is trivial. It follows that γ is injective.

Let τ denote the Toeplitz extension. Then

$$\gamma([\tau]) = \text{ind}\,\tau(z) = \text{ind}\,T_z = -1.$$

Likewise, the extension τ' generated by the backward shift satisfies $\gamma[\tau']) = 1$. Since ± 1 generate $\mathbb{Z}$ as a semigroup, γ is surjective and thus an isomorphism. ∎

IX.4 Positive Maps

A **positive linear map** between C*-algebras is just a linear map which takes positive operators to positive operators. Hence it is a map which preserves the order structure. Let $\mathcal{P}_1(\mathfrak{A}, \mathfrak{B})$ denote the space of all positive unital maps from $\mathfrak{A}$ into $\mathfrak{B}$. If φ is a map between C*-algebras $\mathfrak{A}$ and $\mathfrak{B}$, it induces a map $\varphi^{(n)}$ from $\mathcal{M}_n(\mathfrak{A})$ into $\mathcal{M}_n(\mathfrak{B})$ by

$$\varphi^{(n)}([A_{ij}]) = [\varphi(A_{ij})].$$

Say that φ is **n-positive** if $\varphi^{(n)}$ is positive and **completely positive** is it is n-positive for all $n \geq 1$. These maps were introduced briefly in section II.5. In this section, we will develop some of their basic properties.

A positive map φ on a unital C*-algebra $\mathfrak{A}$ is bounded by $\|\varphi(1)\|$ on $\mathfrak{A}_{sa}$. Indeed every self-adjoint element of norm 1 is the difference of two positive contractions, say $A = P_1 - P_2$. Since $0 \leq \varphi(P_i) \leq \varphi(I) \leq \|\varphi(I)\|I$, it follows that

$$-\|\varphi(I)\|I \leq \varphi(A) \leq \|\varphi(I)\|I.$$

So the norm is always bounded by $2\|\varphi(I)\|$ on $\mathfrak{A}$. In particular, φ is continuous. (In the commutative case, these maps always have norm $\|\varphi(1)\|$; but the 2 is needed in general.)

Proposition IX.4.1 *Let X be a compact Hausdorff space. Every positive map φ from $C(X)$ into a C*-algebra $\mathfrak{B}$ is completely positive.*

Proof. There is a canonical isomorphism between $\mathcal{M}_n(C(X))$ and $C(X, \mathcal{M}_n)$, the space of continuous $\mathcal{M}_n$-valued functions. An element F in $C(X, \mathcal{M}_n)$ is positive if and only if $F(\xi) \geq 0$ for all ξ in X. First note that if $F(\xi) = f(\xi)T$, where f belongs to $C(X)$ and T is a positive scalar matrix in $\mathcal{M}_n$, then

$$\varphi^{(n)}(F) = [\varphi(f)t_{ij}] = \varphi(f)T.$$

In particular, if $f \geq 0$, then $\varphi(f) \geq 0$ and so $\varphi^{(n)}(F) \geq 0$.

Let F in $C(X, \mathcal{M}_n^+)$ and $\varepsilon > 0$ be given. By continuity, there is a finite open cover $\mathcal{O}_1, \ldots, \mathcal{O}_k$ of X such that $\|F(\xi) - F(\zeta)\| < \varepsilon$ when $\xi, \zeta \in \mathcal{O}_i$. Fix a point ξ_i in $\mathcal{O}_i$ for each $1 \leq i \leq k$ and set $T_i = F(\xi_i)$. Let p_i be a partition of unity for this open cover. That is, $p_i \geq 0$, $p_i(\xi) = 0$ for $\xi \notin \mathcal{O}_i$, and $\sum_{i=1}^k p_i = 1$. Then

$$\|F(\zeta) - \sum_{i=1}^k p_i(\zeta)T_i\| \leq \sum_{i=1}^k p_i(\zeta)\|F(\zeta) - T_i\| < \varepsilon$$

because $\|F(\zeta) - T_i\| < \varepsilon$ when $p_i(\zeta) > 0$.

Thus F is uniformly approximated by a sum of the type considered in the first paragraph. As $\varphi^{(n)}$ is continuous, $\varphi^{(n)}(F)$ is uniformly approximated by the positive maps $\sum_{i=1}^k p_i T_i$ and thus $\varphi^{(n)}$ is positive for all $n \geq 1$. ∎

The usefulness of positive maps on $C(X)$ lies in their rigid structure described in the following dilation theorem of **Naimark**. It says that every positive unital map is the corner of a *-representation.

Theorem IX.4.2 *Let φ be a positive unital map of $C(X)$ into $\mathcal{B}(\mathcal{H})$. Then there is a *-representation σ of $C(X)$ on a Hilbert space $\mathcal{K}$ containing $\mathcal{H}$ such that*

$$\varphi(A) = P_{\mathcal{H}}\sigma(A)|\mathcal{H}.$$

The proof follows immediately from the complete positivity of φ and the following more general dilation theorem due to **Stinespring**. The proof is a souped up version of the Gelfand–Naimark Theorem.

IX.4. Positive Maps

Theorem IX.4.3 *Let φ be a unital completely positive map from a unital C*-algebra $\mathfrak{A}$ into $\mathcal{B}(\mathcal{H})$. Then there is a Hilbert space $\mathcal{K}$ containing $\mathcal{H}$ and a $*$-representation σ of $\mathfrak{A}$ on $\mathcal{K}$ such that $\varphi(A) = P_{\mathcal{H}}\sigma(A)|_{\mathcal{H}}$.*

Proof. Form the algebraic tensor product $\mathfrak{A} \otimes \mathcal{H}$. Define a form on it by

$$\langle \sum_{j=1}^{n} A_j \otimes x_j, \sum_{i=1}^{m} B_i \otimes y_i \rangle = \sum_{i=1}^{m}\sum_{j=1}^{n} (\varphi(B_i^* A_j)x_j, y_i).$$

Evidently, this form is sesquilinear. Moreover, it is positive semidefinite because if $\mathbf{x} = (x_1, \ldots, x_n)^t$, then

$$\langle \sum_{j=1}^{n} A_j \otimes x_j, \sum_{i=1}^{n} A_i \otimes x_i \rangle = \sum_{i=1}^{n}\sum_{j=1}^{n} (\varphi(A_i^* A_j)x_j, x_i)$$
$$= (\varphi^{(n)}([A_i^* A_j])\mathbf{x}, \mathbf{x}) \geq 0.$$

The Cauchy–Schwarz inequality implies that $|\langle u, v \rangle| \leq \langle u, u \rangle^{1/2} \langle v, v \rangle^{1/2}$ for all u, v in $\mathfrak{A} \otimes \mathcal{H}$. Thus the set

$$\mathcal{N} = \{v \in \mathfrak{A} \otimes \mathcal{H} : \langle v, v \rangle = 0\}$$
$$= \{v \in \mathfrak{A} \otimes \mathcal{H} : \langle v, u \rangle = 0 \text{ for all } u \in \mathfrak{A} \otimes \mathcal{H}\}$$

is a subspace of $\mathfrak{A} \otimes \mathcal{H}$. Let $\mathcal{K}$ be the Hilbert space completion of $\mathfrak{A} \otimes \mathcal{H}/\mathcal{N}$ in the positive definite inner product induced by $\langle \cdot, \cdot \rangle$. Define a representation of $\mathfrak{A}$ by

$$\sigma(A)[\sum_{i=1}^{n} B_i \otimes y_i + \mathcal{N}] = [\sum_{i=1}^{n} AB_i \otimes y_i + \mathcal{N}].$$

This is well defined because if v is in $\mathcal{N}$, then $\rho_v(X) = \langle Xv, v \rangle$ is positive; whence

$$0 \leq \langle Xv, Xv \rangle = \rho_v(X^* X) \leq \|X\|^2 \rho_v(I) = 0.$$

So $\sigma(\mathfrak{A})\mathcal{N}$ is contained in $\mathcal{N}$.

Clearly σ is linear and multiplicative. To see that it is self-adjoint, compute

$$\langle \sigma(A^*)[\sum_{j=1}^{n} A_j \otimes x_j + \mathcal{N}], [\sum_{i=1}^{n} B_i \otimes y_i + \mathcal{N}] \rangle$$
$$= \langle [\sum_{j=1}^{n} A^* A_j \otimes x_j + \mathcal{N}], [\sum_{i=1}^{n} B_i \otimes y_i + \mathcal{N}] \rangle$$
$$= \sum_{i=1}^{m}\sum_{j=1}^{n} (\varphi(B_i^* A^* A_j)x_j, y_i)$$
$$= \langle [\sum_{j=1}^{n} A_j \otimes x_j + \mathcal{N}], [\sum_{i=1}^{n} AB_i \otimes y_i + \mathcal{N}] \rangle$$

$$= \langle [\sum_{j=1}^{n} A_j \otimes x_j + \mathcal{N}], \sigma(A)[\sum_{i=1}^{n} B_i \otimes y_i + \mathcal{N}] \rangle$$

$$= \langle \sigma(A)^*[\sum_{j=1}^{n} A_j \otimes x_j + \mathcal{N}], [\sum_{i=1}^{n} B_i \otimes y_i + \mathcal{N}] \rangle.$$

So $\sigma(A^*) = \sigma(A)^*$.

To imbed $\mathcal{H}$ into $\mathcal{K}$, define an operator V from $\mathcal{H}$ into $\mathcal{K}$ by $Vx = [I \otimes x + \mathcal{N}]$. This is isometric because

$$\|Vx\|^2 = \langle [I \otimes x + \mathcal{N}], [I \otimes x + \mathcal{N}] \rangle = (\varphi(I)x, x) = \|x\|^2.$$

Moreover,

$$(V^*\sigma(A)Vx, y) = \langle \sigma(A)[I \otimes x + \mathcal{N}], [I \otimes y + \mathcal{N}] \rangle = (\varphi(A)x, y).$$

Thus $V\sigma V^* = \varphi$. ∎

If φ is a positive unital map of a C*-algebra $\mathfrak{A}$ into a quotient algebra $\mathfrak{B}/\mathfrak{J}$, then say that φ has a **positive lifting** if there is a positive unital map ψ of $\mathfrak{A}$ into $\mathfrak{B}$ such that $\varphi = \pi\psi$. We will show that every positive map on $C(X)$ is liftable. We begin with an elementary but somewhat tricky result of general interest.

Lemma IX.4.4 *Let $\mathfrak{J}$ be an ideal in a unital C*-algebra $\mathfrak{A}$, and let π be the quotient map. Let A be a positive element of $\mathfrak{A}$, and let y be an element of $\mathfrak{A}/\mathfrak{J}$ such that $yy^* \leq \pi(A)$. Then there is an element Y in $\mathfrak{A}$ such that $\pi(Y) = y$ and $YY^* \leq A$.*

Proof. Let $a = \pi(A)$, and let Y be an arbitrary lifting of y. Set

$$B = YY^* + |A - YY^*|.$$

Note that $B \geq YY^* + (A - YY^*) = A$, $\quad B \geq YY^*$ and

$$\pi(B) = yy^* + |a - yy^*| = yy^* + a - yy^* = a.$$

Define $Y_n = A^{1/2}(B + \frac{1}{n}I)^{-1/2}Y$. We verify that Y_n is a Cauchy sequence. Set

$$D_{nm} = (B + \tfrac{1}{n}I)^{-1/2} - (B + \tfrac{1}{m}I)^{-1/2}.$$

Then for $m < n$

$$\|Y_n - Y_m\|^2 = \|A^{1/2}D_{nm}YY^*D_{nm}A^{1/2}\| \leq \|A^{1/2}D_{nm}BD_{nm}A^{1/2}\|$$
$$= \|B^{1/2}D_{nm}A^{1/2}\|^2 \quad = \|B^{1/2}D_{nm}AD_{nm}B^{1/2}\|$$
$$\leq \|B^{1/2}D_{nm}BD_{nm}B^{1/2}\| \quad = \|f_{nm}(B)\|$$

IX.4. Positive Maps

where

$$f_{nm}(x) = x^2 \left[(x + \tfrac{1}{n})^{-1/2} - (x + \tfrac{1}{m})^{-1/2} \right]^2$$

$$= \frac{x^2 (\tfrac{1}{m} - \tfrac{1}{n})^2}{(x + \tfrac{1}{n})(x + \tfrac{1}{m}) \left(\sqrt{x + \tfrac{1}{n}} + \sqrt{x + \tfrac{1}{m}} \right)^2}$$

$$< \left(\frac{x}{x + \tfrac{1}{n}} \right) \left(\frac{\sqrt{x}}{\sqrt{x + \tfrac{1}{n}} + \sqrt{x + \tfrac{1}{m}}} \right)^2 \left(\frac{m^{-2}}{x + \tfrac{1}{m}} \right) \leq \frac{1}{4m}.$$

Hence Y_n is Cauchy. Let Y_∞ be the limit. Since

$$Y_n Y_n^* = A^{1/2}(B + \tfrac{1}{n}I)^{-1/2} Y Y^* (B + \tfrac{1}{n}I)^{-1/2} A^{1/2}$$

$$\leq A^{1/2}(B + \tfrac{1}{n}I)^{-1/2} B (B + \tfrac{1}{n}I)^{-1/2} A^{1/2} \leq A,$$

it follows that $Y_\infty Y_\infty^* \leq A$. But $\pi(Y_n) = a^{1/2}(a + \tfrac{1}{n}1)^{-1/2} y$ converges to y. Therefore $\pi(Y_\infty) = y$; and so Y_∞ is the desired lifting. ∎

Corollary IX.4.5 *Suppose that A is a positive element of a C*-algebra $\mathfrak{A}$ with an ideal $\mathfrak{J}$; and that $0 \leq b \leq \pi(A)$ in $\mathfrak{A}/\mathfrak{J}$. Then there is an element B in $\mathfrak{A}$ such that $\pi(B) = b$ and $0 \leq B \leq A$.*

Proof. Take $y = b^{1/2}$, and apply Lemma IX.4.4. ∎

We take the first small step towards lifting positive maps on $C(X)$.

Corollary IX.4.6 *Let $\mathfrak{J}$ be an ideal in a unital C*-algebra $\mathfrak{A}$; let π be the quotient map; and let X be a finite discrete set. Suppose that ρ is a positive unital linear map of $C(X)$ into $\mathfrak{A}/\mathfrak{J}$. Then there is a positive unital linear map σ of $C(X)$ into $\mathfrak{A}$ such that $\rho = \pi \sigma$.*

Proof. Let δ_{x_i} for $1 \leq i \leq n$ denote the minimal idempotents in $C(X)$. Then the positivity of a linear map φ on $C(X)$ is equivalent to $\varphi(\delta_{x_i}) \geq 0$ for $1 \leq i \leq n$. Let $a_i = \rho(\delta_{x_i})$. Since ρ is unital, we have $\sum_{i=1}^n a_i = 1$. It suffices to construct positive operators A_i in $\mathfrak{A}$ such that $\pi(A_i) = a_i$ and $\sum_{i=1}^n A_i = I$.

Proceed by induction. Use the previous corollary to find $0 \leq A_1 \leq I$ so that $\pi(A_1) = a_1$. Then since $a_2 \leq 1 - a_1 = \pi(I - A_1)$, a second use of the corollary yields a positive element $A_2 \leq I - A_1$ such that $\pi(A_2) = a_2$. At the k-th stage, similarly choose a positive element A_k such that $\pi(A_k) = a_k$ and $A_k \leq I - \sum_{i=1}^{k-1} A_i$. For the last term, set $A_n = I - \sum_{i=1}^{n-1} A_i$. Then the map $\sigma(f) = \sum_{i=1}^n f(x_i) A_i$ is the desired lifting. ∎

Next we show that the identity map on $C(X)$ is the limit of (completely) positive maps which factor through finite dimensional C*-algebras. A C*-algebra with this property is **nuclear** (although this is a theorem, not the definition).

Theorem IX.4.7 *Let X be a compact metric space. Then the identity map id is the point-wise limit of a sequence of maps $\alpha_k \beta_k$ where β_k is a $*$-homomorphism of $C(X)$ onto $C(Y_k)$, where Y_k is a finite discrete space, and α_k is a positive unital map of $C(Y_k)$ into $C(X)$.*

Proof. Let Y_k be a finite subset $\{y_{ki} : 1 \leq i \leq n_k\}$ of X such that the balls of radius $\frac{1}{k}$ about the y_{ki} cover X. Let β_k be the restriction homomorphism of $C(X)$ onto $C(Y_k)$. To obtain α_k, construct a partition of unity f_{ki} subordinate to the cover of the $\frac{1}{k}$-balls. Then set

$$\alpha(g) = \sum_{i=1}^{n_k} g(y_{ki}) f_{ki} \quad \text{for} \quad g \in C(Y_k).$$

Clearly this is a unital positive linear map of $C(Y_k)$ into $C(X)$.

If h belongs to $C(X)$ and $\varepsilon > 0$, the uniform continuity of h implies that there is an integer k so that $\text{dist}(x,y) < \frac{1}{k}$ implies $|h(x) - h(y)| < \varepsilon$. Hence

$$|h(x) - \alpha_k \beta_k(h)(x)| = \left| h(x) - \sum_{i=1}^{n_k} h(y_{ki}) f_{ki}(x) \right|$$

$$\leq \sum_{i=1}^{n_k} |h(x) - h(y_{ki})| f_{ki}(x) < \varepsilon$$

since when $f_{ki}(x) > 0$, $\text{dist}(x, y_{ki}) < \frac{1}{k}$ and thus $|h(x) - h(y_{ki})| < \varepsilon$. It follows that $\lim_{k \to \infty} \alpha_k \beta_k(h) = h$ uniformly on X for every h in $C(X)$. ∎

The point-norm topology of pointwise convergence in $\mathcal{P}_1(\mathfrak{A}, \mathfrak{B})$ is metrizable when $\mathfrak{A}$ is separable. Indeed, take any countable dense subset $A_1, A_2, \ldots$ of the unit ball of $\mathfrak{A}$. Then define a metric on $\mathcal{P}_1(\mathfrak{A}, \mathfrak{B})$ by

$$d(\varphi, \psi) = \sum_{n \geq 1} 2^{-n} \|\varphi(A_n) - \psi(A_n)\|.$$

It is clear that a net φ_α in $\mathcal{P}_1(\mathfrak{A}, \mathfrak{B})$ converges in the d metric if and only if $\varphi_\alpha(A_k)$ converges for each k. The boundedness of the net and the density of the sequence A_k implies that this is equivalent to pointwise convergence in the norm topology.

We will be concerned with maps into more than one image C*-algebra. However, if we fix our sequence A_k, there should be no serious confusion if we use d to denote the distance function for all range C*-algebras $\mathfrak{B}$.

Now consider whether it is possible to lift a convergent sequence of positive maps into a quotient algebra to a convergent sequence "upstairs".

Lemma IX.4.8 *Suppose that φ and ψ belong to $\mathcal{P}_1(\mathfrak{A}, \mathfrak{B})$ and $\mathfrak{J}$ is an ideal of $\mathfrak{B}$ with quotient map π of $\mathfrak{B}$ onto $\mathfrak{B}/\mathfrak{J}$. Then*

$$d(\pi\varphi, \pi\psi) = \inf_{\pi\psi' = \pi\psi} d(\varphi, \psi').$$

IX.4. Positive Maps

Proof. It is clear that $d(\pi\varphi, \pi\psi) \leq d(\varphi, \psi')$ whenever ψ' in $\mathcal{P}_1(\mathfrak{A}, \mathfrak{B})$ satisfies $\pi\psi' = \pi\psi$. So consider the other direction. By Theorem I.9.16, there is a quasi-central approximate identity E_λ for $\mathfrak{J}$. Define

$$\psi_\lambda(A) = E_\lambda^{1/2}\varphi(A)E_\lambda^{1/2} + (I - E_\lambda)^{1/2}\psi(A)(I - E_\lambda)^{1/2}.$$

It is routine to verify that this is positive, unital, and that $\pi\psi_\lambda = \pi\psi$.

Notice that for B in $\mathfrak{B}$, the sequence

$$B_\lambda = E_\lambda^{1/2}BE_\lambda^{1/2} + (I - E_\lambda)^{1/2}B(I - E_\lambda)^{1/2}$$

converges to B since $\|B - B_\lambda\|$ is dominated by

$$\|E_\lambda^{1/2}B - BE_\lambda^{1/2}\|\,\|E_\lambda^{1/2}\| + \|(I - E_\lambda)^{1/2}B - B(I - E_\lambda)^{1/2}\|\,\|(I - E_\lambda)^{1/2}\|.$$

and by Exercise II.8 or II.9, $\|E_\lambda^{1/2}B - BE_\lambda^{1/2}\|$ converges uniformly to 0 as a function of $\|E_\lambda B - BE_\lambda\|$. The same is true for $\|(I - E_\lambda)^{1/2}B - B(I - E_\lambda)^{1/2}\|$. So the right hand side converges to 0.

Hence if we set

$$\varphi_\lambda(A) = E_\lambda^{1/2}\varphi(A)E_\lambda^{1/2} + (I - E_\lambda)^{1/2}\varphi(A)(I - E_\lambda)^{1/2},$$

we see that φ_λ converges to φ in the point-norm topology. Thus

$$\inf d(\varphi, \psi_\lambda) \leq \liminf d(\varphi_\lambda, \psi_\lambda).$$

However,

$$d(\varphi_\lambda, \psi_\lambda) = \sum_{n \geq 1} 2^{-n}\|(I - E_\lambda)^{1/2}(\varphi(A_n) - \psi(A_n))(I - E_\lambda)^{1/2}\|.$$

From the proof of Theorem I.5.3, it follows that for every B in $\mathfrak{B}$,

$$\lim_\lambda \|(I - E_\lambda)^{1/2}B(I - E_\lambda)^{1/2}\|$$
$$= \lim_\lambda \|(I - E_\lambda)^{1/2}B - B(I - E_\lambda)^{1/2}\| + \|B(I - E_\lambda)\| = \|\pi B\|.$$

Consequently, if $\varepsilon > 0$, choose N so large that $2^{-N} < \varepsilon/4$. Then choose λ_0 so that for all $\lambda \geq \lambda_0$ and $1 \leq n \leq N$,

$$\|(I - E_\lambda)^{1/2}(\varphi(A_n) - \psi(A_n))(I - E_\lambda)^{1/2}\| < \|\pi(\varphi(A_n) - \psi(A_n))\| + \varepsilon/2.$$

Then it follows that

$$d(\varphi_\lambda, \psi_\lambda) \leq \sum_{n=1}^N 2^{-n}(\|\pi(\varphi(A_n) - \psi(A_n))\| + \varepsilon/2) + \sum_{n > N} 2^{-n}(\|\varphi\| + \|\psi\|)$$
$$\leq \sum_{n=1}^N 2^{-n}\|\pi(\varphi(A_n) - \psi(A_n))\| + \varepsilon/2 + 2^{1-N} < d(\pi\varphi, \pi\psi) + \varepsilon$$

for all $\lambda \geq \lambda_0$. This establishes the non-trivial inequality. ∎

Corollary IX.4.9 *The set of positive unital maps from $\mathfrak{A}$ into $\mathfrak{B}/\mathfrak{J}$ which have unital positive liftings to $\mathfrak{B}$ is closed in the point-norm topology.*

Proof. Suppose that φ_k in $\mathcal{P}_1(\mathfrak{A}, \mathfrak{B}/\mathfrak{J})$ for $k \geq 1$ is a sequence of liftable unital positive maps which converge pointwise to φ. Drop to a subsequence so that $d(\varphi_k, \varphi_{k+1}) < 2^{-k}$ for all $k \geq 1$. Choose any positive lifting ψ_1 in $\mathcal{P}_1(\mathfrak{A}, \mathfrak{B})$ such that $\pi \psi_1 = \varphi_1$. Then use Lemma IX.4.8 to recursively choose ψ_k in $\mathcal{P}_1(\mathfrak{A}, \mathfrak{B})$ such that

$$\pi \psi_k = \varphi_k \quad \text{and} \quad d(\psi_k, \psi_{k+1}) < 2^{-k}.$$

Then ψ_k is Cauchy, and thus has a positive unital limit ψ. Evidently, $\pi \psi = \varphi$. So φ is also liftable. ∎

Now we are ready for the main result on liftings.

Theorem IX.4.10 *Let X be a compact metric space. Every positive unital map of $C(X)$ into a quotient $\mathfrak{B}/\mathfrak{J}$ has a positive unital lifting.*

Proof. Fix a map φ in $\mathcal{P}_1(C(X), \mathfrak{B}/\mathfrak{J})$. Consider the maps $\varphi_k = \varphi \alpha_k \beta_k$ where α_k and β_k are constructed as in Theorem IX.4.7. They converge to φ pointwise. Moreover $\varphi \alpha_k$ are positive unital maps from $C(Y_k)$ into $\mathfrak{B}/\mathfrak{J}$. So by Corollary IX.4.6, there is a positive unital map ψ_k in $\mathcal{P}_1(C(Y_k), \mathfrak{B})$ such that $\pi \psi_k = \varphi \alpha_k$. Since β_k is a unital positive map (in fact a homomorphism), it follows that $\psi_k \beta_k$ is a positive unital lifting of φ_k. Hence by Corollary IX.4.9, φ is also liftable. ∎

IX.5 Ext(X) is a Group

In this section, we establish the non-trivial fact that $\text{Ext}(X)$ is a group. This is now an easy consequence of the material developed about positive maps.

Theorem IX.5.1 *Let X be a compact metric space, and let τ be an extension of $\mathfrak{K}$ by $C(X)$. Then there is another extension σ such that $\tau \oplus \sigma$ is trivial. Hence $[\sigma]$ is an inverse for $[\tau]$ in $\text{Ext}(X)$. Therefore $\text{Ext}(X)$ is a group.*

Proof. Think of τ as a $*$-monomorphism of $C(X)$ into the Calkin algebra $\mathcal{Q}(\mathcal{H})$. This map is, *a fortiori*, a positive unital map. Thus by Theorem IX.4.10, there is a positive linear map $\tilde{\tau}$ of $C(X)$ into $\mathcal{B}(\mathcal{H})$ such that $\tau = \pi \tilde{\tau}$. Then by Naimark's Dilation Theorem IX.4.2, there is a $*$-representation ρ of $C(X)$ on a Hilbert space $\mathcal{K}$ containing $\mathcal{H}$ such that, with respect to the decomposition $\mathcal{K} = \mathcal{H} \oplus \mathcal{H}^\perp$, ρ has the form

$$\rho(f) = \begin{bmatrix} \tilde{\tau}(f) & \rho_{12}(f) \\ \rho_{21}(f) & \rho_{22}(f) \end{bmatrix}.$$

Next we show that $\rho_{12}(f)$ and $\rho_{21}(f)$ are compact. Indeed, since

$$\rho(|f|^2) = \rho(f)\rho(\overline{f})$$

IX.5. Ext(X) is a Group

and τ is a $*$-homomorphism, the 1, 1 entry modulo compacts yields

$$\tau(|f|^2) = \tau(f)\tau(\overline{f}) + \pi(\rho_{12}(f)\rho_{21}(\overline{f})) = \tau(|f|^2) + \pi(\rho_{12}(f)\rho_{12}(f)^*).$$

Thus $\pi\rho_{12}(f) = 0$, whence $\rho_{12}(f)$ is compact. Likewise, $\rho_{21}(f) = \rho_{12}(\overline{f})^*$ is compact. It follows that $\sigma'(f) := \pi\rho_{22}(f)$ is a $*$-homomorphism. Indeed,

$$\sigma'(fg) = \pi\rho_{22}(fg) = \pi(\rho_{22}(f)\rho_{22}(g) + \rho_{21}(f)\rho_{12}(g)) = \sigma'(f)\sigma'(g).$$

Note that it now follows that the trivial extension $\pi\rho$ decomposes as $\tau \oplus \sigma'$.

However, there is no reason that σ' need be a $*$-monomorphism. To ensure that it is injective, we add on a trivial extension γ. So define $\sigma(f) = \sigma'(f) \oplus \gamma(f)$. Since γ is injective, this is a monomorphism into the Calkin algebra. Moreover,

$$\tau \oplus \sigma = \pi\rho \oplus \gamma.$$

This is a trivial extension. Therefore $[\tau] + [\sigma] = 0$. It follows that the semigroup $\mathrm{Ext}(X)$ has inverses, and thus is a group. ∎

If $f : X \to Y$ is a continuous function between compact metric spaces, we may define a map f_* from $\mathrm{Ext}(X)$ into $\mathrm{Ext}(Y)$ as follows. For a monomorphism τ of $C(X)$ into $\mathcal{Q}(\mathcal{H})$, it is natural to consider $\tau(g \circ f)$ for g in $C(Y)$. This is easily seen to be a $*$-homomorphism on $C(Y)$, but in general it is not injective. This is easily remedied by adding on a trivial extension. So we set

$$f_*(\tau)(g) = \tau(g \circ f) \oplus \sigma(g) \quad \text{for all} \quad g \in C(Y)$$

where σ is any trivial extension by $C(Y)$. If τ and τ' are equivalent extensions, then there is a unitary operator U such that $\tau' = \mathrm{Ad}\,\pi U\,\tau$. Likewise, if σ' is another trivial extension by $C(Y)$, then by Theorem IX.2.1 there is a unitary operator V so that $\sigma' = \mathrm{Ad}\,\pi V\,\sigma$. Hence

$$\mathrm{Ad}\,\pi(U \oplus V)(f_*\tau)(g) = \mathrm{Ad}\,\pi(U \oplus V)(\tau(g \circ f) \oplus \sigma(g))$$
$$= (\mathrm{Ad}\,\pi U\,\tau)(g \circ f) \oplus \mathrm{Ad}\,\pi V\,\sigma(g)$$
$$= \tau'(g \circ f) \oplus \sigma'(g) = f_*(\tau')(g).$$

So we have a well defined operation $f_*([\tau]) = [f_*(\tau)]$ on $\mathrm{Ext}(X)$.

Corollary IX.5.2 *Ext is a covariant functor from the category of compact metric spaces to the category of abelian groups.*

Proof. We have seen from Theorem IX.5.1 that $\mathrm{Ext}(X)$ is always an abelian group. Let f be a continuous map from X into Y (a morphism in the category of compact metric spaces). To see that f_* is a group homomorphism (a morphism in the category of abelian groups), notice that

$$(f_*[\tau] + f_*[\tau'])(g) = \tau(g \circ f) \oplus \sigma(g) \oplus \tau'(g \circ f) \oplus \sigma'(g)$$
$$= (\tau \oplus \tau')(g \circ f) \oplus (\sigma \oplus \sigma')(g) = f_*([\tau] + [\tau'])(g).$$

Hence $f_*([\tau] + [\tau']) = f_*[\tau] + f_*[\tau']$.

It is routine to verify that if f is a continuous map from X to Y and g is a continuous map from Y to Z, then $(gf)_* = g_* f_*$. And if id_X is the identity map on X, then id_{X*} is the identity homomorphism on $\mathrm{Ext}(X)$. ∎

IX.6 First Topological Properties

In this section, we establish some important topological properties of Ext. These properties are the beginning of showing that Ext is a homology theory.

Theorem IX.6.1 *Suppose that q is a continuous surjection of X onto Y. Let B be a closed subset of Y containing all points with multiple preimages in X, and let $A = q^{-1}(B)$. Let j denote the injection of A into X, and let q' denote the restriction of q to A.*

$$\begin{array}{ccc} X & \xrightarrow{q} & Y \\ {\scriptstyle j}\uparrow & & \uparrow{\scriptstyle i} \\ A & \xrightarrow{q'} & B \end{array}$$

Then $\ker q_*$ is contained in $j_* \ker q'_*$.

Proof. Suppose that $[\tau]$ in $\mathrm{Ext}(X)$ satisfies $q_*[\tau] = 0$. Because q is surjective, we may define $q_*\tau(g) = \tau(g \circ q)$ without adding a trivial representation as this is already injective on $\mathrm{C}(Y)$. So without loss of generality, $q_*\tau$ is trivial. Thus there is a homomorphism σ of $\mathrm{C}(Y)$ into $\mathcal{B}(\mathcal{H})$ such that $\pi\sigma = q_*\tau$. By Corollary II.4.5, we may suppose that σ is a diagonal representation. Let

$$\mathcal{A} = \{f \in \mathrm{C}(X) : f|_A \text{ is constant}\}.$$

For each f in $\mathcal{A}$, there is a unique function g in $\mathrm{C}(Y)$ such that $f = g \circ q$. Thus we can define a diagonal representation of $\mathcal{A}$ by $\widetilde{\sigma}(f) = \sigma(g)$. This satisfies

$$\tau(f) = \tau(g \circ q) = \pi\sigma(g) = \pi\widetilde{\sigma}(f) \quad \text{for} \quad f \in \mathcal{A}.$$

Let $\mathfrak{D}$ be the diagonal algebra containing the range of σ. There are points ξ_k in X such that $\widetilde{\sigma}(f) = \mathrm{diag}(f(\xi_k))$ for all f in $\mathcal{A}$. Thus the representation $\widetilde{\sigma}$ extends to a representation $\overline{\sigma}$ of the bounded Borel functions on X which are constant on A into the diagonal algebra $\mathfrak{D}$ given by $\overline{\sigma}(h) = \mathrm{diag}(h(\xi_k))$.

Let

$$X_n = \{\xi \in X : \mathrm{dist}(\xi, A) \geq 2^{-n}\} \quad \text{for each } n \geq 1,$$

and set $P_n = \overline{\sigma}(\chi_n)$ where χ_n is the characteristic function of the set X_n.

Now $\tau(f)$ belongs to $\pi\mathfrak{D}$ when f in $\mathcal{A}$. So $\tau(f)$ commutes with πP_n for each $n \geq 1$. We will show that $\tau(f)$ commutes with πP_n for all f in $\mathrm{C}(X)$. To this end, define functions

$$k(t) = 2(t \wedge 1 \vee \tfrac{1}{2}) - 1, \quad t \in \mathbb{R},$$

IX.6. First Topological Properties

and
$$p_n(\xi) = k(2^n \operatorname{dist}(\xi, A)), \quad \xi \in X.$$
Then p_n belongs to $C(X)$, equals 1 on X_n, and is 0 on $X \setminus X_{n+1}$. Notice that
$$\pi P_n = \pi\overline{\sigma}(p_n\chi_n) = \tau(p_n)\pi P_n$$
and
$$\tau(p_n) = \pi\overline{\sigma}(p_n\chi_{n+1}) = \tau(p_n)\pi P_{n+1}.$$
Split f in $C(X)$ as $f = fp_{n+1} + f(1 - p_{n+1})$. Then
$$\tau(f)\pi P_n = \tau(fp_{n+1})\pi P_n + \tau(f(1 - p_{n+1}))\tau(p_n)\pi P_n$$
$$= \pi P_n\tau(fp_{n+1}) + \tau(f(1-p_{n+1})p_n)\pi P_n = \pi P_n\tau(fp_{n+1})$$
$$= \pi P_n\tau(fp_{n+1}) + \pi P_n\tau(p_n)\tau(f(1-p_{n+1})) = \pi P_n\tau(f).$$
Also note that we have shown that $\pi P_n\tau(f) = \pi P_n\tau(fp_{n+1})$.

Let $\{f_i\}$ be a dense subset of the unit ball of $C(X)$. Choose operators T_i so that $\pi T_i = \tau(f_i)$. We have $\pi(T_iP_n - P_nT_i) = 0$, so $T_iP_n - P_nT_i$ is compact for all i and n. Let $E_1 = P_1$ and $E_n = P_n - P_{n-1}$ for $n \geq 2$. Thus $T_iE_n - E_nT_i$ and (from the previous paragraph) $E_n(T_i - \sigma(f_ip_{n+1}))E_n$ are compact. Therefore there are diagonal projections E_n' in $\mathfrak{D}$ of finite codimension in E_n so that
$$\|T_iE_n' - E_n'T_i\| < 2^{-n} \quad \text{and} \quad \|E_n'(T_i - \sigma(f_ip_{n+1}))E_n'\| < 2^{-n}$$
for $1 \leq i \leq n$. Let $P' = \sum_{n \geq 1} E_n'$. Then
$$T_iP' - P'T_i = \sum_{n \geq 1} T_iE_n' - E_n'T_i$$
and
$$P'T_iP' - \sum_{n \geq 1} E_n'\sigma(f_ip_{n+1})E_n'$$
$$= \sum_{n \geq 1} E_n'(T_i - \sigma(f_ip_{n+1}))E_n' + P'\sum_{n \geq 1}(T_iE_n' - E_n'T_i)E_n'$$
are norm convergent sums of compact operators, and thus are compact. We deduce that $\pi P'$ commutes with the range of τ, and $\pi P'\tau(f)$ lies in $\pi P'\mathfrak{D}$.

Define extensions $\tau'(f) = \pi P'\tau(f)|_{P'\mathcal{H}}$ and $\tau_0(f) = \pi P'^{\perp}\tau(f)|_{P'^{\perp}\mathcal{H}}$. It follows that $\tau' = \pi\sigma'$, where $\sigma'(f) = \sum_{n \geq 1} E_n'\sigma(fp_{n+1})E_n'$. Thus τ' is trivial. We next show that τ_0 depends only on $f|_A$. Indeed, let f in $C(X)$ be such that $f|A = 0$, and let $\varepsilon > 0$. Then there is an integer n so that $\|f|_{X \setminus X_n}\| < \varepsilon$. Since
$$\tau(fp_n) = \pi P_{n+1}\tau(fp_n) = \pi P'P_{n+1}\tau(fp_n),$$
it follows that $\tau_0(fp_n) = 0$. Thus
$$\|\tau_0(f)\| = \|\tau_0(f(1-p_n))\| \leq \|f(1-p_n)\| < \varepsilon.$$
Since $\varepsilon > 0$ is arbitrary, it follows that $\tau_0(f) = 0$.

Hence we may define an element $[\rho]$ in $\mathrm{Ext}(A)$ into $\mathcal{B}(P'^{\perp}\mathcal{H} \oplus \mathcal{H})$ by

$$\rho(f) = \tau_0(\tilde{f}) \oplus \mu(f)$$

where $\tilde{f}$ is any continuous extension of f in $C(A)$ to a function in $C(X)$ and μ is a trivial extension of $C(A)$. This definition is well defined because τ_0 depends only on $\tilde{f}|_A = f$. Moreover, it is evident that if μ' is a trivial extension in $\mathrm{Ext}(X)$, then $j_*\rho = \tau_0 \oplus (\mu \oplus \mu')$ is equivalent to $\tau_0 \oplus \tau' = \tau$. ∎

The following important corollary providing a short exact sequence for Ext will be used extensively.

Corollary IX.6.2 *Suppose that A is a closed subset of a compact metric space X. Let j be the canonical injection of A into X, and let p be the quotient map of X onto X/A. Then*

$$\mathrm{Ext}(A) \xrightarrow{j_*} \mathrm{Ext}(X) \xrightarrow{p_*} \mathrm{Ext}(X/A)$$

is exact.

Proof. In Theorem IX.6.1, take B to be the point $\{A/A\}$ in X/A. Since pj is constant, it is evident that $p_*j_* = 0$. Likewise, $p' = p|_A$ is constant, so that $\ker p'_* = \mathrm{Ext}(A)$. Therefore Theorem IX.6.1 shows that $\ker p_*$ is contained in $j_* \mathrm{Ext}(A)$. Thus $\ker p_* = j_* \mathrm{Ext}(A)$. ∎

If X_n is a sequence of compact metric spaces and p_n is a sequence of continuous maps from X_{n+1} to X_n for $n \geq 1$, then the **projective limit** $X = \mathrm{proj}\lim X_n$ is defined as the subset of $\prod_{n \geq 1} X_n$ consisting of those sequences (x_n) such that $p_n(x_{n+1}) = x_n$ for all $n \geq 1$. There are natural maps q_n of X into each X_n via the coordinate maps, and they satisfy $p_n q_{n+1} = q_n$. This space satisfies the universal property that whenever Y is a compact metric space and f_n is a sequence of continuous maps from Y into X_n such that $p_n f_{n+1} = f_n$ for all $n \geq 1$, there is a unique map f of Y into X such that $q_n f = f_n$ for all n. Likewise, we may define the projective limit of groups. In particular, $\{\mathrm{Ext}(X_n), p_{n*}\}$ determines a sequence of groups and group homomorphisms so that $\mathrm{proj}\lim \mathrm{Ext}(X_n)$ is defined. There is a sequence of homomorphisms q_{n*} of $\mathrm{Ext}(X)$ into $\mathrm{Ext}(X_n)$. Thus there is an induced homomorphism κ of $\mathrm{Ext}(X)$ onto $\mathrm{proj}\lim \mathrm{Ext}(X_n)$ such that $q_{n*}\kappa = p_{n*}$. The following result will be helpful.

Theorem IX.6.3 *Suppose that $X = \mathrm{proj}\lim X_n$ is the projective limit of a family $\{X_n, p_n\}$. Then the induced map*

$$\kappa : \mathrm{Ext}(X) \to \mathrm{proj}\lim \mathrm{Ext}(X_n)$$

is surjective.

Proof. First suppose that each p_n is surjective. Let $([\tau_n]_{n \geq 1})$ be an element of the group $\mathrm{proj}\lim \mathrm{Ext}(X_n)$. Choose τ_n be monomorphisms of $C(X_n)$ into $\mathcal{Q}(\mathcal{H})$ such that $p_{n*}\tau_{n+1} = \tau_n$. This is accomplished recursively. Once τ_n is defined,

IX.6. First Topological Properties

choose τ'_{n+1} representing $[\tau_{n+1}]$. Then $p_{n*}\tau'_{n+1}$ is equivalent to τ_n. So there is a unitary operator U_n so that $\operatorname{Ad} \pi U_n\, p_{n*}\tau'_{n+1} = \tau_n$. Thus $\tau_{n+1} = \operatorname{Ad} \pi U_n\, \tau'_{n+1}$ will suffice.

Define τ on the union of the subalgebras $\mathcal{A}_n = \{f \circ q_n : f \in \mathrm{C}(X_n)\}$ of $\mathrm{C}(X)$ by $\tau(f \circ q_n) = \tau_n(f)$. This is well defined because if $f_n \circ q_n = f_m \circ q_m$ for f_n in $\mathrm{C}(X_n)$ and f_m in $\mathrm{C}(X_m)$ and $m < n$, then $g = f_m \circ p_m \circ \cdots \circ p_{n-1}$ belongs to $\mathrm{C}(X_n)$. Since $g \circ q_n = f_m \circ q_m = f_n \circ q_n$, it follows that f_n and g agree on the range of q_n, which is all of X_n by the surjectivity of each p_n. Thus

$$\tau_n(f_n) = \tau_n(f_m \circ p_m \circ \cdots \circ p_{n-1}) = p_{m*}\cdots p_{n-1*}\tau_n(f_m) = \tau_m(f_m).$$

As $\cup_{n\geq 1}\mathcal{A}_n$ is dense in $\mathrm{C}(X)$, the definition of τ extends to $\mathrm{C}(X)$ by continuity.

Clearly τ is a homomorphism since it is a homomorphism on a dense subalgebra. Moreover, on each $\mathcal{A}_n$, the map τ is injective and hence isometric. Therefore τ is isometric and thus injective on the closed union of the $\mathcal{A}_n$'s, $\mathrm{C}(X)$. Finally, it is readily apparent by construction that $q_{n*}\tau = \tau_n$; so $\kappa([\tau]) = ([\tau_n]_{n\geq 1})$.

Now consider the general case. It is easy to construct a continuous surjection from the Cantor set onto any compact metric space. Thus if $Y_n = X_n \vee C_n$ is the disjoint union of X_n and a Cantor set C_n, one may define a surjection p'_n of Y_{n+1} onto Y_n such that $p'_n|_{X_{n+1}} = p_n$. Let $Y = \operatorname{proj lim} Y_n$, and let q'_n be the induced maps of Y onto Y_n.

For each n, let j_n be the injection of X_n into Y_n. Because the j_n commute with the p_n's and q_n's, there is a map $j = \operatorname{proj lim} j_n$ injecting X into Y. In particular, $q_n = q'_n j$ for $n \geq 1$. Also

$$Y_n/X_n = C_n \vee (X_n/X_n) = C_n^+,$$

the disjoint union of C_n and a point, which is totally disconnected. Because $p'_n(X_{n+1})$ is contained in X_n, there are induced maps $\bar{p}_n$ of Y_{n+1}/X_{n+1} onto Y_n/X_n. Therefore

$$Y/X = \operatorname{proj lim} Y_n/X_n.$$

This is homeomorphic to a subspace of $\prod_{n\geq 1} C_n^+$, and thus is totally disconnected.

Corollary IX.3.2 shows that $\operatorname{Ext}(Y_n/\bar{X}_n) = 0$ because Y_n/X_n is totally disconnected. Thus applying Corollary IX.6.2 to the injection j_n yields an isomorphism j_{n*} of $\operatorname{Ext}(X_n)$ onto $\operatorname{Ext}(Y_n)$. Similarly, j_* is an isomorphism of $\operatorname{Ext}(X)$ onto $\operatorname{Ext}(Y)$. Therefore $\overleftarrow{j} = \operatorname{proj lim} j_{n*}$ is an isomorphism of $\operatorname{proj lim} \operatorname{Ext}(X_n)$ onto $\operatorname{proj lim} \operatorname{Ext}(Y_n)$. Thus we have the commutative diagram

$$\begin{array}{ccc} \operatorname{Ext}(X) & \xrightarrow{\kappa_X} & \operatorname{proj lim} \operatorname{Ext}(X_n) \\ {\scriptstyle j_*}\downarrow & & \downarrow{\scriptstyle \overleftarrow{j}} \\ \operatorname{Ext}(Y) & \xrightarrow{\kappa_Y} & \operatorname{proj lim} \operatorname{Ext}(Y_n) \end{array}$$

Since κ_Y is surjective, it follows that $\kappa_X = \overleftarrow{j}^{-1} \kappa_Y j_*$ is also surjective. ∎

One key tool for computing the Ext groups is the Fredholm index map. If πT is an invertible element of the Calkin algebra, then one can compute the Fredholm index $\mathrm{ind}(\pi T) = \mathrm{ind}(T)$. This determines the connected component of $\mathcal{Q}(\mathcal{H})^{-1}$ in which πT lies. Every normal operator N has index zero, and thus πN lies in the connected component of the identity, $\mathcal{Q}(\mathcal{H})_0^{-1}$. Indeed, $\mathrm{ind}(N - \lambda I) = 0$ for every $\lambda \notin \sigma_e(N)$.

Suppose that X is a compact metric space and f is invertible in $C(X)$. Then for $[\tau]$ in $\mathrm{Ext}(X)$, one may compute $\mathrm{ind}(\tau(f))$. This is easily seen to be independent of the choice of representative τ. Indeed, this follows because index is preserved under unitary equivalence; and if σ is a trivial extension, then $\mathrm{ind}(\sigma(f)) = 0$ for every invertible f. As index is a homomorphism, it follows that adding a trivial extension does not change the index. It is also important that index is a homotopy invariant due to the fact that it is a continuous integer valued function. This means that $\mathrm{ind}(\tau(f)) = \mathrm{ind}(\tau(g))$ whenever f and g lie in the same connected component of $C(X)^{-1}$. Thus $\mathrm{ind}\,\tau(f)$ depends only on the equivalence class $[\tau]$ and the homotopy class $[f]$.

The group $C(X)^{-1}/C(X)_0^{-1}$ of homotopy classes of invertible functions is denoted by $\pi^1(X)$. Thus there is a map γ from $\mathrm{Ext}(X)$ into $\mathrm{Hom}(\pi^1(X), \mathbb{Z})$ defined by

$$\gamma[\tau]([f]) = \mathrm{ind}\,\tau(f).$$

It has been shown that this map is well defined. To verify that $\gamma[\tau]$ is a homomorphism, compute:

$$\mathrm{ind}\,\tau(fg) = \mathrm{ind}\big(\tau(f)\tau(g)\big) = \mathrm{ind}\,\tau(f) + \mathrm{ind}\,\tau(g).$$

Finally, we must verify that γ is a homomorphism. This follows from

$$\gamma([\tau_1] + [\tau_2])([f]) = \mathrm{ind}\big(\tau_1(f) \oplus \tau_2(f)\big)$$
$$= \mathrm{ind}\,\tau_1(f) + \mathrm{ind}\,\tau_2(f) = \gamma[\tau_1]([f]) + \gamma[\tau_2]([f]).$$

When necessary, we will write γ_X to indicate that γ is acting on $\mathrm{Ext}(X)$.

A more sophisticated variant makes use of matrix algebras over $C(X)$ and introduces a connection with K-theory. If τ is a $*$-monomorphism of $C(X)$ into $\mathcal{Q}(\mathcal{H})$, then $\tau^{(n)}$ is a monomorphism of $\mathcal{M}_n(C(X))$ into $\mathcal{M}_n(\mathcal{Q}(\mathcal{H}))$ defined by taking an $n \times n$ matrix $[f_{ij}]$ to the matrix $[\tau(f_{ij})]$. Since we may identify $\mathcal{M}_n(\mathcal{Q}(\mathcal{H}))$ with $\mathcal{B}(\mathcal{H}^{(n)})/\mathfrak{K}(\mathcal{H}^{(n)})$ which is the Calkin algebra, this defines an extension of $\mathfrak{K}$ by $\mathcal{M}_n(C(X))$. As above, we may evaluate $\mathrm{ind}\,\tau^{(n)}([f_{ij}])$ for every invertible element of $\mathcal{M}_n(C(X))$. This suggests looking at the group $K_1(C(X))$, which in the commutative case is the group $K^1(X)$. There is a natural imbedding i_n of $\mathrm{GL}_n(X) = \mathcal{M}_n(C(X))^{-1}$ into $\mathrm{GL}_{n+1}(X)$ by sending F to $F \oplus [1]$. This takes the connected component of the identity $\mathrm{GL}_n(X)_0$ into $\mathrm{GL}_{n+1}(X)_0$, and thus induces a map from $\mathrm{GL}_n(X)/\mathrm{GL}_n(X)_0$ into $\mathrm{GL}_{n+1}(X)/\mathrm{GL}_{n+1}(X)_0$. The

group $K^1(X)$ is the direct limit
$$K^1(X) = \varinjlim \mathrm{GL}_n(X)/\mathrm{GL}_n(X)_0.$$

There is an index map γ_X^n from $\mathrm{Ext}(X)$ into $\mathrm{Hom}(\mathrm{GL}_n(X)/\mathrm{GL}_n(X)_0, \mathbb{Z})$ given by
$$\gamma_X^n[\tau]([[f_{ij}]]) = \mathrm{ind}\,\tau^{(n)}([f_{ij}]).$$

This is well defined on equivalence classes, and is a homomorphism by an argument identical to the $n = 1$ case. It is evident that
$$\mathrm{ind}\,\tau^{(n+1)}(F \oplus [1]) = \mathrm{ind}\,\tau^{(n)}(F) + \mathrm{ind}\,\pi I = \mathrm{ind}\,\tau^{(n)}(F).$$

These index maps are compatible in the sense that $\gamma_X^{n+1} i_n = \gamma_X^n$. Therefore there is a direct limit homomorphism $\gamma_X^\infty = \varinjlim \gamma_X^n$ from $\mathrm{Ext}(X)$ into $\mathrm{Hom}(K^1(X), \mathbb{Z})$.

These constructions are functorial. That is, suppose that h is a continuous map of X into Y. Then this induces the endomorphism α_h of $C(Y)$ into $C(X)$ by $\alpha(f) = f \circ h$. Clearly, this takes invertibles to invertibles and maps the connected component of the identity $C(Y)_0^{-1}$ into $C(X)_0^{-1}$. Hence it induces a map h^* of $\pi^1(Y)$ into $\pi^1(X)$. This in turn induces a homomorphism $h_\#$ of $\mathrm{Hom}(\pi^1(X), \mathbb{Z})$ into $\mathrm{Hom}(\pi^1(Y), \mathbb{Z})$ by
$$h_\#(\theta)([f]) = \theta(h^*[f]) = \theta([f \circ h]).$$

This construction is compatible with γ in the sense that
$$\gamma_Y(h_*[\tau]) = h_\# \gamma_X([\tau]).$$

For if $[f]$ is in $\pi^1(Y)$, $[\tau]$ is an element of $\mathrm{Ext}(X)$ and σ is any trivial element of $\mathrm{Ext}(Y)$, then
$$\gamma_Y(h_*[\tau])([f]) = \mathrm{ind}\,\tau(f \circ h) \oplus \sigma(f) = \mathrm{ind}\,\tau(f \circ h)$$
$$= \gamma_X([\tau])([f \circ h]) = h_\#(\gamma_X[\tau])([f]).$$

We will show that γ is injective when X is homeomorphic to a planar set. This will complete the analysis in that case. However, in general, this map has non-trivial kernel. We will see some examples later.

IX.7 Ext for Planar Sets

In this section, we will completely analyze $\mathrm{Ext}(X)$ for X a subset of the plane. The key is to show that the index map γ is an isomorphism for subsets of the plane. This yields a simple, computable set of invariants in this case. For planar sets, $C(X)$ is generated by the identity function z. The group $\pi^1(X)$ has a convenient description in this case.

Theorem IX.7.1 *When X is a subset of the complex plane $\mathbb{C}$, the group $\pi^1(X)$ is the free abelian group generated by $\{[z - \lambda_n]\}$ where $\Lambda = \{\lambda_n\}$ is a set with exactly one point λ_n in each bounded component of $\mathbb{C} \setminus X$.*

Proof. Suppose that X is bounded by finitely many disjoint piecewise smooth Jordan curves $\Gamma_1, \ldots, \Gamma_n$. Given an invertible function f in $\mathrm{GL}(X)$, one can compute the *winding number* $\mathrm{ind}_{\Gamma_i}(f)$ of f about Γ_i for $1 \leq i \leq n$. We assume that the reader is familiar with this concept. There is a continuous logarithm for f if and only if f has winding number 0 about each bounding curve. In this case, one may express f as an exponential e^g; and thus f is in the connected component of the identity. Since winding number is a homotopy invariant in $\mathrm{GL}(X)$, this provides a necessary and sufficient condition for $[f] = 0$. The map ι from $[f]$ to $(\mathrm{ind}_{\Gamma_1}(f), \ldots, \mathrm{ind}_{\Gamma_n}(f))$ in $\mathbb{Z}^n$ injective. So $\pi^1(X)$ is a finitely generated free group. (In fact, deleting those curves forming the exterior boundary of each component will yield an isomorphism.)

Pick a point μ_j in each bounded component $\mathcal{O}_1, \ldots, \mathcal{O}_m$ of $\mathbb{C} \setminus X$. We will show that every function in $\mathrm{GL}(X)$ may be factored as a finite product

$$f = \prod_{j=1}^{m} (z - \mu_j)^{p_j} e^g$$

for a unique choice of integers $p_1, \ldots, p_m$ and a continuous function g in $C(X)$. Hence $[z - \mu_j]$ for $1 \leq j \leq m$ freely generate $\pi^1(X)$. This can be established inductively on the number of bounding curves.

For $n = 1$, the interior of the curve Γ_1 either lies in X or in X^c. In the first case, X is simply connected, there are no bounded components of the complement and $\pi^1(X) = 0$. If the interior $\mathcal{O}$ lies in X^c, then there is one bounded component of the complement and $\mathrm{ind}_\Gamma[z - \mu] = -1$, which freely generates $\mathbb{Z} = \pi^1(X)$.

Assume the result is established for n and consider X with $n + 1$ boundary curves. At least one of these curves, say Γ_1, does not contain any other Γ_j in its interior component $\mathcal{O}$. When $\mathcal{O}$ is contained in X, $Y = \mathcal{O} \cup \Gamma_1$ is a simply connected component of X. Thus $\pi^1(X) = \pi^1(X \setminus Y)$, and $X \setminus Y$ has one fewer boundary component. So by the induction hypothesis, given f in $\mathrm{GL}(X)$, the restriction $f|_{X \setminus Y}$ factors uniquely as $\prod_{j=1}^{m} (z - \mu_j)^{p_j} e^g$. As every function in $\mathrm{GL}(Y)$ is an exponential, the definition of g can be extended to Y to obtain equality on X. The uniqueness of the p_j's follows from the uniqueness on $X \setminus Y$.

Otherwise, $\mathcal{O} = \mathcal{O}_1$ is one of the components of $\mathbb{C} \setminus X$. So $\mathrm{ind}_{\Gamma_1}[z - \mu_1] = 1$ and $\mathrm{ind}_{\Gamma_1}[z - \mu_j] = 0$ for $j > 1$. Given f in $\mathrm{GL}(X)$, let $p_1 = \mathrm{ind}_{\Gamma_1}[f]$. Then the function $h = (z - \mu_1)^{-p_1} f$ has winding number 0 around Γ_1. Hence h may be extended to an invertible continuous function on $Y = X \cup \mathcal{O}_1$. Now Y is bounded by $\Gamma_2, \ldots, \Gamma_{n+1}$ and the components of $\mathbb{C} \setminus Y$ are just $\mathcal{O}_2, \ldots, \mathcal{O}_m$. By the induction hypothesis, there are unique integers $p_2, \ldots, p_m$ so that h factors as $\prod_{j=2}^{m} (z - \mu_j)^{p_j} e^g$. Hence f factors as $\prod_{j=1}^{m} (z - \mu_j)^{p_j} e^g$. Now

$$\mathrm{ind}_{\Gamma_1} \prod_{j=1}^{m} (z - \mu_j)^{q_j} e^g = q_1.$$

IX.7. Ext for Planar Sets

This shows that the integer $p_1 = \text{ind}_{\Gamma_1}[f]$ is uniquely determined as the exponent of $z - \lambda_1$ in the product for f. The uniqueness of the others now follows from the induction hypothesis.

It is easy to show that every planar set X is the intersection of a decreasing sequence of sets X_n which are finitely connected and have piecewise smooth boundary. For example, one may cover X by a finite union of disks of radius 2^{-n} with centres in X. Given f in $\text{GL}(X)$, let $\tilde{f}$ be any extension of f to a continuous function on a large disk containing X. By the uniform continuity of $\tilde{f}$, it will be invertible on X_n for some sufficiently large integer n. Thus f factors as $\prod_{j=1}^{m}(z - \mu_j)^{k_j} e^g$. Let λ_j be the chosen point in the same component $\mathcal{O}_j$ of $\mathbb{C} \setminus X$ as μ_j. Then $\prod_{j=1}^{m}(z - \mu_j)^{k_j}(z - \lambda_j)^{-k_j}$ has winding number 0 around every component of $\mathbb{C} \setminus X$ and thus is an exponential e^h. So

$$f = \prod_{j=1}^{m}(z - \lambda_j)^{k_j} e^{g+h}.$$

This shows that $[z - \lambda_j]$ generate $\pi^1(X)$.

To see that $[z - \lambda_j]$ are free generators, let us suppose that there is a relation $\sum_{j=1}^{m} p_j[z - \lambda_j] = 0$. That is, there is a continuous function g in $C(X)$ so that

$$\prod_{j=1}^{m}(z - \lambda_j)^{p_j} = e^g.$$

As above, there is an integer n sufficiently large so that this relation extends to X_n. But then the freeness of the generators for $\pi^1(X_n)$ shows that $p_j = 0$ for all j. ∎

The main theorem for Ext of subsets of the plane can now be stated.

Theorem IX.7.2 *If X is a compact subset of the plane, then γ is an isomorphism of $\text{Ext}(X)$ onto $\text{Hom}(\pi^1(X), \mathbb{Z})$.*

Before proving it, we note two important operator theoretic corollaries that classify essentially normal operators.

Corollary IX.7.3 *Two essentially normal operators T_1 and T_2 are compalent if and only if $\sigma_e(T_1) = \sigma_e(T_2)$ and $\text{ind}(T_1 - \lambda I) = \text{ind}(T_2 - \lambda I)$ for all $\lambda \notin \sigma_e(T_1)$.*

Proof. The only if direction is easy, so we prove the other direction. Since T_1 and T_2 have the same essential spectrum X, they each determine an extension $\tau_i(f) = f(\pi T_i)$ of $C(X)$ for $i = 1, 2$. Then

$$\gamma[\tau_1]([z - \lambda I]) = \text{ind}(T_1 - \lambda I) = \text{ind}(T_2 - \lambda I) = \gamma[\tau_2]([z - \lambda I]).$$

By Theorem IX.7.1, this implies that $\gamma[\tau_1] = \gamma[\tau_2]$. So by Theorem IX.7.2, τ_1 and τ_2 are equivalent. Thus there is a compact operator K and unitary U so that $T_2 = UT_1U^* + K$ as desired. ∎

Since $\text{ind}(N - \lambda I) = 0$ for every normal operator and $\lambda \notin \sigma_e(T)$, the following is an immediate consequence of the first corollary.

Corollary IX.7.4 *An operator T has the form "normal plus compact" if and only if $T^*T - TT^*$ is compact and $\text{ind}(T - \lambda I) = 0$ for every $\lambda \notin \sigma_e(T)$.*

Another consequence that surprisingly has yet to be given a simple direct proof is the following. Perhaps the reason it is difficult is that the corresponding result for pairs of operators is false (see the examples at the end of the chapter).

Corollary IX.7.5 *The set of all operators of the form "normal plus compact" is a norm closed set.*

Proof. Suppose that T_k are "normal plus compact" and converge to an operator T. Then
$$T^*T - TT^* = \lim_{n \to \infty} T_n^* T_n - T_n T_n^*$$
is compact. Moreover, since the Fredholm index in continuous and the set of Fredholm operators is open, it follows that if $\lambda \notin \sigma_e(T)$, then $\lambda \notin \sigma_e(T_n)$ for large n and
$$\text{ind}(T - \lambda I) = \lim_{n \to \infty} \text{ind}(T_n - \lambda I) = 0.$$
So by the previous corollary, T is also normal plus compact. ∎

In preparation for our proof, we need a few lemmas. The first is a useful partial result.

Lemma IX.7.6 *Suppose that A is a compact subset of a closed interval J. Then the map γ on $\text{Ext}(J/A)$ is injective.*

Proof. It is evident that J/A is homeomorphic to the union X of countably many disjoint smooth curves C_k with disjoint interiors which meet at a common point ξ_0 and have diameters decreasing to 0. Each curve is homeomorphic to a circle or line segment. Let $X_n = \cup_{k=1}^n C_k$ and $Y_n = \cup_{k>n} C_k$. Let i_n and j_n denote the injections of X_n and Y_n into X. Let p_n denote the retraction of X onto X_n obtained by sending Y_n to ξ_0; and similarly, let q_n denote the retraction of X onto Y_n obtained by sending X_n to ξ_0.

$$X_n \underset{p_n}{\overset{i_n}{\rightleftarrows}} X = X_n \cup Y_n \underset{q_n}{\overset{j_n}{\rightleftarrows}} Y_n$$

Fix τ in $\ker \gamma$. Note that $\text{id}_* = (i_n p_n)_* + (j_n q_n)_*$. In particular, consider $n = 1$. Then $p_{1*}[\tau]$ belongs to $\text{Ext}(C_1)$ and
$$\gamma p_{1*}[\tau] = p_{1\#} \gamma[\tau] = 0.$$
By Theorems IX.3.3 and IX.3.1, $\text{Ext}(C_1) = \mathbb{Z}$ or 0 depending on whether C_1 is homeomorphic to a circle or line segment; and γ_{C_1} is an isomorphism in either

IX.7. Ext for Planar Sets

case. So $p_{1*}[\tau] = 0$. Hence

$$[\tau] = i_{1*}p_{1*}[\tau] + j_{1*}q_{1*}[\tau] = j_{1*}[\tau_1]$$

where $[\tau_1] = q_{1*}[\tau]$ belongs to $\mathrm{Ext}(Y_1)$. Therefore there is a faithful representation σ_1 of $C(X)$ such that

$$\tau(f) \simeq \pi\sigma_1(f) \oplus (j_1 q_1)_* \tau(f).$$

Thus there is a projection E_1 so that σ_1 acts on $\mathcal{B}(E_1\mathcal{H})$, and πE_1 commutes with the range of τ.

Repeated use of this argument produces pairwise orthogonal projections E_k and $*$-representations σ_k of $C(X)$ on $\mathcal{B}(E_k\mathcal{H})$ so that

$$\tau(f) = \sum_{k=1}^{n} \oplus \pi\sigma_k(f)E_k \oplus (j_n q_n)_* \tau(f) \Big(\sum_{k=1}^{n} E_k\Big)^\perp$$

on $\mathcal{H} = \sum_{k=1}^{n} \oplus E_k\mathcal{H} \oplus (\sum_{k=1}^{n} E_k)^\perp \mathcal{H}$. Let $\mathcal{A}_n = \{f \circ i_n : f \in C(X_n)\}$. Then it follows that for f in $\mathcal{A}_n$,

$$\tau(f) = \sum_{k=1}^{n} \oplus \pi\sigma_k(f)E_k \oplus f(\xi_0)\Big(\sum_{k=1}^{n} E_k\Big)^\perp.$$

Define

$$\sigma(f) = \sum_{k\geq 1} \oplus \sigma_k(f)E_k \oplus f(\xi_0)\Big(\sum_{k\geq 1} E_k\Big)^\perp.$$

Then $\tau(f) = \pi\sigma(f)$ for all f in $\cup_{n\geq 1}\mathcal{A}_n$, which is dense in $C(X)$. Therefore τ is trivial. ∎

The key step involves cutting the spectrum.

Lemma IX.7.7 *Suppose that X is a compact subset of $\mathbb{C}$, and that $[\tau]$ in $\mathrm{Ext}(X)$ satisfies $\gamma[\tau] = 0$. Let L be a line splitting $\mathbb{C}$ into two half planes H^+ and H^-. Define $X^\pm = X \cap H^\pm$, and let $i^\pm$ be the injections of $X^\pm$ into X. Then there are elements $[\rho^\pm]$ in $\mathrm{Ext}(X^\pm)$ such that $[\tau] = i_*^+[\rho^+] + i_*^-[\rho^-]$ and $\gamma[\rho^\pm] = 0$.*

Proof. Let J be a closed line segment of L sufficiently long to contain the orthogonal projection $\pi_L(X)$ of X onto L. Define injections $j^\pm$ of $X^\pm \cup J$ into $X \cup J$ and j of X into $X \cup J$, and injections $k^\pm$ of $X^\pm$ into $X^\pm \cup J$. Also define a retraction of $X \cup J$ onto $X^+ \cup J$ by

$$r^+(z) = \begin{cases} z & \text{for } z \in X^+ \cup J \\ \pi_L(z) & \text{for } z \in X^- \cup J \end{cases}.$$

The following diagram may be helpful.

$$\begin{array}{ccccc} X^+ & \xrightarrow{i^+} & X & \xleftarrow{i^-} & X^- \\ {\scriptstyle k^+}\downarrow & & {\scriptstyle j}\downarrow & & \downarrow{\scriptstyle k^-} \\ X^+ \cup J & \underset{j^+}{\overset{r^+}{\rightleftarrows}} & X \cup J & \xleftarrow{j^-} & X^- \cup J \end{array}$$

First we show that $[j_*\tau]$ can be decomposed as $j_*[\tau] = j_*^+[\tau^+] + j_*^-[\tau^-]$ for $[\tau^\pm]$ in $\text{Ext}(X^\pm \cup J)$ satisfying $\gamma[\tau^\pm] = 0$. Indeed, $\tau^+ = r_*^+ j_*[\tau]$ lies in $\text{Ext}(X^+ \cup J)$. Let $[\rho] = j_*[\tau] - j_*^+[\tau^+]$. Then

$$r_*^+[\rho] = r_*^+ j_*[\tau] - r_*^+ j_*^+[\tau^+] = (r^+ j)_*[\tau] - (r^+ j^+ r^+ j)_*[\tau] = 0$$

because $r^+ j^+ r^+ = r^+$. Consider the quotient q of $X \cup J$ onto

$$(X \cup J)/(X^- \cup J) = (X^+ \cup J)/J.$$

Since q factors as $q' r^+$, it follows that $q_*[\rho] = 0$. So by Corollary IX.6.2 for the sequence

$$X^- \cup J \xrightarrow{j^-} X \cup J \xrightarrow{q} (X \cup J)/(X^- \cup J)$$

it follows that there is an element $[\tau^-]$ in $\text{Ext}(X^- \cup J)$ such that $[\rho] = j_*^- \tau^-$.

Clearly, $\text{ind}\, \tau^+(z - \lambda) = 0$ for all λ in H^- and $\text{ind}\, \tau^-(z - \lambda) = 0$ for all λ in H^+. So for $\lambda \in H^+ \setminus X \cup L$,

$$\text{ind}\, \tau^+(z - \lambda) = \text{ind}\, j_*^+ \tau^+(z - \lambda)$$
$$= \text{ind}\, \tau(z - \lambda) - \text{ind}\, j_*^- \tau^-(z - \lambda) = 0.$$

Hence $\gamma[\tau^+] = 0$; and similarly, $\gamma[\tau^-] = 0$.

Now consider the sequence

$$X^+ \xrightarrow{k^+} X^+ \cup J \xrightarrow{q^+} (X^+ \cup J)/X^+ = J/(X \cap J).$$

Since $\gamma q_*^+[\tau_+] = q_\#^+ \gamma[\tau^+] = 0$, Lemma IX.7.6 shows that $q_*^+[\tau^+] = 0$. Hence by the short exact sequence of Corollary IX.6.2, there is an element $[\rho^+]$ in $\text{Ext}(X^+)$ such that $[\tau^+] = k_*^+[\rho^+]$. An easy calculation shows that $\gamma[\rho^+] = 0$ as well. Similarly, there is an element $[\rho^-]$ in $\text{Ext}(X^-)$ such that $k_*^-[\rho^-] = [\tau^-]$ and $\gamma[\rho^-] = 0$. So

$$j_*(i_*^+[\rho^+] + i_*^-[\rho^-]) = (j^+ k^+)_*[\rho^+] + (j^- k^-)_*[\rho^-]$$
$$= j_*^+[\tau^+] + j_*^-[\tau^-] = j_*[\tau].$$

IX.7. Ext for Planar Sets

To complete the proof, it must be shown that j_* is injective. Consider

$$\begin{array}{ccccc} X \cap J & \xrightarrow{i_X} & X & \xrightarrow{q} & X/(X \cap J) \\ {\scriptstyle i_J}\downarrow & & {\scriptstyle j}\downarrow & & \parallel \\ J & \xrightarrow{i} & X \cup J & \xrightarrow{p} & (X \cup J)/J \end{array}$$

By Theorem IX.3.1, $\operatorname{Ext}(X \cap J) = 0 = \operatorname{Ext}(J)$. By Corollary IX.6.2, it follows that p_* and q_* are injective. Since $q_* = p_* j_*$, it follows that j_* is also injective. Therefore $i_*^+[\rho^+] + i_*^-[\rho^-] = [\tau]$. ∎

Following the notation of the previous lemma, let $Y = X^+ \vee X^-$ denote the *disjoint union* of X^+ and X^-; and let p denote the natural surjection of Y onto X. In a natural way, $[\rho^+ \oplus \rho^-]$ becomes an element $[\rho]$ in $\operatorname{Ext}(Y)$ such that $p_*[\rho] = [\tau]$ and $\gamma[\rho] = 0$.

We are now prepared to complete the proof of the main result.

Theorem IX.7.8 *For a compact subset X of the plane, the map γ_X is injective.*

Proof. Suppose that $[\tau]$ in $\operatorname{Ext}(X)$ satisfies $\gamma[\tau] = 0$. Think of X as contained in a large square S. Cover the square with a countable number of horizontal and vertical lines L_k which chop S into a grid of smaller rectangles whose diameters decrease to 0 as k tends to infinity. Repeatedly apply the previous lemma to cut the spectrum along L_k for each k. At the k-th stage, let X_k be the disjoint union of the intersection of X with each closed rectangle in the k-line grid, and let p_k be the canonical surjection of X_k onto X_{k-1} (with $X_0 = X$) and let r_k be the surjection of X_k onto X. By the formulation preceding this theorem, there are elements $[\rho_k]$ in $\operatorname{Ext}(X_k)$ such that $\gamma[\rho_k] = 0$ and $p_{k*}[\rho_k] = [\rho_{k-1}]$, whence $r_{k*}[\rho_k] = [\tau]$.

Let Y denote the projective limit of the sequence $\{X_k, p_k\}$ and let q_k denote the maps from Y onto X_k. In particular, q_0 maps Y onto X. Since the diameters of the grids decrease to 0, the space Y is totally disconnected. Thus $\operatorname{Ext}(Y) = 0$ by Theorem IX.3.2. By Theorem IX.6.3, there is an element $[\rho]$ in $\operatorname{Ext}(Y)$ such that $[\rho_k] = q_{k*}[\rho]$ for all $k \geq 0$. Therefore $[\tau] = q_{0*}[\rho] = q_{0*}0 = 0$. ∎

It remains to prove surjectivity. In the case of nice spectrum, we use Toeplitz operators to write down explicit generators. Actually, explicit generators can be given for arbitrary subsets of the plane. However, the proofs are considerably more difficult.

Lemma IX.7.9 *If X is a compact subset of the plane bounded by finitely many disjoint piecewise smooth Jordan curves, then γ_X is surjective.*

Proof. Let $[z - \lambda_k]$ for $1 \leq k \leq m$ be a set of free generators for $\pi^1(X)$ corresponding to the bounded components $\mathcal{O}_k$ of $\mathbb{C} \setminus X$. Thus $\operatorname{Hom}(\pi^1(X), \mathbb{Z})$ is generated by homomorphisms h_j, $1 \leq j \leq m$ such that $h_j[z - \lambda_k] = \delta_{jk}$. Hence it suffices to construct extensions τ_j so that $\operatorname{ind} \tau_j(z - \lambda_k) = \pm \delta_{jk}$.

The boundary of $\mathcal{O}_j$ is given by $\Gamma_0 - \sum_{i=1}^{p} \Gamma_i$, which is a sum of oriented Jordan curves from the boundary of X. It is well known that

$$\text{ind}_{\partial \mathcal{O}_j}(z - \lambda) = \text{ind}_{\Gamma_0}(z - \lambda) - \sum_{i=1}^{p} \text{ind}_{\Gamma_i}(z - \lambda) = \begin{cases} 1 & \lambda \in \mathcal{O}_j \\ 0 & z \in \mathbb{C} \setminus \overline{\mathcal{O}_j} \end{cases}.$$

Let g_0 be a homeomorphism of the unit circle $\mathbb{T}$ onto Γ_0 with positive orientation, and for $1 \leq i \leq p$, let g_i be homeomorphisms of $\mathbb{T}$ onto Γ_i with reversed orientation. Let

$$T_j = \sum_{i=0}^{p} \oplus T_{g_i} \oplus N$$

where T_{g_i} are Toeplitz operators and N is normal with $\sigma(N) = \sigma_e(N) = X$. By Theorem V.1.6, each T_{g_i} is essentially normal with $\sigma_e(T_{g_i}) = \Gamma_i$ and

$$\text{ind}(T_{g_i} - \lambda I) = -\text{ind}_{\mathbb{T}}(g_i - \lambda) = -\text{ind}_{g_i(\mathbb{T})}(z - \lambda).$$

Hence

$$\text{ind}(T_j - \lambda_k I) = -\sum_{i=0}^{p} \text{ind}_{g_i(\mathbb{T})}(z - \lambda_k) = -\text{ind}_{\partial \mathcal{O}_j}(z - \lambda_k) = -\delta_{jk}.$$

As T_j is essentially normal with $\sigma_e(T_j) = X$, we may define an extension τ_j by $\tau_j(f) = f(\pi T_j)$. By the calculation above, $\gamma(\tau_j) = -h_j$ as desired. ∎

The proof of Theorem IX.7.2 can now be completed by establishing surjectivity by a projective limit argument.

Theorem IX.7.10 *If X is a compact subset of the plane, then γ_X is surjective.*

Proof. It is easy to write X an the intersection of a decreasing sequence of closed subsets X_n which are finitely connected and are bounded by a finite number of disjoint piecewise smooth Jordan curves. Let p_n be the injection of X_{n+1} into X_n, and q_n the injection of X into X_n. Then $X = \text{proj}\lim X_n$.

Suppose that h belongs to $\text{Hom}(\pi^1(X), \mathbb{Z})$. Then $h_n = q_{n\#} h$ defines an element of $\text{Hom}(\pi^1(X_n), \mathbb{Z})$ for each n and $p_{n\#} h_{n+1} = h_n$. By Lemma IX.7.9, there is an extension τ_n in $\text{Ext}(X_n)$ such that $\gamma_{X_n}[\tau_n] = h_n$. Also

$$\gamma_{X_n}([\tau_n] - p_{n*}[\tau_{n+1}]) = \gamma_{X_n}[\tau_n] - p_{n\#} \gamma_{X_{n+1}}[\tau_{n+1}]$$
$$= h_n - p_{n\#} h_{n+1} = 0.$$

Since γ_{X_n} is injective by Theorem IX.7.8, this shows that $[\tau_n] = p_{n*}[\tau_{n+1}]$. Thus the sequence $([\tau_k])$ determines an element of $\text{proj}\lim \text{Ext}(X_n)$.

By Theorem IX.6.3, the map κ of $\text{Ext}(X)$ onto $\text{proj}\lim \text{Ext}(X_n)$ is surjective. So we may find $[\tau]$ in $\text{Ext}(X)$ such that $q_{n*}[\tau] = [\tau_n]$ for all n. Finally if f is in

GL(X), extend f to a continuous function F on a large disk containing X_1. Then $f_n = F|_{X_n}$ will belong to GL(X_n) some n sufficiently large. So

$$\gamma_X[\tau](f) = \gamma_X[\tau](f_n \circ q_n) = q_{n\#}\gamma_X[\tau](f_n)$$
$$= \gamma_{X_n} q_{n*}[\tau](f_n) = h_n(f_n) = q_{n\#}h(f_n) = h(f).$$

Hence γ_X is surjective. ∎

IX.8 Quasidiagonality

An operator T is **quasidiagonal** if there is an increasing sequence P_n of finite rank projections tending SOT to the identity such that $\lim_{n\to\infty} \|P_n T - T P_n\| = 0$. In this case, it is possible to drop to a subsequence so that $\|P_n T - T P_n\| < 2^{-n}\varepsilon$ for a given $\varepsilon > 0$. Then the projections $E_n = P_n - P_{n-1}$ form a partition of the identity into finite rank projections such that

$$T = \sum_{n\geq 1} E_n T E_n + K \quad \text{where} \quad K = \sum_{n\geq 1} E_n T P_n^\perp + P_n^\perp T E_n$$

is compact with norm

$$\|K\| \leq \sum_{n\geq 1} \|E_n T P_n^\perp + P_n^\perp T E_n\|$$
$$\leq \sum_{n\geq 1} \max\{\|(P_n T - T P_n)P_n^\perp\|, \|P_n^\perp(P_n T - T P_n)\|\}$$
$$\leq \sum_{n\geq 1} 2^{-n}\varepsilon = \varepsilon.$$

Thus T is a block diagonal operator plus a compact operator. The converse is readily apparent.

Similarly, a subset $\mathcal{S}$ of $\mathcal{B}(\mathcal{H})$ is (jointly) quasidiagonal if there is a sequence $\{P_n\}$ of finite rank projections so that every element of $\mathcal{S}$ is quasidiagonal with respect to $\{P_n\}$. If T is quasidiagonal with respect to a sequence P_n, then so is $T + K$ because $\lim_{n\to\infty} \|KP_n^\perp\| = \|P_n^\perp K\| = 0$ for every compact operator K. Moreover, C*(T) is quasidiagonal. Indeed, write $T = D + K$ where D is block diagonal with respect to a sequence P_n and K is compact. Then

$$\mathrm{C}^*(T) \subset \mathrm{C}^*(D) + \mathfrak{K} \subset \{P_n\}' + \mathfrak{K}.$$

So the sequence P_n implements the quasidiagonality of every element of C*(T).

On the other hand, every semi-Fredholm quasidiagonal operator has index 0. To see this, suppose that $T = D + K$ where $D = \sum_{n\geq 1} \oplus D_n$ is a block diagonal operator and K is compact. Since D_n is finite rank, $\ker D_n$ and $\ker D_n^*$ have the same dimension. The kernels of D and D^* are just the direct sum of these kernels. Since at least one of $\ker D$ or $\ker D^*$ is finite dimensional, only finitely many of $\ker D_n$ and $\ker D_n^*$ are non-zero. Moreover, null(D) = null(D^*); whence ind(T) = ind(D) = 0.

The Weyl–von Neumann–Berg Theorem II.4.2 shows that every normal operator is quasidiagonal. Corollary IX.7.4 shows that the only obstruction to an essentially normal operator being normal plus compact is the Fredholm index. Thus an essentially normal operator is quasidiagonal if and only if it is normal plus compact. We shall see that this is no longer the case for extensions of $C(X)$ when X has higher dimension.

Since quasidiagonality is not affected by compact perturbations, we say that a subalgebra of the Calkin algebra is quasidiagonal if its preimage is quasidiagonal in $\mathcal{B}(\mathcal{H})$. So define an extension $[\tau]$ in $\operatorname{Ext}(X)$ to be **quasidiagonal** if $\pi^{-1}\tau(C(X))$ is quasidiagonal. This is independent of the choice of τ because quasidiagonality is invariant under unitary equivalence. Let $\operatorname{Ext}_{qd}(X)$ denote the set of quasidiagonal extensions in $\operatorname{Ext}(X)$.

By Theorem II.4.1, every trivial extension of $\mathfrak{K}$ by $C(X)$ is quasidiagonal. On the other hand, since quasidiagonal operators have index 0, $\operatorname{Ext}_{qd}(X)$ must lie in $\ker \gamma_X$. As it is easy to verify that if $\mathfrak{A}$ is quasidiagonal, then $\mathcal{M}_k(\mathfrak{A})$ is quasidiagonal for all $k \geq 1$, it also follows that $\operatorname{Ext}_{qd}(X)$ must lie in $\ker \gamma_X^\infty$. So

$$\{0\} \subset \operatorname{Ext}_{qd}(X) \subset \ker \gamma_X^\infty \subset \ker \gamma_X.$$

In fact, we will see in the examples at the end of this chapter that all of these containments may be proper.

Our definition of quasidiagonality of a C*-subalgebra of $\mathcal{B}(\mathcal{H})$ is somewhat different from the definition for the quasidiagonality of an abstract C*-algebra $\mathfrak{A}$ given in section VII.6. However, if σ is a faithful, quasidiagonal representation of $\mathfrak{A}$, then $\sigma^{(\infty)}$ is a faithful, quasidiagonal representation with the additional property that $\sigma^{(\infty)}(\mathfrak{A}) \cap \mathfrak{K} = \{0\}$. By Corollary II.5.6 of Voiculescu's theorem, any other faithful representation ρ of $\mathfrak{A}$ such that $\rho(\mathfrak{A}) \cap \mathfrak{K} = \{0\}$ is approximately unitarily equivalent to $\sigma^{(\infty)}$ relative to $\mathfrak{K}$; and thus is also quasidiagonal. So the quasidiagonality of one faithful representation implies the quasidiagonal of all *essentially* faithful representations.

First we establish a simple lemma that shows that the requirement on increasing sequences is unnecessary.

Lemma IX.8.1 *A separable C*-subalgebra $\mathfrak{A}$ of $\mathcal{B}(\mathcal{H})$ is quasidiagonal if and only if for every $\varepsilon > 0$, finite dimensional projection P and finite subset $\mathcal{A}$ of $\mathfrak{A}$, there is a finite rank projection Q such that $\|PQ^\perp\| < \varepsilon$ and $\|QA - AQ\| < \varepsilon$ for all A in $\mathcal{A}$.*

Proof. If P_n implements the quasidiagonality of $\mathfrak{A}$, then

$$\lim_{n\to\infty} \|PP_n^\perp\| = 0 \quad \text{and} \quad \lim_{n\to\infty} \|P_n A - AP_n\| = 0 \quad \text{for all} \quad A \in \mathfrak{A}.$$

Conversely, suppose that the technical property holds. Choose a sequence A_n dense in the unit ball of $\mathfrak{A}$, and fix a basis $\{e_k\}$ for $\mathcal{H}$. Suppose that we have constructed an increasing set of projections $P_1 < P_2 < \cdots < P_n$ so that $P_i e_i = e_i$

and
$$\|A_k P_i - P_i A_k\| < 2^{-i} \quad \text{for all} \quad 1 \leq k \leq i \quad \text{and} \quad 1 \leq i \leq n.$$

Set $\varepsilon = 2^{-n-1}/5$, $P = P_n \vee e_{n+1}e_{n+1}^*$ and $\mathcal{A} = \{A_1, \ldots, A_{n+1}\}$; and apply the technical property to obtain a finite rank projection Q so that $\|PQ^\perp\| < \varepsilon$ and $\|A_k Q - Q A_k\| < \varepsilon$ for $1 \leq k \leq n+1$.

Following the argument of Lemma III.3.1, there is a projection $P_{n+1} \geq P$ such that $\|P_{n+1} - Q\| < 2\varepsilon$. To recall the ideas, $X = PQP + P^\perp Q P^\perp$ is a positive contraction such that
$$\|X - Q\| = \|PQP^\perp + P^\perp Q P\|$$
$$= \max\{\|PQ^\perp P^\perp\|, \|P^\perp Q^\perp P\|\} \leq \|PQ^\perp\| < \varepsilon.$$

Thus the spectrum of the finite rank operator X is contained in $[0, \varepsilon] \cup [1 - \varepsilon, 1]$. So the projection P_{n+1} obtained from the functional calculus on X for the characteristic function of $[1 - \varepsilon, 1]$ satisfies
$$\|P_{n+1} - Q\| \leq \|P_{n+1} - X\| + \|X - Q\| < 2\varepsilon.$$

Since X commutes with P, so does P_{n+1}. Moreover, $PX = PQP \geq (1 - \varepsilon)P$; whence $P_{n+1} \geq P$. Finally, for $1 \leq k \leq n+1$,
$$\|A_k P_{n+1} - P_{n+1} A_k\| \leq \|A_k Q - Q A_k\| + 2\|A_k\| \|P_{n+1} - Q\|$$
$$< 5\varepsilon = 2^{-n-1}. \qquad \blacksquare$$

There is a natural topology on $\mathrm{Ext}(X)$. This is obtained by taking the topology of pointwise norm convergence on the collection of monomorphisms of $\mathrm{C}(X)$ into $\mathcal{Q}(\mathcal{H})$ and taking the quotient norm on the equivalence classes.

Theorem IX.8.2 *The closure of the zero element in $\mathrm{Ext}(X)$ (in the quotient topology) equals $\mathrm{Ext}_{qd}(X)$.*

Proof. Suppose that τ_k are quasidiagonal monomorphisms of $\mathrm{C}(X)$ into $\mathcal{Q}(\mathcal{H})$ converging pointwise to τ. To verify that the limit τ is quasidiagonal, we apply Lemma IX.8.1. Suppose that $\varepsilon > 0$, a finite rank projection P and finitely many operators $T_1, \ldots, T_n$ in $\pi^{-1}\tau \mathrm{C}(X)$ are given. Let $f_1, \ldots, f_n$ be functions in $\mathrm{C}(X)$ such that $\tau(f_i) = \pi T_i$ for $1 \leq i \leq n$. Choose k sufficiently large that $\|\tau(f_i) - \tau_k(f_i)\| < \varepsilon/3$ for $1 \leq i \leq n$. Then choose operators S_i in $\pi^{-1}\tau_k(f_i)$ so that $\|T_i - S_i\| < \varepsilon/3$. By the quasidiagonality of τ_k, there is a finite rank projection Q such that $\|PQ^\perp\| < \varepsilon$ and $\|S_i Q - Q S_i\| < \varepsilon/3$ for $1 \leq i \leq n$. Then
$$\|T_i Q - Q T_i\| < \|S_i Q - Q S_i\| + 2\|S_i - T_i\| < \varepsilon.$$

Hence τ is quasidiagonal. Therefore $\mathrm{Ext}_{qd}(X)$ is closed. It has already been observed that the zero element of $\mathrm{Ext}(X)$ is quasidiagonal. Hence $\mathrm{Ext}_{qd}(X)$ contains the closure of the zero element.

Now suppose that τ is quasidiagonal, and let P_n be a sequence implementing the quasidiagonality. By dropping to a subsequence if necessary, we may suppose that $\sum_{n\geq 1} \|P_n T - TP_n\| < \infty$ for every T in $\pi^{-1}\tau(C(X))$. Let $E_n = P_n - P_{n-1}$ be the associated partition of the identity. It follows that $T - \sum_{n\geq 1} E_n T E_n$ is compact for all T in $\pi^{-1}\tau(C(X))$. By Theorem IX.4.10, there is a positive linear map φ from $C(X)$ into $\mathcal{B}(\mathcal{H})$ such that $\tau = \pi\varphi$. Replace φ by the positive map $\sum_{n\geq 1} E_n \varphi(\cdot) E_n$. This differs from the original by a map into the compacts. So we assume that φ already has this form. Let $\varphi_n(\cdot) = E_n \varphi(\cdot)|_{E_n \mathcal{H}}$.

By Naimark's Dilation Theorem IX.4.2, each φ_n has a $*$-dilation σ_n into the bounded operators on $\mathcal{H}_n = E_n\mathcal{H} \oplus \mathcal{H}_n'$ such that $\sigma_n(f) = \begin{bmatrix} \varphi_n(f) & \sigma_{12}^n(f) \\ \sigma_{21}^n(f) & \sigma_{22}^n(f) \end{bmatrix}$ with respect to the given decomposition of $\mathcal{H}_n$. Let P_n be the projection of $\mathcal{H}_n$ onto $E_n\mathcal{H}$; and let $P = \sum_{n\geq 1} \oplus P_n$, which is the projection of $\mathcal{K} = \sum_{n\geq 1} \oplus \mathcal{H}_n$ onto $\mathcal{H} = \sum_{n\geq 1} \oplus E_n\mathcal{H}$. Define a representation into $\mathcal{B}(\mathcal{K})$ by $\sigma = \sum_{n\geq 1} \oplus \sigma_n$.

Notice that for any f in $C(X)$, we have

$$\varphi(|f|^2) = P\sigma(|f|^2)P = P\sigma(f)\sigma(f)^*P = \sum_{n\geq 1} \oplus P_n \sigma_n(f)\sigma_n(f)^* P_n$$
$$= \sum_{n\geq 1} \oplus P_n \sigma_n(f) P_n \sigma_n(f)^* P_n + \sum_{n\geq 1} \oplus P_n \sigma_n(f) P_{\mathcal{H}_n'} \sigma_n(f)^* P_n$$
$$= \sum_{n\geq 1} \oplus \varphi_n(f)\varphi_n(f)^* + \sum_{n\geq 1} \oplus P_n \sigma_{12}^n(f) \sigma_{12}^n(f)^* P_n.$$

Therefore,

$$0 = \tau(|f|^2) - \tau(f)\tau(f)^* = \pi\big(\varphi(|f|^2) - \varphi(f)\varphi(f)^*\big)$$
$$= \pi\Big(\sum_{n\geq 1} \oplus P_n \sigma_{12}^n(f) \sigma_{12}^n(f)^* P_n\Big).$$

Consequently, it follows that

$$\lim_{n\to\infty} \|\sigma_{12}^n(f)\| = \lim_{n\to\infty} \|\sigma_{21}^n(f)\| = 0 \quad \text{for all} \quad f \in C(X).$$

To show that $[\tau]$ is in the closure of the trivial elements, consider $\sigma^{(\infty)}$, the direct sum of countably many copies of σ. Clearly, this is a representation of $C(X)$ which contains no compact operators. Moreover, since $\tau(f) = \pi(P\sigma(f)P)$ is a monomorphism, so are $\pi\sigma$ and $\pi\sigma^{(\infty)}$. Hence $\pi\sigma^{(\infty)}$ is a trivial extension. It suffices to show that there are unitary operators U_n so that

$$\lim_{n\to\infty} \|\tau(f) \oplus \pi\sigma^{(\infty)} - \pi U_n \sigma^{(\infty)}(f) U_n^*\| = 0 \quad \text{for all} \quad f \in C(X).$$

IX.8. Quasidiagonality

Notice that

$$\sigma_n^{(\infty)}(f) = \begin{bmatrix} \varphi_n(f) & \sigma_{12}^n(f) & 0 & 0 & 0 & 0 & \cdots \\ \sigma_{21}^n(f) & \sigma_{22}^n(f) & 0 & 0 & 0 & 0 & \cdots \\ 0 & 0 & \varphi_n(f) & \sigma_{12}^n(f) & 0 & 0 & \cdots \\ 0 & 0 & \sigma_{21}^n(f) & \sigma_{22}^n(f) & 0 & 0 & \cdots \\ 0 & 0 & 0 & 0 & \varphi_n(f) & \sigma_{12}^n(f) & \cdots \\ 0 & 0 & 0 & 0 & \sigma_{21}^n(f) & \sigma_{22}^n(f) & \cdots \\ \vdots & \vdots & \vdots & \vdots & \vdots & \vdots & \ddots \end{bmatrix}$$

$$= \begin{bmatrix} \varphi_n(f) & 0 & 0 & 0 & 0 & 0 & \cdots \\ 0 & \sigma_{22}^n(f) & \sigma_{21}^n(f) & 0 & 0 & 0 & \cdots \\ 0 & \sigma_{12}^n(f) & \varphi_n(f) & 0 & 0 & 0 & \cdots \\ 0 & 0 & 0 & \sigma_{22}^n(f) & \sigma_{21}^n(f) & 0 & \cdots \\ 0 & 0 & 0 & \sigma_{12}^n(f) & \varphi_n(f) & 0 & \cdots \\ 0 & 0 & 0 & 0 & 0 & \sigma_{22}^n(f) & \cdots \\ \vdots & \vdots & \vdots & \vdots & \vdots & \vdots & \ddots \end{bmatrix}$$

$$+ \begin{bmatrix} 0 & \sigma_{12}^n(f) & 0 & 0 & 0 & 0 & \cdots \\ \sigma_{21}^n(f) & 0 & -\sigma_{21}^n(f) & 0 & 0 & 0 & \cdots \\ 0 & -\sigma_{12}^n(f) & 0 & \sigma_{12}^n(f) & 0 & 0 & \cdots \\ 0 & 0 & \sigma_{21}^n(f) & 0 & -\sigma_{12}^n(f) & 0 & \cdots \\ 0 & 0 & 0 & -\sigma_{12}^n(f) & 0 & \sigma_{12}^n(f) & \cdots \\ 0 & 0 & 0 & 0 & \sigma_{21}^n(f) & 0 & \cdots \\ \vdots & \vdots & \vdots & \vdots & \vdots & \vdots & \ddots \end{bmatrix}$$

$$\simeq \varphi_n(f) \oplus \sigma_n^{(\infty)}(f) + \varepsilon_n(f).$$

Moreover, $\|\varepsilon_n(f)\| \leq 2\max\{\|\sigma_{12}^n(f)\|, \|\sigma_{21}^n(f)\|\}$ which tends to 0 as n tends to infinity for each f in $C(X)$. Let V_n be the unitary implementing this equivalence, so that

$$V_n \sigma_n^{(\infty)}(f) V_n^* = \varphi_n(f) \oplus \sigma_n^{(\infty)}(f) + \varepsilon_n(f).$$

Then set $W_n = I^{(n)} \oplus \sum_{k>n} \oplus V_k$. Notice that

$$W_n \sigma^{(\infty)}(f) W_n = \sum_{k=1}^n \oplus \sigma_k^{(\infty)}(f) \oplus \sum_{k>n} \oplus V_k \sigma_k^{(\infty)}(f) V_k^*$$

$$= \sum_{k=1}^n \oplus \sigma_k^{(\infty)}(f) \oplus \sum_{k>n} \oplus (\varphi_k(f) \oplus \sigma_k^{(\infty)}(f)) + \varepsilon_k(f)$$

$$\simeq \sum_{k>n} \oplus \varphi_k(f) \oplus \sigma^{(\infty)}(f) + \sum_{k>n} \oplus \varepsilon_k'(f)$$

where the unitary equivalence obtained by shuffling the terms is implemented by a unitary S_n, and $\varepsilon'_k(f) = S_n \varepsilon_k(f) S_n^*$. Then $U_n = S_n W_n$ is a unitary operator such that

$$\sum_{k>n} \oplus \varphi_k(f) \oplus \sigma^{(\infty)}(f) - U_n \sigma^{(\infty)}(f) U_n^* = -\sum_{k>n} \oplus \varepsilon'_k(f).$$

The unitary U_n maps $\sum_{n\geq 1} \oplus \mathcal{H}_n^{(\infty)}$ into $\sum_{k>n} \oplus E_k \mathcal{H} \oplus \sum_{n\geq 1} \oplus \mathcal{H}_n^{(\infty)}$ which is a subspace of $\mathcal{H} \oplus \sum_{n\geq 1} \oplus \mathcal{H}_n^{(\infty)}$ of finite codimension. So we may think of U_n as a Fredholm isometry into this larger space such that

$$\varphi(f) \oplus \sigma^{(\infty)}(f) - U_n \sigma^{(\infty)}(f) U_n^* = \sum_{k=1}^{n} \oplus \varphi_k(f) - \sum_{k>n} \oplus \varepsilon'_k(f).$$

By Corollary IX.2.4, weak equivalence and equivalence coincide for extensions of $C(X)$. Therefore $\pi \operatorname{Ad} U_n \sigma^{(\infty)}$ is a trivial extension. Moreover

$$\|\tau(f) \oplus \pi\sigma^{(\infty)}(f) - \pi \operatorname{Ad} U_n \sigma^{(\infty)}(f)\| = \|\pi \sum_{k>n} \oplus \varepsilon'_k(f)\|$$
$$\leq 2 \max_{k>n} \|\varepsilon_k(f)\|.$$

This tends to 0 pointwise as n increases. Hence τ is the pointwise limit of trivial extensions. ∎

IX.9 Homotopy Invariance

Recall that two maps f and g from X to Y are said to be **homotopic** if there is a continuous function H from $X \times [0,1]$ into Y such that $H(\xi, 0) = f(\xi)$ and $H(\xi, 1) = g(\xi)$. The important topological property of Ext proved in this section is known as **homotopy invariance**, which says that if f and g are homotopic, then $f_* = g_*$.

First we push the quasidiagonality argument a bit further to obtain:

Corollary IX.9.1 *If f_t is a homotopy of maps from X to Y and τ is in $\operatorname{Ext}(X)$, then $[f_{1*}\tau] - [f_{0*}\tau]$ is quasidiagonal.*

Proof. We will show that $f_{1*}\tau$ is the limit of extensions of the form $\operatorname{Ad} U_n f_{0*}\tau$. Let γ be a monomorphism representing the extension $-[\tau]$, and let σ denote a fixed trivial extension by $C(Y)$. Then $\rho_t(g) = \gamma(g \circ f_t) \oplus \sigma(g)$ is an inverse for $f_{t*}\tau$. Moreover, the map taking t to $\rho_t(g)$ is continuous for g in $C(Y)$. Define trivial extensions

$$\sigma_n = \sum_{k=1}^{n} \oplus \big(\rho_{k/n} \oplus f_{k/n*}\tau \big) \quad \text{and} \quad \sigma'_n = \sum_{k=0}^{n-1} \oplus \big(f_{k/n*}\tau \oplus \rho_{k/n} \big).$$

IX.9. Homotopy Invariance

Then
$$(f_{0*}\tau \oplus \sigma_n)(g) = \sum_{k=1}^{n} \oplus \big(f_{(k-1)/n*}\tau \oplus \rho_{k/n}\big)(g) \oplus f_{1*}\tau(g)$$
$$= (\sigma'_n(f) \oplus f_{1*}\tau)(g) + \varepsilon_n(g)$$

where
$$\varepsilon_n(g) \simeq \sum_{k=1}^{n} \oplus \big(\rho_{(k-1)/n}(g) - \rho_{k/n}(g)\big) \oplus 0.$$

By the continuity of $\rho_t(g)$, it follows that $\lim_{n\to\infty} \|\varepsilon_n(g)\| = 0$ for all g in $C(Y)$. This shows that $[f_{1*}\tau]$ is in the closure of $[f_{0*}\tau]$. Hence $[f_{1*}\tau] - [f_{0*}\tau]$ is in the closure of the zero element, and thus is quasidiagonal by Theorem IX.8.2. ∎

The key step is to prove injectivity for a special map.

Lemma IX.9.2 *Let δ_0 be the map of $X \times [0,1]$ into itself given by $\delta_0(\xi,t) = (\xi,0)$. Then δ_{0*} is injective.*

Proof. Suppose that $\delta_{0*}[\tau] = 0$. Then we will show that $[\tau] = 0$. The maps $r_s(\xi,t) = (\xi,\min\{s,t\})$ for $0 \le s \le 1$ define a homotopy from $\delta_0 = r_0$ to the identity map $\text{id} = r_1$. So by Corollary IX.9.1, $[\tau] = \text{id}_*[\tau] - \delta_{0*}[\tau]$ is quasidiagonal. Let P_k be a sequence of finite rank projections implementing the quasidiagonality of τ, and let $E_k = P_k - P_{k-1}$.

As in Theorem IX.8.2, (after replacing the P_k by a subsequence if necessary) there is a positive map $\varphi = \sum_{k \ge 1} \oplus \varphi_k$ of $C(X)$ into $\mathcal{B}(\mathcal{H})$ such that $\tau = \pi\varphi$, where $\varphi_k(f) = E_k\varphi(f)|_{E_k\mathcal{H}}$. Also recall from that proof that there are finite rank linear maps σ_{12}^n and σ_{21}^n such that
$$\varphi_n(fg) - \varphi_n(f)\varphi_n(g) = \sigma_{12}^n(f)\sigma_{21}^n(g)$$
and
$$\lim_{n\to\infty} \|\sigma_{12}^n(f)\| = \lim_{n\to\infty} \|\sigma_{21}^n(f)\| = 0 \quad \text{for all} \quad f \in C(X \times [0,1]).$$

It is worth noting that this limit is uniform on any *compact* set $\mathcal{S}$ of functions because $\|\sigma^n\| = 1$. Indeed, for any $\varepsilon > 0$, there is a finite $\varepsilon/2$ subnet $\{f_i\}_1^N$ of $\mathcal{S}$. For n sufficiently large, both $\|\sigma_{12}^n(f_i)\| < \varepsilon/2$ and $\|\sigma_{21}^n(f_i)\| < \varepsilon/2$ for $1 \le i \le N$. Hence an easy estimate shows that $\|\sigma_{12}^n(f)\| < \varepsilon$ and $\|\sigma_{21}^n(f)\| < \varepsilon$ for all f in $\mathcal{S}$.

Now $r_{s*}\tau(f) = \pi\varphi(f \circ r_s)$ for every $0 \le s \le 1$. Since $\{f \circ r_s : 0 \le s \le 1\}$ is the continuous image of $[0,1]$, it is a compact set. So the previous remarks apply. That is, $\|\sigma_{12}^n(f \circ r_s)\|$ converges to 0 uniformly for s in $[0,1]$.

We will define elements of $\text{Ext}(X \times [0,1])$ by
$$\sigma(f) = \pi \sum_{k \ge 1} \sum_{j=1}^{k} \oplus \varphi_k(f \circ r_{j/k})$$

and
$$\sigma'(f) = \pi \sum_{k \geq 1} \sum_{j=1}^{k} \oplus \varphi_k(f \circ r_{(j-1)/k}).$$

To see that σ is a *-monomorphism, notice that σ is positive and

$$\sigma(fg) - \sigma(f)\sigma(g) = \pi \sum_{k \geq 1} \sum_{j=1}^{k} \oplus \varphi_k(fg \circ r_{j/k}) - \varphi_k(f \circ r_{j/k})\varphi_k(g \circ r_{j/k})$$
$$= \pi \sum_{k \geq 1} \sum_{j=1}^{k} \oplus \sigma_{12}^n(f \circ r_{j/k})\sigma_{21}^n(g \circ r_{j/k}).$$

By the previous paragraph, we see that this sum is a direct sum of finite rank terms converging to zero in norm. Hence it is compact, and thus is annihilated by π. It follows that σ is a homomorphism. It is self-adjoint because of positivity. Injectivity follows from the fact that the restriction to the sum over k of the k-th block of the k-th sum is $\pi \sum_{k \geq 1} \oplus \varphi_k = \tau$ which is already one-to-one.

Next notice that $\sigma' = \sigma$. Indeed,

$$\sigma(f) - \sigma'(f) = \pi \sum_{k \geq 1} \sum_{j=1}^{k} \oplus \varphi_k(f \circ r_{j/k} - f \circ r_{(j-1)/k}).$$

By the uniform continuity of f, $\|f \circ r_{j/k} - f \circ r_{(j-1)/k}\|$ tends to 0 as k increases independent of j. Thus the sum is again a direct sum of finite rank terms tending to 0 in norm, and so is compact and killed by π.

Finally, we will show that $[r_{0*}\tau] + [\sigma] = [\sigma] + [\tau]$. To this end,

$$r_{0*}\tau(f) \oplus \sigma(f) = \pi \sum_{k \geq 1} \oplus \varphi_k(f \circ r_0) \oplus \pi \sum_{k \geq 1} \sum_{j=1}^{k} \oplus \varphi_k(f \circ r_{j/k})$$
$$= \pi \sum_{k \geq 1} \sum_{j=1}^{k} \oplus \varphi_k(f \circ r_{(j-1)/k}) \oplus \pi \sum_{k \geq 1} \oplus \varphi_k(f \circ r_1)$$
$$= \sigma'(f) \oplus \tau(f) = \sigma(f) \oplus \tau(f).$$

Since $\text{Ext}(X)$ is a group, this implies that $[\tau] = [r_{0*}\tau] = 0$ as claimed. ∎

The lemma above handled all the technical difficulties. It remains to mop up in order to establish homotopy invariance.

Theorem IX.9.3 *If f_0 and f_1 are homotopic maps from X into Y, then $f_{0*} = f_{1*}$.*

Proof. Let $j_s(\xi) = (\xi, s)$ be imbeddings of X into $X \times [0,1]$ for $0 \leq s \leq 1$. First let us show that $j_{0*} = j_{1*}$. Indeed, let $[\tau]$ belong to $\text{Ext}(X)$. Note that $\delta_0 j_0 = \delta_0 j_1$. Hence

$$\delta_{0*}(j_{1*}[\tau] - j_{0*}[\tau]) = (\delta_0 j_1)_*[\tau] - (\delta_0 j_0)_*[\tau] = 0.$$

IX.10. The Mayer–Vietoris Sequence

By the lemma above, this shows that $j_{1*}[\tau] = j_{0*}[\tau]$.

More generally, let F be a function from $X \times [0,1]$ to Y implementing the homotopy; that is, $F(\xi, i) = f_i(\xi)$ for $i = 0, 1$. Then if $[\tau]$ is in $\text{Ext}(X)$,

$$f_{0*}[\tau] = F_* j_{0*}[\tau] = F_* j_{1*}[\tau] = f_{1*}[\tau].$$

So homotopy invariance is established. ∎

Recall that a topological space is **contractible** if it is homotopic to a point. The **cone** on X is the space $CX = X \times [0,1]/X \times \{0\}$, which is always a contractible space.

Corollary IX.9.4 *If X is contractible, then $\text{Ext}(X) = 0$. Thus $\text{Ext}(CX) = 0$ for every compact metric space X.*

IX.10 The Mayer–Vietoris Sequence

In this section, we will develop the algebraic topology of Ext a bit more to obtain another computational tool. If f is a continuous map from X to Y, the **mapping cylinder** $Z(f)$ of f is the space $X \times [0,1] \vee Y$ modulo the identifications $(\xi, 1) = f(\xi)$. The **mapping cone** is the space

$$C(f) = Z(f)/(X \times \{0\}).$$

The **suspension** of X is the space $SX = CX/(X \times \{1\})$.

Lemma IX.10.1 *Let f be a continuous map from X into Y, and let i be the injection of Y into $C(f)$. Then there is an exact sequence:*

$$\text{Ext}(X) \xrightarrow{f_*} \text{Ext}(Y) \xrightarrow{i_*} \text{Ext}(C(f)).$$

Proof. Define j to be the inclusion $j(\xi) = (\xi, 0)$ of X into $Z(f)$; q the quotient of $Z(f)$ onto $C(f)$; k to be the injection of Y into $Z(f)$; and p the projection of $Z(f)$ onto Y given by $p(\xi, t) = f(\xi)$ for (ξ, t) in CX and $p(\eta) = \eta$ for η in Y. Consider the diagram

$$X \xrightarrow{j} Z(f) \xrightarrow{q} C(f)$$

with f, p, k, i to Y.

Then $f = pj, i = qk, pk = \text{id}_Y$ and qj is constant. By Corollary IX.6.2,

$$\text{Ext}(X) \xrightarrow{j_*} \text{Ext}(Z(f)) \xrightarrow{q_*} \text{Ext}(C(f))$$

is exact. The map kp is homotopic to $\text{id}_{Z(f)}$ via the map from $Z(f) \times [0,1]$ to $Z(f)$ given by

$$H((\xi, t), s) = (\xi, \max\{s, t\}) \quad \text{for} \quad (\xi, t) \in X \times [0,1]$$
$$H(\eta, s) = \eta \quad \text{for} \quad \eta \in Y.$$

By the homotopy invariance Theorem IX.9.3, $k_*p_* = \mathrm{id}_{Z(f)*}$. As $p_*k_* = \mathrm{id}_{Y*}$, it follows that p_* and k_* are reciprocal isomorphisms. Hence

$$\mathrm{Ext}(X) \xrightarrow{p_*j_*^{-1}} \mathrm{Ext}(Y) \xrightarrow{q_*k_*} \mathrm{Ext}(C(f))$$

is exact. ∎

Lemma IX.10.2 *If A is a contractible closed subset of X, then the quotient map q of X onto X/A induces an isomorphism of $\mathrm{Ext}(X)$ and $\mathrm{Ext}(X/A)$.*

Proof. Consider the sequence

$$\mathrm{Ext}(A) \xrightarrow{j_*} \mathrm{Ext}(X) \xrightarrow{q_*} \mathrm{Ext}(X/A) \xrightarrow{i_*} \mathrm{Ext}(C(q)).$$

This is exact at $\mathrm{Ext}(X)$ by Corollary IX.6.2. Exactness at $\mathrm{Ext}(X/A)$ follows from Lemma IX.10.1. By Corollary IX.9.4, $\mathrm{Ext}(A) = 0$. So it suffices to show that $\mathrm{Ext}(C(q)) = 0$ as well.

There is a natural imbedding k of SA into $C(q)$ given by $k(\alpha, t) = (\alpha, t)$ for α in A and $0 < t < 1$ since this extends by continuity to $k(*, 1) = \xi_0$ where $\xi_0 = A/A$ in X/A. The quotient

$$C(q)/SA = C(X/A)/(\xi_0 \times [0, 1])$$

where $C(X/A)$ is the cone on X/A. Thus both SA and $C(q)/SA$ are contractible. By Corollary IX.6.2,

$$0 = \mathrm{Ext}(SA) \longrightarrow \mathrm{Ext}(C(q)) \longrightarrow \mathrm{Ext}(C(q)/SA) = 0$$

is exact. Therefore $\mathrm{Ext}(C(q)) = 0$ as claimed. ∎

Now we are able to obtain a long exact sequence for Ext.

Theorem IX.10.3 *Suppose that A is a closed subset of X. Then there is a natural long exact sequence*

$$\mathrm{Ext}(A) \xrightarrow{j_*} \mathrm{Ext}(X) \xrightarrow{q_*} \mathrm{Ext}(X/A) \xrightarrow{\partial} \mathrm{Ext}(SA) \xrightarrow{Sj_*} \mathrm{Ext}(SX) \xrightarrow{Sq_*} \cdots$$

Proof. The maps $S^n j$ are the injections of $S^n A$ into $S^n X$, and $S^n q$ is the quotient map of $S^n X$ onto $S^n X / S^n A$ (which is homotopic to $S^n(X/A)$).

The sequence is exact at $\mathrm{Ext}(X)$ by Corollary IX.6.2, and similarly it is exact at $\mathrm{Ext}(S^n X)$ for all $n \geq 1$.

Let $X \cup CA := (X \times \{1\}) \cup CA$ denote the union of X and CA with $A \times \{1\}$ identified with the corresponding subset of $X \times \{1\}$. To define the connecting homomorphism ∂ from $\mathrm{Ext}(X/A)$ to $\mathrm{Ext}(SA)$, consider the quotient maps p from $X \cup CA$ to $(X \cup CA)/X = SA$ and r from $X \cup CA$ onto $(X \cup CA)/CA = X/A$. Since CA is contractible, Lemma IX.10.2 shows that r_* is an isomorphism. Hence we may define $\partial = p_* r_*^{-1}$ from $\mathrm{Ext}(X/A)$ into $\mathrm{Ext}(SA)$.

IX.10. The Mayer–Vietoris Sequence

If i is the injection of X into $X \cup CA$, we have a commutative diagram

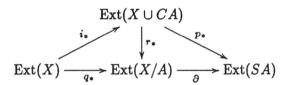

This is exact at $\text{Ext}(X \cup CA)$ by Corollary IX.6.2. Since the diagram commutes and r_* is an isomorphism, it follows that our sequence is exact at $\text{Ext}(X/A)$.

Let $CX \cup CA$ denote the *disjoint union* of these two sets modulo the identification of $A \times \{1\}$ with the corresponding subset of $X \times \{1\}$. Let k be the injection of $X \cup CA$ into $CX \cup CA$; and let s be the quotient map of $CX \cup CA$ onto $(CX \cup CA)/(X \cup CA) = SX$. Also let t denote the quotient of $CX \cup CA$ onto $(CX \cup CA)/CX = SA$. And lastly, we define the flip map of SX onto itself by $f(\xi, t) = (\xi, 1 - t)$. Consider the diagram

$$\begin{array}{ccccc}
\text{Ext}(X \cup CA) & \xrightarrow{k_*} & \text{Ext}(CX \cup CA) & \xrightarrow{s_*} & \text{Ext}(SX) \\
\downarrow r_* & \searrow^{p_*} & \downarrow t_* & & \downarrow f_* \\
\text{Ext}(X/A) & \xrightarrow{\partial} & \text{Ext}(SA) & \xrightarrow{Sj_*} & \text{Ext}(SX)
\end{array}$$

The top row is exact by Corollary IX.6.2. Since CX is contractible, t_* is an isomorphism by Lemma IX.10.2. Since f is a homeomorphism, f_* is an isomorphism. To prove exactness at $\text{Ext}(SA)$, we need to prove that $f_* s_* = S j_* t_*$. This will follow from Theorem IX.9.3 if we can show that fs and $(Sj)t$ are homotopic.

Let ξ_A denote the point A/A and let ξ_X denote X/X. The homotopy H from $(CX \cup CA) \times [0, 1]$ to SX is given by

$$H_{CA}(\alpha, u; v) = (\alpha, (1 - v)u) \qquad \alpha \in A$$
$$H_{CX}(\xi, u; v) = (\xi, 1 - uv) \qquad \xi \in X$$

for $0 \le u \le 1$ and $0 \le v \le 1$. This definition identifies the points $(\alpha, 0)$ with $(\xi_A, 0)$ in CA, the points $(\xi, 0)$ with $(\xi_X, 0)$ in CX and SX, and the points $(\xi, 1)$ with $(\xi_X, 1)$ in SX. Thus

$$H_{CA}(\xi_A, 0; v) = H_{CA}(\alpha, u; 1) = (\xi_A, 0) \qquad \alpha \in A$$
$$H_{CX}(\xi_X, 0; v) = H_{CX}(\xi, u; 0) = (\xi_X, 1) \qquad \xi \in X$$
$$H_{CX}(\xi, 1; 1) = (\xi_X, 0) \qquad \xi \in X.$$

Notice that this is well defined because at $u = 1$,

$$H_{CA}(\alpha, 1; v) = (\alpha, 1 - v) = H_{CX}(\alpha, 1; v) \qquad \alpha \in A.$$

It is evidently continuous even at the endpoints. Moreover, at $v = 0$,

$$H_{CA}(\alpha, u; 0) = (\alpha, u) = (Sj)t(\alpha, u) \qquad \alpha \in A$$
$$H_{CX}(\xi, u; 0) = (\xi_X, 1) = (Sj)t(\xi, u) \qquad \xi \in X$$

and at $v = 1$,

$$H_{CA}(\alpha, u; 1) = (\xi_X, 0) = fs(\alpha, u) \qquad \alpha \in A$$
$$H_{CX}(\xi, u; 1) = (\xi, 1 - u) = fs(\xi, u) \qquad \xi \in X.$$

All the other terms are exact by applying these results to the imbedding of $S^n A$ into $S^n X$ for all $n \geq 0$.

Naturality refers to the fact that the sequence commutes with maps. Suppose that B is a closed subset of another compact metric space Y, and f is a continuous map from X to Y such that $f(A) \subset B$. Then f induces homomorphisms

$$
\begin{array}{ccccccccc}
\mathrm{Ext}(A) & \to & \mathrm{Ext}(X) & \to & \mathrm{Ext}(X/A) & \xrightarrow{\partial} & \mathrm{Ext}(SA) & \to & \mathrm{Ext}(SX) & \to & \cdots \\
\downarrow (f|_A)_* & & \downarrow f_* & & \downarrow \bar{f}_* & & \downarrow (Sf|_A)_* & & \downarrow (Sf)_* & & \\
\mathrm{Ext}(B) & \to & \mathrm{Ext}(Y) & \to & \mathrm{Ext}(Y/B) & \xrightarrow{\partial} & \mathrm{Ext}(SB) & \to & \mathrm{Ext}(SY) & \to & \cdots
\end{array}
$$

To verify that this diagram is commutative, it suffices to verify it on the square involving ∂. This follows from

$$
\begin{array}{ccccc}
X/A & \xleftarrow{r} & X \cup CA & \xrightarrow{p} & SA \\
\downarrow \bar{f} & & \downarrow f \cup Cf|_A & & \downarrow Sf|_A \\
Y/B & \xleftarrow{r} & Y \cup CB & \xrightarrow{p} & SB
\end{array}
$$

As this diagram commutes, it follows that the induced maps commute with ∂. ∎

Remark IX.10.4 This sequence isn't as long as it looks. An important theorem about Ext is **periodicity** which states that there is a natural isomorphism from $\mathrm{Ext}(S^2 X)$ onto $\mathrm{Ext}(X)$. Thus this sequence turns into a 6 term exact cycle. We will not prove periodicity here.

A fairly straightforward consequence of the long exact sequence is the **Mayer–Vietoris** sequence.

Theorem IX.10.5 *Suppose that B and C are two closed subsets of X such that $B \cup C = X$ and $B \cap C = A$. Then there is a natural long exact sequence*

$$\mathrm{Ext}(A) \xrightarrow{(i_{B*}, -i_{C*})} \mathrm{Ext}(B) \oplus \mathrm{Ext}(C) \xrightarrow{j_{B*} + j_{C*}} \mathrm{Ext}(X) \to \mathrm{Ext}(SA) \to \cdots$$

IX.10. The Mayer–Vietoris Sequence

Proof. We denote the injections of A into B and C by i_B and i_C respectively; and the injections of B and C into X by j_B and j_C. Let q be the quotient map of X onto X/B. Then the long exact sequences for the injections of A in C and of B in X are intertwined by the maps induced by j_C:

$$\begin{array}{ccccccccc}
\mathrm{Ext}(A) & \xrightarrow{i_{C*}} & \mathrm{Ext}(C) & \longrightarrow & \mathrm{Ext}(C/A) & \xrightarrow{\partial} & \mathrm{Ext}(SA) & \xrightarrow{Si_{C*}} & \mathrm{Ext}(SC) & \longrightarrow \cdots \\
{\scriptstyle i_{B*}}\downarrow & & {\scriptstyle j_{C*}}\downarrow & & \parallel & & {\scriptstyle Si_{B*}}\downarrow & & {\scriptstyle Sj_{C*}}\downarrow & \\
\mathrm{Ext}(B) & \xrightarrow[j_{B*}]{} & \mathrm{Ext}(X) & \xrightarrow[q_*]{} & \mathrm{Ext}(X/B) & \longrightarrow & \mathrm{Ext}(SB) & \xrightarrow[Sj_{B*}]{} & \mathrm{Ext}(SX) & \xrightarrow[Sq_*]{} \cdots
\end{array}$$

The map from $\mathrm{Ext}(X)$ to $\mathrm{Ext}(SA)$ is defined by ∂q_*.

The exactness of the Mayer–Vietoris sequence at $\mathrm{Ext}(X)$ and onwards to the right now follows from standard "arrow chasing" arguments. For example, consider $\mathrm{Ext}(X)$. It is evident that $\partial q_*(j_{B*} + j_{C*}) = 0$. Suppose that $[\tau]$ in $\mathrm{Ext}(X)$ satisfies $\partial q_*[\tau] = 0$. Exactness at $\mathrm{Ext}(C/A)$ above yields $[\sigma_C]$ in $\mathrm{Ext}(C)$ such that $q_{C*}[\sigma_C] = q_*[\tau]$, where q_C is the quotient of C onto C/A. Since $q_{C*} = q_*j_{C*}$, we see that $q_*([\tau] - j_{C*}[\sigma_C]) = 0$. Hence by exactness at $\mathrm{Ext}(X)$ above, we obtain an element $[\sigma_B]$ in $\mathrm{Ext}(B)$ such that $j_{B*}[\sigma_B] = [\tau] - j_{C*}[\sigma_C]$, whence

$$[\tau] = (j_{B*} + j_{C*})[\sigma_B \oplus \sigma_C].$$

It remains to demonstrate exactness of the diagram at $\mathrm{Ext}(B) \oplus \mathrm{Ext}(C)$. Notice that $j_B i_B = j_A = j_C i_C$, and therefore

$$(j_{B*} + j_{C*})(i_{B*} \oplus -i_{C*}) = j_{A*} - j_{A*} = 0.$$

So we suppose that $[\sigma_B]$ in $\mathrm{Ext}(B)$ and $[\sigma_C]$ in $\mathrm{Ext}(C)$ are extensions satisfying $j_{B*}[\sigma_B] + j_{C*}[\sigma_C] = 0$ in $\mathrm{Ext}(X)$. By Exercise IX.3, there is a natural isomorphism between $\mathrm{Ext}(B) \oplus \mathrm{Ext}(C)$ and $\mathrm{Ext}(B \vee C)$ where $B \vee C$ is the disjoint union. Let $p = j_B \vee j_C$ be the canonical map of $B \vee C$ onto X; and let p' be its restriction to $A \vee A$. Consider the commutative diagram

$$\begin{array}{ccc}
\mathrm{Ext}(A) \oplus \mathrm{Ext}(A) & \xrightarrow{i_{B*} \oplus i_{C*}} & \mathrm{Ext}(B) \oplus \mathrm{Ext}(C) \\
\downarrow & & \downarrow \\
\mathrm{Ext}(A \vee A) & \xrightarrow{i_{B*} \vee i_{C*}} & \mathrm{Ext}(B \vee C) \\
{\scriptstyle p'_*}\downarrow & & {\scriptstyle p_*}\downarrow \\
\mathrm{Ext}(A) & \xrightarrow[i_{A*}]{} & \mathrm{Ext}(X)
\end{array}$$

Since

$$p_*[\sigma_B \vee \sigma_C] = (j_{B*} \vee j_{C*})[\sigma_B \vee \sigma_C] = j_{B*}[\sigma_B] + j_{C*}[\sigma_C] = 0,$$

Theorem IX.6.1 shows that there are extensions $[\rho_B]$ and $[\rho_C]$ in $\mathrm{Ext}(A)$ such that

$$(i_{B*} \vee i_{C*})[\rho_B \vee \rho_C] = [\sigma_B \vee \sigma_C]$$

and
$$0 = p'_*[\rho_1 \vee \rho_2] = [\rho_B] + [\rho_C].$$
Hence $[\rho_C] = -[\rho_B]$. So
$$(i_{B*}, -i_{C*})[\rho_B] = (i_{B*}[\rho_B], i_{C*}[\rho_C]) = ([\sigma_B], [\sigma_C]). \qquad \blacksquare$$

IX.11 Examples

In this section, we will examine some examples of higher dimensional spaces which exhibit interesting topological and analytic aspects that have been alluded to in this chapter.

IX.11.1 Spheres. A direct computation of $\text{Ext}(S^n)$ is difficult because the useful topological sequences we have developed are also valid for homology theories that are not periodic. So we quote the periodicity theorem of Remark IX.10.4 to obtain that $\text{Ext}(S^{2n}) = \text{Ext}(S^0) = 0$ and $\text{Ext}(S^{2n+1}) = \text{Ext}(S^1) = \mathbb{Z}$ for $n \geq 0$.

For the odd spheres, one can use the fact that S^{2n-1} is the unit sphere in $\mathbb{C}^n$ to construct a Toeplitz extension which is a generator for $\text{Ext}(S^{2n-1})$. We consider S^3 here. Another interesting feature of this example is that γ_{S^3} is the zero map, but the higher order map on 2×2 matrices over $C(S^3)$ yields an isomorphism. The reason is the topological fact that $\pi^1(S^3) = 0$, meaning that every invertible function on S^3 is homotopic to the constant function 1. Hence the map γ of $\text{Ext}(S^3)$ into $\text{Hom}(\pi^1(S^3), \mathbb{Z}) = 0$ is necessarily the zero map.

Think of S^3 as $S = \{(z,w) : z, w \in \mathbb{C}, |z|^2 + |w|^2 = 1\}$. Let m be the normalized surface Lebesgue measure on S. Set $H^2(S)$ to be the closure of the polynomials in z and w in the Hilbert space $L^2(m)$, and let P denote the orthogonal projection onto $H^2(S)$. For f in $C(S)$, define a Toeplitz operator by $T_f h = Pfh$. It is easily seen that $T_{\bar{f}} = T_f^*$. Also if f is analytic (uniform limits of polynomials in z and w), then $H^2(S)$ is invariant for T_f. Thus if g is in $C(S)$, then
$$T_g T_f = T_{gf} \quad \text{and} \quad T_{\bar{f}} T_g = T_{\bar{f}g}.$$

Note that $\{z^k w^\ell : k, l \geq 0\}$ is an orthogonal set spanning $H^2(S)$. Thus
$$e_{k\ell} = \|z^k w^\ell\|^{-1} z^k w^\ell = \sqrt{\tfrac{(k+\ell+1)!}{k!\ell!}} z^k w^\ell \quad \text{for } k, \ell \geq 0$$
forms an orthonormal basis for $H^2(S)$. Notice that
$$(\bar{z} z^k w^\ell, z^m w^n) = (z^k w^\ell, z^{m+1} w^n) = \begin{cases} 0 & m \neq k-1 \text{ or } n \neq \ell \\ \|z^k w^\ell\|^2 & m = k-1 \text{ and } n = \ell \end{cases}.$$
Hence
$$P\bar{z} z^k w^\ell = \frac{\|z^k w^\ell\|^2}{\|z^{k-1} w^\ell\|^2} z^{k-1} w^\ell = \frac{k}{k+\ell+1} z^{k-1} w^\ell.$$

IX.11. Examples

(Note that $k = 0$ is a special case.) Compute

$$(T_z^* T_z - T_{|z|^2})e_{k\ell} = P|z|^2 e_{k\ell} - PzP\bar{z}e_{k\ell}$$
$$= \|z^k w^\ell\|^{-1}(P\bar{z}z^{k+1}w^\ell - zP\bar{z}z^k w^\ell)$$
$$= \left(\frac{k+1}{k+\ell+2} - \frac{k}{k+\ell+1}\right)e_{k\ell} = \frac{\ell+1}{(k+\ell+2)(k+\ell+1)}e_{k\ell}$$

Therefore, this commutator is a compact diagonal operator. A similar calculation shows that $T_w^* T_w - T_{|w|^2}$ and $T_z^* T_w - T_{\bar{z}w}$ are compact. It follows that $T_f T_g - T_{fg}$ is compact whenever f and g are uniform limits of polynomials in z, $\bar{z}$, w and $\bar{w}$, which is all of $C(S)$ by the Stone–Weierstrass Theorem.

As in the case of the circle, it is routine to check that $\|T_f\|_e = \|f\|_\infty$. Therefore $\tau(f) = \pi T_f$ is an extension of $\mathfrak{K}$ by $C(S)$. We will show that $[\tau]$ is a generator for $\mathrm{Ext}(S)$. It would be natural to compute $\gamma[\tau]$. However, as noted above, it is a topological fact that $\pi^1(S^3) = 0$, and thus $\gamma_{S^3} = 0$. It is necessary to look at matrix algebras over $C(S)$.

Consider the unitary valued matrix function $F = \begin{bmatrix} z & w \\ -\bar{w} & \bar{z} \end{bmatrix}$. Then

$$A = T_F = \begin{bmatrix} T_z & T_w \\ -T_w^* & T_z^* \end{bmatrix}$$

is essentially unitary ($\pi A^* A = \pi AA^* = I$). We compute $\gamma_2[\tau](F) = \mathrm{ind}\, A$.

First suppose that $\begin{pmatrix} f \\ g \end{pmatrix}$ belongs to $\ker A$. Then

$$\begin{pmatrix} 0 \\ 0 \end{pmatrix} = A\begin{pmatrix} f \\ g \end{pmatrix} = \begin{pmatrix} zf + wg \\ P(-\bar{w}f + \bar{z}g) \end{pmatrix}.$$

Since $zf = -wg$, a consideration of the basis expansion shows that there is a vector h in $H^2(S)$ such that $f = wh$ and $g = -zh$. Thus the second coefficient yields the identity

$$0 = P(-|w|^2 h - |z|^2 h) = -Ph = -h.$$

Therefore $f = g = 0$. So A is injective.

Similarly compute for $\begin{pmatrix} f \\ g \end{pmatrix}$ in $\ker A^*$ to obtain

$$\begin{pmatrix} 0 \\ 0 \end{pmatrix} = A^* \begin{pmatrix} f \\ g \end{pmatrix} = \begin{pmatrix} P\bar{z}f - wg \\ P\bar{w}f + zg \end{pmatrix}.$$

Therefore,

$$g = P(|w|^2 + |z|^2)g = P\bar{w}P\bar{z}f - P\bar{z}P\bar{w}f = (T_z T_w - T_w T_z)^* f = 0.$$

Hence $0 = P\bar{w}f = P\bar{z}f$ which shows that f is constant. In particular, $\begin{pmatrix} 1 \\ 0 \end{pmatrix}$ lies in $\ker A^*$. Hence $\mathrm{ind}\, A = -1$.

This shows that $[\tau] \neq 0$ since $\gamma_2(0) = 0$. It also shows that $[\tau]$ is not a multiple of another extension since -1 is indivisible. As $\mathrm{Ext}(S^3) = \mathbb{Z}$, it follows that $[\tau]$ is a generator.

IX.11.2 The Projective Plane. The real projective plane is introduced to give an example where $\text{Ext}(X)$ has torsion. Clearly the groups $\text{Hom}(\pi^1(X), \mathbb{Z})$ and $\text{Hom}(K^1(X), \mathbb{Z})$ are torsion free. So the homomorphisms γ and γ_∞ cannot detect torsion elements.

The real projective plane $\mathbb{P}^2$ is the 2-sphere S^2 modulo the identification of antipodal points. Perhaps a more useful representation is the unit disk modulo the identification of antipodal points on the boundary circle. We will also see that it is a quotient of the Möbius band $\mathbb{M}$. The Möbius band is the unit square $[0,1]^2$ modulo the identification of $(s,0)$ with $(1-s,1)$ for $0 \leq s \leq 1$.

Consider two imbeddings of the circle $\mathbb{T} = \mathbb{R}/\mathbb{Z}$ into $\mathbb{M}$. Let

$$\alpha(t) = (\tfrac{1}{2}, t) \quad \text{for} \quad 0 \leq t \leq 1$$

and

$$\beta(t) = \begin{cases} (0, 2t) & 0 \leq t \leq \tfrac{1}{2} \\ (1, 2t-1) & \tfrac{1}{2} \leq t \leq 1 \end{cases}.$$

Note that $\alpha(0) = (\tfrac{1}{2}, 0) \sim (\tfrac{1}{2}, 1) = \alpha(1)$ and $\beta(0) = (0,0) \sim (1,1) = \beta(1)$ and $\beta(\tfrac{1}{2}) = (0,1) \sim (1,0)$. These identifications show that α and β are continuous imbeddings of $\mathbb{T}$ into $\mathbb{M}$. There is a retraction of $\mathbb{M}$ onto the centre circle $\alpha(\mathbb{T})$ by $r((s,t)) = (\tfrac{1}{2}, t)$.

Clearly, $r\alpha = \text{id}_\mathbb{T}$; and hence $r_*\alpha_* = \text{id}_*$. Therefore α_* is injective. However, $r\beta(t) = 2t \pmod{1}$ wraps twice around and thus $r_\#\beta_\# = 2\,\text{id}_\#$. As γ is injective on $\text{Ext}(\mathbb{T})$, it follows that $r_*\beta_* = 2\,\text{id}_*$.

Notice that $\mathbb{M}/\alpha(\mathbb{T})$ is homeomorphic to the unit disk $\mathbb{D}$. (The concentric circles given by the first coordinate equal to $\tfrac{1}{2} \pm s$ for $0 < s \leq \tfrac{1}{2}$ have a limit point $\alpha(\mathbb{T})/\alpha(\mathbb{T})$.) Thus by the short exact sequence Corollary IX.6.2, we have

$$\mathbb{Z} = \text{Ext}(\mathbb{T}) \xrightarrow{\alpha_*} \text{Ext}(\mathbb{M}) \xrightarrow{q_*} \text{Ext}(\mathbb{D}) = 0.$$

Therefore α_* is an isomorphism of $\mathbb{Z}$ onto $\text{Ext}(\mathbb{M})$ with inverse r_*. Also

$$\beta_* = (\alpha_* r_*)\beta_* = \alpha_*(2\,\text{id}_*) = 2\alpha_*.$$

Hence the range of β_* is $2\mathbb{Z}$.

Now notice that $\mathbb{M}/\beta(\mathbb{T}) = \mathbb{P}^2$ (as it is a disk with antipodal points on the boundary identified). Thus by the long exact sequence Theorem IX.10.3

$$\begin{array}{ccccccc} \text{Ext}(\mathbb{T}) & \xrightarrow{\beta_*} & \text{Ext}(\mathbb{M}) & \xrightarrow{q_*} & \text{Ext}(\mathbb{P}^2) & \xrightarrow{\partial} & \text{Ext}(S^2) \\ \| & & \| & & \| & & \| \\ \mathbb{Z} & \xrightarrow{2} & \mathbb{Z} & \longrightarrow & \mathbb{Z}/2\mathbb{Z} & \longrightarrow & 0 \end{array}$$

Hence $\text{Ext}(\mathbb{P}^2) = \mathbb{Z}_2$ is a two element group. The non-zero element can be seen to be $q_*\alpha_*(1)$.

IX.11. Examples

Consider the imbedding of $\mathbb{P}^2$ into $\mathbb{C}^2$ defined by sending the unit disk $\mathbb{D}$ of the complex plane into $\mathbb{C}^2$ by $h(z) = (z^2, (1 - |z|^2)z)$. One may readily verify that this is one-to-one on the interior of $\mathbb{D}$ and identifies antipodal points of the boundary. Let (N_1, N_2) be a pair of commuting diagonal operators with joint spectrum equal to $h(\mathbb{D})$. Then they represent the trivial element of $\text{Ext}(\mathbb{P}^2)$. The non-trivial generator of $\text{Ext}(\mathbb{T})$ is given by the Toeplitz extension τ. In this representation, the range of α is homotopic to the map $k(z) = (z, 0)$. So the element $[1]$ in $\text{Ext}(\mathbb{P}^2)$ is generated by

$$k_*(\tau)(z) = (T_z, 0) \oplus (N_1, N_2) = (T_z \oplus N_1, 0 \oplus N_2).$$

In operator theoretic terms, this means that the pair $(T_z \oplus N_1, 0 \oplus N_2)$ is not diagonalizable modulo $\mathfrak{K}$, but the pair $(T_z \oplus T_z \oplus N_1, 0 \oplus 0 \oplus N_2)$ is jointly diagonalizable modulo $\mathfrak{K}$. In fact, the non-zero element of $\text{Ext}(\mathbb{P}^2)$ is not quasidiagonal either. So $\text{Ext}_{qd}(\mathbb{P}^2)$ is properly contained in $\ker \gamma_{\mathbb{P}^2}^\infty$.

IX.11.3 The Suspended Solenoid. Our last example will be used to show several things. It will provide a subset X of $\mathbb{C}^2$ for which $\text{Ext}_{qd}(X) \neq 0$. Hence we will obtain an example that shows that the set of commuting normal plus compact pairs is not closed. It also provides an example in which the surjection κ of Theorem IX.6.3 is not injective.

Consider a solid torus T_1 in $\mathbb{R}^3$ with cross sections of radius 1. Inside T_1, choose another solid torus T_2 which wraps around 3 times inside T_1, and has uniform cross section of radius $1/3$. Continue selecting solid tori T_{n+1} inside T_n wrapping around 3 times, and having uniform cross section of radius 3^{-n}. The intersection T is called a triadic solenoid. The example we want is the suspension ST. It is easy to imbed ST_1 in $\mathbb{C}^2$. So we obtain ST as the intersection of the decreasing sequence ST_n. Each ST_n is homotopic to S^2, and thus $\text{Ext}(ST_n) = 0$. (Alternatively, notice that the injection i of the circle onto the centre circle of the solid torus has a retract r. The solid torus modulo this circle is a solid ball which is contractible. Thus i_* and r_* are reciprocal isomorphisms. The suspended maps yield an isomorphism of $\text{Ext}(S^2) = 0$ and $\text{Ext}(ST_n)$.)

It follows that $\text{proj lim Ext}(ST_n) = 0$ and thus the map κ from $\text{Ext}(ST)$ onto this zero group is the zero map. Thus if $\text{Ext}(ST)$ contains a non-zero element $[\tau]$, we obtain an example where κ is not injective. Suppose that (A_1, A_2) were an essentially normal pair representing $[\tau]$. Then let (N_1, N_2) be a diagonal normal pair with joint spectrum equal to ST. Choose diagonal normal pairs (N_{n1}, N_{n2}) with joint spectrum equal to ST_n converging to (N_1, N_2). Then if j_n is the injection of ST into ST_n, it follows that $j_{n*}\tau$ is trivial. Thus $(A_1 \oplus N_{n1}, A_2 \oplus N_{n2})$ is of the form $(D_{n1}, D_{n2}) + (K_{n1}, K_{n2})$ where (D_{n1}, D_{n2}) is a diagonal normal pair with spectrum ST_n and K_{ni} are compact. Since

$$\lim_{n \to \infty} (A_1 \oplus N_{n1}, A_2 \oplus N_{n2}) = (A_1 \oplus N_1, A_2 \oplus N_1),$$

it follows that τ is quasidiagonal. In fact, this shows that $\text{Ext}(ST) = \text{Ext}_{qd}(ST)$. However, as $[\tau]$ is non-trivial, this is not a diagonalizable pair modulo $\mathfrak{K}$. This shows that the analogue of normals plus compacts being closed fails for jointly normal plus compact pairs, provided that we can find a non-zero element of $\text{Ext}(ST)$.

To facilitate calculations, we will specify the imbeddings of T_{n+1} into T_n more precisely. Parameterize T_n as $\{(z,w) : z \in \mathbb{T}, w \in \mathbb{D}\}$. Define an imbedding j_n of T_{n+1} into T_n by $j_n(z,w) = (z^3, (w+2z)/3)$. Let k_n denote the injection of T into T_n.

We can also express T as a projective limit of circles $\mathbb{T}_n = \mathbb{T}$ via the maps $p_n(z) = z^3$. Denote this projective limit by Z, and let q_n be the induced maps of Z onto $\mathbb{T}_n$. There are maps π_n of T_n onto $\mathbb{T}_n$ by projecting the solid torus onto its centre circle by $\pi_n(z,w) = z$. It is clear from the formulae that we obtain a commutative diagram

$$\begin{array}{ccccccccc}
T_1 & \xleftarrow{j_1} & T_2 & \xleftarrow{j_2} & T_3 & \xleftarrow{} & \cdots & \xleftarrow{} & T_n & \xleftarrow{k_n} & T \\
\pi_1 \downarrow & & \pi_2 \downarrow & & \pi_3 \downarrow & & & & \pi_n \downarrow & & \downarrow \pi \\
\mathbb{T}_1 & \xleftarrow{p_1} & \mathbb{T}_2 & \xleftarrow{p_2} & \mathbb{T}_3 & \xleftarrow{} & \cdots & \xleftarrow{} & \mathbb{T}_n & \xleftarrow{q_n} & Z
\end{array}$$

We will show that map $\pi = \text{proj}\lim \pi_n$ is a homeomorphism. It is easily seen to be surjective. On the other hand, suppose that ξ_1 and ξ_2 are two points in T such that $\pi(\xi_1) = \pi(\xi_2) = (z_n)_{n\geq 1}$ in Z. Then

$$\pi_n k_n(\xi_1) = q_n \pi(\xi_1) = z_n = q_n \pi(\xi_2) = \pi_n k_n(\xi_2) \quad \text{for} \quad n \geq 1.$$

In particular, $q_1(\xi_1)$ and $q_1(\xi_2)$ both lie in the disk $\{z_1\} \times \mathbb{D}$. Inside this disk, both lie in the subset $j_1(\{z_2\} \times \mathbb{D})$ which is a disk of radius of radius $1/3$. Repeating this for the nth disk, both points lie in a common disk of radius 3^{1-n} for all n, and thus $\xi_1 = \xi_2$. So π is injective. A continuous bijection of one compact Hausdorff space onto another is a homeomorphism.

Thus we may identify T with Z. This latter space is the subset of the infinite product group $\mathbb{T}^\infty$ consisting of those sequences $\mathbf{z} = (z_1, z_2, z_3, \ldots)$ such that $z_{n+1}^3 = z_n$. If $\mathbf{z}$ lies in Z, then $-\mathbf{z} = (-z_1, -z_2, -z_3, \ldots)$ also lies in Z. Also notice that each $\mathbf{z}$ has two square roots $\pm \mathbf{z}^{1/2} = \pm(z_1^{1/2}, z_2^{1/2}, \ldots)$ because once a square root $z_1^{1/2}$ is chosen, there is a unique square root of z_2 with cube equal to $z_1^{1/2}$, et cetera.

Let P denote the space obtained from CZ by identifying $(1, \mathbf{z})$ and $(1, -\mathbf{z})$ for all $\mathbf{z}$ in Z. Then there is a non-trivial map of Z into P given by

$$\varphi(\mathbf{z}) = [(1, \pm \mathbf{z}^{1/2})].$$

It is evident that $P/\varphi(Z) = SZ$. Let δ be the quotient map. There is also a map r of P onto the real projective plane $\mathbb{P}^2$ induced by the map r' of CZ onto $\mathbb{D}$ given by $r'(t, \mathbf{z}) = tz_1$. Since $r'(1, \pm \mathbf{z}) = \pm z_1$ which are antipodal points, r is well defined. Define an injection ψ of $\mathbb{T}$ into $\mathbb{P}^2$ by $\psi(z) = [\pm z^{1/2}]$. It is easy to check that $r\varphi = \psi q_1$. We have a commutative diagram

$$\begin{array}{ccccc} Z & \xrightarrow{\varphi} & P & \xrightarrow{\delta} & SZ \\ {\scriptstyle q_1}\downarrow & & \downarrow{\scriptstyle r} & & \\ \mathbb{T} & \xrightarrow{\psi} & \mathbb{P}^2 & & \end{array}$$

Since the map p_n wraps the circle three times around itself, it follows that p_{n*} is multiplication by 3 from $\mathrm{Ext}(\mathbb{T}_{n+1})$ into $\mathrm{Ext}(\mathbb{T}_n)$. Hence the image of $\mathrm{Ext}(\mathbb{T}_{n+1})$ in $\mathrm{Ext}(\mathbb{T}_1)$ is $3^n \mathbb{Z}$. As the map q_{1*} of $\mathrm{Ext}(Z)$ into $\mathrm{Ext}(\mathbb{T}_1)$ factors through p_{n*}, the range of q_{1*} is contained in $3^n \mathbb{Z}$ for all $n \geq 1$ and thus $q_{1*} = 0$. Therefore $r_* \varphi_* = \psi_* q_{1*} = 0$, which means that the range of φ_* is contained in $\ker r_*$. By the short exact sequence Corollary IX.6.2,

$$\mathrm{Ext}(Z) \xrightarrow{\varphi_*} \mathrm{Ext}(P) \xrightarrow{\delta_*} \mathrm{Ext}(SZ)$$

is exact. So if we can find an element $[\sigma]$ in $\mathrm{Ext}(P)$ which is not in $\ker r_*$, then $[\tau] = \delta_*[\sigma] \neq 0$ in $\mathrm{Ext}(SZ)$.

Let $\mathbf{1} = (1, 1, 1, \ldots)$ in Z. Define a map θ of $\mathbb{T}$ into P by

$$\theta(e^{\pi i t}) = [(|t|, \mathrm{sgn}(t)\mathbf{1})] \quad \text{for} \quad -1 \leq t \leq 1.$$

This is well defined because $\theta(e^{\pm \pi i}) = [(1, \pm \mathbf{1})]$ is a single point in P. Notice that $r\theta$ maps $\mathbb{T}$ onto the diameter of $\mathbb{P}^2$. This is the map called $q\alpha$ in the previous example. Therefore $r_*\theta_*(\mathbf{1})$ is the non-zero element of $\mathrm{Ext}(\mathbb{P}^2)$. And consequently $[\sigma] = \theta_*(\mathbf{1})$ does not lie in $\ker r_*$. This completes the construction.

Exercises

IX.1 Suppose that $(\mathfrak{E}_i, \varphi_i)$ represent two extensions of $\mathfrak{K}$ by $C(X)$. Show that the sum of these extensions is given by

$$\left\{ \begin{bmatrix} E_{11} & K_{12} \\ K_{21} & E_{22} \end{bmatrix} : E_{ii} \in \mathfrak{E}_i,\ K_{ij} \in \mathfrak{K},\ \text{and}\ \varphi_1(E_{11}) = \varphi_2(E_{22}) \right\}.$$

IX.2 Show that Corollary IX.9.4 and Corollary IX.6.2 imply homotopy invariance.
HINT: Consider the two injections $j_s(\xi) = (\xi, s)$ for $s = 0, 1$ and the projection p of $X \times [0, 1]$ onto X. Note that $pj_s = \mathrm{id}$ and use the sequence

$$X \xrightarrow{j_{0*}} X \times [0, 1] \longrightarrow CX$$

to show that j_{0*} is surjective.

IX.3 Show that if $X = X_1 \vee X_2$ is the disjoint union of two compact subsets X_1 and X_2, then $\mathrm{Ext}(X) \simeq \mathrm{Ext}(X_1) \oplus \mathrm{Ext}(X_2)$.
HINT: There is a natural map of the sum into $\mathrm{Ext}(X)$. For the inverse,

one may choose a projection corresponding to χ_{X_1} to split an extension τ. The uniqueness of this choice depends on Corollary IX.2.4.

IX.4 Let T be an essentially normal operator with $\sigma_e(T) = X$ corresponding to an extension τ of $\mathfrak{K}$ by $C(X)$. Let A be a closed subset of X, and let p be the quotient of X onto X/A. If $p_*[\tau] = 0$ and $\varepsilon > 0$, show that there is an essentially normal operator R with $\sigma_e(R) = A$ and a diagonal normal operator D with $\sigma(D) = \sigma_e(D) = X$ such that $T \simeq (R \oplus D) + K$ where K is a compact operator with $\|K\| < \varepsilon$.
HINT: Use Corollary IX.6.2 and swallow most of the compact into R by moving a finite projection from D.

IX.5 Suppose that $X = B \cup C$ is the union of closed subsets. Suppose that there is a retract of C onto $A = B \cap C$. Then show that the natural map from $\text{Ext}(B) \oplus \text{Ext}(C)$ into $\text{Ext}(X)$ is surjective.

IX.6 (a) Show that $\text{Ext}(\mathcal{M}_k) = \mathbb{Z}_k$ by lifting matrix units in the Calkin algebra to matrix units in $\mathcal{B}(\mathcal{H})$, and computing the codimension of the lifting of the identity.
(b) Show that $\text{Ext}(\mathcal{M}_{n_1} \oplus \cdots \oplus \mathcal{M}_{n_k}) = \mathbb{Z}_d$ where $d = \gcd(n_1, \ldots, n_k)$.

IX.7 (a) Suppose that j is a unital imbedding of $\mathcal{M}_n$ into $\mathcal{M}_{kn}$, and that τ is an extension of $\mathcal{M}_{kn}$. Show that if σ is a (possibly non-unital) homomorphism of $\mathcal{M}_n$ into $\mathcal{B}(\mathcal{H})$ such that $\pi\sigma(A) = \tau(j(A))$, then there is a homomorphism $\overline{\sigma}$ of $\mathcal{M}_{kn}$ into $\mathcal{B}(\mathcal{H})$ such that $\overline{\sigma}j = \sigma$ and $\pi\overline{\sigma} = \tau$.
HINT: Lift matrix units.
(b) If τ is trivial in part (a) and σ is unital, show that there is a unital lifting $\overline{\sigma}$.
(c) Let $\mathfrak{A}$ be a UHF algebra of type k^∞. Show that the map κ from $\text{Ext}(\mathfrak{A})$ into $\text{proj lim}\,\text{Ext}(\mathcal{M}_{k^n})$ is surjective by repeated use of (a).
(d) Show that κ is an isomorphism by repeatedly using part (b).

IX.8 Show that the Mayer–Vietoris sequence is exact at $\text{Ext}(SA)$.

Notes and Remarks.

The main results of this chapter are due to Brown, Douglas and Fillmore [1973] and [1977]. A nice treatment of the minimal route to the planar case is contained in Davie [1976], as well as the examples at the end of this chapter. Arveson [1974] was the first to point out the role of lifting completely positive maps in this context. Stinespring [1955] generalized Naimark [1943] to the noncommutative setting. Paulsen [1986] provides a nice treatment of completely positive maps. Voiculescu's Theorem II.5.3 shows that the trivial elements form the zero element for every separable C*-algebra. Arveson [1977] shows how to combine Voiculescu's results with a lifting result of Choi and Effros [1976] to obtain inverses for liftable maps. The proof for homotopy invariance was generalized to the non-commutative setting by Salinas [1977], and we follow his approach exploiting quasidiagonality here. O'Donovan [1977] contains a similar argument. Berg and Davidson [1991] provide an operator theoretic proof of the planar case Theorem IX.7.2 using a generalization of Berg's technique.

References

Akemann, C.A. and Pedersen, G.K. [1977], *Ideal perturbations of elements in C*-algebras*, Math. Scand. **41**, 117–139.

Arveson, W.B. [1974], *A note on essentially normal operators*, Proc. Royal Irish Acad. **74**, 143–146.

Arveson, W.B. [1976], *An invitation to C*-algebras*, Grad. Texts Math. **39**, Springer-Verlag, Berlin, New York.

Arveson, W.B. [1977], *Notes on extensions of C*-algebras*, Duke Math. J. **44**, 329–355.

Atiyah, M.F. [1967], *K-theory*, W.A. Benjamin Inc., New York, Amsterdam.

Berg, I.D. [1971], *An extension of the Weyl–von Neumann Theorem to normal operators*, Trans. Amer. Math. Soc. **160**, 365–371.

Berg, I.D. and Davidson, K.R. [1991], *A quantitative version of the Brown-Douglas-Fillmore Theorem*, Acta. Math. **166**, 121–161.

Blackadar, B.E. [1980], *A simple C*-algebra with no non-trivial projections*, Proc. Amer. Math. Soc. **78**, 504–508.

Blackadar, B.E. [1990], *Symmetries of the CAR algebra*, Ann. Math. **131**, 589–623.

Blackadar, B., Bratteli, O., Elliott, G.A. and Kumjian, A. [1992], *Reduction of real rank in inductive limit C*-algebras*, Math. Ann. **292**, 111–126.

Blackadar, B., Dadarlat, M, and Rordam, M. [1991], *The real rank of inductive limit C*-algebras*, Math. Scand. **69**, 211–216.

Blackadar, B. and Kumjian, A. [1985], *Skew products of relations and the structure of simple C*-algebras*, Math. Zeit. **189**, 55–63.

Bratteli, O. [1972], *Inductive limits of finite dimensional C*-algebras*, Trans. Amer. Math. Soc. **171**, 195–234.

Brown, L. G., Douglas, R. G. and Fillmore, P. A. [1973], *Unitary equivalence modulo the compact operators and extensions of C*-algebras*, Proceedings of a conference on operator theory, Halifax, Nova Scotia 1973, Lect. Notes in Math. **345**. (Berlin- Heidelberg-New York: Springer-Verlag), 58–128.

Brown, L. G., Douglas, R. G. and Fillmore, P. A. [1977], *Extensions of C*-algebras and K-homology*, Ann. Math. **105**, 265–324.

Brown, L. G. [1981a], *Universal coefficient theorem for Ext and quasidiagonality*, Proc. Conf. on Operator Theory and Group Representations, Roumania, Pitman Pub. Co., pp. 60–64.

Brown, L.G. [1981b], *Extensions of AF-algebras*, Operator Algebras and Applications, R.V. Kadison (ed.), Proc. Symp. Pure Math. **38**, 175–176, Amer. Math. Soc., Providence.

Brown, L.G. and Pedersen, G.K. [1991], *C*-algebras of real rank zero*, J. Func. Anal. **99**, 131–149.

Bunce, J. and Deddens, J. [1975], *A family of simple C*-algebras related to weighted shift operators*, J. Func. Anal. **19**, 12–34.

Bunce, J. and Salinas, N. [1976], *Completely positive maps on C*-algebras and the left essential matricial spectrum of an operator*, Duke Math. J. **43**, 747–774.

Choi, M.D. and Effros, E.G. [1976], *The completely positive lifting problem for C*-algebras*, Ann. Math. **104**, 585–609.

Choi, M.D. [1979], *A simple C*-algebra generated by two finite order unitaries*, Can. J. Math. **31**, 867–880.

Choi, M.D. [1980], *The full C*-algebra of the free group on two generators*, Pacific J. Math. **87**, 41–48.

Choi, M.D. [1983], *Lifting projections from quotient C*-algebras*, J. Operator Thy. **10**, 21–30.

Coburn, L. [1967], *The C*-algebra of an isometry*, Bull. Amer. Math. Soc. **73**, 722–726.

Cohen, J.M. [1979], *C*-algebras without idempotents*, J. Func. Anal. **33**, 211–216.

Cohen, J.M. and Figà-Talamanca, A. [1988], *Idempotents in the reduced C*-algebra of a free group*, Proc. Amer. Math. Soc. **103**, 779–782.

Connes, A. [1986], *Non-commutative differential geometry*, Publ. Math. IHES **62**, 257–360.

Corach, G. and Larotonda, A.R. [1984], *Stable range in Banach algebras*, J. Pure Appl. Algebra **32**, 289–300.

Cuntz, J. [1977], *Simple C*-algebras generated by isometries*, Comm. Math. Phys. **57**, 173–185.

Cuntz, J. [1981], *K-theory for certain C*-algebras*, Ann. Math. **113**, 181–197.

Davidson, K.R. [1984], *Berg's technique and irrational rotation algebras*, Proc. Royal Irish Acad. **84A**, 117–123.

Davie, A.M. [1976], *Classification of essentially normal operators*, Lect. Notes Math. **512**, pp. 31–55, Springer Verlag, Berlin, New York.

Dixmier, J. [1964], *Les C*-algèbres et leurs représentations*, Gauthier–Villars, Paris.

Douglas, R.G. [1972], *Banach algebra techniques in operator theory*, Academic Press, New York and London.

Effros, E.G., Handelman, D.E. and Shen, C.L. [1980], *Dimension groups and their affine transformations*, Amer. J. Math. **102**, 385–402.

Effros, E.G. [1989], *Why the circle is connected: an introduction to quantized topology*. Math. Intell. **11**, 27–34.

Elliott, G.A. [1976], *On the classification of inductive limits of sequences of semi-simple finite dimensional algebras*, J. Algebra **38**, 29–44.

Elliott, G.A. [1978], *On totally ordered groups and K_0*, Proc. Ring Theory conf., Waterloo, D. Handelman and J. Lawrence (eds.), Lect. Notes Math. **734**, 1–49, Springer–Verlag, New York, 1978.

Elliott, G.A. and Evans, D.E. [1993], *The structure of the irrational rotation C*-algebra*, Ann. Math. **138**, 477–501.

Fuchs, L. [1965], *Riesz groups*, Ann. Scuola Norm. Pisa **19**, 1–34.

Fukamiya, M. [1952], *On a theorem of Gelfand and Neumark and the B*-algebra*, Kumamoto J. Sci. **1**, 17–22.

Gelfand, I.M. [1941], *Normierte Ringe*, Mat. Sbornik **9**, 3–24.

Gelfand, I.M. and Naimark, M. [1943], *On the imbedding of normed rings into the ring of operators in Hilbert space*, Mat. Sbornik **12**, 197–213.

Glimm, J. [1960a], *On a certain class of operator algebras*, Trans. Amer. Math. Soc. **95**, 318–340.

Glimm, J. [1960b], *A Stone–Weierstrass theorem for C*-algebras*, Ann. Math. **72**, 216–244.

Hadwin, D. [1977], *An operator valued spectrum*, Indiana Univ. Math. J. **26**, 329–340.

References

Hadwin, D. [1980], *Nonseparable approximate equivalence*, Trans. Amer. Math. Soc. **266**, 203–231.

Ji, R. [1992], *Smooth dense subalgebras of reduced C*-algebras, Schwartz cohomology of groups, and cyclic cohomology*, J. Func. Anal. **107**, 1–33.

Kadison, R.V. [1957], *Irreducible operator algebras*, Proc. Nat. Acad. Sci. U.S.A. **43**, 273–276.

Kadison, R.V. and Pedersen, G.K. [1985], *Means and convex combinations of unitary operators*, Math. Scand. **57**, 249–266.

Kaplansky, I. [1949], *Normed algebras*, Duke J. Math. **16**, 399–418.

Kaplansky, I. [1951], *A theorem on rings of operators*, Pacific J. Math, **1**, 227–232.

Kumjian, A. [1988], *An involutive automorphism of the Bunce–Deddens algebra*, C.R. Math. Rep. Acad. Sci. Canada **10**, 217–218.

Murray, F.J. and von Neumann, J. [1936], *On rings of operators*, Ann. Math. **37**, 116–229.

Naimark, M.A. [1943], *On a representation of additive operator set functions* (in Russian), C.R. (Doklady) Acad. Sci. URSS **41**, 359–361.

O'Donovan, D. [1977], *Quasidiagonality in the Brown–Douglas–Fillmore theory*, Duke J. Math. **44**, 767–776.

Paschke, W. and Salinas, N. [1979], *Matrix algebras over $\mathcal{O}_n$*, Mich. J. Math. **26**, 3–12.

Paterson, A.L. [1988], *Amenability*, Math. Surveys and Monographs **29**, Amer. Math. Soc., Providence.

Paulsen, V.I. [1986], *Completely bounded maps and dilations*, Pitman Res. Notes Math. **146**, Longman Sci. Tech. Harlow.

Pedersen, G.K. [1979], *C*-algebras and their automorphism groups*, Academic Press, London.

Pimsner, M. [1983], *Embedding some transformation groups C*-algebras into AF algebras*, Ergodic Thy. Dynam. Sys. **3**, 613–626.

Pimsner, M. and Popa, S. [1978], *Ext groups of some C*-algebras considered by J. Cuntz*, Rev. Roum. Pures Appl. **23**, 1069–1076.

Pimsner, M. and Voiculescu, D.V. [1980a], *Imbedding the irrational rotaation algebras into an AF algebra*, J. Operator Thy. **4**, 201–210.

Pimsner, M. and Voiculescu, D.V. [1980b], *Exact sequences for K-groups and Ext-groups of certain crossed products of C*-algebras*, J. Operator Thy. **4**, 93–118.

Pimsner, M. and Voiculescu, D.V. [1982], *K-groups of reduced crossed products by free groups*, J. Operator Thy. **8**, 131–156.

Power, S.C. [1978], *Simplicity of C*-algebras of minimal dynamical systems*, J. London Math. Soc. (2) **18**, 534–538.

Powers, R.T. [1975], *Simplicity of the C*-algebra associated with the free group on two generators*, Duke J. Math. **42**, 151–156.

Putnam, I. [1989], *The C*-algebras associated with minimal homeomorphisms of the Cantor set*, Pacific J. Math. **136**, 329–353.

Putnam, I. [1990a], *On the topological stable rank of certain transformation group C*-algebras*, Ergodic Thy. Dynam. Sys. **10**, 197–207.

Putnam, I. [1990b], *The invertibles are dense in the irrational rotation C*-algebras*, J. reine angew. Math. **410**, 160–166.

Rieffel, M.A. [1981], *C*-algebras associated with irrational rotations*, Pacific J. Math. **93**, 415–429.

Rieffel, M.A. [1983a], *Dimension and stable rank in the K-theory of C*-algebras*, Proc. London Math. Soc. (3) **46**, 301–333.

Rieffel, M.A. [1983b], *The cancellation theorem for projective modules over irrational rotation algebras*, Proc. London Math. Soc. (3) **47**, 285–302.

Rosenberg, J. [1987], *Quasidiagonality and nuclearity*, J. Operator Thy. **18**, 15–18.

Russo, B. and Dye, H.A. [1966], *A note on unitary operators in C*-algebras*, Duke J. Math. **33**, 413–416.

Segal, I.E. [1947], *Irreducible representations of operator algebras*, Bull. Amer. Math. Soc. **53**, 73–88.

Segal, I.E. [1949], *Two sided ideals in operator algebras*, Ann. Math. **50**, 856–865.

Stinespring, W. [1955], *Positive functions on C*-algebras*, Proc. Amer. Math. Soc. **6**, 211–216.

Vershik, A.M. [1981], *Uniform approximation of shift and multiplication operators*, Sov. Math. Dokl. **24**, 97–100.

Voiculescu, D.V. [1976], *A non-commutative Weyl–von Neumann Theorem*, Rev. Roum. Pures Appl. **21**, 97–113.

von Neumann, J. [1929], *Zur Algebra der Funktionaloperationen und Theorie der normalen Operatoren*, Math. Ann. **102**, 370–427.

von Neumann, J. [1931], *Uber Funktionen von Funktionaloperationen*, Ann. Math. **32**, 191–226.

Zeller–Meier, G. [1968], *Produits croisés d'une C*-algèbre par une group d'automorphismes*, J. Math. Pures Appl. **47**, 101–239.

Zhang, S. [1990], *A property of purely infinite C*-algebras*, Proc. Amer. Math. Soc. **109**, 717–720.

Index

AF algebra, 75, 235
   characterization, 82
   ideals, 85
algebraically irreducible, 26
amenable group, 185
approximate identity, 11
approximate unitary equivalence, 57
approximate unitary equivalence, relative to $\mathfrak{K}$, 57
approximately finite dimensional, 75
approximately inner automorphism, 116
approximately inner derivation, 96
automorphisms of $\mathfrak{K}$, 151

Berg's technique, 174, 239
Bratteli diagram, 76
Bunce–Deddens algebra, 139, 162, 231, 238, 246, 247

C*-dynamical system, 216
cancellation, 99, 246
canonical anticommutation relations, 87
canonical shuffle, 140
CAR algebra, 76, 87, 106
Cauchy-Schwarz inequality, 28
Cesàro sums, 223
character of a group, 184
classical dynamical system, 223
Coburn's Theorem, 137
commutant, 19
compalence, 72
compalent, 252
completely positive, 65, 259
concrete C*-algebra, 1
cone, 101, 129
cone on X, 289
continuous functional calculus, 8
contractible C*-algebra, 129

contractive homomorphism of ordered groups, 103
covariant representation, 216
crossed product, 217
crystallographic group, 193
Cuntz algebra, 144, 219, 234
current algebra, 89
cyclic representation, 46
cyclic vector, 46

derivation, 96
diagonal representation, 62
dimension group, 102
discrete Heisenberg group, 200
Double Commutant Theorem, 19

Effros–Handelman–Shen Theorem, 123
Elliott's Theorem, 109
enveloping C*-algebra, 217
equivalent extension, 151, 253
equivalent idempotents, 97
∗-equivalent projections, 97
ergodic transformation, 224
essential spectrum, 252
essentially normal, 252
expectation, 145, 167, 206, 222
Ext$(\mathcal{O}_n)$, 154
Ext(X), 254
extension, 91, 150, 252

Fibonacci algebra, 106, 116
finite C*-algebra, 101
finite dimensional C*-algebra, 74
free group, 203

gauge automorphisms, 89
gauge invariant, 89
Gelfand–Naimark Theorem, 33
GICAR algebra, 89, 108

Glimm's Lemma, 64
GNS construction, 29
Grothendieck group, 100
group C*-algebra, 184

Haar measure, 182
hereditary, 13, 85, 102
homotopic maps, 286
homotopy invariance, 98, 286
hull, 191
hull–kernel topology, 191

ideal, 12
index map, 272
inductive, 84
infinite C*-algebra, 147
initial projection, 23
inner automorphism, 116
inner derivation, 96
inner function, 162
inverse closed, 15
irrational rotation algebra, 167, 218, 234, 244
irreducible, 26

Jacobson radical, 33
Jordan decomposition, 43

K_0, 101
K_1, 233
Kadison Transitivity Theorem, 27
Kadison–Kaplansky conjecture, 200
Kaplansky Density Theorem, 20

$L^1(G)$, 182
L^∞ functional calculus, 51
lattice ordered group, 119
left regular representation, 184
limit circle algebra, 159, 242, 246
long exact sequence for Ext, 290

m-times around imbedding, 141
mapping cone, 127, 289
mapping cylinder, 289
Markov–Kakutani Theorem, 185
matrix units, 74
maximal abelian algebra, 48
Mayer–Vietoris sequence, 292
mean, invariant, 185
minimal Cantor system, 235
minimal dynamical system, 225
modular function, 182
multiplicity free representation, 48

multiplicity function, 56

n-positive, 65, 259
Naimark Dilation Theorem, 260
non-degenerate representation, 34
normal, 2

odometer transformation, 230
operator monotone, 40
order, 9
order ideal, 112
order unit, 101

$\pi^1(X)$, 273
P–V short exact sequence, 234
partial isometry, 23
partial multiplicities, 75
partition, 235
periodic weighted shift, 137
periodicity, 292
polar decomposition, 23
positive, 2
positive definite function, 187
positive homomorphism, 103
positive lifting, 262
positive linear functional, 27
positive map, 65, 259
prime ideal, 191
primitive ideal, 190
primitive ideal space, 190
projectionless C*-algebra, 124, 200, 210
projective limit, 270
projective plane, 296
proper isometry, 132
properly infinite, 147
pure state, 30
purely infinite, 149

quasi-similarity, 72
quasicentral approximate unit, 35
quasidiagonal C*-algebra, 204, 209
quasidiagonal extension, 282
quasidiagonal operator, 281
quotient algebra, 13

range projection, 23
real rank zero, 156
reduced crossed product, 218
reduced group C*-algebra, 184
Rieffel projections, 170
Riesz group, 118

Index

Rohlin's Lemma, 225
Russo–Dye Theorem, 25

scale, 102
self-adjoint, 2
semi-commutator, 134
semi-simple, 33
separating vector, 48
simple dimension group, 114
spectral measure, 51
Spectral Theorem, 47
spectrum, 3
spectrum of a C*-algebra, 191
stably equivalent, 99
stably finite, 101
state, 27
state of a dimension group, 114
state space, 30
Stinespring Dilation Theorem, 260
strong operator topology, 16
strongly continuous, 20
subrepresentation, 44
supernatural number, 86, 142
suspended solenoid, 297
symmetry, 19

2-adic numbers, 124
Toeplitz operator, 133
topological stable rank, 244
topologically irreducible, 26
trace, 42, 114
trace class, 41
translation invariant measure, 224
triadic solenoid, 297
trivial extension, 153, 254

UHF algebra, 86, 219, 238
uniform multiplicity, 54
uniformly hyperfinite, 86
unilateral shift, 132
unimodular, 182
unitarily equivalence, 46
unitary, 2
unitary representation, 182
universal C*-algebra, 166
universal representation, 33
unperforated, 118

vector state, 33
Voiculescu's Theorem, 68
von Neumann algebra, 19

weak approximate unitary equivalence, 57
weak operator topology, 16
weak-∗ topology, 42
weakly compalent, 252
weakly equivalent extension, 152, 254
weighted unilateral shift, 137
Weyl–von Neumann–Berg Theorem, 59
Wold decomposition, 136